ABOUT ISLAND PRESS

Island Press is the only nonprofit organization in the United States whose principal purpose is the publication of books on environmental issues and natural resource management. We provide solutions-oriented information to professionals, public officials, business and community leaders, and concerned citizens who are shaping responses to environmental problems.

In 2006, Island Press celebrates its twenty-first anniversary as the leading provider of timely and practical books that take a multidisciplinary approach to critical environmental concerns. Our growing list of titles reflects our commitment to bringing the best of an expanding body of literature to the environmental community throughout North America and the world.

Support for Island Press is provided by the Agua Fund, The Geraldine R. Dodge Foundation, Doris Duke Charitable Foundation, The William and Flora Hewlett Foundation, Kendeda Sustainability Fund of the Tides Foundation, Forrest C. Lattner Foundation, The Henry Luce Foundation, The John D. and Catherine T. MacArthur Foundation, The Marisla Foundation, The Andrew W. Mellon Foundation, Gordon and Betty Moore Foundation, The Curtis and Edith Munson Foundation, Oak Foundation, The Overbrook Foundation, The David and Lucile Packard Foundation, The Winslow Foundation, and other generous donors.

The opinions expressed in this book are those of the author(s) and do not necessarily reflect the views of these foundations.

ABOUT THE SOCIETY FOR ECOLOGICAL RESTORATION INTERNATIONAL

The Society for Ecological Restoration International (SER) is an international nonprofit organization comprising members who are actively engaged in ecologically sensitive repair and management of ecosystems through an unusually broad array of experience, knowledge sets, and cultural perspectives.

The mission of SER is to promote ecological restoration as a means of sustaining the diversity of life on Earth and reestablishing an ecologically healthy relationship between nature and culture.

SER, 285 W. 18th Street, #1, Tucson, AZ 85701. Tel. (520) 622-5485, Fax (520) 622-5491, e-mail, info@ser.org, www.ser.org.

FOUNDATIONS OF RESTORATION ECOLOGY

Society for Ecological Restoration International (SER)

The Science and Practice of Ecological Restoration
James Aronson, editor
Donald A. Falk, associate editor

Wildlife Restoration: Techniques for Habitat Analysis and Animal Monitoring,
by Michael L. Morrison

Ecological Restoration of Southwestern Ponderosa Pine Forests,
edited by Peter Friederici, Ecological Restoration Institute
at Northern Arizona University

Ex Situ *Plant Conservation: Supporting the Survival of Wild Populations*,
edited by Edward O. Guerrant Jr., Kayri Havens, and Mike Maunder

Great Basin Riparian Ecosystems: Ecology, Management, and Restoration,
edited by Jeanne C. Chambers and Jerry R. Miller

Assembly Rules and Restoration Ecology: Bridging the Gap Between Theory and Practice,
edited by Vicky M. Temperton, Richard J. Hobbs, Tim Nuttle, and Stefan Halle

The Tallgrass Restoration Handbook: For Prairies, Savannas, and Woodlands,
edited by Stephen Packard and Cornelia F. Mutel

The Historical Ecology Handbook: A Restorationist's Guide to Reference Ecosystems,
edited by Dave Egan and Evelyn A. Howell

Foundations of Restoration Ecology,
edited by Donald A. Falk, Margaret A. Palmer, and Joy B. Zedler

Foundations of Restoration Ecology

Edited by
Donald A. Falk, Margaret A. Palmer, and Joy B. Zedler

Foreword by
Richard J. Hobbs

Society for Ecological Restoration International

Washington · Covelo · London

Library of Congress Cataloging-in-Publication Data
Foundations of restoration ecology / edited by Donald A. Falk, Margaret
A. Palmer, and Joy B. Zedler ; foreword by Richard J. Hobbs.
p. cm. — (Science and practice of ecological restoration)
Includes bibliographical references and index.
ISBN 13: 978-1-59726-017-6 (pbk.: alk. paper)
ISBN 10: 1-59726017-7 (pbk.: alk. paper)
1. Restoration ecology. I. Falk, Donald A. II. Palmer, Margaret A.
III. Zedler, Joy B. IV. Series.
QH541.15.R45F68 2006
577—dc22

2005026650

British Cataloguing-in-Publication Data available

Printed on recycled, acid-free paper 

Manufactured in the United States of America

10 9 8 7 6 5

CONTENTS

FOREWORD

If you have built castles in the air, your work need not be lost; that is where they should be. Now put the foundations under them.
Henry David Thoreau (1854)

It is axiomatic that any building needs firm foundations on which to rest, to ensure its strength and persistence. The taller the building the deeper and more secure its foundations must be. So, too, with fields of endeavor: lasting success and influence depends on the development of firm foundations on which the field can develop and grow.

For restoration ecology, these foundations are being put in place now. The importance of restoration ecology as the science that informs the practice of ecological restoration has increased dramatically over a relatively short period of time. Although the need for environmental repair has been recognized for a long time, and there have been attempts to restore different types of ecosystem for many decades, the science underpinning these attempts is still young and in a formative stage.

Restoration ecology has been raised within other parent disciplines, particularly ecology, which have their own long and venerable traditions. It is in these disciplines that restoration ecology finds the principles and ideas to guide its development. The building blocks with which the scientific foundations of restoration ecology are being constructed come from a broad range of topics and approaches within theoretical and applied ecology and other fields. These building blocks are not necessarily taken unchanged into restoration ecology; rather, they often need to be reworked in recognition of the fact that the science interacts strongly with the practicalities of doing restoration. Theory has to make sense in practice. Indeed, the practice of restoration should ideally serve as a testing ground, which can help inform and improve ecological theory, as has been eloquently argued in the past (Bradshaw 1987).

At the same time, the success and effectiveness of restoration practice can be significantly enhanced by ensuring that ideas and approaches being used are based on an up-to-date understanding of how ecosystems are put together and function, and by being able to learn from successes and failures in other systems and other parts of the world. The practice of ecological restoration has frequently been based largely on local understanding of how particular ecosystems work, without any real reference to a recognized body of theory or generalized framework. Some of these local restoration projects have been spectacularly successful in achieving their goals, due to the insight and energy of those involved. These are the castles that have been built in the air already, which stand as examples of what can be achieved. We can learn from these and aim to emulate the successes elsewhere.

Others have been less successful and have not met their restoration goals, because they have failed to reverse ecosystem degradation or to reconstruct the desired ecosystem in terms of species composition or structure. The lack of success has sometimes stemmed from poor planning or execution, but at other times has been due to an incomplete or erroneous understanding of how ecological systems work. These castles have tumbled. We can also learn from these. What went wrong? What lessons are there for future projects elsewhere? The failures are not necessarily the fault of the people involved and may have more to do with a general shift in the past few decades in the way we perceive ecological systems. While restoration ecology has been growing up, ecology has also been developing, and ecologists have shifted ground on many basic concepts of how communities and ecosystems work. In particular, there has been a shift from viewing ecosystems as static equilibrium entities in which simple linear causation can be detected, to viewing them as complex dynamic systems in which nonlinearity, historical dependence, and unpredictability are often major features. These shifts have dramatic implications for how we manage and restore systems (Botkin 1990; Wallington et al. 2005).

These shifts obviously have important implications for how we choose and assemble the building blocks for the foundations of restoration ecology. If we build the foundation on outdated or erroneous building blocks, we can expect to see many more castles tumble as people repeat the mistakes of the past. If, however, we build the foundations on the best and sturdiest building blocks we have available, we have the chance to build even bigger and better castles. Developing a sound foundation based on current ideas and concepts provides restoration ecology with the opportunity to start being able to provide generalizable approaches that can be tried and tested in different systems and different places. This is why this book is so important. Certainly, we will find that not every idea presented here will stand the test of time or work in practice everywhere. But the editors and authors are building a foundation on which we can build and move forward. It's time for restoration ecology to have a solid foundation and to develop as the enabling science behind the truly awe-inspiring restoration activities needed in this increasingly human-damaged world.

Richard J. Hobbs, Perth, Australia

Literature Cited

Botkin, D. B. 1990. *Discordant harmonies. A new ecology for the twenty-first century*. Oxford, UK: Oxford University Press.

Bradshaw, A. D. 1987. Restoration: An acid test for ecology. In *Restoration ecology: A synthetic approach to ecological research*, ed. W. R. Jordan, M. E. Gilpin, and J. D. Aber, 23–30. Cambridge, UK: Cambridge University Press.

Thoreau, H. D. 1854. *Walden; or, Life in the woods*. Boston: Ticknor and Fields.

Wallington, T. J., R. J. Hobbs, and S. A. Moore. 2005. Implications of current ecological thinking for biodiversity conservation: A review of the salient issues. *Ecology and Society* 10:15. http://www.ecologyandsociety.org/vol10/iss1/art15/.

ACKNOWLEDGMENTS

Books such as this one become a reality only with the support and involvement of many people. The Editors wish to thank some of those who have contributed to our work over a period of several years in fulfillment of our vision to bring this book together.

First, we express our thanks to Paul Zedler and Bill Halvorson (ESA and SER program chairs respectively) and symposium reviewers for accepting our original symposium at the joint meeting of those two great organizations, where we first tried out some of the ideas represented in this book. In addition, we thank other participants in the original ESA–SER symposium for contributing their ideas, including Edie Allen, Sara Baer, Steve Handel, Alan Hastings, Bruce Pavlik, Charles Peterson, and Joan Roughgarden.

As the book developed, we were consistently astounded at the remarkable insights and contributions of the authors of the chapters in this book. Collectively they are a remarkable group of creative scientists, and this book truly stands on their shoulders.

All of us have had the benefit of discussions about restoration ecology over many years. The Editors wish to thank all of our colleagues who have generously shared their ideas and perspectives on a range of topics in restoration ecology, especially Dave Allan, Craig Allen, Edith Allen, James Aronson, Emily Bernhardt, Julio Betancourt, John Callaway, Brad Cardinale, André Clewell, Wally Covington, Meo Curtis, Sheila David, Suzanne van Drunick, Walter Dunn, Dave Egan, Mima Falk, Alan Watson Featherstone, Pete Fulé, George Gann, Ed Guerrant, Bill Halvorson, Steve Handel, Dan Harper, Emily Heyerdahl, Eric Higgs, Laura Jackson, William Jordan III, Kathryn Kennedy, Mary Kay LeFevour, Don McKenzie, Guy McPherson, Carol Miller, Gary Nabhan, Peggy Olwell, Steve Packard, LeRoy Poff, Peter Raven, Rob Robichaux, Bill Romme, Melissa Savage, Tom Sisk, Sean Smith, Nate Stephenson, Chris Swan, Tom Swetnam, Keith VanNess, Cameron Weigand, Peter White, Steve Windhager, Truman Young, and Paul Zedler. We also thank our lab members, graduate students, and postdoctoral collaborators, many of whom have carried ideas from the classroom into important research and practice.

We are grateful to Barbara Dean, Barbara Youngblood, and the staff of Island Press for their encouragement and commitment to excellence in publishing in restoration ecology. James Aronson supported inclusion of this work in the Island Press–SER series, *Science and Practice of Ecological Restoration*, where we always thought this work belonged. Susan

Halverson and Emily Sievers did yeoman work at the University of Wisconsin compiling material for the final manuscript submission.

Finally, we offer thanks to our home institutions for providing logistic and financial support during our work on this book. We hope that this work will contribute to the stature of our universities as places where scientists and policy-makers can come together to envision a sustainable future for the planet.

D.A.F., M.A.P., J.B.Z

Chapter 1

Ecological Theory and Restoration Ecology

MARGARET A. PALMER, DONALD A. FALK, AND JOY B. ZEDLER

Ecological restoration has been practiced in some form for centuries. For instance, many indigenous peoples tended lands to sustain natural ecosystem services, such as production of basket-weaving materials, food crops, or forage for game animals, and they continue to do so (Stevens 1997). Today, the practice of ecological restoration is receiving immense attention because it offers the hope of recovery from much of the environmental damage inflicted by misuse or mismanagement of the Earth's natural resources, especially by technologically advanced societies (*Economist* 2002; Malakoff 2004).

Strictly speaking, ecological restoration is an attempt to return a system to some historical state, although the difficulty or impossibility of achieving this aim is widely recognized. A more realistic goal may be to move a damaged system to an ecological state that is within some acceptable limits relative to a less disturbed system (Figure 1.1). In this sense of the term, ecological restoration can be viewed as an attempt to recover a natural range of ecosystem composition, structure, and dynamics (Falk 1990; Allen et al. 2002; Palmer et al. 2005). Correspondingly, restoration ecology is the discipline of scientific inquiry dealing with the restoration of ecological systems.

The simplest restorations involve removing a perturbation and allowing the ecosystem to recover via natural ecological processes. For example, a small sewage spill to a large lake might correct itself, if microorganisms can decompose the organic matter and the added nutrients do not trigger algal blooms. Locally extirpated species can recolonize sites as habitat quality improves, and the physical structure of communities can begin to resemble the predisruption condition.

More often, however, restoration requires multiple efforts, because multiple perturbations have pushed ecosystems beyond their ability to recover spontaneously. For example, restoring streams affected by urbanization often requires new stormwater infrastructure to reduce peak flows, followed by channel regrading and riparian plantings (Brown 2000). For coastal marshes that have been dredged for boat traffic, restoration might involve removing fill, recontouring intertidal elevations, amending dredge spoil substrates, and introducing native plants. In some cases, "restoration" *sensu latu* is never finished, as some level of maintenance is always needed (e.g., in wetlands dominated by invasive species). Full restoration means that the ecosystem is once again resilient—it has the capacity to recover from stress (SERI 2002; Walker et al. 2002). Yet it is rarely possible to achieve the self-sustaining state

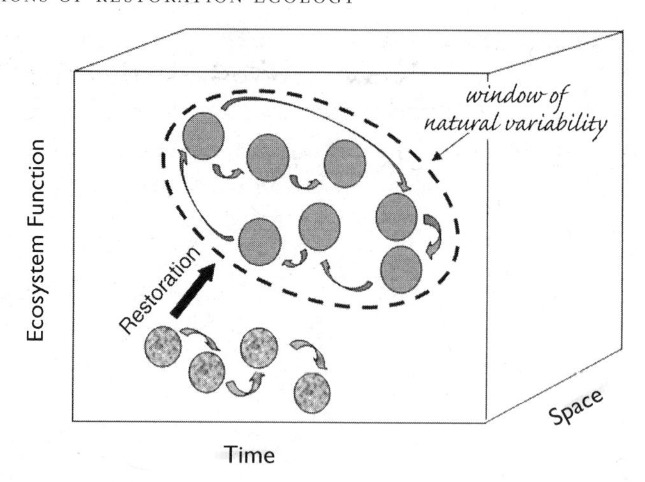

FIGURE 1.1 Ecological systems are highly dynamic entities. Thus, all attributes of natural systems, including levels of ecosystem processes (dark grey spheres), vary over time and space within a natural window of variability (dashed oval line). Restoration should be attempted when the system attribute moves outside that window of natural variability (mottled grey spheres). Once "restored," the system is unlikely to be exactly where it was predisturbance. Although this figure is drawn in three dimensions, the true assessment of both reference and degraded conditions is likely to be multivariate. Illustration motivated by Walker and Boyer (1993).

because degraded ecosystems typically lack natural levels of environmental variability (Baron et al. 2002; Pedroli et al. 2002) and their resilience is no longer recoverable (Suding et al. 2004).

While restoration is sometimes considered an art or a skill that is honed by practice and tutelage (Van Diggelen et al. 2001), science-based restorations are those projects that benefit from the infusion of ecological theory and application of the scientific method. Science-based restorations follow (1) explicitly stated goals, (2) a restoration design informed by ecological knowledge, and (3) quantitative assessment of system responses employing pre- and postrestoration data collection. Restoration becomes adaptive when a fourth step is followed: (4) analysis and application of results to inform subsequent efforts (Zedler and Callaway 2003). Analogous to adaptive management, the corrections that are made to the restoration process should be guided by sound theory and experimentation, not just trial and error.

An unfortunate aspect of ecological restoration as it is commonly practiced today is that the results of most efforts are not easily accessible to others. Despite pleas to report long-term responses (Zedler 2000; Lake 2001), most projects are not monitored postrestoration (NRC 1992; Bernhardt et al. 2005). Informing later efforts is in many ways the most critical element—science, in its simplest form, is the sequential testing of ideas that over time leads to a better understanding of nature.

Ecological Experimentation in a Restoration Context

The focus of this book is the mutual benefit of a stronger connection between ecological theory and the science of restoration ecology. Ecological restoration provides exciting opportunities to conduct large-scale experiments and test basic ecological theory, both of which have the potential to build the science of restoration ecology (Figure 1.2). A fundamental premise of this book is that the relationship of restoration ecology to ecological theory works in both directions: restoration ecology benefits from a stronger grounding in basic theory, while ecological theory benefits from the unique opportunities for experimentation in a restoration context (Palmer et al., 1997). Many examples of this reciprocity are found throughout this book.

Although ecology overall lacks a general unified theory, the field has developed a strong and diverse body of theory addressing nearly every aspect of ecological interactions (Weiner 1995; McPherson and DeStefano 2003). As evidenced throughout the book, this body of theory is highly relevant to both the science of restoration ecology and the practice of ecological restoration (Table 1.1). While ecological restoration has scientific underpinnings, the integration of ecological theory and restoration has been uneven, despite recognition that the practice could be enhanced by such integration (Young et al. 2005).

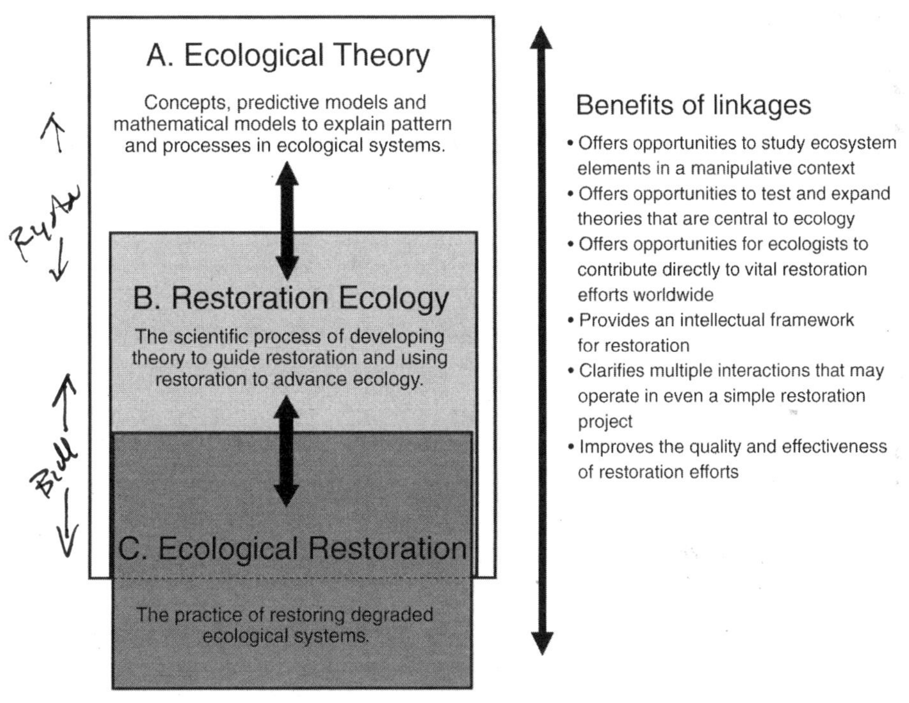

FIGURE 1.2 The relationship between ecological theory, restoration ecology, and ecological restoration can be viewed in a hierarchical fashion. While there is a very large body of ecological theory (A, unfilled box), only some of it can be directly applied to restoration ecology at the present time (B, grey box). There is thus a demand to extend and develop theory, and the benefits of doing so extend in both directions. Ecological science benefits from the linkage, as does restoration ecology and ecological restoration. There is also a large part of ecological restoration that will never be guided by restoration ecology (C, black box); instead, contextual constraints and societal objectives, such as co-opting natural resources or modifying ecological systems for human use, will determine restoration objectives and potential much of the time.

TABLE 1.1

Broad areas of ecological theory that are foundational to the science of restoration ecology and are covered in the book.

Relevant ecological theory	Examples of ecological restoration questions	Examples of current themes, issues, and models	Contributors
Population and ecological genetics	Which propagule sources and numbers should be introduced?	Bottlenecks and founder events, drift in small populations, locally adapted genotypes, within- and among-population genetic diversity, inbreeding and outbreeding effects, genetic neighborhoods and spatial genetics, effective population size, gene flow	Falk, Richards, Montalvo, and Knapp (*Chapter* 2)
Ecophysiological and functional ecology	What are the potential physiological challenges in the restored environment?	Stress tolerance, physiological limits of survival and reproduction, adaptation to novel environments, phenotypes tolerant of unusual conditions	Ehleringer and Sandquist (*Chapter* 3)
Demography, population dynamics, metapopulation ecology	How can we tell if populations will persist?	Population dynamics, demographic transition matrices, seed dormancy and germination, population persistence and resilience, population spatial structure, age structure and density dependence, dispersal among subpopulations, metapopulation dynamics	Maschinski (*Chapter* 4)
Community ecology	What assemblages will persist in each part of the restored site? In what order should they be introduced?	Community composition, coexistence of species, assembly theory, alternative successional pathways, sensitivity to initial conditions, predation, trophic structure, dispersal, environmental filters, disturbance regimes, mutualism	Menninger and Palmer (*Chapter* 5)
Evolutionary ecology	How will organisms adapt to novel restored environments?	Evolutionary environment, adaptation to novel environments, trait selection, metapopulations, genetic diversity, evolutionary potential, landscape genetics	Stockwell, Kinnison, and Hendry (*Chapter* 6)
Fine-scale heterogeneity	How can sites be modified to enhance diversity?	Spatial heterogeneity of resources and ecosystem functionality, spatial and temporal variation at individual or community level, coexistence of multiple species at multiple spatial scales	Larkin, Vivian-Smith, and Zedler (*Chapter* 7)
Food webs	Do interacting species need to be introduced?	Trophic cascades, bottom-up/top-down dynamics, food-web networks, productivity and food-web structure, plant-herbivore interactions, predator-prey theory, indirect interactions	Vander Zanden, Olden, and Gratton (*Chapter* 8)
Ecological dynamics and trajectories	How will the restored system develop?	Trajectories of ecosystem degradation and recovery, natural variability, linear and nonlinear dynamics, multiple stable states vs. ordered succession, resilience, multiple equilibria, ecological thresholds	Suding and Gross (*Chapter* 9)

TABLE 1.1 (*continued*)

Broad areas of ecological theory that are foundational to the science of restoration ecology and are covered in the book.

Relevant ecological theory	Examples of ecological restoration questions	Examples of current themes, issues, and models	Contributors
Biodiversity and ecosystem functioning	Can a single restoration site maximize species richness and ecosystem functions?	Diversity-stability relationships, functional diversity, functional equivalence, redundancy, interface between community and ecosystem ecology, ecological insurance and ecosystem reliability	Naeem (*Chapter 10*)
Modeling and simulations	How predictable are restoration outcomes?	Stochastic influences on deterministic processes, uncertainty, natural range of variability, spatial interactions, heuristic and simulation models, multivariate statistics	Urban (*Chapter 11*)
Invasive species and community invasibility	How should sites be managed to exclude undesired species?	Properties of invasive species, community invasibility, alteration of ecosystem processes, plant community responses, resistance and resilience, competition, top-down and bottom-up control, disturbance theory	D'Antonio and Chambers (*Chapter 12*)
Research design and statistical analysis	How can we design restoration experiments and analyze the resulting data?	Replication, power analysis, sample size, general statistical framework, time series and repeated measures, chronosequence analysis, multivariate characterization, estimating effect size, BACI designs	Osenberg, Bolker, White, St. Mary, and Shima (*Chapter 13*)
Macroecology	How does the larger spatial context influence an individual restored site?	Large-scale ecological processes, species and population migrations over time and space, ecosystem size and community diversity/structure, cross-system fluxes	Maurer (*Chapter14*)
Paleoecology, climate change	Can restoration be planned within the context of expected global change?	Climatic cycles, climate-vegetation relationships and migration of vegetation, vegetation-climate (dis)equilibrium, natural variability, temporal variation	Millar and Brubaker (*Chapter 15*)

There is also great potential to enhance understanding of the basic structure and function of ecological systems by using restoration settings to develop and test theory (Bradshaw 1987; Jordan et al. 1987; Palmer et al. 1997; Hobbs and Harris 2001; Perrow and Davy 2002). Indeed, restored sites, or those that are soon to be restored, represent virtual playgrounds for asking how well ecological theories can predict the responses of natural systems.

The opportunity to test ecological theory in restoration sites is exciting; at the same time, ecologists and evolutionary biologists are challenged to use theory to devise experiments that can be conducted in restoration settings. We do not think this limits our inquiry to a reductionist paradigm: as with ecology itself, understanding can progress even when formal hypotheses cannot be framed (Pickett et al. 1994). Even more difficult is the challenge of designing experiments that are workable within a project's spatial extent, timing constraints, and resources. Finding suitable sites, receptive managers, interested researchers, appropriate ideas to test, and funding to test them—all at the same time and place—is challenging, but feasible and worth the effort. The payoff for the practice of ecological restoration comes in learning how to improve approaches, how to correct errors, how to accomplish desired outcomes, and how to plan future projects.

Can basic ecological abstractions of nature and mathematical models be used to inform restoration practice, given that ecological responses are often context-dependent? We think so. Every step in the restoration process can be informed by existing ecological theory (Table 1.1); however, every attempt to state predictions from theory also indicates the need to expand theory itself. Thus, we ask: Under what circumstances can we grow the science of restoration ecology using existing ecological theory? What issues or settings require an extension of our theories and models or even the development of theories de novo?

The Imperative to Advance Theory

Experience indicates that restoration follows multiple pathways, which means that outcomes are difficult to predict. Part of the difficulty is that restoration takes place across a multidimensional spectrum of specific sites within various kinds of landscapes, and where goals range from highly specific (e.g., enhance the population of one rare animal species) to general (e.g., encourage vegetation to cover bare substrate). The task of developing theory that offers a high level of predictability is akin to figuring out how to grow myriad crops across a heterogeneous continent. If we consider the centuries it has taken agriculturalists to optimize the crops that farmers should grow in one field in one region (e.g., alternate corn and beans or alfalfa within the cornbelt using specified soil amendments, planting, and harvesting protocols), the difficulty of reproducing entire ecosystems on demand becomes understandable. It could take much longer for the science of restoration to achieve predictable results, because there are more ecosystem types and a wider variety of tools. We assert that these conditions create a great need for guidance from ecological theory. For some ecosystems, ecological theory needs to be melded with physical science theory; for example, river restoration must be informed by geomorphic, hydrological, and ecological theory (Wohl et al. 2005; Palmer et al. 2006).

The need to develop a sound theoretical base for ecological restoration is imperative for at least three reasons. First, restoration is a booming business that requires the support of a knowledge base and research innovations (*Economist* 2002). Billions of dollars are spent annually to restore polluted and sediment-clogged streams (Bernhardt et al. 2005; Hassett et al. 2005) and to reforest lands that have been degraded and fragmented (Lamb and Gilmore 2003). Yet many restoration efforts are still trial-and-error improvisations. For example, every new biological invasion prompts a series of attempts to reduce or eradicate populations that increasingly damage native communities. Systematic evaluations of multiple tools in a common site come only after long delays in recognizing the magnitude of the problem and obtaining the resources to fund appropriate research.

Second, the stakes are far too high *not* to develop a stronger theory for restoration ecology. As the global human population continues to expand, vital resources, such as fresh water and arable soils, are threatened and depleted (Gleick 2003; McMichael et al. 2003; Stocking 2003). Obviously, conservation of resources prior to their degradation is desirable, but our crowded planet's current rate of resource consumption suggests that we must do more than hold the line (Sugden et al. 2003; Palmer et al. 2004). Where conservation has failed to sustain crucial ecological services, ecological restoration should be the option of choice (Dobson et al. 1997; Young 2000; Ormerod 2003). Given the state of our environment, restoration must use ecologically designed solutions (Pimm 1996; Palmer et al. 2004); our only other recourse is technological fixes to maintain ecosystem processes, an expensive and often ineffective option. Admittedly, some ecological technology (e.g., waste treatment) can improve people's lives, but many problems (e.g., spatially distributed water shortages) cannot be solved by technology, at least not affordably (Gleick 2003). Furthermore, technological fixes lack the aesthetic appeal of restored ecosystems and the species they support.

A third reason to enhance the linkage between ecological theory and restoration is to grow the field of ecology. Regardless of their specialty, ecologists can benefit greatly by testing theory in a restoration context (Palmer et al. 1997; Young et al. 2001). As Bradshaw (1987) noted, restoration is the "acid-test of ecological theory." If we cannot predict the development of a community at a restored or managed site based on knowledge of species and their interactions, then perhaps we can make use of what we observe to refine our theories and predictions and improve their predictive power (Zedler 2000; Hobbs and Harris 2001).

Origins and Structure of This Book

The fields of *ecological restoration* and *restoration ecology* have been well served by two journals of those same names for many years. Since their inception, these journals have published hundreds of articles on topics ranging from tools, techniques, research ideas, results, and philosophy. Today, articles on restoration also appear in mainstream ecological journals (e.g., *Ecological Applications, Journal of Applied Ecology, Science*). Yet, despite years of intellectual development, restoration ecology remains to be defined as a field of scientific endeavor and its conceptual foundations articulated. This realization ultimately is what led us to create this book.

Initially, we organized a symposium (Palmer et al. 2002) for the 2002 joint meeting of the Ecological Society of America (ESA) and the Society for Ecological Restoration International (SERI). In some respects, the 2002 symposium was a follow-up to a previous (1996) meeting of ecologists and land managers at the National Center for Ecological Analysis and Synthesis (NCEAS) to discuss the conceptual basis of restoration ecology (Allen et al. 1997). This culminated in a series of journal articles (Allen et al. 1997; Ehrenfeld and Toth 1997; Michener 1997; Montalvo et al. 1997; Palmer et al. 1997; Parker 1997; White and Walker 1997) devoted to identifying the conceptual framework for restoration ecology and outlining critical research questions that offer unique opportunities to couple basic research with the practical needs of restorationists. Our hope was to move both ecology and the field of restoration ecology forward.

For the 2002 symposium, we asked scientists well versed in ecological theory—but not necessarily active in restoration work—to present their most creative ideas on the linkage (real or potential) between ecological theory and restoration ecology. We also asked scientists actively involved in restoration research to illustrate how ecological theory has been coupled with restoration efforts and/or how they have tested ecological theory in a restoration context. This emphasis on two-way communication of ideas between ecological theorists and restoration ecologists is carried forward in this volume.

Selecting the topics to include in this book was not easy. We have used the word *theory* broadly to include ecological and evolutionary concepts, predictive models, and mathematical models. We organized the book around the ecological concepts and principles that are fundamental to restoration. Our goals were to provide comprehensive overview of the theoretical foundations of restoration ecology, and to identify critical areas in which new theory is needed, existing theory needs to be tested, and new and exciting cross-disciplinary questions need to be addressed.

Each chapter in this book addresses a particular area of ecological theory. Some of these (e.g. population genetics, demography, community ecology) are traditional levels of biological hierarchy, while others (species interactions, fine-scale heterogeneity, successional trajectories, invasive species ecology, ecophysiology, and functional ecology) explore specific topics of central relevance to the challenges of restoration ecology. Several chapters focus on research tools (research design, statistical analysis, modeling, and simulations), or place restoration ecology research in a larger context (macroecology, paleoecology and climate change, evolutionary ecology). Some areas merit more specific coverage, including ecosystem processes (e.g., restoration of biogeochemical processes) and landscape-level spatial ecology, both of which are highly relevant to restoration and merit further work. Other important areas fell outside the scope of this book, and we urge readers to consult other sources for information on the economics of ecological restoration; on sociological issues, such as stakeholder "buy-ins" that often determine the success of a project; and on engineering principles and technical issues that are required for some types of restoration.

We have organized the book into parts reflecting three general areas of ecological theory (levels of biological hierarchy, restoring ecological functions and processes, and the macroecological context). Each part is introduced briefly by the Editors. The chapters follow a common structure designed to assist the reader, particularly the student new to the field. After a brief introduction to the general area and its significance within ecological research, each chapter summarizes the body of theory most relevant to restoration ecology, including its central concepts and models, current issues, and front lines of research. The authors then discuss the application of this body of theory to restoration ecology as specifically as possible, with references to the restoration literature, where possible. The chapters end with perspectives on (1) tests of ecological theory research that could help build and strengthen restoration ecology, and (2) how restoration offers opportunities to test ideas in basic ecology.

This book is meant to provide a scientific framework for restoration ecology that can be used to inform ecological restoration as well as stimulate advances in our understanding of nature. As you read, bear in mind that the implementation of ecological restoration is not only escalating at an astounding rate, but also that it remains the most ecologically viable and aesthetically appealing remedy for mending Earth's ever-increasing number and scale of degraded ecosystems.

Literature Cited

Allen, C. D., M. Savage, D. A. Falk, K. F. Suckling, T. W. Swetnam, T. Schulke, P. B. Stacey, P. Morgan, M. Hoffman, and J. T. Klingel. 2002. Ecological restoration of southwestern Ponderosa pine ecosystems: A broad perspective. *Ecological Applications* 12 (5): 1418–1433.

Allen, E., W. W. Covington, and D. A. Falk. 1997. Developing the conceptual basis for restoration ecology. *Restoration Ecology* 5 (4): 275–276.

Baron, J. S., N. L. Poff, P. L. Angermeier, C. N. Dahm, P. H. Gleick, N. G. Hairston, R. B. Jackson, C. A. Johnston, B. D. Richter, and A. D. Steinman. 2002. Meeting ecological and societal needs for freshwater. *Ecological Applications* 12 (5): 1247–1260.

Bernhardt, E. S., M. A. Palmer, J. D. Allan, G. Alexander, S. Brooks, S. Clayton, J. Carr, C. Dahm, J. Follstad-Shah, D. L. Galat, S. Gloss, P. Goodwin, D. Hart, B. Hassett, R. Jenkinson, G. M. Kondolf, S. Lake, R. Lave, J. L. Meyer, T. K. O'Donnell, L. Pagano, P. Srivastava, and E. Sudduth. 2005. Restoration of U.S. rivers: A national synthesis. *Science* 308:636–637.

Bradshaw, A. D. 1987. Restoration: An acid test for ecology. In *Restoration ecology: A synthetic approach to ecological research*, ed. W. R. Jordan, M. E. Gilpin, and H. J. D. Aber, 23–29. Cambridge, UK: Cambridge University Press.

Brown, K. 2000. *Urban stream restoration practices: An initial assessment*. Ellicott City, MD: Center for Watershed Protection. http://www.cwp.org/stream_restoration.pdf.

Dobson, A., A. D. Bradshaw, and A. J. M. Baker. 1997. Hopes for the future: Restoration ecology and conservation biology. *Science* 277:515–522.

Economist, The. 2002. Restoration drama. August 8.

Ehrenfeld, J. G., and L. A. Toth. 1997. Restoration ecology and the ecosystem perspective. *Restoration Ecology* 5 (4): 307–317.

Falk, D. A. 1990. Discovering the past, creating the future. *Restoration and Management Notes* 8 (2): 71–72.

Gleick, P. H. 2003. Global freshwater resources: Soft-path solutions for the 21st century. *Science* 302: 1524–1527.

Hassett, B., M. A. Palmer, E. S. Bernhardt, S. Smith, J. Carr, and D. Hart. 2005. Status and trends of river and stream restoration in the Chesapeake Bay watershed. *Frontiers in Ecology and the Environment* 3:259–267.

Hobbs, R. J., and J. A. Harris. 2001. Restoration ecology: Repairing the Earth's ecosystems in the new millennium. *Restoration Ecology* 9:209–219.

Jordan, W. R., M. E. Gilpin, and H. J. D. Aber, editors. 1987. *Restoration ecology: A synthetic approach to ecological research*. Cambridge, UK: Cambridge University Press.

Lake, P. S. 2001. On the maturing of restoration: Linking ecological research and restoration. *Ecological Management and Restoration* 2:110–115.

Lamb, D., and D. Gilmour. 2003. *Rehabilitation and restoration of degraded forests*. Washington, DC: World Wildlife Fund.

Malakoff, D. 2004. The river doctor. *Science* 305:937–939.

McMichael, A. J., C. D. Butler, and C. Folke. 2003. New visions for addressing sustainability. *Science* 302:1919–1921.

McPherson, G. R., and S. DeStefano. 2003. *Applied ecology and natural resource management*. Cambridge, UK: Cambridge University Press.

Michener, W. K. 1997. Quantitatively evaluating restoration experiments: Research design, statistical analysis, and data management considerations. *Restoration Ecology* 5 (4): 324–337.

Montalvo, A. M., S. L. Williams, K. J. Rice, S. L. Buchmann, C. Cory, S. N. Handel, G. P. Nabhan, R. Primack, and R. H. Robichaux. 1997. Restoration biology: A population biology perspective. *Restoration Ecology* 5 (4): 277–290.

National Research Council (NRC). 1992. *Restoration of aquatic ecosystems*. Washington, DC: National Academy Press.

Ormerod, S. J. 2003. Restoration in applied ecology: Editor's introduction. *Journal of Applied Ecology* 40:44–50.

Palmer, M. A., R. F. Ambrose, and N. L. Poff. 1997. Ecological theory and community ecology. *Restoration Ecology* 5 (4): 291–300.

Palmer, M. A., J. B. Zedler, and D. A. Falk. 2002. Ecological theory and restoration ecology. *Ecological Society of America (ESA) 2002 Annual Meeting Abstracts*. http://abstracts.co.allenpress.com/pweb/esa2002.

Palmer, M. A., E. Bernhardt, E. Chornesky, S. Collins, A. Dobson, C. Duke, B. Gold, R. Jacobson, S. Kingsland, R. Kranz, M. Mappin, M. L. Martinez, F. Micheli, J. Morse, M. Pace, M. Pascual, S. Palumbi, O. J. Reichman, A. Simons, A. Townsend, and M. Turner. 2004. Ecology for a crowded planet. *Science* 304:1251–1252.

Palmer, M. A., E. Bernhardt, J. D. Allan, G. Alexander, S. Brooks, S. Clayton, J. Carr, C. Dahm, J. Follstad-Shah, D. L. Galat, S. Gloss, P. Goodwin, D. Hart, B. Hassett, R. Jenkinson, G. M. Kondolf, S. Lake, R. Lave, J. L. Meyer, T. K. O'Donnell, L. Pagano, P. Srivastava, and E. Sudduth. 2005. Standards for ecologically successful river restoration. *Journal of Applied Ecology* 42:208–217.

Palmer, M. A., and E. S. Bernhardt. 2006. Scientific pathways to effective river restoration. *Water Resources Research*, 42(3): W03507.

Parker, V. T. 1997. The scale of successional models and restoration objectives. *Restoration Ecology* 5 (4): 301–306.

Pedroli, B., G. de Blust, K. van Looy, and S. van Rooij. 2002. Setting targets in strategies for river restoration. *Landscape Ecology* 17:5–18.

Perrow, M. R., and A. J. Davy, editors. 2002. *Handbook of ecological restoration*. Volumes 1 and 2. Cambridge, UK: Cambridge University Press.

Pickett, S. T. A., J. Kolasa, and C. D. Jones. 1994. *Ecological understanding: The nature of theory and the theory of nature*. San Diego: Academic Press.

Pimm, S. L. 1996. Designer ecosystems. *Nature* 379:217–218.

Society for Ecological Restoration International (SERI). 2002. *The SERI primer on ecological restoration*. Tucson: SERI Science and Policy Working Group. www.seri.org/.

Stevens, S., editor. 1997. *Conservation through cultural survival: Indigenous peoples and protected areas*. Washington, DC: Island Press.

Stocking, M. A. 2003. Tropical soils and food security: The next 50 years. *Science* 302:1356–1359.

Suding, K. N., Gross, K. L., and D. R. Housman. 2004. Alternative states and positive feedbacks in restoration ecology. *Trends in Ecology & Evolution* 19:46–53.

Sugden, A., C. Ash, B. Hanson, and J. Smith. 2003. Where do we go from here? *Science* 302:1906.

Van Diggelen, R., A. P. Grootjans, and J. A. Harris. 2001. Ecological restoration: State of the art or state of the science? *Restoration Ecology* 9:115–118.

Walker, J. L., and W. D. Boyer. 1993. An ecological model and information needs assessment for longleaf pine ecosystem restoration. In *Silviculture: From the cradle of forestry to ecosystem management*, comp. L. H. Foley, 138–144. General Technical Report SE-88. Southeastern Forest Experiment Station: USDA, Forest Service.

Walker, B., S. Carpenter, J. Anderies, N. Abel, G. S. Cumming, M. Janssen, L. Lebel, J. Norberg, G. D. Peterson, and R. Pritchard. 2002. Resilience management in social-ecological systems: A working hypothesis for a participatory approach. *Conservation Ecology* 6:14.

Weiner, J. 1995. On the practice of ecology. *Journal of Ecology* 83:153–158.

White, P. S., and J. L. Walker. 1997. Approximating nature's variation: Selecting and using reference sites and reference information in restoration ecology. *Restoration Ecology* 5:338–349.

Wohl, E., P. L. Angermeier, B. Bledsoe, G. M. Kondolf, L. MacDonnell, D. M. Merritt, M. A. Palmer, N. L. Poff, and D. Tarboton. 2005. River restoration. *Water Resources Research* (41)10:W10301.

Young, T. P. 2000. Restoration ecology and conservation biology. *Biological Conservation* 92:73–82.

Young, T. P., J. M. Chase, and R. T. Huddleston. 2001. Community succession and assembly: Comparing, contrasting and combining paradigms in the context of ecological restoration. *Ecological Restoration* 19:5–18.

Young, T. P., D. A. Petersen, and J. J. Clary. 2005. The ecology of restoration: Historical links, emerging issues and unexplored realms. *Ecology Letters* 8:662–673.

Zedler, J. B. 2000. Progress in wetland restoration ecology. *Trends in Ecology & Evolution* 15:402–407.

Zedler, J. B., and J. C. Callaway. 2003. Adaptive restoration: A strategic approach for integrating research into restoration projects. In *Managing for Healthy Ecosystems*, ed. D. J. Rapport, W. L. Lasley, D. E. Rolston, N. O. Nielsen, C. O. Qualset, and A. B. Damania, 164–174. Boca Raton: Lewis Publishers.

PART ONE

Ecological Theory and the Restoration of Populations and Communities

For decades, biologists have conceived of life in terms of nested levels of biological organization, from subcellular structures to processes that extend across and among whole ecosystems. This hierarchy obscures much of reality: what happens to the whole organism ramifies inevitably to the life of the cell, and no organism exists independently of interactions with other organisms or the physical environment. Nonetheless, levels of biological organization have proven to be a powerful conceptual tool for understanding how the processes of life are distributed among many components.

It is this understanding of interacting processes at different levels of organization that makes the biological hierarchy model relevant to restoration ecology. Inevitably, the restoration ecologist initiates or observes changes across multiple levels of organization. For example, setting a prescribed fire in a tallgrass prairie changes soil, water, and nutrient levels and availability; changes whole-plant water status and mycorrhizal function; alters the competitive balance among species; influences the demography of populations through size and age-dependent mortality; modifies seed germination, dispersal, and establishment rates; and redirects the flow of carbon, water, and energy through the ecosystem. None of these outcomes is fully independent of any other. And, yet, to advance restoration ecology, we need to probe inside such complex cross-scale processes in order to understand the mechanisms at work.

While the hierarchy of life extends from the infinitely small to the globally large, the genetic composition of individual organisms and populations is a convenient and logical base. Falk, Richards, Montalvo, and Knapp explore the importance of genetic variation to restoration ecology. Genetic variation provides both the potential for, and limitations of, organism responses to novel environments. In addition to defining the envelope of individual response, the distribution of genetic variability within and among populations can be crucial to restoration efforts, particularly where the practical goal of restoration is to promote the establishment of self-sustaining populations. Genetic variation is thus the (often unseen) foundation of the biological outcomes of restoration, and it merits increased attention in both practice and research.

The response of organisms to both degraded and restored ecosystems is mediated largely through physiological processes: how the bodies of organisms work. Ehleringer and Sandquist examine this important area for restoration ecology, using the example of how plant ecophysiology can govern the outcomes of restoration experiments. Whether by design or accident, the restorationist alters the local distribution of light and energy aboveground, and water and chemistry belowground. Postrestoration conditions may favor plants with different photosynthetic pathways, as well as species that can tolerate novel microclimatic conditions of temperature and humidity. Belowground, plant tolerance for changes in macro- and micronutrients, as well as exposure to increased concentrations of toxic metals and changes to soil pH, salinity, and water, can be critical to the outcome of restoration. The altered hydrological environment can influence plant rooting depth, root/shoot allocation, and the reliance on mycorrhizal symbioses. Restoration exposes plants to a wide range of physiological stressors, and outcomes will depend on the ability of species to tolerate altered environments.

Plant or animal populations are frequently the focal point of restoration research and practice. Population size is a basic metric of restoration success or failure, as is the variability of populations in space and time, particularly in uncertain environments. Maschinski describes some of the contemporary tools of population biology that are most relevant to restoration ecology, especially population viability analysis, which incorporates the effects of population size and demographic and environmental stochasticity. Elasticity and sensitivity analysis, which measure the effects on population growth rates of changes in vital rates (birth, death, growth) are particularly promising tools for restoration research. Beyond the individual population, metapopulation theory has high potential for helping restorationists design long-term strategies. The reason is simple: a completely isolated population is unlikely to survive for long in a variable environment. Exchange of genes and individuals among subpopulations is a fundamental dynamic in a natural landscape, and no less so as we try to restore sustainable populations in fragmented and altered landscapes.

A complex community with multiple coexisting and persisting species is a common restoration objective. As Menninger and Palmer observe, however, assemblages of species and population levels are rarely stable over space and time. Community-level restoration ecology must thus address both the emergent patterns of species coexistence and the underlying processes that govern community composition. Three levels of community function are relevant to restoration ecology: Regional processes, which determine species composition through the regional species pool, dispersal, and colonization dynamics, operate within and among sites. Environmental and habitat attributes constitute a set of biotic and abiotic filters that govern which species are likely to establish and persist. And, finally, biotic interactions are highly variable, ranging from directly competitive to mutualistic and potentially varying with circumstances. A deeper understanding of all of these community-level processes will be fundamental to restoration ecology.

While restoration projects typically span a few months or years, restoration ecology needs to extend to time frames that include slow processes and very long-term outcomes. Stockwell, Kinnison, and Hendry address a challenging and little-explored aspect of restoration ecology, the evolutionary perspective. Both degraded (prerestoration) and restored environments may represent novel circumstances for many species. Where the local environment is significantly outside the envelope of conditions to which species are adapted, strong selection can alter gene frequencies rapidly, leading to the emergence of novel character distributions and even new ecotypes. The rate at which this occurs is influenced by the degree of genetic variability within and among populations, as well as dispersal, colonization, and survival. By manipulating these variables (for example, by moving propagules to a new location), the restoration ecologist inevitably influences processes of adaptation and species distributions. The evolutionary response to restoration thus integrates all levels of biological organization, from genes and ecophysiology to the structure and dynamics of metapopulations and the interaction of species in complex communities.

Chapter 2

Population and Ecological Genetics in Restoration Ecology

DONALD A. FALK, CHRISTOPHER M. RICHARDS, ARLEE M. MONTALVO, AND ERIC E. KNAPP

Genetic diversity serves as the basis for adaptive evolution in all living organisms. Heritable differences among individuals influence how they interact with the physical environment and other species, and how they function within ecosystems. Genetic composition affects ecologically important forms and functions of organisms, including body size, shape, physiological processes, behavioral traits, reproductive characteristics, tolerance of environmental extremes, dispersal and colonizing ability, the timing of seasonal and annual cycles, disease resistance, and many other traits (Lewontin 1974; Hedrick 1985; Booy et al. 2000; Lowe et al. 2004). Genetic diversity within a species thus provides the means for responding to environmental uncertainty and forms the base of the biodiversity hierarchy (Stebbins 1942; Crow 1986; Noss 1990; Hartl 1997; Reed and Frankham 2003). To overlook genetic variation is to ignore a fundamental force that shapes the ecology of living organisms.

Restoration ecologists are often faced with practical consequences of this variation when selecting plant and animal materials for restoration projects. Ecological genetics are thus fundamental to the design, implementation, and expectations of any restoration project, whether or not consideration of the genetic dimension is explicit. For these and many other reasons, genetic variability merits increased attention in restoration practice and research (Falk and Holsinger 1991; Fenster and Dudash 1994; Havens 1998; Young 2000; Rice and Emery 2003; Schaal and Leverich 2005).

In this chapter we outline some genetic considerations important to the design, implementation, and long-term success of populations in natural habitats. We begin by reviewing the fundamental importance of genetic variation in population ecology. We then discuss how genetic variation is measured and assessed at the levels of individuals and populations. We conclude by examining how genetic information can be used in restoration ecology and ecological restoration practice.

Why Is Genetic Variation Important to Restoration Ecology?

We begin our examination of genetic variation in restoration ecology by focusing on two aspects that are likely to be encountered in the restoration process: its importance in providing the basis for adaptation of organisms to changing environments, and its role in preventing or ameliorating deleterious effects of inbreeding in small or isolated populations.

Genetic Diversity Is the Primary Basis for Adaptation to Environmental Uncertainty

Genetic variation holds the key to the ability of populations and species to persist through changing environments over evolutionary time (Frankel 1974; Lewontin 1974; Freeman and Herron 1998; Stockwell et al. 2003). The magnitude and pattern of adaptive variation is critical for the long-term persistence of a species, whether endangered or widespread (Booy et al. 2000; Reed and Frankham 2003; Rice and Emery 2003).

Environments that vary in time and over space are often described in terms of the *natural* or *historic range of variability* in weather, disturbance events, resource availability, population sizes of competitors, and so forth (Morgan et al. 1994; White and Walker 1997; Swetnam and Betancourt 1998). In a completely stable physical and biological environment, a species might benefit more by maintaining a narrow range of genotypes adapted to prevailing conditions, and allele frequencies might eventually attain equilibrium (Rice and Emery 2003). By contrast, if the environment is patchy, unpredictable over time, or includes a wide and changing variety of diseases, predators, and parasites, then subtle differences among individuals increase the probability that some individuals and not others will survive to reproduce—that is, individuals will vary in *fitness* when traits influencing survival or reproduction are exposed to selection.

For example, Knapp et al. (2001) found that while individual blue oak trees flowered for less than ten days, different trees in the population initiated flowering over a period of a month in the spring. Such variability is potentially adaptive, since at least some trees in the population will flower during warm sunny periods when wind pollination is most successful and an acorn crop is more likely. Because differences among individuals are often determined at least in part by genes that are under selection, population genetic theory predicts that a broader range of genetic variation (higher heterozygosity) will persist in variable environments (Cohen 1966; Chesson 1985; Tuljapurkar 1989). For instance, within-population variability is central to the adaptation of desert annuals to uncertain precipitation regimes (Adondakis and Venable 2004).

On a longer time scale, during periods of rapid climate change or increased climatic variability, the zone of suitable climate for a species may shift in latitude and elevation. Populations with individuals containing different genes for adaptation to new climatic conditions are more likely to persist, and if their seeds are dispersed into the new location the population can "migrate" across the landscape over generations (Ledig et al. 1997). By contrast, populations with a narrower range of genotypes (more phenotypically uniform) may fail to survive and reproduce as conditions become less locally favorable. Such populations are more likely to become *extirpated* (locally extinct). The challenge to restoration ecology is to utilize sufficient diversity to allow adaptation to new circumstances, while avoiding the adverse effects of introducing genotypes that are poorly adapted to the environment (Rice and Emery 2003; Gustafson et al. 2004a, 2004b).

Genetic Diversity Within Populations Reduces Potentially Deleterious Effects of Inbreeding.

In addition to its adaptive value at the population level, genetic variation (or its lack) can affect the survival and performance of individuals. In a diploid organism, when an individual is *homozygous* at a gene or locus, mutations that are not beneficial to survival or reproduction are more likely to be expressed. This can lead to developmental, physiological, or behavioral problems of genetic origin, such as malformed physical structure, poor biochemical balance, improper organ formation and function, altered social behavior, and susceptibility to disease (Barrett and Kohn 1991; Hartl and Clark 1997; Schaal and Leverich 2005).

Homozygosity at key gene loci is a common result of *inbreeding*, which is mating among closely related individuals. Populations that are small, isolated, or subdivided into small groups because of restricted dispersal can be particularly susceptible to inbreeding and *inbreeding depression*, reduced overall fitness of organisms with low heterozygosity. In small or highly inbred populations, *genetic drift*, the chance selection of genotypes, can cause deleterious alleles to become either fixed or purged from the population (Templeton 1991; Husband and Schemske 1996; Keller and Waller 2002). If populations that have been fixed for different alleles are crossed, *heterosis* (increased vigor of hybrids) in progeny may indicate inbreeding depression (Keller and Waller 2002). Such increases in fitness are known as *genetic rescue*, which occurs when new genetic material is added to inbred populations (Hedrick 1995; Richards 2000; Ingvarsson 2001; Tallmon et al. 2004).

It is in a restoration context critical to distinguish the *census population* (the number of individuals counted) from the *effective population size* (Lande and Barrowclough 1987). For any population, the effective population size (N_e) is the number of individuals that contribute genes to succeeding generations. This number is typically smaller than the number of individuals in a population census, because not all individuals reproduce, and progeny numbers vary among individuals. Plants with a high proportion of self-fertilization, or those with unequal sex ratios can have much lower N_e than the census number, as do animal populations where breeding success is determined by behavior and social interactions. Obligate outcrossing species (where individuals cannot self-fertilize) are especially vulnerable to the effects of small N_e (Barrett and Kohn 1991; Schaal and Leverich 2005). Effective population size—and the components of breeding systems that influence it—are important considerations in collection strategies and the management of small remnant or restored populations (Barrett and Kohn 1991; Ryman and Lairke 1991; Allendorf 1994; Nunney and Elam 1994; Newman and Pilson 1997).

Heterozygosity is not always beneficial, nor does inbreeding always have adverse effects on the fitness of populations (Waser 1993; Byers and Waller 1999). In some circumstances, a population may be so well adapted to local circumstances that introducing alleles from other populations actually reduces its performance once populations hybridize (the extrinsic, or ecological form of *outbreeding depression*) (Templeton 1991; Waser 1993; Tallmon et al. 2004). Alternatively, isolated populations may have diverged and become so different, even if adapted to similar environmental conditions, that they suffer chromosomal mismatches or cytoplasmic incompatibilities when hybridized, reflecting loss of co-adaptation (the intrinsic, or genetic form of outbreeding depression) (Templeton 1994; Fenster and Galloway 2000; Montalvo and Ellstrand 2001; Galloway and Etterson 2004). The outcome can depend on multiple factors, including relatedness of hybridizing populations, their genetic architectures, and the immediate environment (Edmands and Timmerman 2003; Rogers and Montalvo 2004). If crossed populations are highly differentiated, then the deleterious effects of outcrossing may outweigh the beneficial effect of reducing the adverse effects of inbreeding.

Restoring populations that balance the adverse effects of excess inbreeding and outbreeding can thus be challenging (Havens 1998). An understanding of how genetic diversity is distributed within and among populations can provide clues to achieving such a balance. Studies that examine the relative fitness of translocated populations and effects of mating among individuals within and between populations can help inform decisions about populations used for restoration (Keller et al. 2000; Montalvo and Ellstrand 2001; Hufford and Mazer 2003).

How Is Genetic Variation Detected and Measured?

A fundamental problem in ecological genetics is to distinguish between variations caused by differences in genotype and those attributable to environment. The measurable outward appearance of a trait (phenotype) can be subdivided broadly into genetic and environmental (nongenetic) components (Lynch and Walsh 1998). Partitioning the observed trait variance permits evaluation of the relative importance of heritable variation in shaping morphology and other complex traits. Many organisms, however, can display a wide variety of phenotypic responses due to the environment that are ecologically important but not heritable (*phenotypic plasticity*). This flexibility in phenotype may in itself be adaptively important and may have a genetic basis (Scheiner 1993). The underlying variation and genetic basis of traits is often referred to as *genetic architecture*, and a variety of methods exist for assessing genetic variation, classified broadly as quantitative (biometric) and molecular.

Biometric Studies

Variation for heritable traits is often evaluated in common garden studies, by planting individuals from different source populations in a common environment so that genetic differences among individuals can be revealed (Clausen et al. 1940; Erickson 2004; Rogers and Montalvo 2004). Source populations and test locations often focus on the range of habitat variation within a geographic region of interest. Plantings are ideally *reciprocal*, such that every accession is planted at every test location.

Common garden experiments can reveal adaptive differentiation among ecotypes by partitioning the variance in trait values. For instance, a two-way analysis of variance model can estimate the significance of the genotype–environment ($G \times E$) interaction in a common garden trial (Comstock and Moll 1963; Lynch and Walsh 1998). Reciprocal transplants and common garden experiments allow the researcher to evaluate the relative performance of genotypes in different environments and look for the signature of home site advantage indicative of ecotypes, or to differentiation on opposing ends of an environmental cline (Montalvo and Ellstrand 2000).

In restoration projects, data from genotypes grown in multiple environments can help predict the success (or risk) of seed translocation, transplanting, or augmentation of declining populations from different sources. In addition, the risk of inbreeding and outbreeding depression and the potential to adapt to climate change can be evaluated simultaneously if progeny of controlled population crosses are incorporated into common garden experiments (Fenster and Galloway 2000; Keller et al. 2000; Montalvo and Ellstrand 2001; Etterson 2004a, 2004b; Rogers and Montalvo 2004).

Molecular Marker Variation

Genetic diversity can be measured directly using *molecular markers* (Schaal et al. 1991; Hartl and Clark 1997; Petit et al. 1998). Marker-based estimates of diversity have a number of properties that make them particularly useful in estimating the breeding structure of populations, population bottlenecks, and the biogeographic structure of species (Schoen and Brown 1993; Roy et al. 1994; Hedrick 1999). The expansion of molecular approaches has been influenced by enormous increases in computing speed, cost-effectiveness of implementing molecular genotyping, and improvements in technologies for sequencing DNA markers with high variability (Lowe et al. 2004) (Table 2.1 and Box 2.1).

Quantitative Variation in Phenotype, Genotype, and Genes

Interest in the architecture of complex traits, their adaptive value, and the distribution of quantitative genetic variation in the wild has been longstanding (Clausen et al. 1940; Stebbins 1950; Mather and Jinks 1982; Slate 2005; Tonsor et al. 2005). Many traits important for restoration, such as biomass, flower number, water-use potential, timing of seed flush, seed weight, and photosynthetic efficiency, are influenced by multiple genes that display continuous rather than discrete variation for trait values. The measurement of such heritable phenotypes is directly applicable to defining seed transfer zones, where local ecotypes may play an important role in population persistence.

Quantitative traits are generally thought to be influenced by multiple loci, and may thus reflect functional variation in several portions of the genome (Table 2.1 and Box 2.1). The identification of genes that influence quantitative characters focuses on *quantitative trait loci* (QTLs). With recent technological advancements and the availability of numerous fully sequenced genomes, emphasis is shifting from the analysis of marker genes toward analyses on the genomic scale (Black et al. 2001; Luikart et al. 2003).

Data for "the genes that matter" may ultimately play an important role in the management of germplasm resources and restoration ecology by determining the heritable component of ecologically important traits such as growth rates and tolerance for drought or extreme temperatures (Mitchell-Olds 1995; van Tienderen et al. 2002; Howe et al. 2003). Quantitative genetic approaches can generate useful and testable predictions for the evolutionary dynamics of phenotypes subject to selection under changing environmental conditions, such as may be encountered in both disturbed or restored ecosystems (Stockwell et al., this volume).

Spatial and Temporal Dimensions of Diversity and Divergence

In addition to genetic differences among individuals, genetic differences also exist among populations of most species. A *population* is defined as a group of potentially interbreeding individuals that share a common gene pool. The genetic profile of populations typically varies from place to place across a species range. Differences among populations may arise as the result of chance occurrences, such as the genetic composition of dispersing individuals that create a new population (*founder effect*), or changes in allele frequencies that result from chance mating and reproductive success in very small populations (genetic drift) (Primack and Kang 1989; Templeton 1991; Eckert et al. 1996; Paland and Schmid 2003). Differences among

TABLE 2.1

A comparison of marker systems for evaluation of genetic variation.

In this table, *genic* refers to markers located in defined regions such as a particular gene, whereas *anonymous* markers are uncharacterized, with an unknown distribution in the genome. *Dominance* of the marker system determines whether (as in a diploid organism) the allelic state of both parental alleles (codominant) or just one (dominant) can be retrieved. *Transferability* denotes how readily markers developed in one species can be used in another. The *potential for estimating selection* is related to whether the marker affects a character under selection (as distinguished from neutral variation). Many marker systems exploit variation in noncoding (presumably neutral or nearly so) regions of the genome, so their variation is independent of selective traits. The *information* criterion of a marker indicates whether the marker gives information about the genotype (both alleles); the haplotype, which is the linear ordered arrangement of alleles found on one (haploid) chromosome; or phenotype. The *development* of these markers varies from technically difficult (requiring specialized equipment and significant investment) to logistically difficult (requiring field space and plot management). *Genomic coverage* denotes the number of loci that can be reasonably handled in a single study. Some markers are used primarily to examine variation at a single locus (*) but can be scaled up to encompass many loci. See Box 2.1 for additional details.

Feature	Allozymes	RFLP	AFLP	SSR	SNP	Quantitative traits
Source of marker information	Genic protein	Anonymous or genic DNA	Anonymous DNA	Anonymous or genic DNA	Genic DNA	Multigenic morphology and physiology
Dominance	Codominant	Codominant	Dominant (in practice)	Codominant	Codominant	Variable
Transferability	High	Moderate	High	Variable	Variable	High
Potential for estimating selection	Limited	Limited	Limited	Limited	Moderate to high	High
Information	Molecular phenotype	Genotype	Molecular phenotype	Genotype	Genotype, haplotype	Phenotype
Ease of development	Moderate	Technically difficult	Moderate	Technically difficult	Technically difficult	Logistically difficult
Genomic coverage	Low	Moderate	High	Moderate	Low*	High

Box 2.1

Sources of Information About Genetic Variation

A wide variety of techniques are used in population and ecological genetics. Much of the resulting information is potentially relevant to restoration ecology, but it is important to understand how data are generated, and the limitations of various techniques. Because each method looks into the genome in a particular way, various methods may provide significantly different pictures of genetic diversity and structure (Table 2.1).

While it is beyond the scope of this chapter to review the many tools used in population and ecological genetics in detail, fortunately a number of excellent texts and summary papers are available. A good starting point is a recent synthesis by Lowe et al. (2004). Many of the techniques currently in use can be found there with key supporting references. In addition to technical caveats, the authors also point out the suitability and limitations of various marker systems for questions relevant to the ecological genetics of restoration. Similar overviews of genetic tools and techniques are available in several other works (Brown et al. 1990; Schaal et al. 1991; Hedrick and Miller 1992; Hartl and Clark 1997).

Allozymes are allelic variants of the same enzyme protein (isozyme) that can be separated on starch gels using electrophoresis (Hartl and Clark 1997). Isozyme analysis has been a workhorse marker of studies in plant population genetics (Brown 1978; Clegg 1990). While the application of these markers is relatively inexpensive and requires simple equipment, the number of loci that can be analyzed and their variability is more limited than for direct assays of DNA variation. Moreover, allozymes reflect variability in gene products (proteins), not genes themselves, and thus provide only an indirect view of genetic structure and variability. Nonetheless, given its wide application over many years, isozyme studies still constitute a primary source of information for many species of restoration interest (Hamrick and Godt 1990; Knapp and Rice 1998; Hedrick 2001).

A variety of techniques are used currently to evaluate variation at the level of DNA itself, including *restriction fragment length polymorphisms* (RFLP) (Botstein et al. 1980), *amplified fragment length polymorphisms* (AFLP) (Vos et al. 1995), *simple sequence repeats* (SSR) or *microsatellites* (Weber and May 1989; Theil et al. 2003), and *single nucleotide polymorphisms* (SNP) (Black et al. 2001). Many of these techniques use the *polymerase chain reaction* (PCR) to "amplify" a target locus by making multiple copies of a specific DNA sequence. This amplified DNA can then be used for molecular analysis (Strand et al. 1997).

Molecular techniques have revolutionized population, ecological, and evolutionary genetics and play an increasingly important role in conservation biology (Young 2000; Hedrick 2001). Molecular genetics also show great promise for restoration ecology because of the detailed view they provide of variation among individuals and populations (Excoffier et al. 1992; Hedrick 1999; Black et al. 2001; Hedrick 2001; Holsinger et al. 2002; Luikart et al. 2003; Ryder 2005).

The most direct phenotypically based information on genetic variation comes from measuring *quantitative morphological* or *physiological variation* in organism traits (Lewontin 1984; Storfer 1996; Frankham 1999). Unlike classic Mendelian variation (for instance, eye color = brown or blue), many ecologically important traits, such as height, mass, growth rate, reproductive output, seed weight, and drought or disease resistance, are continuously variable.

Quantitative traits relating to growth, morphology, phenology, and fitness have been used to assess spatial patterns of adaptive differentiation in a number of restoration contexts. For instance, Howe et al. (2003) found that quantitative traits in forest trees revealed population differentiation for characters associated with adaptation to cold and cold hardiness. Quantitative traits can be used in evaluating seed transfer zones or guidelines and assessing standing levels of genetic diversity (Storfer 1996; Burton and Burton 2002; Hufford and Mazer 2003; Erickson et al. 2004; Johnson et al. 2004; Rogers and Montalvo 2004). These results suggest that quantitative trait analysis may be of central importance for restoration ecology.

populations can also arise deterministically (i.e., by natural selection), especially if the environment exposes individuals to different selection pressures for survival and reproduction.

Populations often diverge in their genetic composition, particularly when there is little gene flow between populations (e.g., limited dispersal of seeds, vegetative propagules, or pollen, or limited movement of animals across physiographic barriers). Indeed, "populations" are defined as much by patterns of mating and gene flow as by the physical distribution of individuals, although the two are often closely related (Levin 1981; Slatkin 1987; Knapp and Rice 1996; Neigel 1997; Manel et al. 2003).

Patterns of genetic diversity thus reflect both biology and history (Wright 1965; Nei 1975). For example, nearby populations of plants that are cross-pollinated by bees may share many alleles because genes (packaged in pollen grains) can flow easily between sites that are within dispersal distances of pollinators. Seeds of species that are dispersed by birds or wind may disperse many kilometers. Populations of species without significant barriers to dispersal tend to be genetically similar and may have fewer unique alleles, especially when they are geographically close. By contrast, there may be less gene flow among populations of species that self-fertilize or are pollinated by ground-dwelling flightless beetles, or plants whose heavy fruits fall to the ground in the vicinity of the parent plant (Manel et al. 2003).

Differences among populations are commonly quantified by the use of statistics such as Wright's inbreeding coefficient (F_{ST}) and Nei's coefficient of gene variation (G_{ST}) (Figure 2.1, Table 2.2). These indices reflect how heterozygosity is partitioned within and among populations, based on differences in allele frequencies (Wright 1969; Nei 1975; Wright 1978; McKay and Latta 2002). Genomic-level analyses have led to increased understanding of genetic structure and the development of new analytical methods (Black et al. 2001; Luikart et al. 2003).

Quantitative traits can also be examined to reveal hierarchical structure (i.e., within and among populations). The proportion of quantitative trait variance that occurs among populations relative to that of the total population is called Q_{ST} (in all of these hierarchical measures [Q_{ST}, F_{ST}, and G_{ST}], the subscript ST indicates the variation in subpopulations compared to all populations taken together). A value of *zero* means that variation is distributed randomly in space—that is, all of the variation observed is due to differences among individuals within populations, and none to differences among populations. In contrast, the maximum value of *one* means that all the variation is due to differences among populations and that individuals within populations are similar to each other (Figure 2.1).

In general, the distribution of genetic variation within and among populations is linked strongly to life-history traits, particularly dispersal and reproductive mode (Hamrick et al.

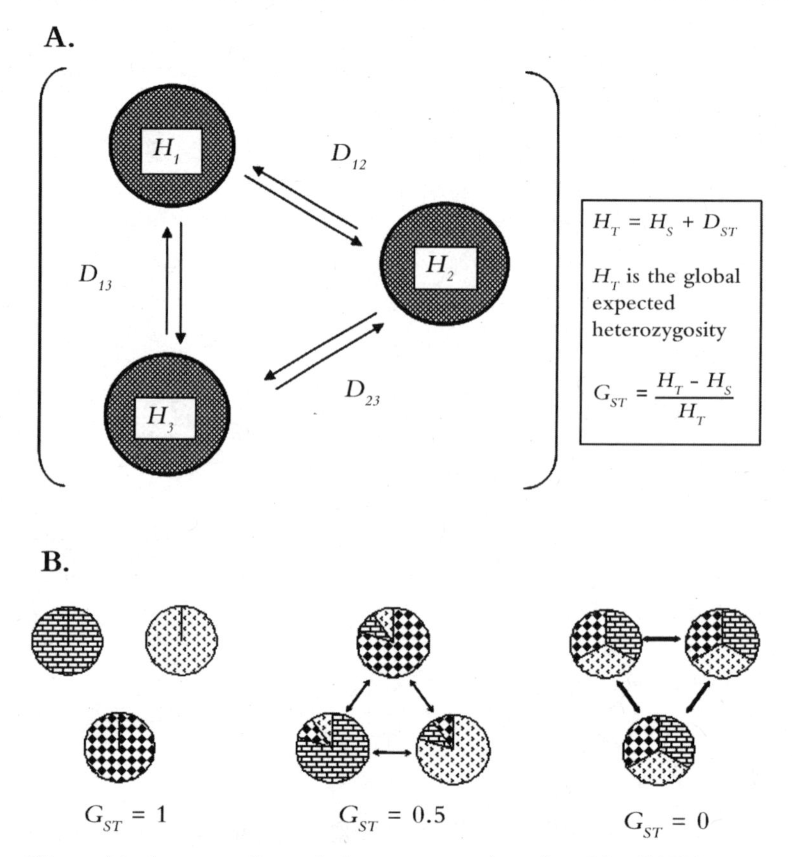

FIGURE 2.1 Hierarchical nature of population structure, based on Nei (1973).
A. The observed heterozygosity (H_i) for each of three subpopulations is used to calculate the gene diversity (D_{ij}), also sometimes called among population divergence, for each pair of subpopulations. Mean population divergence is D_{ST}. The average expected heterozygosity of the subpopulations is H_S, and the expected heterozygosity for the pooled subpopulations is H_T. Adapted with permission from Meffe and Carroll (1994).
B. Resulting population structure for three hypothetical populations based on one locus with three alleles with no gene flow, low gene flow, and high gene flow (left to right). Most studies would utilize data from multiple loci.

1979; Hamrick and Godt 1990; Hamrick et al. 1991). Species that disperse genes (in plants, this includes both pollen and seeds) widely and frequently tend, other things being equal, to have lower G_{ST} (i.e., populations will be relatively similar). Even a moderate rate of gene movement among populations (one individual every a few generations) can link the gene pools of two populations.

Restoration ecology has considerable potential to contribute to understanding of population genetics, by treating (re)introductions as experimental populations (Newman and Pilson 1997; Gustafson et al. 2004b). For instance, Williams and Davis (1996) found significantly reduced genetic diversity (percentage of polymorphic loci, allele richness, expected and observed heterozygosity) in transplanted eelgrass beds (*Zostera marina*) compared to naturally occurring locations. Their data point to small founder populations with limited initial diver-

TABLE 2.2

Common statistical measures of genetic diversity within and among populations.

G and Q_{ST} are based on quantitative traits. The remaining measures are based on discrete trait variation, usually molecular markers. All models except for Q_{ST} assume selectively neutral variation. Most of these measures apply to the average of all pairs of gene copies from the population level specified.

Measure	Description	Basis of value
Genetic diversity within populations		
G	Genetic component of phenotypic variation in quantitative traits	The portion of phenotypic variance controlled by genes
h	Broad sense heritability	Percent of total phenotypic variance due to all forms of genetic variance
H^2	Narrow sense heritability	Percent of total phenotypic variance that is additive (measures ability to respond to selection)
H_o	Average observed heterozygosity	Observed fraction of heterozygotes, averaged over all sample loci
H_e	Average expected heterozygosity	Expected fraction of heterozygotes based on allele frequencies, averaged over sample loci
P	Percentage of polymorphic loci	Percent of all loci with > one allele
A	Average alleles per locus	Number of alleles/locus averaged over all sample loci
F	Inbreeding coefficient (the probability of alleles being identical, that is, or probability of homozygosity)	The difference between H_o and H_e relative to H_e: $[F = (H_e - H_o)/H_e]$
Genetic diversity among populations		
Q_{ST}	Proportion of quantitative trait variance among populations	Relative to total phenotypic variance measured over all populations
F_{ST}, G_{ST}	Proportion of total molecular marker variation among populations, averaged over loci	Relative to variation measured over all populations
K_{ST}	Proportion of mean substitutions per nucleotide site within populations, averaged over sites	Relative to mean substitutions over all populations
Genetic distance	Fraction of alleles and frequencies not shared among pair of populations	Pair-wise comparison among populations
Genetic similarity	Fraction of alleles and their frequencies shared among a pair of populations	Pair-wise comparison among populations

Modified from Rogers and Montalvo (2004).

sity as a likely explanation. Using AFLPs, Smulders et al. (2000) found reduced genetic variation among reintroduced populations of two meadow species in the Netherlands when compared to source populations, again likely because of a small number of founders. By contrast, using a combination of allozymes, RAPDs, and competition experiments, Gustafson et al. (2002, 2004b) found no significant genetic differences between several remnant and restored tallgrass prairie populations, although geographic variation was evident.

Geographic variation for ecologically significant traits such as drought tolerance may not be distributed within and among populations in the same way as allozymes and molecular markers. Nonetheless, recent developments in technical and analytical tools have moved the fields of genetics and ecology closer together. From a restoration perspective, what may be most important is to recognize that data on neutral and adaptive variation are likely complementary (Merlia and Crnokrak 2001; McKay and Latta 2002). An understanding of the forces that shape the distribution of genetic diversity can help to strengthen restoration ecology (Hedrick 2001).

Applying Population and Ecological Genetics in Restoration Ecology

Ecological restoration varies widely in its objectives and applications. Its focus can include *introduction*, *reintroduction*, or *augmentation* of populations (Table 2.3) as well as the restoration of communities and ecosystems. Restoration can also involve a variety of spatial contexts with different potentials for influencing resident populations. For example, the genetic composition and scale of restoration materials relative to that of resident populations can influence the success of both restored and resident populations profoundly (Figure 2.2).

A significant literature in applied restoration genetics is developing to guide both practical and fundamental research questions (Knapp and Rice 1994; Hufford and Mazer 2003; Guerrant et al. 2004; Rogers and Montalvo 2004). In this section we focus on three recurrent topics in restoration ecology where genetics may play a large role: identifying the goals of restored populations, selecting source populations to obtain material for restoration, and the process of collecting such material for restoration from source populations.

TABLE 2.3

Types of population-level restoration.

Restoration materials can be native to a project site or brought in from elsewhere. If a species is not native to a project site, genetic appropriateness of the plant material can differ, compared to when a species is resident or connected to nearby resident populations by gene flow. Introduction, reintroduction, and augmentation may involve both rare and common species. See Falk and Holsinger (1991), Gordon (1994), and Rogers and Montalvo (2004) for discussion.

Term	Definition
Type of restoration	
Introduction	Species or genotypes not presently at the project site, and not known to have existed there previously, are established at a site. Species may or may not be native to broader geographic area
Reintroduction	Reestablishment of species or genotypes not presently at the project site, but that did occur there in the past (population was extirpated and reestablished)
Augmentation	Individuals of a species are added to a site where the species occurs presently (also called *restocking*)
Type of restoration material	
Resident	Species, populations, or genotypes native to a local site. These can be extracted from a local site for onsite restoration or augmentation
Translocated	Genotypes collected offsite for planting or release at a project site within the natural range of the species. Differs from usage in Gordon (1994)
Introduced	Species, populations, or genotypes collected offsite and introduced to a project site outside their historical range.

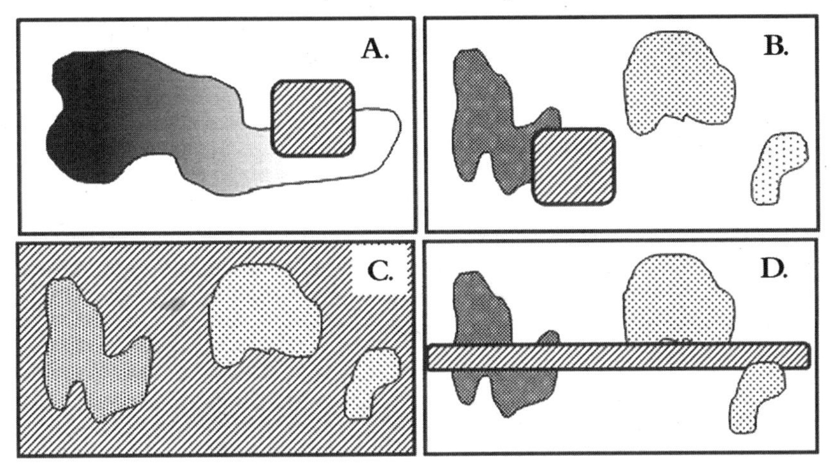

FIGURE 2.2 Spatial context and scale of restoration projects and natural populations. The potential for introduced genotypes (striped areas) to interact with resident genotypes (dotted or grated areas) differs in A–D. Dotted/graded areas are the natural range of a species and the striped area is the restoration project area. In A, gene frequencies of the resident population vary continuously across a gradient, and the restoration area is small relative to the resident area. The subpopulations in B–D have become genetically differentiated in isolation. Restoration could affect resident populations by providing a bridge for gene flow among isolated populations. In C, once isolated resident populations have become connected by gene flow, but are also potentially swamped by introduced genotypes. In D, the restoration area is long and narrow, as in a road right-of-way, creating a dispersal corridor that connects isolated resident populations.

Articulating Goals for Restored Populations

The starting pool of genetic variation is a critical element in design and implementation of a restoration project, but opinions differ on how much explicit attention should be given to the genetic component. For example, the primary goal for a population-level restoration project may be variously: (1) reintroduction of a species that has been extirpated; (2) restoration of critical habitat components (such as nesting structure or food plants) for a species of interest; or (3) demographic or genetic augmentation of an existing but reduced population.

Since few reintroductions can replicate past populations exactly, questions about genetics are often pragmatic: How similar is this source population to the population we wish to augment? Should we combine material from multiple source populations? Which is the more immediate threat: genetic, demographic, or environmental stochasticity? These are difficult questions with few general answers (Clewell 2000). In the case of a wilderness manager restoring a high-quality reference site, the goal might be complete fidelity to the historic distribution of genotypes. This standard can be difficult to achieve, however, and also ignores the possibility that populations are not at genetic equilibrium. By contrast, the highest priority for a restoration project in a severely degraded ecosystem may be to establish a functional plant community for which tolerance of extreme conditions may be paramount (Stockwell et al., this volume).

Geographic Location of Source Material for Restoration

Restorationists may be concerned not only with the overall degree of variability but also its particular geographic distribution and phylogenetic lineage. The most common approach is to specify a geographic (and often elevation) range within which source material should be collected (Fenster 1991). This approach is based on the assumption that populations near one another and growing in similar conditions will be more similar genetically due both to ecotypic variation and the effects of gene flow (Govindaraju 1990). If a population is "genetically local" to a site, it would presumably be adapted to the site and compatible with existing populations of the same species at the site (Campbell 1986; Gustafson et al. 2004b; Rogers and Montalvo 2004).

Geographic distance is indeed sometimes a reasonable first approximation of genetic distance. For instance, on public lands in the western United States, commonly applied guidelines suggest that material for outplanting be collected from within 300 m (1,000 ft) elevation bands and 160 km (100 miles) lateral distance of the planting area (elevation bands vary from 150–900 m [500–3,000 ft] for different agencies) (Johnson et al. 2004).

Geographic proximity and genetic similarity, however, are not always highly correlated (Knapp and Rice 1998). Some geographic areas (e.g., California and the Sky Island bioregion of southwestern North America) are highly heterogeneous in topography, soils, and climate at relatively small spatial scales, while other areas (e.g., shortgrass prairies and high plains) are more homogeneous over large spatial scales. Montalvo and Ellstrand (2001) found that the cumulative fitness of crossed populations was affected significantly by genetic distance and environmental factors, but not strongly correlated with geographic distance in this heterogeneous landscape (Figure 2.3).

Climatic zones or measures of environmental distance may be better predictors of fitness than genetic distance or geographic distance, especially if there is a clinal variation in an adaptive trait (Knapp and Rice 1998; Montalvo and Ellstrand 2000, 2001; ONPS 2001). Furthermore, there are no simple distance rules that apply equally to all species, because species vary in gene flow among populations, population size, and the resulting distribution of diversity (G_{ST}). For some species (e.g., self-fertilizing plants in small, isolated patches of habitat, or fishes in isolated stream reaches), each site may reflect a unique local adaptation, and the geographic range of suitable genotypes can be very small (a few km^2). Other species (for example, those with wind-dispersed pollen and seeds, higher rates of gene flow, and larger effective populations) are generally less differentiated over the landscape and can be collected across wider ranges.

Moreover, many species have dispersal curves that are *leptokurtic* (i.e., with long tails), meaning that a few seeds or offspring in each generation may travel far beyond the majority (Clark 1998). While few in number, these long-range dispersers may play a critical role in helping species to adjust their ranges in periods of climate and vegetation change (Schwartz 1992, 1993; Higgins et al. 2003). For each species, the restorationist must ask how widely the species disperses its genes under normal conditions, and what factors (distance, geographic barriers, habitat types) influence where genes can spread.

Of course, it is often possible to mix genotypes from different populations, and let selection sort out the variation: this is, after all, exactly what nature does. Some geneticists advocate the use of regional mixtures, creating composite collections of genotypes, all of which

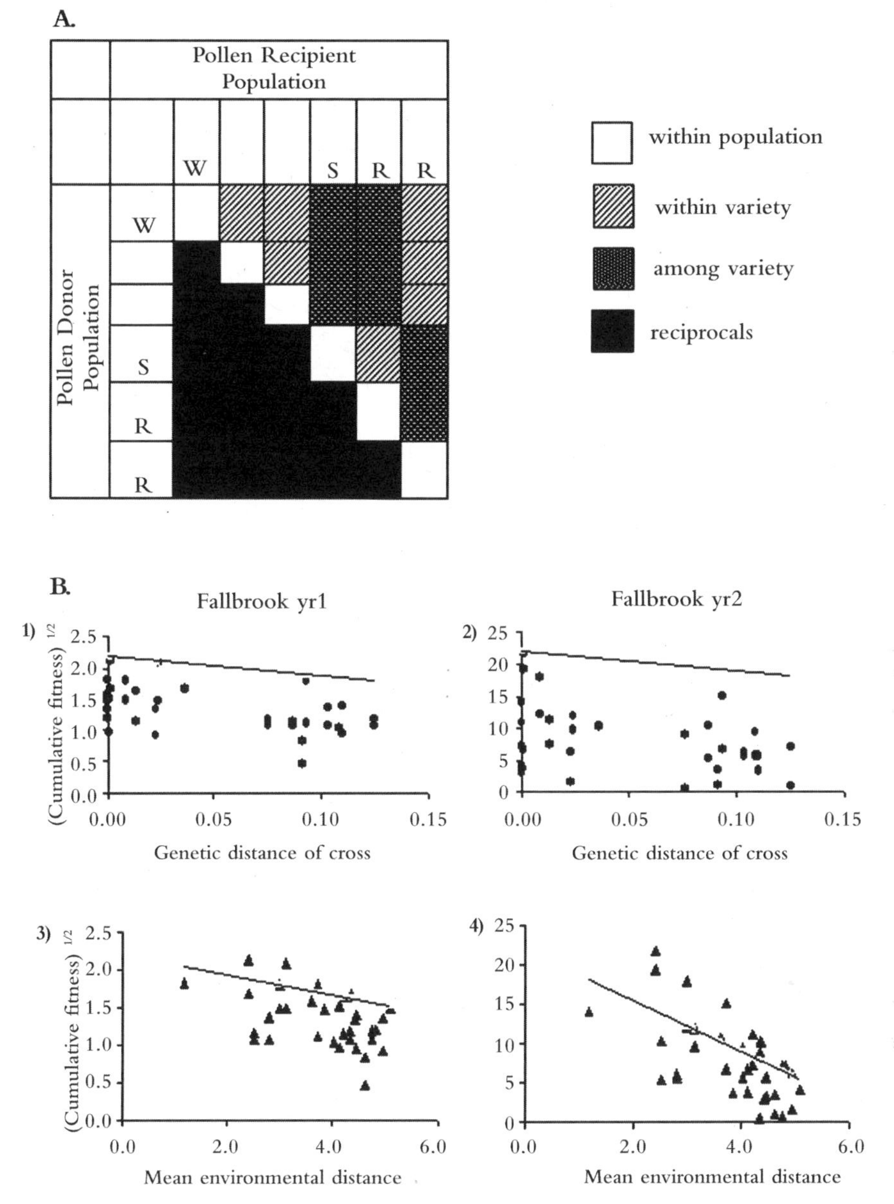

FIGURE 2.3 Outbreeding depression and local adaptation in *Lotus scoparius*.
A. Experimental crossing design used to test for outbreeding depression. Plants from populations listed across the top of the diagram served as mothers, while other plants on the vertical axis served as fathers. The cells below the diagonal represent the reciprocal of those crosses shown above the diagonal. Column and row headings are site name abbreviations.
B. Cumulative fitness of progeny planted at one of two common gardens, plotted as a function of genetic distance of the crossed parental populations (1 and 2), and as a function of the mean environmental distance of the parental source sites to the common garden site (3 and 4). For each variable, data are shown for the juvenile phase (1 and 3) and at maturity (2 and 4). In this study, genetic distance and mean environmental distance significantly predicted success of hybrids, whereas geographic distance (not shown) was not a significant predictor of fitness. Figures from Montalvo and Ellstrand (2001).

are at least moderately well adapted to the general environment and within a broad geographic zone (Knapp and Dyer 1997; Lesica and Allendorf 1999). Mixing populations sampled from within a local ecoregion may be one way of striking a balance that assures input of genetic variation while limiting extreme differentiation (balancing inbreeding and outbreeding).

Since many adaptive traits have a genetic basis, restoration material that performs well may come from a site with similar ecological conditions that is not necessarily close geographically (Knapp and Rice 1998; Procaccini and Piazzi 2001). If ecosystem functionality is the main objective, and there are no resident populations that could hybridize with the restoration materials, then a range of regional genotypes can be introduced, allowing selection (differential survival and reproduction) to sort out the best genotypes for the site.

Mixing genotypes may be particularly suitable if existing "populations" are in fact isolated and reduced remnants of formerly widespread and interconnected groups (Maschinski, this volume). Recombining populations fragmented by human alteration of the landscape may be appropriate provided that the geographic scale is ecologically realistic and it is not done indiscriminately. Mixing too broadly (i.e., combining genotypes from very diverse ecological settings) can result in many individuals that are poorly adapted to the new environment (high genetic load). Care must also be taken to assure that populations that appear similarly adapted do not differ in other aspects that could result in a breakup of intrinsic *co-adaptation* (favorable interactions among alleles).

The gene pools of many remnant native populations have been seriously eroded, so that what persists today is often a small remnant of the original diversity. Small gene pools are more prone to inbreeding, as well as random genetic change from drift. Populations that formerly exchanged genes regularly may have also become genetically isolated by habitat fragmentation (Schwartz 1993; Young and Clarke 2000; Gustafson et al. 2004b; Uesugi et al. 2005). In such cases, a credible argument can be made to bring together genetic material from several populations, in effect replacing the natural (but now disrupted) processes of gene flow. In addition, some restoration sites may be so heavily disturbed (i.e., mine tailing reclamation areas) that the most geographically local population is no longer the one best adapted to the new growing environment (Stockwell et al., this volume). In such circumstances, a wide diversity of genotypes may increase the chances that at least some plants will survive.

If remnant native populations of the species being introduced already exist at a site, additional genetic considerations reinforce the importance of the scale of collection. It is possible, for instance, that genotypes from outside the apparent current range may swamp the local genotype, even if they do not outperform locally adapted varieties. Introduced populations may hybridize with the existing native population, introducing new genes (genetic pollution) and potentially affecting genetic integrity adversely (Rieseberg 1991; Glenne and Tepedino 2000). If a few poorly adapted individuals recruit into the existing native population, natural selection may eventually remove the deleterious genes, but this could take many generations depending on initial establishment success and reproductive factors.

The issue is partly one of numbers; with commercial seed production of many native species, restorationists now have the tools to introduce large volumes of seed into ecosystems. If the number of poorly adapted, nonlocal propagules is large in relation to the number of local native types, the chance of mating between poorly and well-adapted plants will increase (especially in cross-pollinating species), potentially swamping the native population and diluting the local genes. Progeny from such matings may experience outbreeding depression (i.e., poor survival and growth in relation to the parents). Carefully chosen introductions, with genotypes similar to the existing native population, can avoid these negative impacts.

Sampling the Diversity of Source Populations

The genetic diversity of a restoration project is limited initially by the diversity of the original sample. While other alleles may enter the project area over time (by migration of individuals, dispersal of gametes, or additional reintroduction measures), the starting pool of genetic diversity may govern the performance of a reintroduced population for a long time. Restored populations almost inevitably represent a genetic subset of the available source populations. Consequently, many restored populations are subject to some form of the founder effect, the result of an initially limited gene pool due to a relatively small number of founders (Carr et al. 1989; Eckert et al. 1996; Montalvo et al. 1997; Knapp and Connors 1999). Hence, maximizing the diversity of the original source collection within an ecologically meaningful portion of the species range is a critical consideration in restoration.

A variety of guidelines have been developed for sampling wild populations of plants and animals for breeding and reintroduction (CPC 1991; Guerrant 1992; BGCI 1994; Guerrant 1996; IUCN/SSC 1998; Guerrant et al. 2004; Rogers and Montalvo 2004). These and other collecting guidelines vary in their purposes and conclusions; some focus on seed collection for long-term banking, while others address the needs of plant material for reintroduction of populations or restoration of habitats. Many of these guidelines derive their sampling size estimates in part from early work on this subject by Marshall and Brown (1975). There have been many variations over the last thirty years emphasizing different aspects of the collection, such as multilocus diversity and efficiency in collection cost for return in diversity captured (Falk 1991; Lockwood et al. 2005). Most strategies seek to capture all alleles in a population with a frequency greater than some value (commonly 5%) with a probability of 95%.

Although the purposes of collections vary, most guidelines address certain common sampling issues:

1. How many individuals will be sampled from each population?
2. How many populations will be sampled to create the source pool?
3. What is the probability of a collected sample surviving to establishment?

Number of Individuals to Sample Within Populations

The underlying theoretical basis for sampling multiple individuals within a population is that populations are rarely truly *panmictic* (that is, with completely randomized breeding). In plants, a large proportion of mating occurs between neighboring individuals, even when pollination occurs via an animal vector. In animal populations, a wide range of behavioral adaptations commonly concentrates breeding success in a few individuals. The result is that populations are not genetically homogeneous; to capture their genetic diversity adequately, multiple individuals need to be sampled.

In their original analysis of sample size, Marshall and Brown (1975) present the following fundamental expression:

$$P[A_1, A_2] = 1 - (1 - p_1)^n - (1 - p_2)^n - (1 - p_1 - p_2)^n$$

This equation says that the probability of capturing two alleles at a single locus ($P[A_1, A_2]$) is one minus the probability of *not* selecting either or both alleles in n tries, where A_1 and A_2 are the two alleles in question, and p_1 and p_2 are the frequencies of those alleles in the sampled population. Lawrence et al. (1995) extend this general framework to accommodate different outcrossing rates and multiple loci.

Lockwood et al. (2005) recently reviewed these and other sampling strategies published over the past 35 years, and found general agreement when sampling goals are taken into account (e.g., level of genetic diversity desired, allele frequencies, probability of capture). For instance, Marshall and Brown (1975 et seq.) estimate a sample of 50–100 seeds collected from separate individuals per site, whereas, using a different theoretical model, Lawrence et al. conclude that collection of 172 seeds from separate individuals will meet sampling objectives. These sampling strategies represent a minimum collection, however, and the restorationist must also take into consideration the viability of field-collected material through to the stage of reintroduction (see below). For species that are locally rare, large seed collections may not be advised if they would potentially interfere with the dynamics of the source population.

Perhaps the most important insight from population genetics with regard to restoration sampling strategies, however, is the influence of ecological and life-history variation on the distribution of genetic diversity within and among populations (Hamrick and Godt 1996; Lockwood et al. 2005). As we have stressed above, processes such as population size, dispersal rates and distances, strength of local selection, breeding strategy (clonal, selfing, outcrossing), and metapopulation dynamics all have important effects on gene flow and diversity, and thus on the amount and distribution of genetic variation within a species (Maschinski, this volume). Average expected heterozygosity within a population (H_e, Table 2.2) may provide an indication of the mean genetic difference among individuals. For restoration ecology, it is clearly most important to understand the ecology, genetics, and evolutionary biology of a target species first, and then to craft a collecting strategy accordingly.

Number of Populations to Sample

In most species, the cumulative amount of genetic variation captured increases as successive populations are added to the sample. However, since populations have some degree of similarity ($0 < G_{ST} < 1$), each additional population added to a sample collects some alleles that are new to the sample, and some that are already present from previous samples. As the number of populations sampled increases, the marginal diversity rate decreases (that is, fewer and fewer novel alleles are captured), and the cumulative diversity function approaches an asymptote (Figure 2.4). For a pool of populations sampled at random, there comes a point at which further sampling across populations provides little or no additional genetic benefit (Falk 1991; Neel and Cummings 2003).

The number of populations at which this occurs is related strongly to the measure of differentiation among populations, G_{ST}. When G_{ST} is high, populations are more differentiated

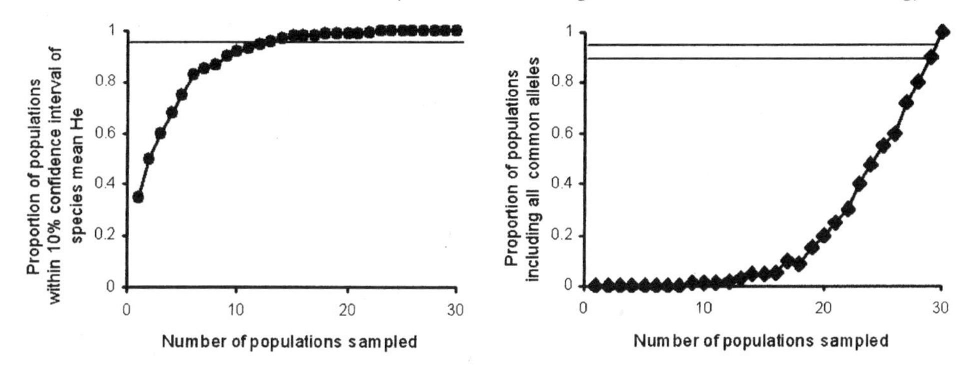

FIGURE 2.4 Sampling for *in situ* conservation of the rare species *Erigeron parishii*. Neel and Cummings (2003) used a resampling procedure to determine the proportion of 1,000 samples required to reach known levels of expected heterozygosity (left) and to include all common alleles (right) for a finite number of populations of four rare plant species. Original sample data included 30 populations and up to 30 samples pêr population. For 14 isozyme loci, there were 49 common alleles (frequency of alleles >0.05 in at least one population) and $H_e = 0.19$. The point at which 95% of samples were within ± 10% is indicated with a horizontal line (left figure). Levels corresponding to the 90% and 95% probabilities of all common alleles are indicated with horizontal lines (right figure).

from one another, so more populations need to be sampled to capture the maximum total diversity. When G_{ST} is low, populations are relatively similar, so sampling from only a few will capture most of the diversity that exists.

Beyond these general patterns, and given the great variability within and among organisms, there are few absolute rules for the number of populations to be sampled that apply to all taxa of restoration interest. Moreover, different components of the genome may accumulate at different rates; for example, as successive populations are sampled, rare or uncommon alleles accumulate at a slower rate than common alleles. Thus, if it is suspected that important adaptations occur at low frequency (for instance, because of recent environmental change), then it may be advisable to sample a larger number of sites. Brown (Marshall and Brown 1975; Brown and Briggs 1991; Brown and Hardner 2000) has argued that collections for diversity should span the environmental range of the focal species. It is important to realize that many of these sampling schemes seek to maximize the capture of genetic variation within a taxon for *ex situ* conservation or agriculture. Sampling for reintroduction to a particular restoration site may be restricted to sites that are within gene flow range or located in ecologically similar habitats.

One other exception to the general null model is for populations that are strongly differentiated along habitat lines (ecotypes or opposite ends of clines). If the reintroduction area is unusual habitat for a species, then it is worth sampling other populations that occur in similar conditions to increase the likelihood of capturing useful adaptive variation. This follows from the observation that habitat characteristics (and not simple geographic proximity) may best predict quantitative variation that is under the most direct selection and thus of the greatest ecological and adaptive significance.

Restoration experiments may prove extremely useful in testing the effect of combining propagules from multiple populations. For instance, Smulders et al. (2000) found that genetic diversity in reintroduced populations of two meadow species was increased when propagules were drawn from two or more source populations, even though there was only moderate differentiation among populations. Similar results have been obtained in other studies (Gustafson et al. 2002, 2004b).

Probability of a Collected Sample Surviving to Establishment

In the end, what counts in a reintroduction is the number of individuals in the new population, as well as their diversity (Menges 1991; Guerrant 1996; Primack 1996; Montalvo et al. 1997). Almost certainly, less than 100% of the samples (seeds, cutting, eggs, adults) collected in the field will survive to establishment on the restoration site. Attrition occurs at every step along the way: during the collecting process, transportation, storage, propagation/curation, and outplanting/release (Guerrant 1992, 1996). High initial mortality rates are frequently observed in reintroduced populations, often continuing for several years (Brown and Briggs 1991; Howald 1996).

A simple "back of the envelope" calculation can help estimate the compensation for attrition of collected samples and reintroduced progeny. Let P_s represent the probability that a collected sample s will survive to establishment through the restoration process, and N, the number of individuals desired in the final restored population. Then we account for attrition during the reintroduction process by collecting N/P_s samples in the field. For example, suppose that plants dug for transplant have an ultimate survival rate of 40% from initial collecting to the third year of a reintroduction project and that we want to establish a population of 50 individuals. Correcting for attrition, an initial collection of (50/.40), or 125 individuals, will have a reasonable chance of providing the final population size required.

The processes of collecting and reintroduction set the stage for the future performance of the population. For instance, attrition in an outplanted or released population has genetic as well as demographic consequences (Maschinski, this volume). If a large proportion of individuals fail to survive, the restored or reintroduced population may be subject both to the founder effect and then further genetic bottlenecks as effective population size diminishes due to mortality (Newman and Pilson 1997; Robichaux et al. 1997). If the initial outplanting or release is genetically uniform to begin with (e.g., hundreds or thousands of plants produced by cloning), the resulting gene pool in the field can be quite narrow. Such apparently large populations are, in genetic terms, very small (Lande and Barrowclough 1987; Barrett and Kohn 1991; Nunney and Elam 1994).

Of course, genetically limited populations occur in nature, often resulting from processes similar to those that shape restored populations. Founder events resulting from long-distance dispersal of a small number of colonizing individuals are genetically narrow almost by definition. The effects of genetic homogeneity in a reintroduced population may not be immediately evident, but over a period of years the population may have lower rates of growth, survival, and reproduction and may persist less successfully through periods of natural environmental variability.

Summary

Genetic variation is often the "invisible dimension" of ecological restoration (Rice and Emery 2003). Understandably, restoration clients, practitioners, and even researchers may be immediately concerned with more readily apparent variables, such as the size and location of a restored or reintroduced population, or the overall composition of a restored community. Nonetheless, we argue that genetic variation (or its absence) is a keystone factor in the outcome of restoration experiments, especially in the long term.

In the most general terms, restorationists (practitioners or scientists) should be aware of the degree of genetic diversity with which they are working. Restoration researchers, planners and managers must understand how the plants and animals they use were generated, by asking the collector, propagator, or breeder how the material was obtained, and what steps were taken to assure the presence of a suitably wide range of genotypes. While flats of thousands of identical plants produced by cloning may appear to offer short-term advantages of predictable response to current growing conditions (as they do for agricultural crops), in the long run (again as with crops) such populations may be less likely to persist and succeed in the face of disease, competition, and environmental variability. Of course, knowing the methods by which individuals or material were produced provides genetic information only by inference; it is unfortunately uncommon to have actual genetic data ahead of time for reintroduction efforts.

The range of genetic tools and information available to the restoration ecologist is vastly greater than even a decade ago. New techniques for DNA sequencing and quantitative variation in particular have opened up a detailed view of the genomes of many species. Even if a species of restoration interest has not been sequenced or studied, it is increasingly likely that congeners or other species with similar life histories have been. Partnerships between restoration ecologists and population geneticists can be fruitful in the interpretation and application of observed patterns, and restoration ecology stands to benefit from such interactions.

The differentiation of populations across the range of a species remains at the heart of the genetic equation for restoration ecology. Despite the search for universal and simple rules, population genetics shows us that species vary in their dispersal rates and distances, and hence in rates of gene flow and the degree of genetic differentiation among populations. These differences are often correlated closely with life-history attributes particular to each species. This leads us to conclude that the most relevant guidelines for restoration ecology will often be species-specific, although some general patterns appear robust. If populations are highly divergent, reflecting either the neutral effects of isolation and small population size, or the diversifying effects of selection, the restoration ecologist must seek to understand this variation.

Geographic distance is a reasonable but potentially misleading means of matching genotypes to the appropriate environment. In a disrupted and fragmented landscape, nearby remnant populations may be genetically depauperate and may lack genotypes that correspond to the conditions under restoration. If populations are strongly selected for local habitat conditions, then habitat similarity may outweigh geographic or physical distance as a criterion for choosing propagules for restoration. Local and regional climatic and soil zones may be more useful criteria for obtaining material for reintroduction, since these will often represent organisms more closely adapted to conditions at the restoration site.

As sources of material for reintroduction, large, genetically diverse populations may be preferable to small, limited populations even when the latter are closer to the restoration site, although small populations may also contain valuable variation. It may also be desirable in some circumstances to combine material from several suitable sites, to capture a wider array of genotypes that can succeed in the new location. Such "regional mixtures" should draw on populations within a similar ecological zone (climate, soil) to the restoration site, to avoid too large a proportion of individuals that are not well adapted to the new conditions. However, such mixtures also increase the risk of outbreeding depression through breakup of local adaptation and co-adaptation, and should be monitored closely. Small local populations can be "swamped" by non-native genotypes if a species already exists at a restoration site. If existing populations are to be augmented, the number of individuals introduced from other locations should not be so large as to overwhelm the local gene pool, particularly if there is evidence of local adaptation.

We have argued that population and ecological genetics are important sources of ideas and information for restoration. We also maintain that restoration ecology has a great deal of unique value to offer in return. It should be evident from this chapter that restoration ecology is fertile ground for testing many basic ideas in ecological and population genetics. Restoration ecology is particularly well configured to contribute empirical tests of genetic drift, founder events, breakup of co-adapted gene complexes and maladaptation, inbreeding and outbreeding depression, reduced gene flow, and small effective population size, all of which are possible outcomes of the restoration process itself. These are all promising areas of research in restoration genetics.

Restoration involves, almost by definition, the movement of propagules and consequently the dispersal of genes. Restoration frequently includes the establishment of new populations or the augmentation of existing ones, where initial population size is known (something the evolutionary biologist or ecological geneticist rarely knows from nature). With a bit of added effort, the diversity and even the specific genomic content of reintroduced populations could be documented routinely. Restoration experiments can provide valuable baselines for tracking temporal change in population genetics, something that is very difficult using naturally occurring populations exclusively (Newman and Pilson 1997; Gustafson et al. 2004b).

Given the thousands of reintroduction and restoration projects currently underway worldwide, the opportunities for empirical tests of basic concepts in population and ecological genetics are enormous. Today, much of that potential is wasted, as both practical and research restoration projects are implemented with no genetic baseline. There could be a leap of orders of magnitude in scientific understanding were we to bring basic questions from population genetics to the restoration ecologist's doorstep.

Acknowledgments

Work on this manuscript was supported by our home institutions, for which the authors express their gratitude. An earlier version of this material was prepared under the auspices of the Science & Policy Working Group of the Society for Ecological Restoration. We also thank Andy Clewell, David Egan, Emilia Falk, Joy Zedler, and two anonymous reviewers for comments on drafts of this chapter.

Literature Cited

Adondakis, S., and D. L. Venable. 2004. Dormancy and germination in a guild of Sonoran Desert annuals. *Ecology* 85:2582–2590.

Allendorf, F. W. 1994. Genetically effective sizes of grizzly bear populations. In *Principles of conservation biology*, ed. G. K. Meffe and C. R. Carroll, 155–156. Sunderland, MA: Sinauer Associates.

Barrett, S. C. H., and J. R. Kohn. 1991. Genetic and evolutionary consequences of small population size in plants: Implications for conservation. In *Genetics and conservation of rare plants*, ed. D. A. Falk and K. E. Holsinger, 3–30. New York: Oxford University Press.

Black, W. C., C. F. Baer, M. F. Antolin, and N. M. DuTeau. 2001. Population genomics: Genome-wide sampling of insect populations. *Annual Review of Entomology* 46:441–469.

Booy, G., R. J. J. Hendriks, M. J. M. Smulders, J. M. V. Groenendael, and B. Vosman. 2000. Genetic diversity and the survival of populations. *Plant Biology* 2:379–395.

Botanic Gardens Conservation International (BGCI). 1994. *A handbook for botanic gardens on the reintroduction of plants to the wild*. Richmond, Surrey, UK: BGCI, in association with the IUCN Species Survival Commission, Reintroduction Specialist Group 42.

Botstein, D., R. L. White, M. Skolnick, and R. W. Davis. 1980. Construction of a genetic linkage map in man using restriction fragment length polymorphisms. *American Journal of Human Genetics* 32:314–331.

Brown, A. H. D., M. T. Clegg, A. L. Kahler, and B. S. Weir. 1990. *Plant population genetics, breeding and genetic resources*. Sunderland, MA: Sinauer Associates.

Brown, A. H. D. 1978. Isozymes, plant population genetic structure and genetic conservation. *Theoretical and Applied Genetics* 52:145–157.

Brown, A. H. D., and J. D. Briggs. 1991. Sampling strategies for genetic variation in *ex situ* collections of endangered plant species. In *Genetics and conservation of rare plants*, ed. D. A. Falk and K. E. Holsinger, 99–119. New York: Oxford University Press.

Brown, A. H. D., and C. M. Hardner. 2000. Sampling the gene pools of forest trees for *ex situ* conservation. In *Forest conservation genetics: Principles and practice*, ed. A. Young, T. J. B. Boyle, and D. Boshier, 185–196. Wallingford, UK: CABI Publishing.

Burton, P. J., and C. M. Burton. 2002. Promoting genetic diversity in the production of large quantities of native plant seed. *Ecological Restoration* 20 (2): 117–123.

Byers, D. L., and D. M. Waller. 1999. Do plant populations purge their genetic load? Effects of population size and mating history on inbreeding depression. *Annual Review of Ecology and Systematics* 30:479–513.

Campbell, R. K. 1986. Mapped genetic variation of Douglas-fir to guide seed transfer in southwest Oregon. *Silvae Genetica* 35:2–3.

Carr, G. D., R. H. Robichaux, M. S. Witter, and D. W. Kyhos. 1989. Adaptive radiation of the Hawaiian Silversword alliance (Compositae-Madiinae). In *Genetics, speciation, and the founder principle*, ed. L. V. Giddings, K. Y. Kaneshiro, and W. W. Anderson, 79–97. Oxford, UK: Oxford University Press.

Center for Plant Conservation (CPC). 1991. Genetic sampling guidelines for conservation collections of endangered plants. In *Genetics and conservation of rare plants*, ed. D. A. Falk and K. E. Holsinger, 225–238. New York: Oxford University Press.

Chesson, P. L. 1985. Coexistence of competitors in spatially and temporally varying environments: A look at the combined effects of different sorts of variability. *Theoretical Population Biology* 28:263–287.

Clark, J. S. 1998. Why trees migrate so fast: Confronting theory with dispersal biology and the paleorecord. *American Naturalist* 152 (2): 204–224.

Clausen, J., D. D. Keck, and W. M. Hiesey. 1940. *Experimental studies on the nature of species. I. Effect of varied environments on western North American plants*. Publication 520. Washington, DC: Carnegie Institution of Washington.

Clegg, M. T. 1990. Molecular diversity in plant populations. In *Plant population genetics, breeding, and genetic resources*, ed. A. H. D. Brown, M. T. Clegg, A. L. Kahler, and B. S. Weir, 98–115. Sunderland, MA: Sinauer Associates.

Clewell, A. F. 2000. Restoring for natural authenticity. *Ecological Restoration* 18 (4): 216–217.

Cohen, D. 1966. Optimising reproduction in a randomly varying environment. *Journal of Theoretical Biology* 12:119–129.

Comstock, R. E., and R. H. Moll. 1963. Genotype-environment interactions. In *Statistical genetics and plant breeding*, ed. W. D. Hanson and H. F. Robinson, 164–196. Resources Council Publication No. 982. Washington, DC: National Academy of Sciences.

Crow, J. F. 1986. *Basic concepts in population, quantitative, and evolutionary genetics*. New York: W. H. Freeman & Co.

Eckert, C. G., D. Manicacci, and S. C. H. Barrett. 1996. Genetic drift and founder effect in native versus introduced populations of an invading plant, *Lythrum salicaria* (Lythraceae). *Evolution* 50 (4): 1512–1519.

Edmands, S., and C. C. Timmerman. 2003. Modeling factors affecting the severity of outbreeding depression. *Conservation Biology* 17:883–892.

Erickson, V. J., N. L. Mandel, and F. C. Sorensen. 2004. Landscape patterns of phenotypic variation and population structuring in a selfing grass, *Elymus glaucus* (blue wildrye). *Canadian Journal of Botany* 82:1776–1789.

Etterson, J. R. 2004a. Evolutionary potential of *Chamaecrista fasciculata* in relation to climate change. I. Clinal patterns of selection along an environmental gradient in the Great Plains. *Evolution* 58:1446–1458.

Etterson, J. R. 2004b. Evolutionary potential of *Chamaecrista fasciculata* in relation to climate change. II. Genetic architecture of three populations reciprocally planted along an environmental gradient in the Great Plains. *Evolution* 58:1459–1471.

Excoffier, L., P. E. Smouse, and J. M. Quattro. 1992. Analysis of molecular variance inferred from metric distances among DNA haplotypes: Applications to human mitochondrial DNA restriction data. *Genetics* 131:479–491.

Falk, D. A. 1991. Joining biological and economic models for conserving plant diversity. In *Genetics and conservation of rare plants*, ed. D. A. Falk and K. E. Holsinger, 209–223. New York: Oxford University Press.

Falk, D. A., and K. E. Holsinger, editors. 1991. *Genetics and conservation of rare plants*. New York: Oxford University Press.

Fenster, C. B. 1991. Gene flow in *Chamaecrista fasciculata* (Leguminosae) I. Gene dispersal. *Evolution* 45:398–408.

Fenster, C. B., and M. R. Dudash. 1994. Genetic considerations for plant population restoration and conservation. In *Restoration of Endangered Species*, ed. M. L. Bowles and C. J. Whelan, 34–62. Cambridge, UK: Cambridge University Press.

Fenster, C. B., and L. F. Galloway. 2000. Inbreeding and outbreeding depression in natural populations of *Chamaecrista fasciculata* (Fabaceae). *Conservation Biology* 14:1406–1412.

Frankel, O. H. 1974. Genetic conservation: Our evolutionary responsibility. *Genetics* 78:53–65.

Frankham, R. 1999. Quantitative genetics in conservation biology. *Genetical Research* 74:237–241.

Freeman, S., and J. C. Herron. 1998. *Evolutionary analysis*. Upper Saddle River, NJ: Prentice-Hall.

Galloway, L. F., and J. R. Etterson. 2004. Population differentiation and hybrid success in *Campanula americana*: Geography and genome size. *Journal of Evolutionary Biology* 18:81–89.

Glenne, G., and V. J. Tepedino. 2000. *A Clark County dilemma: Native bees as agents of rare plant pollination and hybridization*. Third Southwestern Rare and Endangered Plant Conference, The Arboretum at Flagstaff, AZ, USDA Agricultural Research Service, Bee Biology & Systematics Laboratory, Utah State University, Logan.

Govindaraju, D. R. 1990. Gene flow, spatial patterns and seed collection zones. *Forest Ecology and Management* 35:291–302.

Guerrant Jr., E. O. 1992. Genetic and demographic considerations in the sampling and reintroduction of rare plants. In *Conservation biology: The theory and practice of nature conservation, preservation and management*, ed. P. L. Fiedler and S. K. Jain, 322–344. New York: Chapman & Hall.

Guerrant Jr., E. O. 1996. Designing populations: Demographic, genetic, and horticultural dimensions. In *Restoring diversity: Strategies for reintroduction of endangered plants*, ed. D. A. Falk, C. I. Millar, and M. Olwell, 171–208. Washington, DC: Island Press.

Guerrant Jr., E. O., K. Havens, and M. Maunder, editors. 2004. *Ex situ plant conservation: Supporting species survival in the wild*. Science and Practice of Ecological Restoration. Washington, DC: Island Press.

Gustafson, D. J., D. J. Gibson, and D. L. Nikrent. 2002. Genetic diversity and competitive abilities of *Dalea purpurea* (Fabaceae) from remnant and restored grasslands. *International Journal of Plant Science* 163 (6): 979–990.

Gustafson, D. J., D. J. Gibson, and D. L. Nikrent. 2004a. Competitive relationships of *Andropogon gerardii* (big bluestem) from remnant and restored native populations and select cultivated varieties. *Functional Ecology* 18 (3): 451–457.

Gustafson, D. J., D. J. Gibson, and D. L. Nikrent. 2004b. Conservation genetics of two co-dominant grass species in an endangered grassland ecosystem. *Journal of Applied Ecology* 41:389–397.

Hamrick, J. L., and M. J. Godt. 1990. Allozyme diversity in plant species. In *Plant population genetics, breeding, and genetic resources*, ed. A. H. D. Brown, M. T. Clegg, A. L. Kahler, and B. S. Weir, 43–63. Sunderland, MA: Sinauer Associates.

Hamrick, J. L., and M. J. W. Godt. 1996. Effects of life history traits on genetic diversity in plant species. *Philosophical Transactions of the Royal Society of London, Series B* 351:1291–1298.

Hamrick, J. L., M. J. W. Godt, D. A. Murawski, and M. D. Loveless. 1991. Correlations between species traits and allozyme diversity: Implications for conservation biology. In *Genetics and conservation of rare plants*, ed. D. A. Falk and K. E. Holsinger, 75–86. New York: Oxford University Press.

Hamrick, J. L., Y. B. Linhart, and J. B. Mitton. 1979. Relationships between life history characteristics and electrophoretically detectable variation in plants. *Annual Review of Ecology and Systematics* 10: 173–200.

Hartl, D. L., and A. G. Clark. 1997. *Principles of population genetics*. Sunderland, MA: Sinauer Associates.

Havens, K. 1998. The genetics of plant restoration: An overview and a surprise. *Restoration & Management Notes (Ecological Restoration)* 16 (1): 68–72.

Hedrick, P. 1985. *Genetics of populations*. Portola Valley: Jones & Bartlett Publishers.

Hedrick, P. W. 1995. Gene flow and genetic restoration: The Florida panther as a case study. *Conservation Biology* 9:996–1007.

Hedrick, P. W. 1999. Perspective: Highly variable loci and their interpretation in evolution and conservation. *Evolution* 53:313–318.

Hedrick, P. W. 2001. Conservation genetics: Where are we now? *Trends in Ecology & Evolution* 16:629–636.

Hedrick, P. W., and P. S. Miller. 1992. Conservation genetics: Techniques and fundamentals. *Ecological Applications* 2:30–46.

Higgins, S. I., J. S. Clark, R. Nathan, T. Hovestadt, F. Schurr, J. Fragoso, M. Aguiar, E. Ribbens, and S. Lavorel. 2003. Forecasting plant migration rates: Managing uncertainty for risk assessment. *Journal of Ecology* 91 (3): 341–347.

Holsinger, K. E., P. O. Lewis, and D. K. Dey. 2002. Bayesian approach to inferring population structure from dominant markers. *Molecular Ecology* 11:1157–1164.

Howald, A. M. 1996. Translocation as a mitigation strategy: Lessons from California. In *Restoring diversity: Strategies for reintroduction of endangered plants*, ed. D. Falk, C. I. Millar, and M. Olwell, 293–330. Washington, DC: Island Press.

Howe, G. T., S. N. Aitken, D. B. Neale, K. D. Jermstad, N. C. Wheeler, and T. H. H. Chen. 2003. From genotype to phenotype: Unraveling the complexities of cold adaptation in forest trees. *Canadian Journal of Botany* 81:1247–1266.

Hufford, K. M., and S. J. Mazer. 2003. Plant ecotypes: Genetic differentiation in the age of ecological restoration. *Trends in Ecology & Evolution* 18:147–154.

Husband, B. C., and D. W. Schemske. 1996. Evolution of the magnitude and timing of inbreeding depression in plants. *Evolution* 50 (1): 54–70.

Ingvarsson, P. K. 2001. Restoration of genetic variation lost—The genetic rescue hypothesis. *Trends in Ecology & Evolution* 16:62–63.

IUCN/SSC.1998. *IUCN guidelines for re-introductions*. Gland, Switzerland and Cambridge, UK: International Union for Conservation of Nature (IUCN)/Species Survival Commission (SSR) Reintroduction Specialist Group.

Johnson, G. R., F. C. Sorensen, J. B. St. Clair, and R. C. Cronn. 2004. Pacific Northwest forest tree seed zones: A template for native plants? *Native Plants Journal* 5:131–140.

Keller, L. F., and D. W. Waller. 2002. Inbreeding effects in wild populations. *Trends in Ecology & Evolution* 17:230–241.

Keller, M., J. Kollmann, and P. J. Edwards. 2000. Genetic introgression from distant provenances reduces fitness in local weed populations. *Journal of Applied Ecology* 37:647–659.

Knapp, E. E., and P. G. Connors. 1999. Genetic consequences of a single-founder population bottleneck in *Trifolium amoenum* (Fabaceae). *American Journal of Botany* 86 (1): 124–130.

Knapp, E. E., and A. R. Dyer. 1997. When do genetic considerations require special approaches to ecological restoration? In *Conservation biology for the coming decade*, ed. P. L. Fiedler and P. Kareiva, 344–363. New York: Chapman & Hall.

Knapp, E. E., M. A. Goedde, and K. J. Rice. 2001. Pollen-limited reproduction in blue oak: Implications for wind pollination in fragmented populations. *Oecologia* 128:48–55.

Knapp, E. E., and K. J. Rice. 1994. Starting from seed: Genetic issues in using native grasses for restoration. *Restoration and Management Notes* 12:40–45.

Knapp, E. E., and K. J. Rice. 1996. Genetic structure and gene flow in *Elymus glaucus* (blue wildrye): Implications for native grassland restoration. *Restoration Ecology* 4:1–10.

Knapp, E. E., and K. J. Rice. 1998. Comparison of isozymes and quantitative traits for evaluating patterns of genetic variation in purple needlegrass (*Nassella pulchra*). *Conservation Biology* 12 (5): 1031–1041.

Lande, R., and G. Barrowclough.1987. Effective population size, genetic variation, and their use in population management. In *Viable populations for conservation*, ed. M. E. Soulé, 87–124. Cambridge, UK: Cambridge University Press.

Lawrence, M. J., D. F. Marshall, and P. Davies. 1995. Genetics of genetic conservation. II. Sample size when collecting seed of cross-pollinating species and the information that can be obtained from the evaluation of material held in gene banks. *Euphytica* 84:101–107.

Ledig, F. T., V. Jacob-Cervantes, P. D. Hodgskiss, and T. Eguiluz-Piedra. 1997. Recent evolution and divergence among populations of a rare Mexican endemic, Chihuahua spruce, following Holocene climatic warming. *Evolution* 51 (6): 1815–1827.

Lesica, P., and F. W. Allendorf. 1999. Ecological genetics and the restoration of plant communities: Mix or match? *Restoration Ecology* 7:42–50.

Levin, D. A. 1981. Dispersal *versus* gene flow in plants. *Annals of the Missouri Botanical Garden* 68:232–253.

Lewontin, R. C. 1974. *The genetic basis of evolutionary change*. New York: Columbia University Press.

Lewontin, R. C. 1984. Detecting population differences in quantitative characters as opposed to gene frequencies. *American Naturalist* 123 (1): 115–124.

Lockwood, D. R., G. M. Volk, and C. M. Richards. 2005. A review of wild plant sampling strategies: The roles of ecology and evolution. *Plant Breeding Reviews* (forthcoming).

Lowe, A., S. Harris, and P. Ashton. 2004. *Ecological genetics: Design, analysis and application*. Oxford, UK: Blackwell Publishing.

Luikart, G., P. R. England, D. Tallmon, S. Jordan, and P. Taberlet. 2003. The power and promise of population genomics: From genotyping to genome typing. *Nature Reviews Genetics* 4:981–994.

Lynch, M., and B. D. Walsh. 1998. *Genetics and analysis of quantitative traits*. Sunderland, MA: Sinauer Associates.

Manel, S., M. K. Schwartz, G. Luikart, and P. Taberlet. 2003. Landscape genetics: Combining landscape ecology and population genetics. *Trends in Ecology & Evolution* 18:189–197.

Marshall, D. R., and A. H. D. Brown. 1975. Optimum sampling strategies in genetic conservation. In *Crop genetic resources for today and tomorrow*, ed. O. H. Frankel and J. G. Hawkes, 53–80. Cambridge, UK: Cambridge University Press.

Mather, K., and J. L. Jinks. 1982. *Biometrical genetics*. New York: Chapman & Hall.

McKay, J. K., and R. G. Latta. 2002. Adaptive population divergence: Markers, QTL and traits. *Trends in Ecology & Evolution* 17:285–291.

Meffe, G. K., and C. R. Carroll. 1994. *Principles of conservation biology*. Sunderland, MA: Sinauer Associates.

Menges, E. S. 1991. The application of minimum viable population theory to plants. In *Genetics and conservation of rare plants*, ed. D. A. Falk and K. E. Holsinger, 45–61. New York: Oxford University Press.

Merlia, J., and P. Crnokrak. 2001. Comparison of genetic differentiation at marker loci and quantitative traits. *Journal of Evolutionary Biology* 14:892–903.

Mitchell-Olds, T. 1995. The molecular basis of quantitative variation in natural populations. *Trends in Ecology & Evolution* 10:324–328.

Montalvo, A. M., and N. C. Ellstrand. 2000. Transplantation of the subshrub *Lotus scoparius:* Test of the home site advantage hypothesis. *Conservation Biology* 14:1034–1045.

Montalvo, A. M., and N. C. Ellstrand. 2001. Nonlocal transplantation and outbreeding depression in the subshrub *Lotus scoparius* (Fabaceae). *American Journal of Botany* 88:258–269.

Montalvo, A. M., S. L. Williams, K. J. Rice, S. L. Buchmann, C. Cory, S. N. Handel, G. P. Nabhan, R. Primack, and R. H. Robichaux. 1997. Restoration biology: A population biology perspective. *Restoration Ecology* 5 (4): 277–290.

Morgan, P., G. H. Aplet, J. B. Haufler, H. C. Humphries, M. M. Moore, and W. D. Wilson. 1994. Historical range of variability: A useful tool for evaluating ecosystem change. *Journal of Sustainable Forestry* 2 (1–2): 87–111.

Neel, M., and M. P. Cummings. 2003. Effectiveness of conservation targets in capturing genetic diversity. *Conservation Biology* 17 (1): 219–229.

Nei, M. 1973. Analysis of gene diversity in subdivided populations. *Proceedings of the National Academy of Science, USA* 70:3321–3323.

Nei, M. 1975. *Molecular population genetics and evolution*. Amsterdam: North-Holland.

Neigel, J. E. 1997. A comparison of alternative strategies for estimating gene flow from genetic markers. *Annual Review of Ecology and Systematics* 28:105–128.

Newman, D., and D. Pilson. 1997. Increased probability of extinction due to decreased genetic effective population size: Experimental populations of *Clarkia pulchella*. *Evolution* 51:354–362.

Noss, R. F. 1990. Indicators for monitoring biodiversity: A hierarchical approach. *Conservation Biology* 4:355–364.

Nunney, L., and D. R. Elam. 1994. Estimating the effective population size of conserved populations. *Conservation Biology* 8:175–184.

Oregon Native Plant Society (ONPS). 2001. *Guidelines on use of native plants for gardening*. Portland. 3 pp. http://www.npsoregon.org/publica.htm.

Paland, S., and B. Schmid. 2003. Population size and the nature of genetic load in *Gentianella germanica*. *Evolution* 57:2242–2251.

Petit, R. J., A. E. Mousadik, and O. Pons. 1998. Identifying population for conservation on the basis of genetic markers. *Conservation Biology* 12:844–855.

Primack, R. B. 1996. Lessons from ecological theory: Dispersal, establishment, and population structure. In *Restoring diversity: Strategies for reintroduction of endangered plants*, ed. D. A. Falk, C. I. Millar, and M. Olwell, 209–234. Washington, DC: Island Press.

Primack, R. B., and H. Kang. 1989. Measuring fitness and natural selection in wild populations. *Annual Review of Ecology and Systematics* 20:367–396.

Procaccini, G., and L. Piazzi. 2001. Genetic polymorphism and transplanting success in the Mediterranean Seagrass, *Posidonia oceanica* (L.) Delile. *Restoration Ecology* 9 (3): 332–338.

Reed, D. H., and R. Frankham. 2003. Correlation between fitness and genetic diversity. *Conservation Biology* 17:230–237.

Rice, K. J., and N. C. Emery. 2003. Managing microevolution: Restoration in the face of global change. *Frontiers in Ecology* 1 (9): 469–478.

Richards, C. M. 2000. Inbreeding depression and genetic rescue in a plant metapopulation. *American Naturalist* 155:383–394.

Rieseberg, L. H. 1991. Hybridization in rare plants: Insights from case studies in *Cercocarpus* and *Helianthus*. In *Genetics and conservation of rare plants*, ed. D. A. Falk and K. E. Holsinger, 171–181. New York: Oxford University Press.

Robichaux, R. H., E. A. Friar, and D. W. Mount. 1997. Molecular genetic consequences of a population bottleneck associated with reintroduction of the Mauna Kea Silversword (*Argyroxiphium sandwicense* ssp. *sandwicense* [Asteraceae]). *Conservation Biology* 11 (5): 1140–1146.

Rogers, D. L., and A. M. Montalvo. 2004. *Genetically appropriate choices for plant materials to maintain biological diversity*. Report to the USDA, Forest Service, Rocky Mountain Reigon. Lakewood, CO: University of California. 343 pp. http://www.fs.fed.us/r2/publications/botany/plantgenetics.pdf.

Roy, M. S., E. Geffen, D. Smith, E. A. Ostrander, and R. K. Wayne. 1994. Patterns of differentiation and hybridization in North American wolflike canids, revealed by analysis of microsatellite loci. *Molecular Biology and Evolution* 11 (4): 553–570.

Ryder, O. A. 2005. Conservation genomics: Applying whole genome studies to species conservation efforts. *Cytogenetics and Genome Research* 108:6–15.

Ryman, N., and L. Lairke. 1991. Effects of supportive breeding on the genetically effective population size. *Conservation Biology* 5 (4): 325–329.

Schaal, B. A., and W. J. Leverich. 2005. Conservation genetics: Theory and practice. *Annals of the Missouri Botanical Garden* 92 (1): 1–11.

Schaal, B. A., W. J. Leverich, and S. H. Rogstad. 1991. Comparison of methods for assessing genetic variation in plant conservation biology. In *Genetics and conservation of rare plants*, ed. D. A. Falk and K. E. Holsinger, 123–134. New York: Oxford University Press.

Scheiner, S. M. 1993. Genetics and evolution of phenotypic plasticity. *Annual Review of Ecology and Systematics* 34:35–68.

Schoen, D. J., and A. H. D. Brown. 1993. Conservation of allelic richness in wild crop relatives is aided by assessment of genetic markers. *Proceedings of the National Academy of Science, USA* 90:10623–10627.

Schwartz, M. W. 1992. Potential effects of global climate change on the biodiversity of plants. *Forestry Chronicle* 68:462–471.

Schwartz, M. W. 1993. Modeling effects of habitat fragmentation on the ability of trees to respond to climatic warming. *Biodiversity and Conservation* 2:51–61.

Slate, J. 2005. Quantitative trait mapping in natural populations: Progress caveats and future direction. *Molecular Ecology* 14:363–379.

Slatkin, M. 1987. Gene flow and population structure of natural populations. *Science* 236:787–792.

Smulders, M. J. M., J. van der Schoot, R. H. E. M. Geerts, A. G. Antonisse-de Jong, H. Korevaar, A. van der Werf, and B. Vosman. 2000. Genetic diversity and the reintroduction of meadow species. *Plant Biology* 2:447–454.

Stebbins, G. L. 1942. The genetic approach to problems of rare and endemic species. *Madroño* 6:241–272.

Stebbins Jr., G. L. 1950. *Variation and evolution in plants*. New York: Columbia University Press.

Stockwell, C. A., A. P. Hendry, and M. T. Kinnison. 2003. Contemporary evolution meets conservation biology. *Trends in Ecology & Evolution* 18:94–101.

Storfer, A. 1996. Quantitative genetics: A promising approach for the assessment of genetic variation in endangered species. *Trends in Ecology & Evolution* 11:343–348.

Strand, A. E., J. Leebens-Mack, and B. G. Milligan. 1997. Nuclear DNA-based markers for plant evolutionary biology. *Molecular Ecology* 6:113–118.

Swetnam, T. W., and J. L. Betancourt. 1998. Mesoscale disturbance and ecological response to decadal climatic variability in the American Southwest. *Journal of Climate* 11:3128–3147.

Tallmon, D. A., G. Luikart, and R. S. Waples. 2004. The alluring simplicity and complex reality of genetic rescue. *Trends in Ecology & Evolution* 19:489–496.

Templeton, A. R. 1991. Off-site breeding of animals and implications for plant conservation strategies. In *Genetics and conservation of rare plants*, ed. D. A. Falk and K. E. Holsinger, 182–194. New York: Oxford University Press.

Templeton, A. R. 1994. Coadaptation, local adaptation, and outbreeding depression. In *Principles of conservation biology*, ed. G. K. Meffe and C. R. Carroll, 152–153. Sunderland, MA: Sinauer Associates.

Theil, T., W. Michalek, R. K. Varshney, and A. Graner. 2003. Exploiting EST databases for the development and characterization of gene-derived SSR-markers in barley (*Hordeum vulgare* L.). *Theoretical and Applied Genetics* 106:411–422.

Tonsor, S. J., C. Alonso-Blanco, and M. Koorneef. 2005. Gene function beyond the single trait: Natural variation, gene effects and evolutionary ecology in *Arabidopsis thaliana*. *Plant, Cell and Environment* 28:2–20.

Tuljapurkar, S. 1989. An uncertain life: Demography in random environments. *Theoretical Population Biology* 35:227–294.

Uesugi, R., N. Tani, K. Goak, J. Nishihiro, Y. Tsumura, and I. Wishitani. 2005. Isolation and characterization of highly polymorphic microsatellites in the aquatic plant, *Nymphoides peltata* (Menyanthaceae). *Molecular Ecology Notes* 5:343–345.

van Tienderen, P. H., A. A. de Haan, C. G. van der Linden, and B. Vosman. 2002. Biodiversity assessment using markers for ecologically important traits. *Trends in Ecology & Evolution* 17:577–582.

Vos, P., R. Hogers, M. Bleeker, et al. 1995. AFLP: A new technique for DNA fingerprinting. *Nucleic Acids Research* 23:4407–4414.

Waser, N. M. 1993. Population structure, optimal outbreeding, and assortative mating in angiosperms. In *The natural history of inbreeding and outbreeding: Theoretical and empirical perspectives*, ed. N. W. Thornhill, 173–199. Chicago: University of Chicago Press.

Weber, J. L., and P. E. May. 1989. Abundant class of human polymorphisms which can be typed using polymerase chain reaction. *American Journal of Human Genetics* 44:388–396.

White, P. S., and J. L. Walker. 1997. Approximating nature's variation: Selecting and using reference information in restoration ecology. *Restoration Ecology* 5 (4): 338–349.

Williams, S. L., and C. A. Davis. 1996. Population genetic analyses of transplanted eelgrass (*Zostera marina*) beds reveal reduced genetic diversity in southern California. *Restoration Ecology* 4 (2): 163–180.

Wright, S. 1965. The interpretation of population structure by F-statistics with special regard to systems of mating. *Evolution* 19:395–420.

Wright, S. 1969. *The theory of gene frequencies*. Chicago: University of Chicago Press.

Wright, S. 1978. *Variability within and among natural populations*. Chicago: University of Chicago Press.

Young, A. G., and G. M. Clarke, editors. 2000. *Genetics, demography, and viability of fragmented populations*. Cambridge, UK: Cambridge University Press.

Young, T. P. 2000. Restoration ecology and conservation biology. *Biological Conservation* 92:73–83.

Chapter 3

Ecophysiological Constraints on Plant Responses in a Restoration Setting

JAMES R. EHLERINGER AND DARREN R. SANDQUIST

Plant restoration activities can be positively or negatively impacted by changes in the abiotic environment, such as changes in aboveground microclimate, soil structure, or soil nutrients, from that of the predisturbance condition. Through an understanding of the ecophysiological and biochemical mechanisms of adaptation that describe the potential for a plant to persist in a habitat or location, one can better assess the impact of an altered environment on future plant performance and restoration outcomes. This feature of plants is often referred to as *tolerance*. Plant species differ in their capacities to tolerate different biotic and abiotic stressors and this tolerance can be the basis for why some species are capable of reestablishing themselves quickly in a restoration setting, whereas the reestablishment of other species proceeds at a much slower rate, if at all. This chapter focuses on two basic ecophysiological themes that relate to the capacity to become reestablished and tolerate variations in abiotic conditions: (1) light and energy relations in aboveground processes, and (2) water and nutrient relations in belowground processes. We describe the basic requirements of plants, as well as the types of stressors and plant responses associated with these themes. We also describe specific examples that relate to needs required for mitigation in a restoration context, where mitigation could refer to improving factors that impact the reestablishment of particular species or of an ecosystem process.

It is important to recognize that in some cases the physical environment (aerial microclimate or soil conditions) may have been so extensively modified by a previous disturbance or land-use activity that plant reestablishment may not be possible in the short term, because anthropogenic activity has irreversibly altered the environment (Suding et al. 2004). In such cases, reestablishment of historically present species may not be possible or practical. An example of such an extreme would include reestablishment of particular native species in portions of the north central United States (e.g., Iowa and Indiana) where extensive belowground tile systems were installed a century ago to convert swamp regions into fertile agricultural lands (Prince 1997). A second example is the portions of Australia where historical conversion of forest to agricultural lands has resulted in salt migration to the soil surface that makes plant establishment difficult (Cocks 2003; Eberbach 2003). Lastly, tailings that have accumulated from mining activities in the western United States, as well as other places in the world, have resulted in soils that either are so contaminated with toxic elements that no species can persist or simply lack the basic soil structure that allows plants to get established

(Shaw 1990). Nonetheless, efforts to restore vegetation to these significantly altered ecosystems can and should benefit from understanding the ecophysiological principles that allow tolerance of stressors associated with these altered systems.

Light and Energy Balance: Aboveground Processes

Photosynthesis is the basic process whereby the simultaneous capture of carbon dioxide from the atmosphere and photons from the sun results in the formation of the organic compounds used as the building blocks of growth in plants. In general, neither of these two essential substrates for photosynthesis differs in concentration between pristine habitats and those disturbed sites undergoing restoration. What may differ, though, is the light profile within the vegetation, which becomes relevant if plant species vary in their tolerances of light levels. In this regard, it is prudent to recognize that different species have quite different tolerances in the degree to which their leaves will persist when exposed to full sunlight conditions. In addition, plants with different photosynthetic pathways may have a differential capacity to utilize light resources for photosynthetic carbon gain, especially when considering a restoration setting involving herbaceous species. Three major photosynthetic pathways exist: C_3, C_4, and Crassulacean Acid Metabolism (CAM) (Farquhar et al. 1989; Sage and Monson 1999; Taiz and Zieger 1999). However, owing to slow growth rates and relatively low abundances of CAM species worldwide, only C_3 and C_4 photosynthesis are particularly relevant to restoration activities in most cases. These two pathways share similar biochemical and structural features to capture the sun's photons, and as a result produce ATP and NADPH to drive the photosynthetic reduction of CO_2 to form sugars. Where the two pathways differ is in how carbon dioxide is fixed, which results in C_4 taxa typically having a greater capacity to fix carbon than C_3 taxa in the same environment, particularly in warm and high-light conditions.

C_3 Versus C_4 Photosynthesis

C_3 photosynthesis is the ancestral pathway common to all taxonomic lines (Ehleringer and Monson 1993; Sage and Monson 1999). During photosynthesis, carbon dioxide diffuses into leaves through stomata and then diffuses to chloroplasts where it combines with ribulose bisphosphate (a 5-C molecule) via ribulose bisphosphate carboxylase (Rubisco) to form two molecules of phosphoglycerate (a 3-C molecule), which can then be transformed into usable sugar molecules. However, Rubisco can also combine ribulose bisphosphate with atmospheric oxygen to form one molecule of phosphoglycerate and one molecule of glycolate. The glycolate produced cannot be directly transformed into a usable sugar and must thereby be processed through a biochemical "salvage" pathway referred to as *photorespiration* that results ultimately in the generation of carbon dioxide. Although Rubisco has a much greater affinity for CO_2 than O_2, photorespiration reduces overall photosynthetic carbon gain in C_3 plants in proportion to the ratio of ambient CO_2 vs. O_2 (the two competing substrates for Rubisco). Under current atmospheric conditions of 0.037%, CO_2 and 21% O_2, the reduction of net carbon gain is about 35% and this inefficiency increases even more as temperatures increase.

C_4 photosynthesis appears to have evolved multiple times and most likely as a result of low carbon dioxide conditions (Ehleringer et al. 1997; Sage and Monson 1999). It is a modification of the C_3 pathway that spatially restricts the C_3 photosynthetic cycle to the interior portions of a leaf, such as bundle sheath cells as shown in Figure 3.1. In the outer cells of C_4 pathway leaves, PEP carboxylase takes up carbon dioxide (actually bicarbonate) at a high rate to produce oxaloacetate (a 4-C molecule). The 4-C molecules diffuse into the interior bundle sheath cells where a decarboxylation reaction occurs and the resulting carbon dioxide is fixed into organic matter using the C_3 photosynthetic cycle. Because of the greater enzymatic activity of PEP carboxylase relative to Rubisco, the PEP carboxylation activity results in a pump-like mechanism that creates high carbon dioxide concentrations at the leaf interior where the Rubisco portion of the photosynthetic cycle takes place. Thus, photorespiratory carbon dioxide loss does not occur in C_4 plants.

C_4 plants tend to have a higher photosynthetic rate relative to C_3 plants because they lack photorespiratory activity. They usually also have higher growth rates, particularly in warm climates. Not surprisingly, many of the most common invasive species on disturbed sites in temperate to tropical regions possess C_4 photosynthesis. Indeed, some of the world's worst weeds are C_4 taxa (Sage and Monson 1999). Thus, in any restoration activity, it is important to recognize the often-superior competitive ability of C_4 taxa, especially if the objective is to reestablish C_3, nonwoody vegetation. This competitive advantage comes from the ability of C_4 plants to take advantage of today's relatively low carbon dioxide atmosphere. Additionally, on open, disturbed sites, warm microclimatic conditions, especially during the summer, also favor C_4 taxa over C_3 taxa, because the high temperatures at the soil surface tend to increase photorespiration and reduce net photosynthetic carbon gain in C_3 taxa. Ironically, with human burning of fossils fuels resulting in increased atmospheric carbon dioxide levels, it is pos-

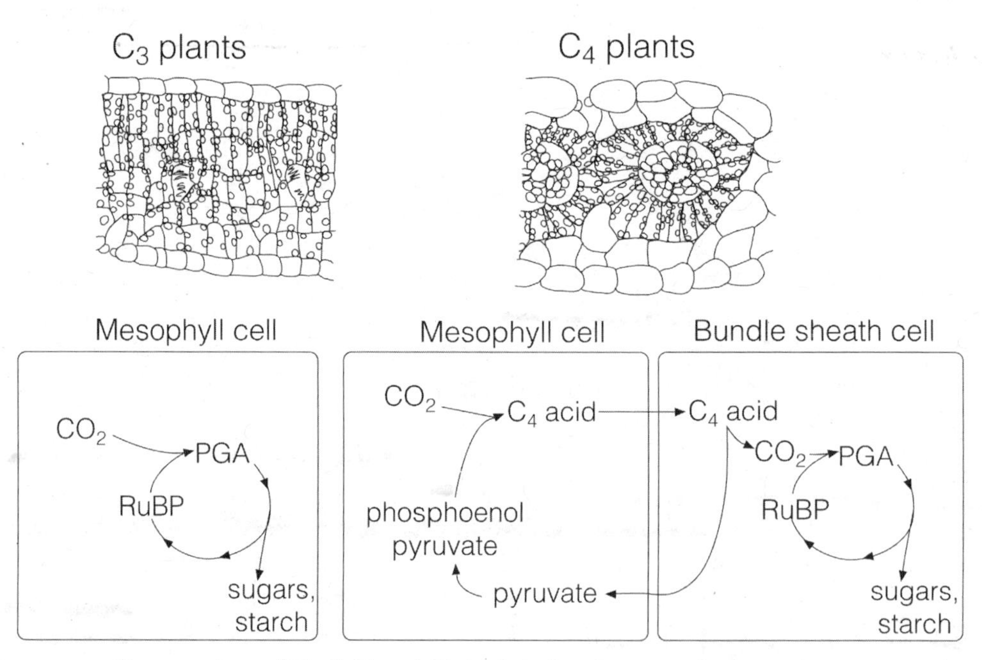

FIGURE 3.1 Cross-sections of C_3 (left) and C_4 (right) plant leaves and schematic representations for each photosynthetic pathway showing the basic differences in carbon dioxide fixation. Note that C_4 plants have the C_3 photosynthetic pathway, but it is restricted to interior cells.

sible that the competitive advantage of C_4 over C_3 taxa today will diminish in the next century (Ehleringer et al. 2004).

High Light as a Stressor

Photosynthetic uptake of carbon dioxide in leaves of both C_3 and C_4 taxa increases with increasing sunlight. This is expected since two of the essential substrates for photosynthetic carbon gain are the ATP and NADPH generated by the light reactions of photosynthesis. Eventually a plateau is reached where there is no further increase in photosynthetic carbon dioxide uptake with increasing light levels (Figure 3.2). While several factors determine the light level at which photosynthesis does not increase further, the two most common features not associated with the light reactions of photosynthesis are stomatal conductance and leaf protein content (typically estimated by leaf nitrogen content). Each of these factors responds to the plant's growth environment, with the upper limits often well correlated with leaf life expectancy (Reich et al. 1999). Stomatal conductance is a measure of how wide open the stomatal pores are that allow the inward diffusion of carbon dioxide for photosynthesis. Water stress (described below) tends to result in reduced stomatal conductance, reduced photosynthetic rates, and for the light saturation point of photosynthesis to occur at lower light levels.

The same response applies for protein content. Since the majority of leaf protein is associated with photosynthetic activity, reduction in leaf protein content will reduce photosynthetic rates, particularly under water stress. The successful establishment of plants in a

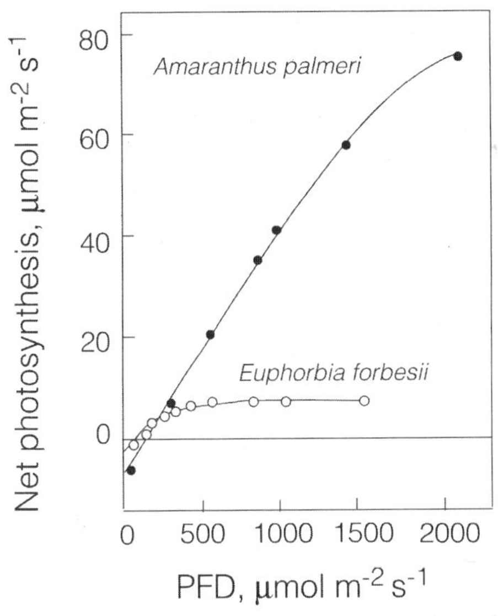

FIGURE 3.2 The response of photosynthesis (as measured by carbon dioxide fixation rate) to changes in the sunlight (photon fluxdensity, PFD) for two C_4 species adapted to different light conditions. *Amaranthus palmeri* is a desert annual, adapted to high-light environments. *Euphorbia forbesii* is a shade-adapted species from the forests of Hawaii. Note the correlation between maximum photosynthetic rate and sunlight level at which photosynthesis saturates. Modified from Pearcy and Ehleringer (1984).

restoration setting will depend on both a sufficient supply of nutrient resources to build plant tissues and support photosynthetic activities and on sufficient amounts of water supplied to leaves to maintain stomatal conductance for the inward diffusion of carbon dioxide.

Exposure to light levels far in excess of those experienced during development, such as for greenhouse plants transplanted to the field or for shade plants exposed to higher light levels than they might be exposed to under more natural conditions, can create a significant challenge for plants in a restoration context. That is, once photosynthetic light saturation is achieved, as shown in Figure 3.2, high light levels can become a stressor (Demmig-Adams 1998), inhibiting plant establishment and potentially causing leaf mortality. Photosynthetic light saturation can occur at light levels that are as little as 5%–20% of midday sunlight for leaves of understory plants or shade leaves of large trees. High light can also become a stressor if the photosynthetic apparatus generates too much ATP and NADPH (the products of the light reactions) and thus exceeds the ability of these substrates to be utilized in the dark reactions. In this case, carbon dioxide availability for the dark reactions generally limits overall photosynthesis (often due to stomatal diffusion limitation).

Another type of light stress is photoinhibition, a process that can occur when leaves are exposed to sunlight levels that are above the light saturation point, as shown in Figure 3.2 (Adir et al. 2003; Demmig-Adams 2003). The excess energy from the light reactions of photosynthesis oxidizes cellular components unless a mechanism is available to dissipate this energy. The effects of photoinhibition can include a reduction in photosynthetic capacity and loss of chlorophyll (bleaching). There are protective mechanisms within a leaf to minimize the potential damage caused by excess light availability (Demmig-Adams 1998; Adir et al. 2003; Adams et al. 2004), including the xanthophyll cycle, where excessive energy is dissipated without causing chlorophyll pigment loss (Demmig-Adams 2003). However, sometimes the light level exposure is too high for protection to be effective, as for plants that naturally grow in shade but are exposed to high light during transplanting, or plants exposed to water stresses and high temperature conditions. Here, exposure to high light does constitute a stress that results in photoinhibition and degradation of protein components of the photosynthetic apparatus (Demmig-Adams 1998; Jiao et al. 2004). When leaves of low-light adapted or acclimated plants are exposed to high light levels, the photoinhibitory effects often result in a photosynthetic light response curve (e.g., Figure 3.2) in which photosynthetic rates actually decline at higher light levels, such as in the case of some tropical tree species (Langenheim et al. 1984).

Microclimatic Stressors

Microclimate variation contributes greatly to the small-scale topographic heterogeneity that plays an important role in the ecology of both plant and animal systems (Larkin et al., this volume). For example, plants may experience a microclimate in which the air and leaf temperatures in the 0.5 m above the soil surface can be significantly hotter during the day and significantly cooler at night than those experienced at greater heights. Microclimatic conditions such as these can be considered stressors because they can result in tissue desiccation, protein degradation, high respiration, and other biochemical dysfunctions. The vicinity of the soil surface is the harshest of environments because plant tissues are potentially exposed to both contrasting stressors, especially during the sensitive period of plant establishment.

During the day, the sun's energy is absorbed by the soil surface, potentially raising the temperature of the soil surface to particularly high levels on sunny days. A portion of the surface heat is transferred to the air by convection, raising the air temperature nearest the surface, and creating an air temperature profile that is hottest near the soil surface (Figure 3.3). Now consider two plants of differing heights with leaves in the same microclimate profile. Metabolic rates, such as rates of photosynthesis, respiration, and transpiration, are a function of leaf temperature. So, we would expect the highest transpiration rates and respiration rates to occur in leaves nearest the soil surface. This poses a thermal stress, especially during seedling establishment, since the rooting depths, water transport capacities, and carbon reserves are likely to be lower in young, establishing plants than in mature, established plants. The impact of a soil-surface microclimate stress can be even greater under certain conditions and ultimately result in mortality. This is because leaf temperatures often can be elevated 1°–10°C above air temperatures.

The difference between leaf and air temperatures will depend on the net leaf energy balance, qualitatively described as:

absorbed solar + infrared radiation = infrared reradiation + convection + transpiration,

where the solar and infrared radiation absorbed by a leaf represent the energy gained by a leaf that must, in turn, be dissipated through reradiation, convection, and transpiration. Leaf temperatures will rise until the energy absorbed by a leaf equals the amount of energy dissipated by these three processes. Thus, leaf temperatures will normally exceed air temperatures by an amount reflecting the net energy gain. If leaves are able to transpire at a high rate or if leaves are small so that convection rates are potentially high, then leaf temperatures may be similar to air temperatures. However, seedlings with large leaves near the surface, or leaves not able

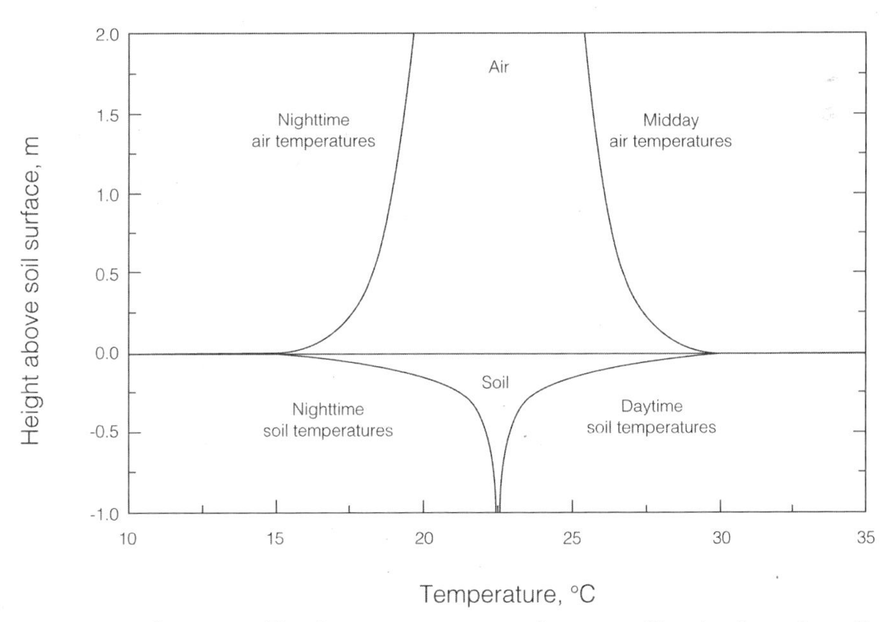

Figure 3.3 Microclimate profile of air temperatures, a function of height above the soil surface during midday and nighttime conditions.

to dissipate heat through transpirational cooling, will have higher temperatures than that of the adjacent air. Over time, these elevated leaf temperatures can result in dehydration and leaf mortality of young seedlings getting established. Not surprisingly there is strong selective pressure for leaf tissues to grow beyond the soil surface in order to reduce transpiration rates and reduce respiratory carbon losses. Having leaves even 2–3 cm above the soil surface is enough to greatly reduce transpiration rates. This is one reason why screens and other shading structures are so important in promoting establishment of seedlings in a restoration setting: they reduce the net energy load incident on the seedling.

At the critical stage of seedling establishment, spring nighttime conditions at the soil surface in some habitats can also represent a thermal stress. This is because at night the coldest part of the microclimatic profile on a bare surface is at the soil surface (Figure 3.3). Here energy is lost by reradiation; the radiative loss from the soil is greater than the absorption of infrared radiation from a nighttime sky, resulting in falling soil and leaf temperatures at the soil surface. During early spring conditions in temperate regions, frost develops at the soil surface as a result of this thermal imbalance. Again, emerging seedling tissues at the soil surface are most vulnerable to this freezing stress, which can often be avoided by leaf and bud tissues that are elevated 5–10 cm above the soil surface. Increasing wind speed (and therefore convective heat transfer to the surface) will reduce the magnitude of the cooling effect in the microclimatic profile, but often the air is most stable at night and so wind speeds are typically low.

Water and Nutrients: Belowground Processes

Many terrestrial restoration settings are associated with significantly altered soil conditions. These include disturbances that have completely replaced surface soils (such as in mining operations); changes in the composition of soil components (such as the addition of clay particles or the loss of organic matter); altered bulk soil densities (such as those associated with compaction); losses of microbes (such as mycorrhizae and nitrogen-fixing bacteria); and the addition of contaminants (such as heavy metals from a smelting operation). Such activities may affect both the availability and distribution of belowground resources and alter the ability of plants to acquire the critical resources essential for growth. The extent to which the belowground environment has changed can strongly dictate the potential for recovery owing to highly sensitive nutrient and water dependencies of vegetation. In many cases belowground alterations, even subtle structural changes in soil compaction or increases in some elements, will preclude native ecosystem restoration. In contrast to these terrestrial cases, the restoration challenge of wetland ecosystems is often increased surface salinities that alter plant/water relations, thereby reducing the likelihood of plant establishment (Handa and Jeffrie, 2000; Zedler et al. 2003).

Tolerances Associated with Minerals in Soils

Most plants take up nutrients through their roots, specifically through single-celled roots that probe the aqueous soil environment surrounding a root. A common practice in restoration settings is to surface supply some of the critical macronutrients for plant growth—particularly calcium, iron, magnesium, nitrogen, phosphorus, and sulfur—as fertilizer (Bloomfield et al. 1982; Cione et al. 2002; Bradshaw 2004). Mineral nutrients, such as nitrate and ammonium, are highly soluble in soil water and have a relatively high diffusion rate in a water solution, facilitating their uptake. The uptake of nutrients by roots is an active, energy-dependent process, in contrast to the uptake of water, which is a largely passive process. The uptake of minerals is facilitated by their solubilities, but this also makes these same minerals highly leachable from soils, especially in high-precipitation environments. Of the mineral elements extracted from the soil, nitrogen is the element needed in highest concentration within leaves as an essential component of proteins, pigments, and nucleic acids, which explains why high additions of nitrogen are particularly important (Bradshaw 1983, 1984). Some pioneer species that readily establish in restoration settings have the ability to produce their own nutrient nitrogen, avoiding the requirement that nitrogen be supplied in the soil. These plants are known as nitrogen-fixers, and they accumulate organic nitrogen through nitrogen fixation in association with a bacterial symbiont.

Often the surface area and lateral extension of a root hair are inadequate to provide sufficient exposure for roots to all essential nutrients available in the soil. This is particularly true for phosphorus, an essential element that has a low solubility and low diffusivity in the soil water solution. Thus, fungal associations are essential to establishment and nutrient uptake by most higher plants (Lambers et al. 1998; Chapin et al. 2002; Fitter and Hay 2002). Fungal hyphae are able to extend up to several orders of magnitude farther away from the root than can root hairs, creating such an effective mineral-uptake situation that many plants do not grow or have significantly reduced growth rates in the absence of their symbiotic mycorrhizal partners. Reclamation studies have provided some of the strongest evidence of the critical roles of mycorrhizal associations for the establishment of plants in a restoration setting (Allen 1991; Caravaca et al. 2003; Querejeta et al. 2003). Disturbance processes (e.g., strip mining activities) that precede the restoration phase often kill or remove mycorrhizal spores, requiring that seeds or transplanted seedlings on restoration sites be provided a fungal inoculum.

Terrestrial restoration settings often differ from more natural habitats by an abundance of toxic elements in the soil (Bradshaw 1983, 1984, 2004; Shaw 1990). The three most common mineral-related challenges to restoration are highly saline soils (discussed later), soils with altered pH levels, and high-metal-toxicity soils. The physiological impacts of these three stressors on plants are as different as the solutions applied in a restoration setting. Interestingly, there are often populations or taxa adapted to these unusual soil regimes, with the tolerance mechanism being either as accumulators or excluders. Studies of genetic variation for tolerance to heavy metals have also been extremely insightful. For example, the ability of different grass species to invade and colonize mine spoils is related to genetic variation in non-related features (Shaw 1990; MacNair 1993).

Altered soil pH levels have multiple effects on plant roots and tolerances are fairly general. Directly, pH can have a negative impact through the effect of excess H^+ or OH^- on membrane integrity and ion uptake systems. Indirectly, pH can influence the solubility of metals that are toxic to plants. In contrast, heavy-metal tolerance in plants is often fairly specific and limited to a single metal, rather than species being tolerant of a wide range of heavy metals (Shaw 1990). For instance, aluminum toxicity (Al^{3+}) occurs in acidic soils and is a major constraint on plant growth in all but calcifuge ("chalk-escaping," "acid-loving") species, which hyper-accumulate aluminum (Jansen et al. 2002). The presence of Al^{3+} generally reduces root elongation and uptake rates of essential cations such as calcium and magnesium

(Fuente-Martinez and Herrera-Estrella 1999). Zinc, cadmium, copper, iron, and other metals can have negative effects on plant metabolism when present in the soil in high concentrations (Shaw 1990; Rout and Das 2003). While much is known about whether tolerant species accumulate or avoid these metals, much less is known about the specific mechanisms of adaptation. In many cases, genotype-specific tolerance has been identified (MacNair 1993). However, there is limited molecular-level information on the mechanisms of tolerance and sensitivity to the metal among species (Pollard et al. 2002; Assuncao et al. 2003). This is a promising area of restoration ecology research (Fuente-Martinez and Herrera-Estrella 1999; Prasad and Freitas 2003). With the advances in genetics and the rise of new molecular tools such as micro-array analyses, it may soon be possible to pinpoint the specific molecular mechanism(s) that underlie the differing capabilities of species to persist in restoration settings with high soil metal concentrations.

Water Availability and Acquisition

As with nutrients, the acquisition of water via belowground plant structures may be significantly altered in a restored habitat, owing to effects on both water availability and plant function (i.e., uptake and transport). The former is primarily a hydrological issue, influenced by soil properties, soil salinity, and climate (Sperry 2000). However, ecological effects, such as competition for water by neighboring plants (Ehleringer et al. 1991) and hydraulic redistribution of water from deep to shallow depths (Burgess et al. 1998), can also play an important role in altering the abundance of water resources. At the plant functional level, basic water uptake via roots is generally similar among most species, but the degrees of sensitivity to water limitation or water excess result in strongly varied responses. In natural systems, these differences can determine species distributions, and in restored systems may dictate a plant's ability to survive (Lambers et al. 1998). Basic rooting zones for water uptake differ between juvenile and adult plants for many perennial species (Donovan and Ehleringer 1992, 1994). Water acquisition can also be increased by mycorrhizal associations (found in many species) and by specific plant adaptations, such as hydraulic redistribution (generally defined as the movement of soil water through root systems from areas of high water availability to areas with lower water availability). Direct interception of moisture, such as fog, may also be critical to the establishment and maintenance of both tree and understory species in restoration of maritime terrestrial ecosystems (Burgess and Dawson 2004). Facilitating the maintenance or recovery of these biotic contributions to resource enhancement may be particularly crucial to restoration. For example, mycorrhizae abundance and their association with plants have been shown to be sensitive to nutrient supplementation (e.g., Egerton-Warburton and Allen 2000; Corkidi et al. 2002), a common practice in many restoration projects.

In a natural setting, plant species within the community often exhibit pronounced differences in effective rooting depth, with root density and effective rooting depth for water uptake varying within the soil profile (Dawson and Ehleringer 1998). This is illustrated in a study from the southwestern United States (Figure 3.4), which showed that the hydrogen isotope ratio of xylem water quantitatively reflects the depth in the soil from which water was derived. Summertime precipitation events resulted in upper-surface soil layers having a hydrogen isotope ratio of ca. –20%, whereas wintertime precipitation events had ratios closer to ca. –90%. Following a summer rain event, a large fraction of the vegetation did not use that

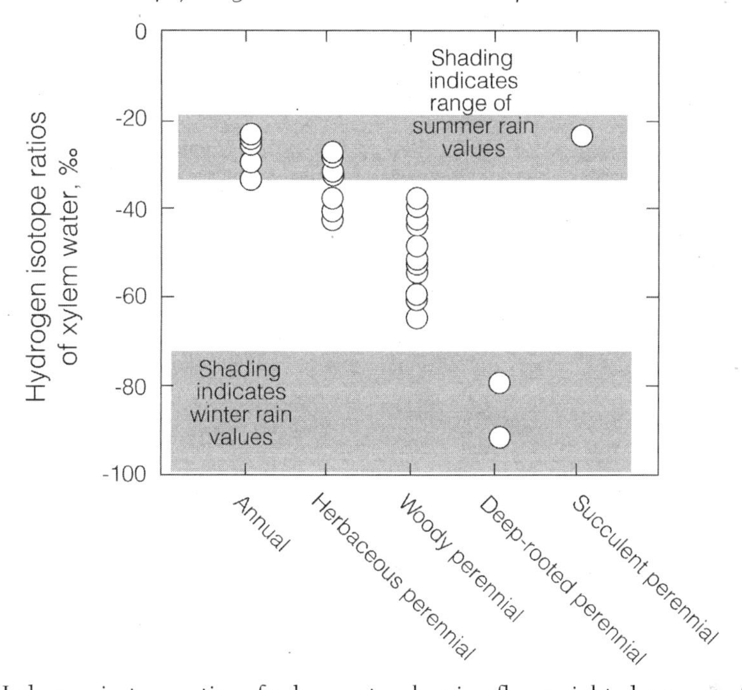

FIGURE 3.4 Hydrogen isotope ratios of xylem water showing flux-weighted sources of water used by different annual and perennial species following strong summer rains in an arid land community of the Colorado Plateau in the southwestern United States. Modified from Ehleringer et al. (1991).

resource at all, or used only a portion of that new moisture as a transpiration source. Some species appeared to utilize soil water from surface and deeper soil layers equally, while others utilized moisture from recent summer rains or moisture from stored winter rains. Uptake of nitrogen from the soil and uptake of water need not come from the same rooting depths, as roots often exhibit resource acquisition specialization (Gebauer and Ehleringer 2000; Gebauer et al. 2002). Therefore, to increase the probability of restoration, it becomes critical to know both the actual rooting distributions as well as the depths from which plants extract nutrient and water resources.

Water Limitation Stressors and Biotic Feedbacks

Many studies underscore the importance of water availability and water acquisition in sustaining community properties and ecosystem processes (Chapin et al. 2002). Stresses owing to the lack of water constitute the majority of these examples, in part because water limitation, rather than water excess, is more prevalent as a stress and has a greater overall negative impact on productivity. This is due in great part to reduced carbon gain owing to stomatal closure, but water stress also decreases cellular and biochemical functions, and may negatively affect production by altering structural integrity. Thus, it comes as no surprise that reductions in water availability are easily capable of altering entire communities and ecosystems (such as found in stream diversions) (Naeem, this volume).

Following uptake, the transport of water through a plant is achieved by the presence of a water potential gradient from the site of water uptake (the soil) to the site of water loss (air). Commonly referred to as the soil-plant-air continuum (SPAC), this water transport mechanism is largely passive and driven by leaf-level transpiration, but because transport depends on the maintenance of this gradient, it is critical that management of each end-member (soil and air) accompany restoration of the transport medium (plant). Although the SPAC gradient is passively derived, the actual water fluxes are regulated by biotic factors, such as stomatal function and hydraulic architecture, and environmental factors, such as the leaf-to-air vapor pressure difference (Sperry 2000; Sperry et al. 2002).

Over the past decade or so, it has become clear that plant hydraulic architecture plays a fundamental role in governing the flow of water through plants. Given that water in the xylem is held under tension, low soil moisture availability and high evaporation demand can cause xylem within plant stems and roots to lose its conductive ability (i.e., cavitate), resulting in a disruption of water flow from the soil to the transpiring leaf surfaces. Different plant species have contrasting "vulnerability" curves, which describe the relationship between the plant water potential (a measure of water stress) and xylem cavitation (a measure of the plant's ability to move water between roots and leaves) (Figure 3.5). The xylem tissues transporting water between roots and shoots of plant species from more mesic habitats tend to cavitate at higher plant water potentials. The steep changes in cavitation that can occur over a narrow plant water potential range underscore the importance of maintaining adequate soil moisture, especially during the development and establishment of plants in a restored community.

Leaf stomata have the greatest effect on regulating water fluxes from plants (Boyer 1985). Stomata are sensitive to both plant water status and relative humidity, and generally close during periods of water stress (Kozlowski and Pallardy 2002; Sperry 2002). There is not a single stomatal response exhibited by all plants to humidity and water-deficit stresses, but rather stomatal pores of different species exhibit a wide range of sensitivities (Schulze and Caldwell 1994; Flexas et al. 2004). To the extent that plants in a restoration setting are influenced by

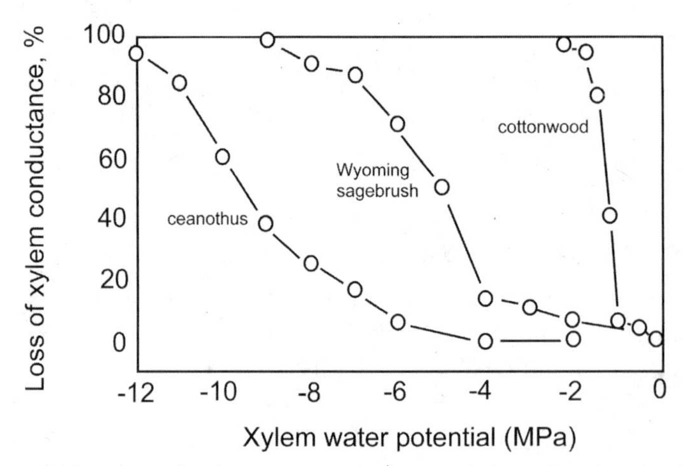

FIGURE 3.5 Vulnerability curves for three contrasting perennial species, showing loss of xylem conductance (xylem embolism) within the plant hydraulic system as a function of plant water potential. Modified from Sperry (2000).

pronounced diurnal or seasonal changes in humidity, both transpiration and photosynthesis rates will exhibit similar dynamics.

The differential rooting depths and sensitivities of stomata to humidity describe a fundamental water-relations challenge in restoring species within arid ecosystems. However, facilitation by shading to alter the microclimate can be a viable mechanism permitting species with differing rooting depths to become established. Maestre et al. (2001) described establishment of three desired shrub species (*Medicago arborea*, *Quercus coccifera*, and *Pistacia lentiscus*) in a Mediterranean restoration setting using the tussock of *Stipa tenacissima* (alpha grass) to facilitate establishment. It is likely that the differential use of soil moisture in surface and deeper soil layers by the grass and shrubs species, respectively, afforded an opportunity to both reduce the energy load on developing perennials and avoid competition for water at depth. With better knowledge of the differential rooting depths for water uptake of juvenile and adult perennials, it is possible to devise irrigation routines that increase the probability that perennials will become established in a restoration setting.

Variation of water availability, uptake, and transport, and the factors that affect them in restoration settings should follow patterns similar to those found under natural conditions. However, in light of the altered soil conditions typical of most projects, future restoration efforts would benefit from designs that explicitly incorporate the ecological importance of water relations, especially if the restoration objectives include efforts to recover some semblance of a normal or sustainable hydrological cycle. Indeed, because water availability is found repeatedly to be the resource most limiting to plant and ecosystem production (Chapin et al. 2002), recognizing the factors that govern water acquisition and transport is critical to restoration programs. What we don't know is which water-use traits are most difficult to recover in restoration programs and the degree to which water limitation influences the success of such efforts.

In recent years, the integration of ecophysiology and ecosystem ecology has promoted greater recognition of biotic feedbacks and their importance in sustaining water resources of ecosystems. The feedback framework has proved useful in elucidating the functional importance of species within a community, and identifying which species should be considered keystone from a functional perspective (Naeem, this volume). For example, fog-water interception and subsequent fog drip caused by redwood trees in the coastal forests of northwestern California have been shown to contribute substantial portions of the monthly water consumption by understory species (Dawson 1998). In the absence of these tall trees, summer soil moisture input for understory and shrub species would be nil since rainfall is absent during the summer in this ecosystem. Similarly, belowground water redistribution (hydraulic lift) by key tree species within eastern deciduous forests can enhance water availability in the upper soil layers, not only to the tree species itself, but also to many forb and herbaceous species in the tree's immediate proximity (Dawson 1993, 1996).

The absence of certain canopy trees has also been shown to increase leaf-to-air vapor pressure difference (recall Figure 3.4), which leads to increased transpiration of remaining plants and hastens drought and water stress in the system. This negative feedback can lead to slow but pronounced changes in species function and composition, ultimately resulting in type conversion to a relatively more xerophytic flora. The danger of such wholesale conversions is the possibility of a system reverting to an alternative state that may be resilient to restoration (Suding et al. 2004; Suding and Gross, this volume).

Salinizing and Groundwater Shifts at Landscape Scales

Human activities affect the hydrology of entire river basins, and modern high-nutrient agricultural systems leach nutrients into groundwater systems. Perhaps nowhere is the hydrologically driven restoration challenge made clearer than in riparian zones. Existing ecological and agroecological systems can alter the salinity and water relations of a region such that restoration activities become challenged by an altered groundwater system (David et al. 1997; Kozlowski 2002; Cocks 2003; Eberback 2003). Conversely the tiling of waterlogged soils to drain marshes and other wetlands can alter the surface-groundwater dynamics irreversibly, to the extent that reestablishment of native communities becomes challenging, at best, or even impossible in the foreseeable future (Prince 1997).

One extreme example is the hydrologically driven restoration effort being made in western Australia, where both the challenges and the needs for restoration come face to face. Here the emergence of an extensive and highly productive wheat agriculture economy over the past century has replaced trees as the dominant plant type across the landscape (Bell 1999; Eberbach 2003). The forest-to-crop conversion during this time period has resulted in lower overall transpiration water fluxes across the landscape and, consequently, the rise of a saline groundwater table that now has the potential to jeopardize the stability of the wheat-driven economy. The rising saline water table also is likely to impact nearby salt-intolerant ecosystems (Cramer and Hobbs 2002).

In this instance, restoration for a mix of stable agricultural and natural vegetation requires lowering of the groundwater by the reintroduction of tree and perennial shrub species capable of drawing down the water table with their higher-than-wheat transpiration rates (Bell 1999). This would seem initially possible given the higher and more continuous transpiration rates by a dense forest stand. Cocks (2003) suggests that native biodiversity could be maintained by planting native trees with higher water-flux rates, which might be harvested periodically for bioenergy, wood products, and fuel. Yet salinity is also likely to reduce transpiration rates by perennial species in the short term, through its impact on water potential and stomatal conductance. Lefroy and Stirzaker (1999) predict that an extensive agroforestry effort would be required to manage the rising saline water tables. However, Hatton et al. (2003) recently concluded that even with extensive revegetation of the landscape, this effort might be inadequate to achieve a stable hydrogeochemical state given the magnitude of the historical crop-related activities. Clearly the situation in western Australia presents an exciting opportunity for restoration ecology, ecophysiological, and agricultural interests to work together to identify feasible alternatives and implement measures that will offer a sustainable future for the region.

Summary

In the most ideal research, plant ecophysiological performance in a restored setting should be compared to that of reference plants in a natural ecosystem. Such studies provide the best opportunity for identifying performance expectations and ultimately attaining restoration goals. Thus, field-based comparative experiments are likely to offer the greatest insights for restoration, but in the past, this research tended to be time-intensive and technologically expensive—burdens that often precluded adequate sample sizes. However, improved technological capabilities over the past two decades, such as lightweight, portable, gas-exchange systems and compact data loggers, have made field ecophysiological assessments much more rapid and tractable. In addition, the use of proxies, such as stable isotopes that correlate well with long-term, integrated, ecophysiological function, provide a relatively easy means by which to monitor plant performance and predict restoration outcomes. Careful selection of which ecophysiological variables to monitor, and on which species, also helps to refine such studies—the variables should be based on the stresses that are expected to have the greatest impact on plant survival (e.g., water potential in an arid system or light response in a high-light environment) and for those species that best represent the reference ecosystem. Simple proxies, such as leaf area and stem elongation, can provide a decent integrated evaluation of stress response, but if certain ecosystem functions, such as water or carbon fluxes, are an objective of restoration, more sophisticated measurements may be necessary. In all cases, however, ecophysiological trait values that match the expected ranges seen in reference plants should be included in the performance standards of a restoration project.

It is clear that not all plant species exhibit the same sets of physiological response curves or stress tolerances. Thus, changes in the state of aboveground microclimate conditions and belowground resource states are likely to produce different species responses that might be predictable once the basic ecophysiological characteristics of the key species are understood. Restoration involves not only an understanding of the role of the physical environment as a driver of plant performance, but also an appreciation of the biotic feedbacks that influence plant performance directly. An appreciation of these basic ecophysiological mechanisms of adaptation and physiological environmental responses can shed fundamental insights that inform the practice of ecological restoration, as well as help guide restoration ecology research and restoration experiments. Furthermore, because restoration settings often pose unique environmental challenges to plants, ecophysiological studies in these settings may also provide significant new insights about plant ecophysiological function.

Acknowledgments

We thank Don Falk and Joy Zedler for helpful comments on early drafts of this chapter. Support for chapter preparation was provided by the Office of Basic Science at the U.S. Department of Energy, Contract No. DE-FG02-DOER63012, and the National Science Foundation, Grant No. DEB-0129326.

Literature Cited

Adams, W. W., C. R. Carter, V. Ebbert, and B. Demmig-Adams. 2004. Photoprotective strategies of overwintering evergreens. *BioScience* 54:41–49.

Adir, N., H. Zer, S. Shochat, and I. Ohad. 2003. Photoinhibition—A historical perspective. *Photosynthesis Research* 76:343–370.

Allen, E. B., M. E. Allen, L. Egerton-Warburton, L. Corkidi, and A. Gomez-Pompa. 2003. Impacts of early- and late-seral mycorrhizae during restoration in seasonal tropical forest, Mexico. *Ecological Applications* 13:1701–1717.

Allen, M. F. 1991. *The ecology of mycorrhizae*. Cambridge, UK: Cambridge University Press.

Assuncao, A. G. L., H. Schats, and M. G. M. Aarts. 2003. *Thlaspi caerulescens*, an attractive model species to study heavy metal hyperaccumulation in plants. *New Phytologist* 159:351–360.

Bell, D. T. 1999. Australian trees for the rehabilitation of waterlogged and salinity-damaged landscapes. *Australian Journal of Botany* 47:697–716.

Bloomfield, H. E., J. F. Handley, and A. D. Bradshaw. 1982. Nutrient deficiencies and the aftercare of reclaimed derelict land. *Journal of Applied Ecology* 19:151–158.

Boyer, J. S. 1985. Water transport. *Annual Review of Plant Physiology* 36:473–516.

Bradshaw, A. D. 1983. The reconstruction of ecosystems. *Journal of Applied Ecology* 20:1–7.

Bradshaw, A. D. 1984. Ecological principle and land reclamation practice. *Landscape Planning* 11:35–48.

Bradshaw, A. D. 2004. The role of nutrients and the importance of function in the assembly of ecosystems. In *Assembly rules and restoration ecology: Bridging the gap between theory and practice*, ed. V. M. Temperton, R. J. Hobbs, T. Nuttle, and S. Halle, 325–340. Washington, DC: Island Press.

Burgess, S. S. O., M. A. Adams, N. C. Turner, and C. K. Ong. 1998. The distribution of soil water by tree root systems. *Oecologia* 115:306–311.

Burgess, S. S. O., and T. E. Dawson. 2004. The contribution of fog to the water relations of *Sequoia sempervirens* (D. Don): Foliar uptake and prevention of dehydration. *Plant Cell and Environment* 27: 1023–1034.

Caravaca, F., J. M. Barea, J. Palenzuela, D. Figueroa, M. M. Alguacil, and A. Roldan. 2003. Establishment of shrub species in a degraded semiarid site after inoculation with native or allochthonous arbuscular mycorrhizal fungi. *Applied Soil Ecology* 22:103–111.

Chapin III, F. S., P. A. Matson, and H. A. Mooney. 2002. *Principles of terrestrial ecosystem ecology*. New York: Springer-Verlag.

Cione, N. K., P. E. Padgett, and E. B. Allen. 2002. Restoration of a native shrubland impacted by exotic grasses, frequent fire, and nitrogen deposition in southern California. *Restoration Ecology* 10:376–384.

Cocks, P. S. 2003. Land-use change is the key to protecting biodiversity in salinising landscapes. *Australian Journal of Botany* 51:627–635.

Corkidi, L., D. L. Rowland, N. C. Johnson, and E. B. Allen. 2002. Nitrogen fertilization alters the functioning of arbuscular mycorrhizae at two semiarid grasslands. *Plant and Soil* 240:299–310.

Cramer, V. A., and R. J. Hobbs. 2002. Ecological consequences of altered hydrological regimes in fragmented ecosystems in southern Australia: Impacts and possible management responses. *Austral Ecology* 27:546–564.

David, M. B., L. E. Gentry, D. A. Kovaic, and K. M. Smith. 1997. Nitrogen balance in and export from an agricultural watershed. *Journal of Environmental Quality* 26:1038–1048.

Dawson, T. E. 1993. Hydraulic lift and water use by plants: Implications for water balance, performance, and plant-plant interactions. *Oecologia* 95:565–574.

Dawson, T. E. 1996. Determining water use by trees and forests from isotopic, energy balance and transpiration analyses: The roles of tree size and hydraulic lift. *Tree Physiology* 16:263–272.

Dawson, T. E. 1998. Fog in the California redwood forest: Ecosystem inputs and use by plants. *Oecologia* 117:476–485.

Dawson, T. E., and J. R. Ehleringer. 1998. Plants, isotopes and water use: A catchment-scale perspective. In *Isotope tracers in catchment hydrology*, ed. C. Kendall and J. J. McDonnel, 165–202. Amsterdam: Elsevier.

Demmig-Adams, B. 1998. Survey of thermal energy dissipation and pigment composition in sun and shade leaves. *Plant and Cell Physiology* 39:474–482.

Demmig-Adams, B. 2003. Linking the xanthophyll cycle with thermal energy dissipation. *Photosynthesis Research* 76:73–80.

Donovan, L. A., and J. R. Ehleringer. 1992. Contrasting water-use patterns among size and life-history classes of a semi-arid shrub. *Functional Ecology* 6:482–488.

Donovan, L. A., and J. R. Ehleringer. 1994. Water-stress and use of summer precipitation in a Great Basin shrub community. *Functional Ecology* 8:289–297.

Eberbach, P. L. 2003. The eco-hydrology of partly cleared, native ecosystems in southern Australia: A review. *Plant and Soil* 257:357–369.

Egerton-Warburton, L. M., and E. B. Allen. 2000. Shifts in arbuscular mycorrhizal communities along an anthropogenic nitrogen deposition gradient. *Ecological Applications* 10:484–496.

Ehleringer, J. R., T. E. Cerling, and M. D. Dearing, editors. 2004. *A history of atmospheric CO_2 and its effects on plants, animals, and ecosystems*. New York: Springer-Verlag.

Ehleringer, J. R., T. E. Cerling, and B. R. Helliker. 1997. C-4 photosynthesis, atmospheric CO_2 and climate. *Oecologia* 112:285–299.

Ehleringer, J. R., and R. K. Monson. 1993. Evolutionary and ecological aspects of photosynthetic pathway variation. *Annual Review of Ecology and Systematics* 24:411–439.

Ehleringer, J. R., S. L. Phillips, W. F. S. Schuster, and D. R. Sandquist. 1991. Differential utilization of summer rains by desert plants: Implications for competition and climate change. *Oecologia* 88:430–444.

Farquhar, G. D., J. R. Ehleringer, and K. T. Hubick. 1989. Carbon isotope discrimination and photosynthesis. *Annual Review of Plant Physiology and Molecular Biology* 40:503–537.

Fitter, A. H., and R. K. M. Hay. 2002. Environmental physiology of plants. Academic Press, London.

Flexas, J., J. Bota, G. Cornic, and T. D. Sharkey. 2004. Diffusive and metabolic limitations to photosynthesis under drought and salinity in C-3 plants. *Plant Biology* 6:269–274.

Fuente-Martinez, J. M., and L. Herrera-Estrella. 1999. Advances in the understanding of aluminum toxicity and the development of aluminum tolerant transgenic plants. *Advances in Agronomy* 66:103–120.

Gebauer, R. L. E., and J. R. Ehleringer. 2000. Water and nitrogen uptake patterns following moisture pulses in a cold desert community. *Ecology* 81:1415–1424.

Gebauer, R. L. E., S. Schwinning, and J. R. Ehleringer. 2002. Interspecific competition and resource pulse utilization in a cold desert community. *Ecology* 83:2602–2616.

Handa, I. T., and R. L. Jeffries. 2000. Assisted revegetation trials in degraded salt-marshes. *Journal of Applied Ecology* 37:944–958.

Hatton, T. J., J. Ruprecht, and R. J. George. 2003. Preclearing hydrology of the Western Australia wheatbelt: Target for the future? *Plant and Soil* 257:341–356.

Jaio, S. X., H. Emmanuel, and J. A. Guikema. 2004. High light stress inducing photoinhibition and protein degradation of photosystem I in *Brassica rapa*. *Plant Science* 167:733–741.

Jansen, S., M. R. Broadley, E. Robbrecht, and E. Smets. 2002. Aluminum hyperaccumulation in angiosperms: A review of its phylogenetic significance. *Botanical Review* 68:235–269.

Kozlowski, T. T. 2002. Physiological ecology of natural regeneration of harvested and disturbed forest stands: Implications for forest management. *Forest Ecology and Management* 158:195–221.

Kozlowski, T. T., and S. G. Pallardy. 2002. Acclimation and adaptive responses of woody plants to environmental stresses. *Botanical Review* 68:270–334.

Lambers, H., H. Poorter, and M. M. I. Van Vuven. 1998. *Inherent variation in plant growth*. Leiden: Bachuys Publishers.

Langenheim, J. H., C. B. Osmond, A. Brooks, and P. J. Ferrar. 1984. Photosynthetic responses to light in seedlings of selected Amazonian and Australian rain forest tree species. *Oecologia* 63:215–224.

Lefroy, E. C., and R. J. Stirzraker. 1999. Agroforestry for water management in the cropping zone of southern Australia. *Agroforestry Systems* 45:277–302.

MacNair, M. R. 1993. The genetics of metal tolerance in vascular plants. *New Phytologist* 124:541–559.

Maeste, F. T., S. Bautista, J. Cortina, and J. Bellot. 2001. Potential for using facilitation by grasses to establish shrubs on a semiarid degraded steppe. *Ecological Applications* 11:1641–1655.

Pearcy, R. W., and J. R. Ehleringer. 1984. Comparative ecophysiology of C_3 and C_4 plants. *Plant Cell and Environment* 7:1–13.

Poff, N. L., J. D. Allan, M. B. Bain, J. R. Karr, K. L. Prestegaard, B. D. Richter, R. E. Sparks, and J. C. Stromberg. 1997. The natural flow regime: A paradigm for river conservation and restoration. *BioScience* 47:769–784.

Pollard, A. J., K. D. Powell, F. A. Harper, and J. A. C. Smith. 2002. The genetic basis of metal hyperaccumulation in plants. *Critical Reviews in Plant Sciences* 21:539–566.

Prasad, M. N. V., and H. M. D. Freitas. 2003. Metal hyperaccumulation in plants—Biodiversity prospecting for phytoremediation technology. *Electronic Journal of Biotechnology* 6:285–321.

Prince, H. C. 1997. *Wetlands of the American midwest: A historical geography of changing attitudes*. Chicago: University of Chicago Press.

Querejeta, J. I., J. M. Barea, M. F. Allen, F. Caravaca, and A. Roldan. 2003. Differential response of $\delta^{13}C$ and water use efficiency to arbuscular mycorrhizal infection in two aridland woody plant species. *Oecologia* 135:510–515.

Reich, P. B., D. S. Ellsworth, M. B. Walters, J. M. Vose, C. Gresham, J. C. Volin, and W. D. Bowman. 1999. Generality of leaf trait relationships: A test across six biomes. *Ecology* 80:1955–1969.

Rout, G. R., and P. Das. 2003. Effect of metal toxicity on plant growth and metabolism: I. Zinc. *Agronomie* 23:3–11.

Sage, R. F., and R. K. Monson, editors. 1999. *C_4 plant biology*. San Diego: Academic Press.

Schulze, E.-D., and M. M. Caldwell, editors. 1994. *Ecophysiology of photosynthesis*. New York: Springer-Verlag.

Shaw, A. J. 1990. *Heavy metal tolerance in plants*. Boca Raton: CRC Press.

Sperry, J. S. 2000. Hydraulic constraints on plant gas exchange. *Agricultural and Forest Meteorology* 104:13–23.

Sperry, J. S., U. G. Hacke, R. Oren, and J. P. Comstock. 2002. Water deficits and hydraulic limits to leaf water supply. *Plant Cell and Environment* 25:251–263.

Suding, K. N., K. L. Gross, and G. R. Houseman. 2004. Alternative states and positive feedbacks in restoration ecology. *Trends in Ecology & Evolution* 19:46–53.

Taiz, L., and E. Zeiger. 1998. *Plant physiology*. Sunderland, MA: Sinauer Associates.

Zedler, J. B., H. Morzaria-Luna, and K. Ward. 2003. The challenge of restoring vegetation on tidal, hypersaline substrates. *Plant and Soil* 253:259–273.

Chapter 4

Implications of Population Dynamic and Metapopulation Theory for Restoration

JOYCE MASCHINSKI

For many who have tried to restore viable self-sustaining populations to the wild, there has come a great sense of humility and wonder at the complexity of systems that have a deceiving appearance of simplicity. Many early attempts at restoring populations met with low success, because of this naïve perception that it would be simple and quick (examples in Falk et al. 1996). But experience has taught us otherwise.

Understanding how populations change in the face of spatial and temporal variation or in response to environmental, genetic, and demographic uncertainty is central to planning any restoration effort or recovering any rare species. Interactive factors influencing population persistence are complex and much more research is needed in this area. Even basic data for estimating birth rates, death rates, rates of population increase, and habitat occupation is often lacking, yet it is essential for developing effective, reliable recovery and restoration plans (Schemske et al. 1994).

Theoretical population viability and metapopulation models have become integral to the legal protection of rare species and habitats (Schemske et al. 1994; ESA 1996; Dreschler and Burgman 2004). They have become widely used for conservation management, and they also have relevance for restoration planning (Possingham et al. 2000). Because they are being used to document the compliance of restoration projects with mitigation laws, developing reliable models capable of detecting "real" change is essential (Dreschler and Burgman 2004). Although some argue that widespread application may be inappropriate because the assumptions of the models do not hold for all natural populations, patches, or metapopulations (Doak and Mills 1994), models can help guide management actions for conservation of species and can substantiate the success or failure of restoration projects.

Theoretical constructs from population viability and metapopulation analyses can help provide testable hypotheses for restoration projects, which may in turn help refine theory. In the face of increasing threats of habitat fragmentation and climate change, it is critical that restoration efforts include a research agenda and an experimental component.

Here, I briefly highlight key principles of the major theories of relevance to restoration and note the challenges of each. I review several studies that have tested hypotheses either supporting or refuting theory. Using case studies, I illustrate some of the limitations and potential of using population dynamics and metapopulation theory in restoration. My examples include plant populations, particularly rare species, and animal populations. Using these theoretical constructs for restoration planning is akin to unraveling the woven rug to learn how it was constructed. The theoretical threads can trigger the questions, but only experimentation can provide the evidence needed to move the science forward.

Population Viability Analysis

Population viability analysis (PVA) has become fundamental to understanding and predicting the persistence of populations. Rather than predicting the absolute fate of a population, PVA can best be used as a heuristic tool for estimating the *relative* viability of populations under variable management or experimental regimes (Possingham et al. 2000; Brigham and Thomson 2003), in natural versus restored habitats (Bell et al. 2003), under scenarios of projected future conditions (Maschinski et al. 2006), or for risk assessment (e.g., Madden and Van den Bosch 2002). In that sense PVA provides a perfect theoretical platform from which restoration plans can be derived.

Approaches to PVA can be deterministic or stochastic, analytical or simulation based, spatially structured or nonspatial (Beissinger and Westphal 1998). Although the most commonly used modeling approach is the stage or age-structured matrix model, other options are available (e.g., Dennis et al. 1991; Meir and Fagan 2000; Holmes 2004). Models are becoming more sophisticated and challenge researchers to hone approaches to understanding species' biology. (See reviews by Menges 2000; Beissinger and McCullough 2002; Reed et al. 2002; Brigham and Schwartz 2003; Lande et al. 2003.)

PVA modeling can help address restoration questions about the size of site needed for the founding population, the type of propagule that should be used, and whether the restored population is sustainable. PVA forms a formal framework for exploring potential effects of different restoration strategies (Ball et al. 2003). Below I address several components of PVA that have relevance to restoration.

Minimum Viable Population (MVP)

The concept of minimum viable population size required for conserving a species in a particular location and time (Soulé 1987) arose from population viability analysis. Early ideas of MVP were based upon the size of population needed to sustain a genetic condition necessary for short-term survival, continued adaptation to environmental change, and continued evolution to new forms. Early estimates of MVP ranged from 50 to 500 (Franklin 1980; Soulé 1980). Later empirical and theoretical work incorporating risks of deleterious mutations increased the suggested effective population size from 10^3 to 10^6 (Shaffer 1987; Menges 1991; Lande 1995). For many rare species, actual population sizes are much less than one thousand individuals (Wilcove et al. 1998), which begs the question of whether their populations have a chance of maintaining evolutionary potential and long-term genetic viability.

For reintroductions of rare species, simulation models have been used to determine feasibility and the MVP (smallest founding group) that could result in a sustainable population with an acceptable probability of persistence. Simulations using varying initial population size generate trajectories of populations over time. If simulation models also incorporate environmental and genetic components, they offer more robust predictions.

Uncertainty

Although early models were deterministic, population viability models now incorporate uncertainty or stochasticity (Lande 2002) and analysis of extinction probability or quasi extinction, defined as a population dropping below an arbitrary threshold assigned by the modeler (Burgman et al. 1993). Shaffer (1987) first introduced and defined categories of uncertainty that influence population viability: *demographic uncertainty*, resulting from random events in the survival and reproduction of individuals; *environmental uncertainty*, caused by unpredictable changes in weather, resource supply, and populations of predators, competitors, and so forth; *natural catastrophes*, such as floods, fires, and droughts, which are extreme manifestations of a fluctuating environmental (Lande 1993); and *genetic uncertainty*, caused by random changes in the genetic makeup of populations due to inbreeding, genetic drift, or founder effects. All of these factors have a stronger influence on small populations than on large populations (Shaffer 1987; Menges 1998), because larger populations are better buffered against stochasticity.

Several authors have made relative comparisons of the type of stochasticity that has the greatest import to population viability. Most agree that, of these factors, environmental stochasticity is likely to have the most important effects on populations, while demographic and genetic stochasticity will likely play their greatest roles in small populations (Shaffer 1987; Lande 1993, but see Lande 1995 and Kendall and Fox 2002). Populations with modest growth rates and delayed reproduction tend to be influenced more strongly by demographic stochasticity (Menges 1998). Populations that have a mean per capita growth rate larger than variance will have greater persistence under environmental stochasticity (Lande 1993). Regardless of initial population size, a population with negative long-run growth rates will have high probabilities of extinction (Lande 1993).

Knowing that uncertainty plays an important role in population dynamics calls for testing the performance of founding populations with varying genetic diversity and densities in habitats that vary in quality, level of disturbance, or other factors (e.g., Kephart 2004).

Elasticity Analysis

Elasticity analyses measure the proportional change in population growth (λ) given small changes in stage-specific vital rates (Caswell 2001). Elasticity values can be used to identify the life stage that has the strongest influence on the population viability model; matrix elements with the highest elasticity contribute the most to the overall population growth (e.g., Silvertown et al. 1996). Thus, elasticity analysis can be used to determine what life stage will have greatest promise for building and sustaining a restored population. Recently, alternative methods to elasticity analyses have been reviewed or presented in the literature (Horvitz et al. 1997; Caswell 2000; Grant and Benton 2000; Wisdom et al. 2000; Caswell 2001).

It is important to be aware of underlying assumptions that generate elasticities and recognize that elasticity analyses may not always provide accurate predictions for restorations. Vital rates with high elasticities do not necessarily correspond to the life history stages that are currently limiting population growth or that are the most productive targets for management (Brigham and Thomson 2003; Schwartz 2003). Models assume that transition elements are independent while, in reality, transitions may be correlated. Improving reproduction may negatively impact survival or growth, for example. Some elements may vary widely across years, such that selecting the highest value life stage for a restoration in one year may not hold for another year. Elements with high elasticities tend to have low variation across years or sites (Pfister 1998, but see Pico et al. 2003); therefore, changes in these life stages may have little impact on population growth. For example, long-lived species often have the largest elasticity values for surviving adults, yet the major threat for species' persistence may result from lack of recruitment (Schwartz 2003). Restoring a population with adults may not only be the technically most difficult and resource-intensive approach, but it also may not improve the species' conservation status in the long term.

Challenges of Population Viability Modeling

Population viability models require long-term data sets and a good understanding of the species' biology. Ideally, long-term data sets capture the range of variation in vital rates of the species. Long-lived species present a special problem in that they have very slow responses, low mortality, low turnover, and extremely episodic recruitment (Schwartz 2003). These become especially problematic for measuring population growth of restored populations using PVA (e.g., Bell et al. 2003).

Just as population behavior is controlled by "weakest" links in time (Menges 1998), the models we are able to construct are limited by the "weakest" link(s) in our data. Notoriously large information gaps that directly affect the quality of the models we are able to build are dormancy, seed or egg banks, survival of dispersed young, and the effects of uncertainty on all model parameters. Often empirical data needed to quantify or estimate accurately the transition stages for these parameters are not available or would require a great deal of time to collect.

Data required to make a good prediction about dormancy, such as long-term variability in reproductive success, may be no faster or easier to collect than actual vital rates for seeds (Doak et al. 2002). Further, vital rate measurements in a greenhouse or laboratory setting, such as percent germination or percent survival, may not translate to the field (e.g., Dudash 1990). Even small changes in estimates of annual seed or egg survival, or annual seed germination or egg hatching, can result in dramatically different estimates of population extinction risk, and these effects are most important under conditions of highly variable environments (Aikio et al. 2002; Doak et al. 2002).

For example, a 21-year data set demonstrated that two co-occurring *Daphnia* species had dramatically different rates of dormancy and hatching (Caceres 1997) (Figure 4.1). *Daphnia galeata mendotae* failed to reproduce in 8 out of 21 years sampled in Oneida Lake, New York. Multiyear dormancy allowed this weak competitor to persist at this location. In contrast, the co-occurring *Daphnia pulicaria* did not rely on egg storage for persistence and had consistently high recruitment. Thus, the importance of the dormant stage had markedly different influences on population growth rates depending upon species-specific responses to environmental variation. Similarly, co-occurring plant species may germinate from seed banks at different rates under varying environmental conditions, thereby enabling coexistence of species (Baskin and Baskin 1998; Pake and Venable 1996) and maintaining variation in population genetic structure (Levin 1990; McCue and Holtsford 1998). Variable rates of germination from dormant stages reduce the ability to generalize seed or egg bank parameters for a PVA across co-occurring species.

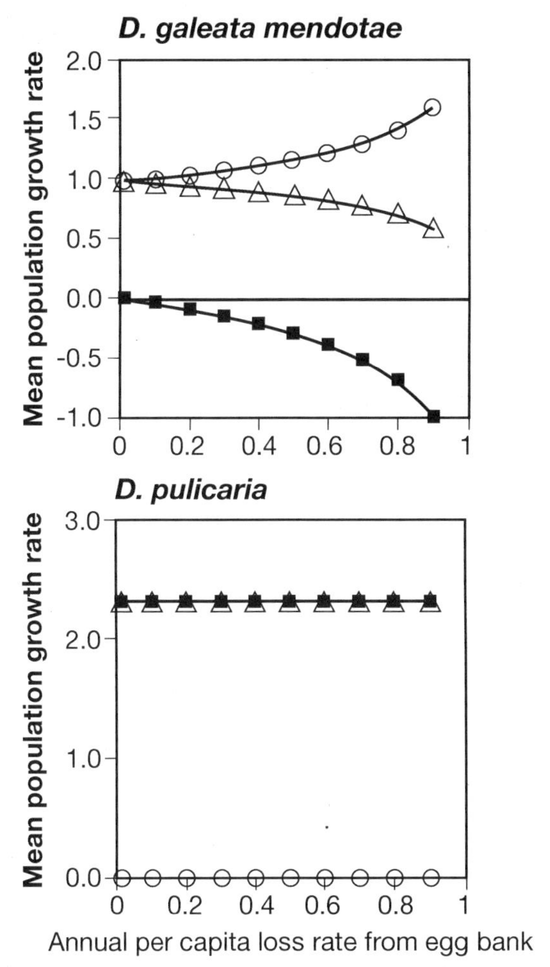

FIGURE 4.1 Example of two co-occurring *Daphnia* species with dramatically different egg dormancies. Mean population growth rates of two *Daphnia* species in a homogeneous egg bank are represented. The portion of the population growth rate that does not include recruitment variation is represented by black squares; the portion of the population growth rate that is due to annual variation in recruitment is represented by open circles; the boundary growth rate, represented by open triangles, is the sum of each circle/square pair. For *D. galeata mendotae*, all squares fall below the zero line, indicating that this species could not persist without variable recruitment and reestablishment from the egg bank. In contrast, *D. pulicaria* persistence did not depend upon egg bank storage. Redrawn from Caceres (1997).

Quantifying the survival or mortality of individuals that disperse from a study site is also problematic for building PVA models. This is especially true for pelagic or wind-dispersed organisms whose young may settle far from the parent (Crowder et al. 1994). Assessment of juvenile dispersal and survival may be compromised by sampling techniques (Szacki 1999), accurate identification of young produced within a sample area (Sork et al. 2002), and the patterns of movement across season and location (Szacki 1999; Ball et al. 2003). These parameters may be unknown, even in organisms that have been studied extensively.

PVA predictions are most accurate for short time intervals (10–20 years), because models will become increasingly imprecise as time from field data collection increases (Ellner et al. 2002). Models are sensitive to estimated mean vital rates, which can vary with stochastic factors, time, and space.

Metapopulation Analysis

Spatial and temporal variation in population demography is critical to the long-term population dynamics and persistence of a species across its range (Gilpin and Soulé 1986; Pulliam 1988; Lande 1993; references in Kauffman et al. 2003). Metapopulation models link population ecology (local abundance) with biogeography (regional occurrence) and provide a useful framework for understanding correlative and experimental data on population distribution and abundance (Gotelli 1991). A metapopulation approach is likely to provide useful tools for developing restoration strategies for optimizing among-population processes critical for the persistence of many natural systems (Thrall et al. 2000).

Hanski and Gyllenberg (1993) considered two theoretical models for metapopulation analysis as extremes of a continuum (Figure 4.2). The mainland-island model, based on the equilibrium theory of island biogeography (MacArthur and Wilson 1967), assumes a large and invulnerable source population on the "mainland," from which individuals migrate to smaller habitat patches ("islands") with more transient populations. Levins's model (1969, 1970) assumes a set of equally large habitat patches, or islands, with local populations frequently going extinct and vacated patches recolonized from the currently occupied set of patches. Most species occur intermediate to these extremes, where there is significant spatial variation in habitat patch sizes, even if there is no true "mainland" invulnerable to extinction (Harrison 1991).

Kareiva (1990) reviewed models that describe spatial organization in heterogeneous environments: island models, where populations are subdivided; stepping-stone models, where patches have explicit spatial dimensions; and reaction-diffusion models, which assume a ho-

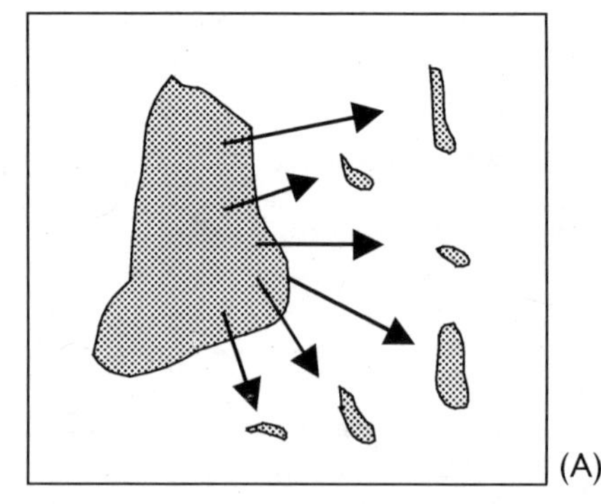

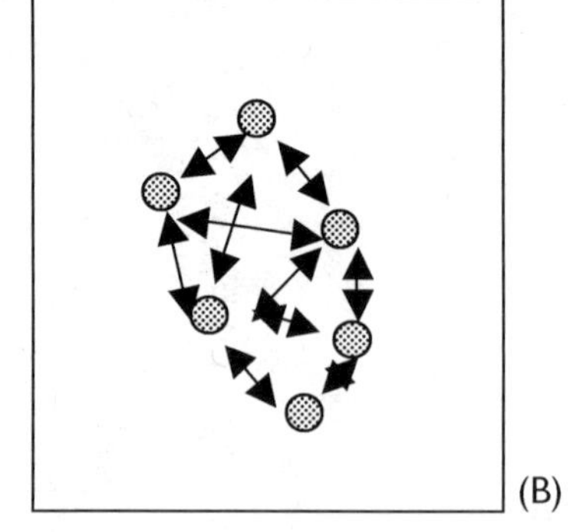

FIGURE 4.2 (A) *Mainland-Island model* (adapted from Hanski and Gyllenberg 1993); (B) *Levins model* (Levins 1969, 1970). In the Mainland-Island model, individuals migrate from the large, invulnerable source population to the smaller habitat patches. The smaller patches are vulnerable to extinction and can be recolonized. The Levins model assumes equal-sized patches with multidirectional dispersal, and all patches are capable of extinction and recolonization.

mogeneous environment and provide a null model describing spatial patterns that arise from random motion and population growth alone (Figure 4.3). All require good information about dispersal, whether or not dispersal depends upon density and whether direction of movement is influenced by the quality of habitats. All depend on knowing the spatial scale over which population dynamics are considered.

Several generalities arise from metapopulation theory (see Box 4.1). The theory predicts that a threshold number of suitable patches is required for large-scale metapopulation persistence; immigration and colonization must be greater than extinction. Non-equilibrium metapopulations are destined for ultimate extinction, with the time to extinction of the metapopulations being the same as the time to extinction of the largest populations (Hanski 1999). Spatially explicit models allow for inclusion of the areas and spatial locations of the patches (Hanski 1994, 1999) and are becoming more widely used for conservation and

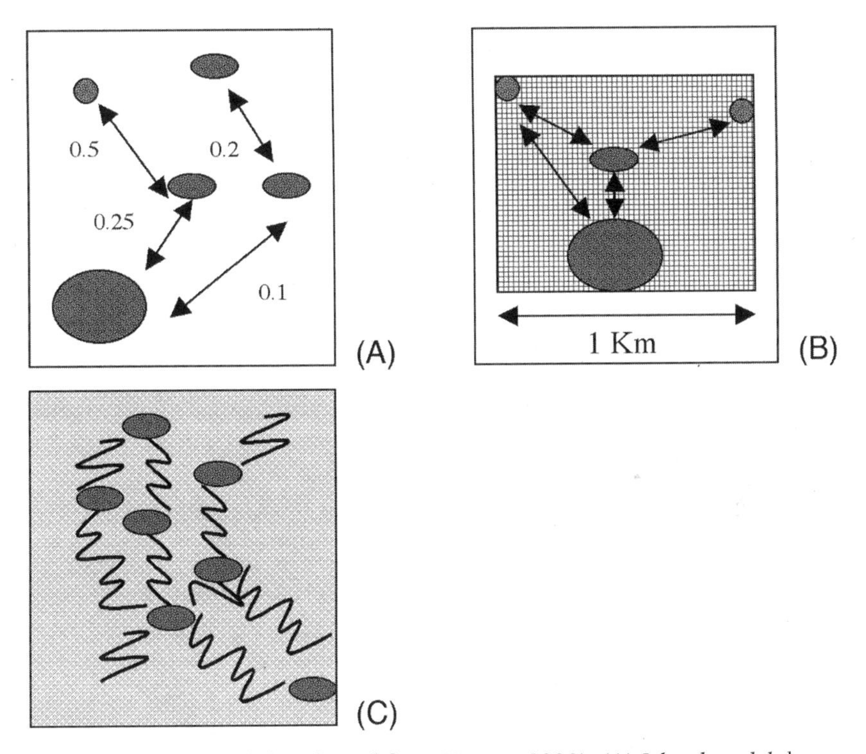

FIGURE 4.3 Types of spatial models (adapted from Kareiva 1990): (A) *Island models* have a collection of patches coupled by one common pool of dispersers; all patches are equally accessible; there is no explicit spatial dimension; and dispersal rates are fractions of individuals that move without regard to distance moved. Island models can be used to investigate how spatial subdivision (or fragmentation) alters the metapopulation behavior. (B) *Stepping-stone models* have the same qualities as Island models except that patches have fixed spatial coordinates. These can be used to examine consequences of long-range versus short-range dispersal. (C) *Reaction-diffusion models* assume a homogeneous environment, and use standard continuous-time Lotka-Volterra representations of local dynamics and a constant, random rate of dispersal. These models can examine the consequences of population density and habitat quality on metapopulation structure.

Box 4.1
Key Principles of Metapopulation Theory

1. The probability of extinction decreases as average population or patch size increases (Shoener and Spiller 1987; Hanski 1991), as the fraction of large patches increases (Hanski and Gyllenberg 1993), and as the total number of patches increase (Simberloff 1976; Hanski and Gyllenberg 1993). The largest patches have the lowest extinction risk, and these determine estimates of time to extinction of the metapopulation.
2. Persistence of metapopulation is possible only if recolonization exceeds extinction (Hanski 1991). Preserving a metapopulation requires either increasing colonization or reducing extinction.
3. As maximum reproductive rate increases within a patch, the probability of extinction decreases (Earn et al. 2000).
4. The "rescue effect" occurs when increasing the number of immigrants increases patch occupancy and decreases the risk of extinction (Brown and Kodric-Brown 1977).
5. The "establishment effect" occurs when increasing the proportion of suitable habitats occupied by a species increases the rate of successful colonization through dispersal and augmentation (Lande et al. 2003).
6. Heavy emigration will make local populations smaller and hence more vulnerable to extinction (Hess 1996a; Maschinski 2001; Menges et al. 2004).
7. The closer the proximity of patches, the higher the migration between the patches and the greater the likelihood of recolonization of vacant patches (e.g., Ovaskainen and Hanski 2003).
8. Larger patches have a greater probability of contributing migrants to a metapopulation; therefore the genetic composition of the largest population influences that of the entire metapopulation (Hanski and Gyllenberg 1993).
9. Patch arrangement and corridor quality can influence metapopulation size. Landscapes with greater interior patches will support larger metapopulations than those with more peripheral patches. Increasing the number of high-quality corridors (those that allow for greater survival after dispersal) will increase the metapopulation size (Anderson and Danielson 1997).

restoration applications to help determine the value of patches for species' persistence (e.g., Hanski and Ovaskainen 2003; Ovaskainen and Hanski 2003). Models that consider heterogeneous habitat, such as stochastic patch occupancy models (SPOM) are helping to develop more spatially realistic metapopulation theory (Hanski 2001).

Several studies have shown the impact of habitat fragmentation and/or disturbance on metapopulation persistence (e.g., Collingham and Huntley 2000; Hanski and Ovaskainen 2003). Using a site occupancy and recruitment model of eighteen plant species growing in grasslands of Scandinavia, Eriksson and Kiviniemi (1999) found a significant relationship between species diversity, availability of suitable habitat, and ability to colonize roadsides. Generalist grassland species with good dispersal ability were predicted to increase or remain stable, while species with limited seed dispersal, low seed set and/or low disturbance tolerance had higher risk of extinction. In fragmented Douglas-fir forests of the U.S. Pacific Northwest, populations of seed predators increased three- to fourfold, thereby increasing seed predation and the extinction risk of *Trillium ovatum* (Tallmon et al. 2003). Where habitats are fragmented, networks of small habitat patches can serve as stepping stones connecting and facilitating migration between landscape patches (Huntley 1991a, 1991b), and larger patches are better. Models indicate that corridors are important for migrating species, but not necessarily sessile organisms (Collingham and Huntley 2000). Due to assumptions of the models, metapopulation models may not be suitable for all restoration systems, but they may be useful as heuristic tools to help identify risks of either removing populations or creating new populations (e.g., Lande 1988).

Minimum Viable Metapopulation Size

Hanski et al. (1996) defined minimum viable metapopulation (MVM) size as the minimum number of interacting local populations necessary for long-term persistence of a metapopulation in balance between local extinctions and recolonizations. The minimum amount of suitable habitat (MASH) was defined as the minimum density (or number) of suitable habitat patches necessary for metapopulation persistence. In order for a metapopulation to persist, there must be a balance between local extinctions and recolonizations of empty but suitable habitat patches. In general, 15 to 20 well-connected patches are required for MVM.

With high turnover rates and habitat destruction, a metapopulation is not at equilibrium and is destined for extinction. Chance variation in the number of extant populations is analogous to demographic stochasticity and can lead to extinction, especially when metapopulations are small. Metapopulations consisting of a small number of local populations, each with a high risk of extinction, are not likely to persist long. Hanski et al. (1996) concluded that many rare and endangered species fall below the minimum viable metapopulation size and may already be headed toward extinction, unless the fragmentation of their habitat is reversed.

Sources, Sinks, Population Regulation, and Habitat Selection

In complex habitat mosaics, individuals may be distributed among habitats and have variable or habitat-specific demographic rates (i.e., different life spans, developmental rates, birth and death rates) (Pulliam 1988; Pulliam and Danielson 1991). Surplus individuals from highly productive "source" habitats may immigrate to less productive "sink" habitats, where within-habitat reproduction fails to keep pace with within-habitat mortality (Pulliam 1988). Habitat selection based on differences in habitat quality from source habitats can maintain large sink populations (Pulliam 1988). If good breeding sites in the source habitat are rare, and poor sites in the sink are relatively common, a large population may occur in the sink. Populations can be sustained by immigration from more productive source habitats.

Pulliam and Danielson (1991) modeled differences in λ two habitat types: source habitat that produces a surplus of individuals available for dispersal, and sink habitat that cannot produce enough young to meet even its own losses. The models indicate that the effect of habitat loss depends upon the ability of the individuals to sample the environment and the quality of the habitat (m). If 90% of poor-quality habitat was removed, the population was reduced by only a moderate amount because the high-quality habitat was so productive that it continued to produce offspring to saturate both habitats (Figure 4.4). By contrast, removing good-quality habitat had a much greater impact on population size and caused extinction when site selection was low. Habitat fragmentation increased distance between successive sites on the dispersal path. If dispersal was limited, then there was reduced sampling (m), which could cause extinction. Coupled effects of decreased sampling and habitat loss are the key to extinction, rather than either factor singly.

Habitat-specific demographic rates may be more important ecologically than age-specific demographic rates. Because species may occur commonly (and breed successfully) in sink habitats, populations need to be studied in the landscape context to understand how habitat heterogeneity influences population dynamics (Pulliam 1988). Several studies have shown that population dynamics are affected by disturbance, spatial variation, and environmental

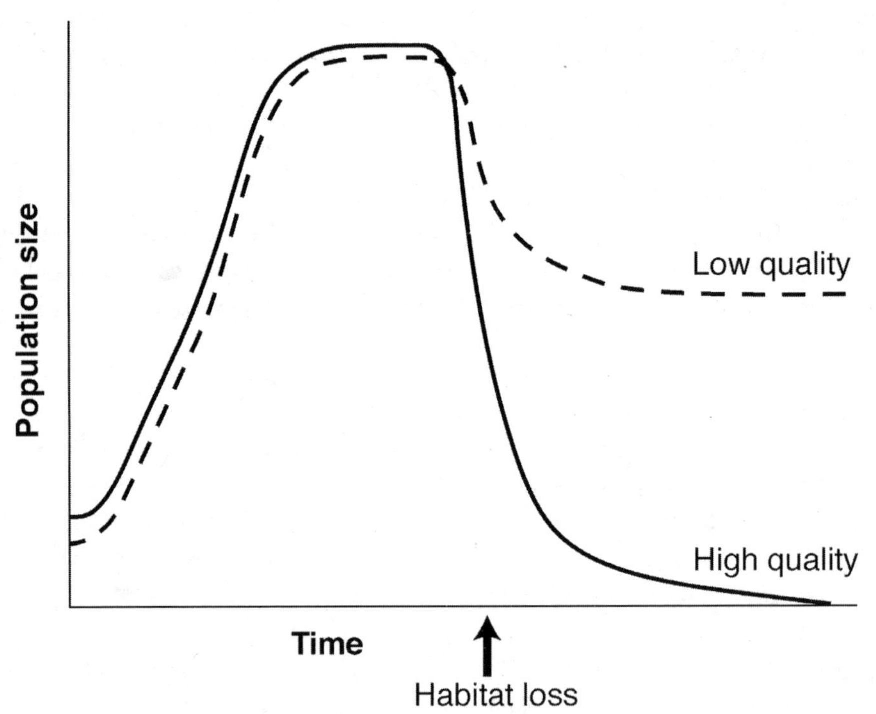

FIGURE 4.4 Comparison of projected population size when 90% of low-quality and high-quality habitat is removed (adapted from Pulliam and Danielson 1991). Loss of good-quality habitat had a more devastating effect on the population size and caused extinction when site selection was low. Loss of poor-quality habitat had less impact, because immigration from the high-quality habitat maintained populations at a moderate level.

heterogeneity (references in Menges and Quintana-Ascencio 2003). In large restoration sites, population size and growth rate may be a function of the relative proportion of the habitat types provided.

Although source-sink models apply to highly mobile species with the ability to select the best habitats and influence their own reproduction, the models offer insights to restoration planners, who are the "habitat selectors." If *m* is translated into reintroduction attempts, it may be possible for populations to increase to equilibrium and stabilize with enough attempts. Because not all utilized habitats are necessarily valuable to the species, restoration using multiple habitats may help to determine where the species' demographic vital rates will be highest. Estimates of the relative contributions of different habitat types to population size and growth will help to determine which habitats are the most valuable to the species. Source-sink models allow us to determine the effect that a given change in the availability of high- and low-quality habitats will have on the global population (Pulliam and Danielson 1991). When a species with very low dispersal ability is to be managed, patch size and spatial dispersion may be more important than the total amount of habitat that is preserved. It is important to determine the ratio of good to poor habitat that will have the best chance of maintaining a viable population of a species. The key is to understand the extent to which the persistence of the population depends upon immigration (Pulliam and Danielson 1991).

Challenges of Metapopulation Theory

Metapopulation models have inherent limitations that present challenges to practitioners. As is true of PVA, metapopulation models need all of the data required for a simple demographic model, including means and variances of all vital rates. In addition, metapopulation models depend upon good estimates of migration, dispersal, or colonization, for which accurate and relevant data can be difficult to obtain (e.g., Freckleton and Watkinson 2002). Few ecological studies of any species occur at the spatial or temporal scale required to obtain this data (Doak and Mills 1994). Empirical estimates of the model parameters will significantly influence the results and their interpretation; these will be magnified by increasingly complex models (Doak and Mills 1994). Therefore, comparative simulations rather than absolute risk assessments are recommended (Melbourne et al. 2004). Restoration experiments provide opportunities for testing population and metapopulation theory through long-term data sets that will aid our refinement of the theory and our understanding of biological patterns in species with varying life histories.

Indirect evidence can be used to infer how plants colonize vacant sites. For instance, colonization is a function of distance from the nearest existing population (Harrison et al. 2000; Jacquemyn et al. 2003). It may be unclear whether a new population occurrence is a true colonization or simply an awakened dormant seed bank (van der Meijden et al. 1992). Although a state variable can depict dispersal through time, this is rarely included in PVA and metapopulation models (Freckleton and Watkinson 2002).

Similarly, accurate estimates of animal dispersal are difficult to obtain. For example, migration of bank voles (*Clethrionomys glareolus*) and yellow-necked mice (*Apodemus flavicollis*) varied with season, sex, resource abundance, and nature of the matrix between patches (Szacki 1999), making it difficult to generalize across sites, years, or species for metapopulation modeling.

Addressing Restoration Questions Using Population and Metapopulation Theory

In this section, I illustrate how theory has been applied to restoration planning and implementation. In some cases, empirical tests have given insight into limitations of theory or have not been done, while others show support for the generalizations of theory (see Box 4.1). Incorporating experimentation in restorations offers opportunities to test and refine theory.

How Many Individuals Should Be Reintroduced?

Theory indicates that the smaller the population, the larger the influence of stochastic factors and the larger the extinction risk. Using as large a founding population as is practical should increase chances of reestablishing the species.

Bell et al. (2003) compared the number of propagule types that would be required to create an MVP with less than 5% extinction probability in the next one hundred years. They found that more than 400 transplants of one-year rosettes, or 1,600 seedlings, or 250,000 seeds would be required to create a viable restored population of *Cirsium pitcheri*. These numbers may not be reasonable to achieve with extremely rare populations, especially if removing substantial propagules from a wild population for restoration to another location may endanger the wild population (Menges et al. 2004).

Researchers simulated the founding population size needed for reintroducing capercaillie (*Tetrao urogallus*) in Scotland (Marshall and Edwards-Jones 1998). They found that a minimum of 60 individuals would be required in 5,000 ha to have greater than 95% probability of a population establishing for 50 years. However, if populations were supplemented every five years with two birds genetically unrelated to the founding population, the MVP was reduced, and genetic health, as measured by levels of heterozygosity, was maintained (Figure 4.5). Supplying new individuals to the population allows for smaller initial populations to be used in a restoration. To date, empirical evidence of capercaillie reintroduction success is not available.

How Large Should Patches Be?

Theory indicates that larger patches have lower extinction risk. The largest patches within a metapopulation will determine the persistence of the metapopulation. Restoration sites may be limited; however, practitioners can choose the largest patches in the best available habitat for reintroducing target species. For example, in southern Scotland, sites selected for capercaillie reintroduction were forest blocks with the largest area and most suitable annual precipitation (Marshall and Edwards-Jones 1998). (See also the *Jacquemontia reclinata* case study later in this chapter.)

The degree of isolation interacts with population size to determine population fate. In a metapopulation analysis of the federally endangered wireweed (*Polygonella basiramia*), Boyle et al. (2003) found that populations growing in small isolated gaps in Florida scrub were more likely to go extinct than those growing in larger gaps. Gaps created by fires are critical to the species' persistence.

Patch size and patch isolation influence reproductive success. Large outcrops of serpentine morning glory (*Calystegia collina*) had greater flower and fruit production and greater

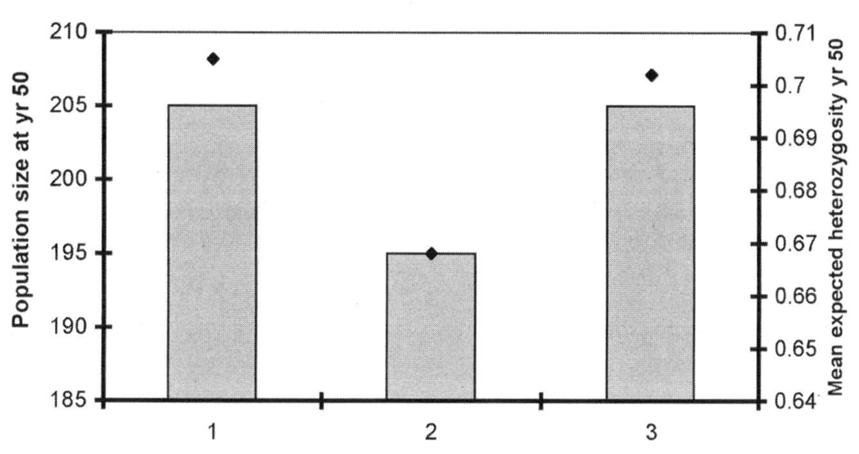

FIGURE 4.5 Population size and mean expected heterozygosity of capercaillie after 50 years following releases into habitats with K = 300. The x-axis represents three scenarios: (1) Initial population is 30 and is supplemented with two unrelated individuals every 5 years. Models indicate100% probability of survival to year 50. (2) Initial population is 30, and no supplementation occurs. Models indicate 88% probability of survival to year 50. (3) Initial population is 60, and no supplementation occurs. Models indicate 88% probability of survival to year 50. Therefore, with supplementation of new individuals, a lower number of founders can be used. Redrawn from Marshall and Edwards-Jones (1998).

densities of flowering patches (Wolf and Harrison 2001). Pollination was positively correlated with flower and patch density due to higher quality of pollen being transferred.

What Developmental Stage Should Be Used?

Elasticity analyses indicate the life stages that have the greatest influence on population growth rate and can guide choice of propagule for use in restoration. Using population viability models of three plants with varying life histories, Guerrant (1996) simulated variation of quasi-extinction rate and final population size as a function of founding population size. Larger initial propagule sizes or stage classes could reduce extinction probability and demographic cost of restoration plantings regardless of the life history of the species he tested. High seedling mortality was partially responsible for this pattern. For one annual species, only those populations founded by whole plants exceeded the original cohort size after 10 years.

Elasticity predictions may miss important aspects of restored population behavior, which can be different from that of the natural population. Plants may be altered by the transplanting or may face environmental conditions unlike any they have seen in natural habitats (Bell et al. 2003; Maschinski et al. 2004). For example, elasticity analyses of the federally endangered Upper Sonoran shrub (*Purshia subintegra*) indicated that vegetative and reproductive adult survival had the greatest contributions to population growth (Table 4.1) (Maschinski et al. 2006) suggesting that adult plants would be the best life stage to use for reintroduction. However, reintroductions comparing whole plants and seeds showed seeds to be the most successful propagule after five years (Figure 4.6) (Maschinski et al. 2004; Maschinski et al.

TABLE 4.1

Elasticities for *Purshia subintegra* for transition years 2001–2002 and 2002–2003 in dry sites. Note that stasis in vegetative and reproductive adults have the highest elasticities.

	2001–2002			λ 0.961	
	Seedbank	Seedling	Juvenile	Veg. Adult	Rep. Adult
Seedbank	0	0	0	0	0
Seedling	0	0	0	0	0
Juvenile	0	0	0	0	0
Veg.	0	0	0	0.998	0.0011
Rep.	0	0	0	0.001	0
	2002–2003			**λ 0.779**	
	Seedbank	Seedling	Juvenile	Veg. Adult	Rep. Adult
Seedbank	0	0	0	0	0
Seedling	0	0	0	0	0.0028
Juvenile	0	0.0028	0.0008	0	0
Veg.	0	0	0.0028	0.188	0.1845
Rep.	0	0	0	0.187	0.4313

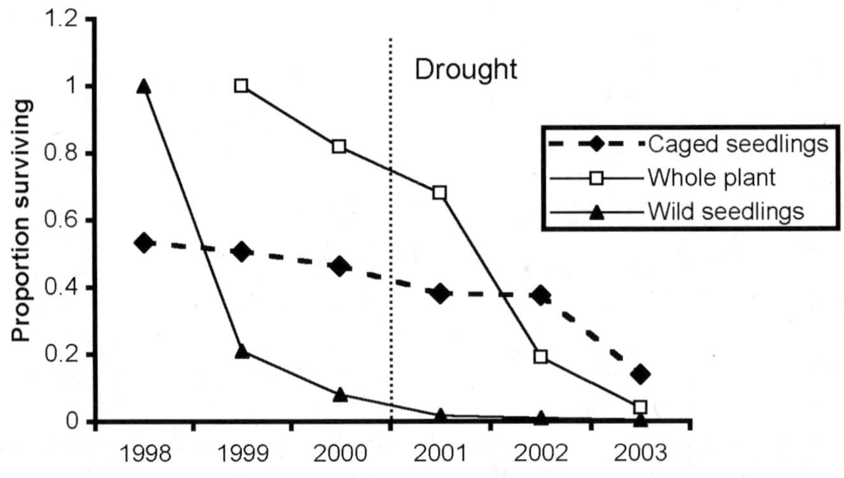

FIGURE 4.6 Proportion of surviving *Purshia subintegra* in reintroduction trials and in the natural population in Verde Valley, Arizona, measured from 1998–2003 (adapted from Maschinski et al. 2004 and Maschinski et al. 2002). Reintroduced whole plants were propagated from cuttings taken from the Verde Valley population and watered for five months after transplanting to the site. Caged seedlings arose from seeds introduced into cages; the survival of seedlings is reported. No supplemental water was given to seeds or seedlings. The survival of naturally occurring wild seedlings was assessed in 30 demographic plots.

2006). Several factors may have been responsible for the improved survival of the experimental versus natural seedlings, including the loosening of the soil during sowing that may have allowed better root development and the caging that provided protection from predation and desiccation. Propagule selection has logistic considerations also. Introducing 4,800 *P. subintegra* seeds was less expensive than propagating, growing, planting, and maintaining 450 whole plants. In the end, elasticity analyses can provide insight to current natural population behavior, but there is no replacement for field testing (Lande et al. 2003) or for the experience of the practitioner.

What Is Suitable Habitat?

What constitutes a site suitable for colonization is sometimes unclear (e.g., Watkinson et al. 2000; Freckleton and Watkinson 2002). Experimental tests have demonstrated the existence of suitable, yet unoccupied, habitat (e.g., Quintana-Ascencio et al. 1998). Simply knowing that a species historically occurred at a site may not be an indication that the site still represents suitable habitat, especially if the factor that caused extirpation from the site has not been removed. Metapopulation models generally assume that patches in which extinction occurs are immediately available for recolonization (Hess 1996a), when in reality extinction may have been caused by habitat destruction or some other alteration that permanently changes the suitability of the habitat. Removing the cause of local extinction will make the largest contribution toward reducing extinction rate (Hess 1996a).

When selecting suitable sites for restoration, one must consider physical, biological, logistical, and historical criteria. However, even with such data, suitable sites are not always evident (Fiedler and Laven 1996). Population dynamic and metapopulation theory suggest that the spatial arrangement of restored patches, their size, and distance from one another are equally important considerations.

Two major factors complicate the selection of areas of suitable habitat for restoration. First, habitats may change over time due to anthropogenic or natural factors, making the establishment of "suitability" less certain. Second, it is often difficult to identify what factors make a site suitable. Evaluating the quality of habitat using fitness attributes of the population (such as mean annual population growth rate) must be done over a long enough time scale to average performance of the organism over good and bad years. Local extirpation of a population may have been due to the marginal suitability of the site, to the lack of recolonization or recruitment in the site, or to some anthropogenic disturbance that permanently altered it. Sowing experiments have been used to test whether habitat is suitable, but unoccupied (Quintana-Ascencio et al. 1998; Eriksson and Kiviniemi 1999; Ehrlen and Eriksson 2000).

As an example, little is known about the natural habitat requirements of the endangered hoary pea, *Tephrosia angustissima* var. *corallicola*. Its only extant population in the continental United States is in a cultivated field in south Florida that has been mowed for 50 years or more. Herbarium records indicate that the species once grew in pine rockland habitat. Although there are several herbarium specimens from Cuba, the status of populations and their ecology there is unknown. To increase the total number of U.S. populations and improve the species' conservation status, Fairchild Tropical Botanic Garden researchers transplanted replicate genotypes into each of three habitats within the pine rockland: along a

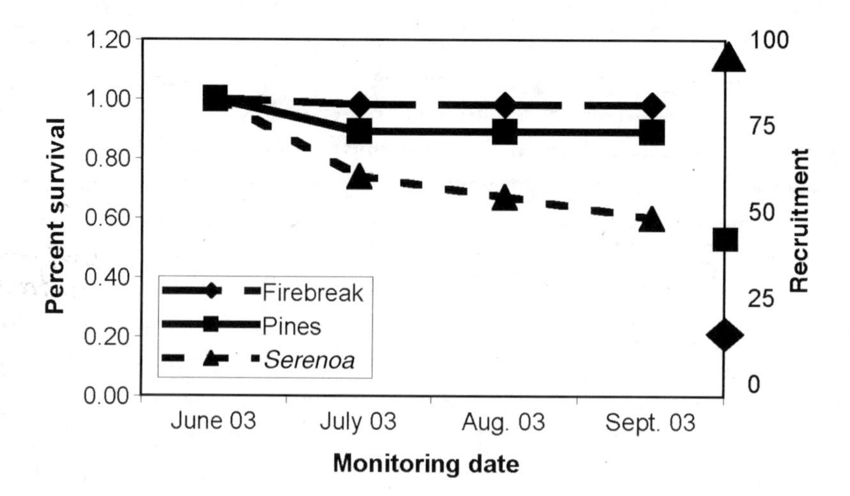

FIGURE 4.7 Percent survival of *Tephrosia angustissima* var. *corallicola* introduced as whole plants and their seedling recruits into three habitat types within pine rockland in south Florida: roadside, under *Pinus elliottii* var. *densa* canopy, and in sand pockets adjacent to *Serenoa repens*. Whole plants were propagated from cuttings taken from the mowed population (adapted from Wendelberger et al. 2003).

firebreak road, under the canopy of *Pinus elliotti* var. *densa*, and in sand pockets adjacent to *Serenoa repens* (Wendelberger et al. 2003). Survival of transplants varied across microsites (Figure 4.7). More surprising was that within three months after transplanting, seedlings recruited into all three sites. Thus, this experimental reintroduction greatly increased knowledge of the species' biology, while increasing population numbers and decreasing the risk of extinction.

Suitability of habitat may also change temporally with the advent of disturbance events. A comparison of *Eucalyptus cladocalyx* in its native habitat in the southern Flinders Ranges of South Australia and in southwestern Australia, where it is invasive, indicated that fire frequency is a primary force influencing the population dynamics of the species (Ruthrof et al. 2003). Following fire, the species has mass seedling recruitment. Because the introduced site has burned more frequently in the last 15 years than the native site, *E. cladocalyx* has become invasive. Several studies have shown that time-since-fire affects the demography of plant species dramatically (Platt et al. 1988; Satterthwaite et al. 2002; Quintana-Ascencio et al. 2003; Menges and Quintana-Ascencio 2004; Suding and Gross, this volume).

Can Increasing Dispersal and Colonization Improve Metapopulation Persistence?

Restoration practitioners can tip the scales toward greater metapopulation persistence by repeatedly collecting and introducing new individuals (rescue effect and establishment effect) and by modifying the habitat to increase likelihood of survival (e.g., Maschinski et al. 2004). For example, Kauffman et al. (2003) used population modeling to evaluate how past management activities influenced endangered peregrine falcons (*Falco peregrinus anatum*) in California. The authors used habitat-specific fecundity trends and estimated survival rates in a time-varying matrix model to estimate population growth rates of peregrine falcons in rural and urban habitats. In urban habitats, population growth rates were 29% per year, birds had higher fecundity, and survival rates of first-year birds were higher than in rural habitats, where $\lambda = 0.99$ and there was slower improvement in eggshell thickening through the 1980s. The models indicated that introductions were pivotal in recovering the rural population.

Colonization and dispersal processes must be preserved for any species to have a long-term future (Thrall et al. 2000). To restore dispersal and colonization to land fills, Robinson and Handel (2000) created habitat islands of trees and shrubs believed to be good bird attractors and good sources of clonal or seed propagules. After five years, avian dispersers facilitated the introduction of 26 other species to the woodland, most of which were from sources in close proximity to the experimental site.

How Should Restoration Sites Be Spatially Arranged?

Selecting the spatial arrangement of restored patches can be critical to the long-term persistence of the metapopulation. Clustering of patches can benefit species retention due to increased dispersal opportunities (Kareiva and Wennergren 1995) and pollinator visitation (Groom 2001). But there is a possibility for the unintended effects of connecting or enhancing the dispersal of undesirable organisms (such as pathogens, predators, exotics) that may have adverse effects on target species (Noss 1987; Simberloff and Cox 1987; Hess 1996b). Increasing connectivity may also synchronize local population fluctuations and increase extinction risk (Earn et al. 2000). Further, recent studies indicate that the interpatch matrix greatly influences dispersal between patches (Ricketts and Morris 2001). The authors suggest that modifying the matrix could reduce patch isolation and extinction risk of populations in fragmented landscapes.

For practitioners, these diametrically opposed views can make the decision about spatial arrangement of patches in the complex landscape matrix problematic. Knowledge of the species' biology is essential for assessing risks and benefits of spatial structure. This area has great potential for experimental design in restoration that can contribute to theory.

Theory predicts that the closer restored populations are to intact habitats and populations, the greater the opportunity for dispersal leading to colonization and persistence. In a study comparing geographical locations of forest patches on total species colonization, Jacquemyn et al. (2003) found that total species richness was higher for non-isolated than for isolated patches. Because only presence/absence data were collected, it was not possible to assess whether patch isolation influenced the viability of the colonizing species. The authors expressed a concern that forest restoration might be pointless if there are no intact habitats nearby to serve as a propagule source. Others have argued that although larger tracts are better, even small tropical forest fragments, especially if close to intact forest, can support a reasonable array of species (Turner and Corlett 1996). Isolation is less critical for species with long-lived seed banks (Piessens et al. 2005).

Clustering patches can either reduce or increase extinction risk, depending upon the species' ecology. In a review of 25 species of Sonoran Desert freshwater fishes, successful recolonization of empty habitat was significantly related to clustering and the occasional long-distance dispersal during periods of high discharge (Fagan et al. 2002). Groom (2001) experimentally manipulated (1) patch isolation while maintaining constant size; (2) patch spatial arrangement (whether clustered or single); and (3) pollen availability to examine whether levels of isolation influenced pollination, herbivory, population growth, and persistence. In the six-year study, she found more pollen limitation and less herbivory in small isolated populations of *Clarkia concinna concinna* than in clustered patches. Small isolated patches also had lower population growth and more extinctions. Although patch size interacted with isolation to influence pollinator behavior, Groom (1998, 2001) recommended clustering subpopulations of insect pollinated plants to enhance long-term population growth (but see objections to unintentional results of connectivity in Hess 1996a, 1996b, and Earn et al. 2000).

In a comparison of 57 species at 81 sites, Dupre and Ehrlen (2002) found that habitat quality, especially pH, was more important for the incidence of species than habitat configuration. Patch area and isolation significantly affected only 11 and 4 species, respectively. Species favored by larger area were also disadvantaged by greater isolation. The importance of habitat configuration varied with life history. Habitat specialists and clonal perennials that produced few seeds were more negatively affected by patch isolation, while animal-dispersed species were more negatively affected by small stand size.

Using decision analysis methods, Dreschsler et al. (2003) examined four hypothetical management scenarios where patches of larval host plants for the Glanville fritillary butterfly (*Melitaea cinxia*) in southwestern Finland were removed (Figure 4.8). The models indicated that the removal of the small, dispersed, stepping-stone patches would have the least detrimental effect on the butterfly metapopulation, while eliminating the largest, most closely clustered patches would have the greatest impact. This finding is in agreement with metapopulation theory, but it does not take into account the risks associated with connectivity nor the possible genetic cost of losing rare alleles from small isolated populations at the edge of the species' range.

How Many Propagules Should Be Moved from One Patch to Another to Sustain a Metapopulation? Which Patches or Populations Should Be Augmented?

Metapopulation theory indicates that large patches have lower extinction rates than small patches. Large patches in high-quality habitat should have highest reproduction and they may serve as source populations for the metapopulation.

To test whether human-assisted dispersal could enhance persistence of an endangered plant, I used metapopulation modeling. In a northern New Mexico canyon, nearly 80% of the population of the federally endangered Holy Ghost ipomopsis (*Ipomopsis sancti-spiritus*), a short-lived monocarpic perennial, grows along a 4.5 km stretch of a steep road. Colonization is constrained by the canyon topography; most migration between patches probably results from water carrying seeds downhill. With current population vital rates and no human-assisted dispersal, metapopulation analysis of average Leslie matrices over six transition years predicted that *I. sancti-spiritus* had a high probability of extinction; 60% of the demographic transects had negative growth rates, occupancy of *I. sancti-spiritus* in any transect had a probability of zero within 44 years (Figure 4.9a) and within the next 10 years only two of the ten transects would be occupied (Maschinski 2001). Persistence improved with human-assisted dispersal uphill from the more fecund patches in the lower part of the canyon (source popu-

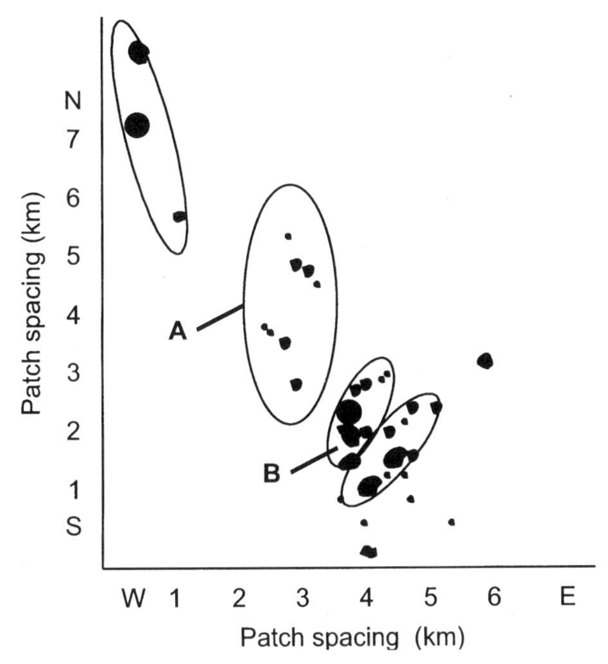

FIGURE 4.8 The network of habitat patches of the Glanville fritillary (*Melitaea cinxia*) in southwestern Finland used for a management decision exercise to determine which type of site (indicated by the numbered ellipses) could be removed with the least impact on the metapopulation. The coordinates on the axes give the east-west and north-south locations of the patches in the landscape, scaled in kilometers. The sizes of the dots indicate the sizes of the patches, with the largest patch (A) having an area of 0.91 ha and the smallest one (B) having an area of 0.01 ha. (redrawn from Drechsler et al. 2003).

lations) to less fecund areas at the upper part of the canyon (sink populations) (Figure 4.9b). However, there was a cost of dispersing seeds away from the downhill patches, as removing 45% of the seeds decreased the patch longevity (Figure 4.9b).

A potential strategy for conserving the species is to disperse seeds from source to sink patches. But better options for preserving the metapopulation would be to augment patches from *ex situ* seed sources, or to improve the habitat quality in the upper part of the canyon by increasing light and decreasing competition from aggressive herbaceous competitors where *I. sancti-spiritus* grows. Even with such actions, the species remains at high risk of extinction.

Jacquemontia reclinata: A Case Study

The federally endangered coastal perennial vine, beach jacquemontia, *Jacquemontia reclinata* is endemic to the southeastern coast of south Florida. Intensive development has left nine populations, which are declining in isolated habitat fragments (Figure 4.10). Five population extinctions have been documented. The Florida Multispecies Recovery Plan (USFWS 1999) calls for the establishment of new populations and augmentation of existing populations, an idea supported by many coastal public land managers, who are working to preserve and restore coastal dune habitat.

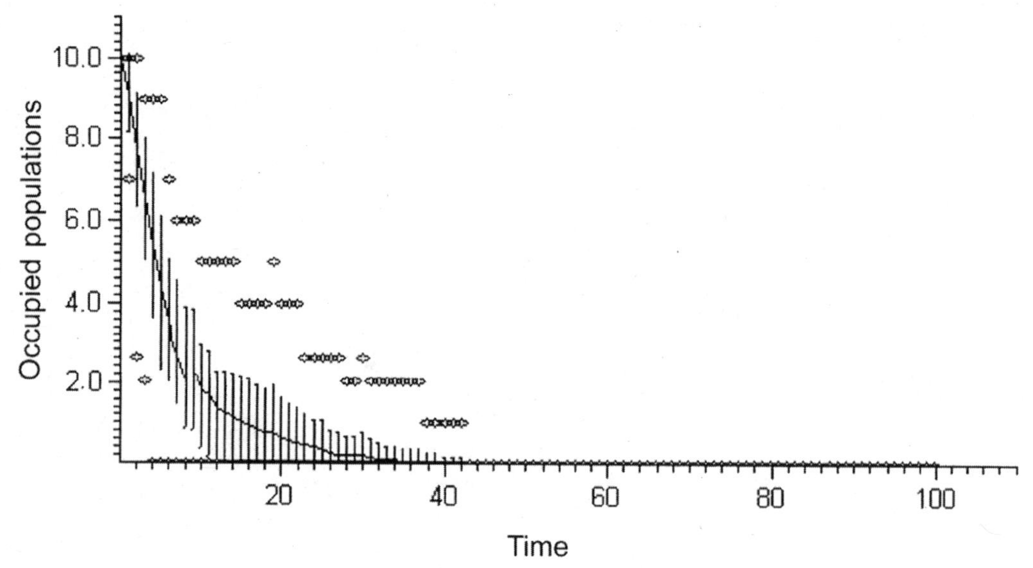

FIGURE 4.9A Expected occupancy (number of patches) within metapopulation of *Ipomopsis sancti-spiritus* in northern New Mexico over the next 100 years. The average ± 1 standard deviation (vertical bars) and minimum and maximum number of extant patches (diamonds) are indicated. The model used average matrices and assumed downhill dispersal only to adjacent patches.

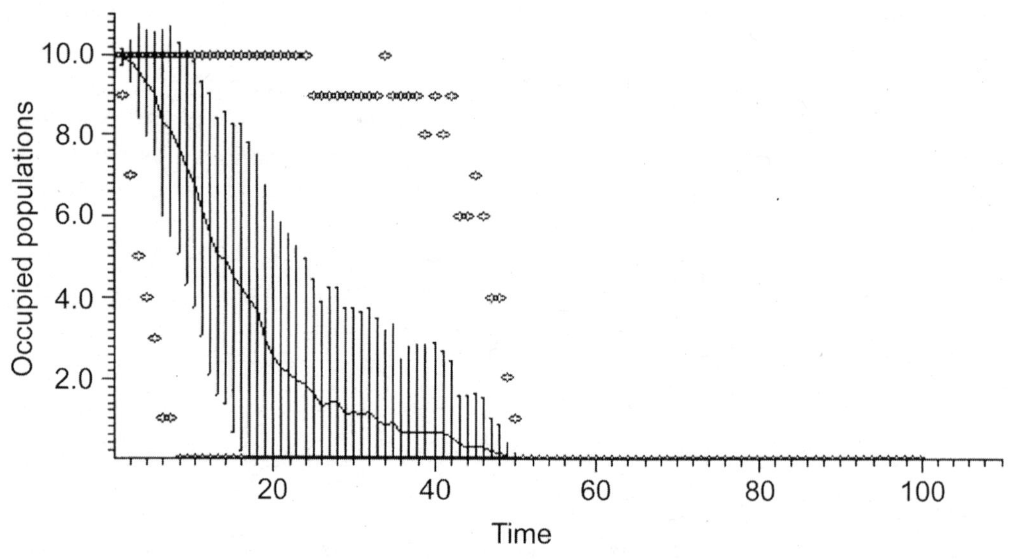

FIGURE 4.9B Expected occupancy (number of patches) within metapopulation of *Ipomopsis sancti-spiritus* in northern New Mexico over the next 100 years, assuming human-aided dispersal uphill to all patches from the two patches at the bottom of the canyon. The average ± 1 standard deviation (vertical bars) and minimum and maximum number of extant patches (diamonds) are indicated. The model used average matrices and assumed human-aided dispersal uphill to all transects. Note the greater variance and range than in Figure 4.9a.

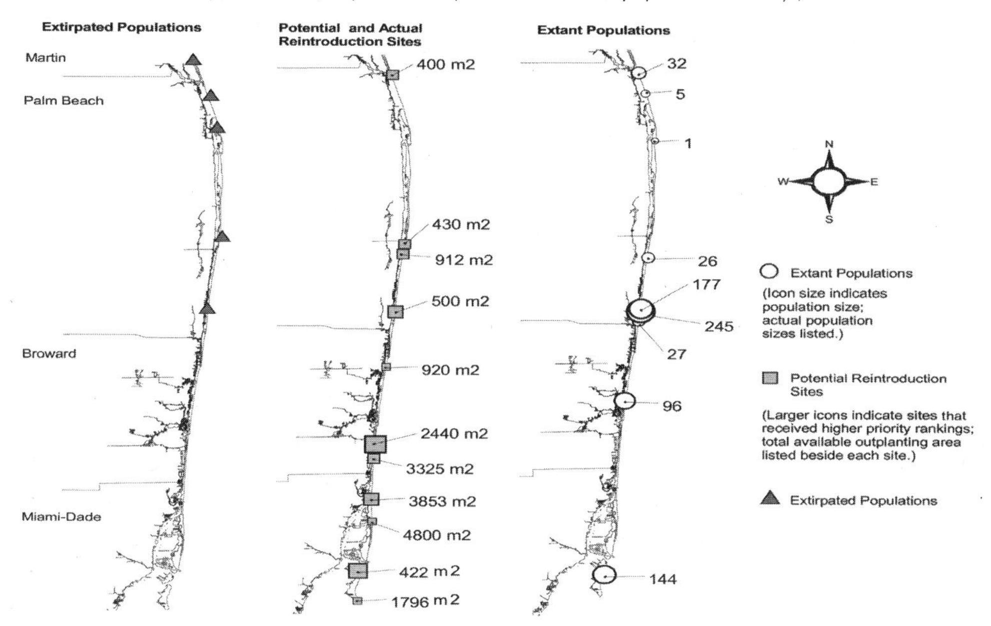

FIGURE 4.10 Map of extirpated, extant, and potential reintroduction sites for *Jacquemontia reclinata* along the eastern coast of south Florida. Population sizes are indicated for extant sites. Area of potential reintroduction sites is indicated.

Because 66% of the natural *J. reclinata* populations have fewer than 100 individuals, and no population has more than 250 individuals (Figure 4.10), all populations will require active intervention to ensure their long-term survival. Although all *J. reclinata* populations fall below the minimum viable population size predicted by theory (Shaffer 1987; Lande 1995), the northern populations have the smallest numbers of plants and highest risk of extinction (Figure 4.10). Genetic diversity was positively correlated with population size; the two largest populations had the highest genetic diversity and the smallest populations in the north had relatively low genetic diversity (Thornton 2003a, 2003b). Thus, the smallest populations have high demographic and genetic risk.

The total number of populations falls below the minimum viable metapopulation size of 15 to 20 well-connected patches (Hanski et al. 1996) and some extant patches are separated by greater than 64 km. Random Amplified Polymorphic DNA (RAPD) analysis of extant *J. reclinata* populations revealed that migration rates were very high between populations (m = 4.05) (Thornton 2003b). Despite habitat fragmentation and large distances between some populations, Thornton (2003a) suggested that hurricane dispersal of seeds may mix genotypes across the geographic range of the species.

How can population dynamic and metapopulation theory help structure the experimental restoration of *J. reclinata* along the southeastern coast of Florida? Wright and Thornton (2003) identified eleven sites that have characteristics suitable for *J. reclinata* introductions (Figure 4.10). These habitats range in size from 422 m^2 to 4,800 m^2 and are dispersed throughout the extant range of the species. Some are clustered and near source populations, whereas others are more isolated. Some are near extirpated populations, but none are historic sites.

Theory predicts that increases in immigration and gene flow can increase the probability of metapopulation persistence. Essentially, restorationists can be the artificial "rescue effect" that moves migrants into the system and the artificial "establishment effect" that increases colonization until the metapopulation stabilizes. Researchers could create eleven new populations in "suitable" habitat that should set the species on the course to recovery.

According to theory, the highest priority sites for introduction are those with the largest area closest to the extant populations, where natural dispersal and recolonization could have the highest probability of occurring (i.e., sites with 2,440 m^2, 3,335m^2, and 3,853 m^2), while lowest priority would be assigned to the isolated small northern populations. Northern habitats may lack the resources required for optimal reproduction of the species, therefore investing effort for their augmentation would not improve the species' persistence, whereas augmenting the southern larger populations, which are potentially the source populations, would improve the probability of species' persistence from both demographic and genetic perspectives. Unfortunately, foregoing any augmentation in the north would most likely ensure the extirpation of those small isolated populations and would greatly contract the range of the species, and is not recommended.

Areas of Research Need and Opportunity

Several guidelines are available for restoration of populations of rare species, communities, and ecosystems (Falk et al. 1996; ANPC 1998; IUCN 1998; Clewell et al. 2000; SERI 2002). Most critical is the identification and correction of key factors causing population decline (Groom 1998, 2001), as restoring sites with ongoing threats will be a losing proposition (D'Antonio and Chambers, this volume). Still, there remain many unanswered questions and opportunities for utilizing population and metapopulation theory in restoration planning.

Basic research is needed on demography of young (seeds and dispersed juveniles), seed and egg banks, dormancy in nature, and dispersal and colonization across populations and patches. This will best be done in heterogeneous environments over enough time to assess variability and microsites needed for optimal population growth. Research directed toward these topics will improve our understanding of species' biology, will enable better models to be constructed, and will improve their application toward conserving populations and metapopulations.

Restorations can be used to test the generalities of PVA and metapopulation theory. This will require that restoration experimental designs incorporate founding population size, spatial arrangement, restoration site and matrix quality, connectivity, migration, and establishment. These studies will require large spatial scale and monitoring over many years, and researchers should expect possible failure in some locations.

Examining population demography and metapopulation structure is most helpful if it is integrated with population genetics (Falk et al., this volume) and ecological assessment of habitat heterogeneity (Larkin et al., this volume) over the entire range of the species. This is a tall order. However, such correlative information will be essential for understanding the underlying causes (and effects) of variation in demographic rates and will be critical for guiding restoration decisions. Models can provide tools for exploring assumptions about restoration activities or about changes under future conditions. By integrating connectivity of landscapes at a biogeographic scale with species' demography, it may become possible to examine potential demographic consequences of climate change and/or habitat destruction, and to develop conservation strategies across whole landscapes (e.g., Opdam and Wascher 2004; Millar and Brubaker, this volume). It will be important to test the efficacy of corridors and variable nonhabitat mosaics for propagule movement and gene flow (e.g., Opdam and Wascher 2004; Ricketts and Morris 2001) and to determine the best spatial context for optimal population growth of restored patches. Studies encompassing the entire range of a species may require cooperation across several land-managing jurisdictions and incorporation of the social and cultural contexts in which they occur.

But perhaps a more pressing dilemma is whether it is worthwhile to restore populations that are small, lack genetic diversity, and have little habitat for expansion. Holsinger (2000) warned that populations with negative long-term growth rates will require constant management and frequent supplementation to prevent their extinction. Many of the world's rare species have these attributes. It is clear that without long-term commitments and investments of time and human action, small populations may not have much chance of long-term persistence. Those committed to preserving biodiversity are building long-term data sets and preliminary results indicate that human action can improve the persistence of populations. Most restoration projects strive toward the goal of self-sustaining systems, but the untested question is whether the populations will have a chance, even with such commitment.

Because it is impractical to develop PVA or metapopulation models for every species of concern, several practitioners are expanding the application of the models to develop rules of thumb, decision analyses, and risk assessments, which can be applied to a wide range of ecological conditions and species for management purposes (Frank and Wissel 1998; Frank 2004). This is a worthy new direction in the field that opens opportunities for empirical tests within restoration contexts.

Acknowledgments

I am grateful to the editors for giving me the opportunity to contribute to this volume, and to J. Baggs, C. Sacchi, H. Thornton, S. Wright, K. Wendelberger, J. Possley, A. Frances, J. Pascarella, E. Guerrant, P. Quintana-Ascencio, and J. Richards, whose research and discussions with me have greatly influenced my thoughts on this subject. The U.S. Fish and Wildlife Service, Arizona Department of Transportation, Center for Plant Conservation, Florida Department of Agriculture and Consumer Services Division of Plant Industry, Fairchild Tropical Botanic Garden, The Arboretum at Flagstaff, Frances McAllister, Otto and Gallina Franz, and Ron Horn generously supported the research. M. Petersen, Kai, D. Warr, C. McDonald, W. Albrecht, J. Busco, D. Callihan, B. Cammack, J. Campbell, J. and N. DeMars, K. Evans, M. Gibson, J. Giguere, K. Hueston, A. Levine, T. Martinez, E. Menges, S. Murray, N. Nahstoll, B. Phillips, P. Quintana-Ascencio, G. Seymour, W. Hines, P. West, M. Brown, B. Camp, A. Chitty, G. Cronk, C. Delfino, E. Doll, M. Fellows, A. Frances, T. Greenwood, T. Hedges, M. Hoover, D. LaPuma, T. Magellan, J. Possley, C. Rosell, R. Savary, M. Wright, Boy Scout Troop 333, and students from Boca Raton High School kindly assisted with field research. Land managers who have been very supportive of our research related to the conservation and management of *Jacquemontia reclinata* in south Florida include G. Atkinson, S. Bass, S. Corrada, P. Davis, J. Duquesnel, D. Farmer, J. Fernandez, R. Garcia, J. Higgins, F. Griffiths, L. Golden, P. Howell, M. Knight, D. Jennings, E. Lynk, J. Maguire, D. Martin,

G. McCausland, L. McDonald, C. Morgenstern, G. Powell, R. Skinner, T. Simmons, S. Thompson, K. Volker, and J. Weldon. R. Redsteer crafted the figures.

Literature Cited

Aikio, S., E. Ranta, V. Kaitala, and P. Lundberg. 2002. Seed bank in annuals: Competition between banker and non-banker morphs. *Journal of Theoretical Biology* 217:341–349.

Anderson, G. S., and B. J. Danielson. 1997. The effects of landscape composition and physiognomy on metapopulation size: The role of corridors. *Landscape Ecology* 12:261–271.

Australian Network for Plant Conservation (ANPC). 1998. *Guidelines for the translocation of threatened plants in Australia*. Canberra, Australia.

Ball, S. J., D. B. Lindenmayer, and H. P. Possingham. 2003. The predictive accuracy of population viability analysis: A test using data from two small mammal species in a fragmented landscape. *Biodiversity and Conservation* 12:2393–2413.

Baskin, C. C., and J. M. Baskin. 1998. *Seeds: Ecology, biogeography, and evolution of dormancy and germination*. San Diego: Academic Press.

Beissinger, S. R., and M. I. Westphal. 1998. On the use of demographic models of population viability in endangered species management. *Journal of Wildlife Management* 62:821–841.

Beissinger, S. R., and D. R. McCullough, editors. 2002. *Population viability analysis*. Chicago: University of Chicago Press.

Bell, T. J., M. L. Bowles, and A. K. McEachern. 2003. Projecting the success of plant population restoration with viability analysis. In *Population viability in plants*, ed. C. A. Brigham and M. W. Schwartz, 313–348. Berlin: Springer-Verlag.

Boyle, O. D., E. S. Menges, and D. M. Waller. 2003. Dances with fire: Tracking metapopulation dynamics of *Polygonella basiramia* in Florida scrub (USA). *Folia Geobotanica* 38:255–262.

Brigham, C. A., and M. W. Schwartz. 2003. *Population viability in plants*. Berlin: Springer-Verlag.

Brigham, C. A., and D. M. Thomson. 2003. Approaches to modeling population viability in plants: An overview. In *Population viability in plants*, ed. C. A. Brigham and M. W. Schwartz, 145–172. Berlin: Springer-Verlag.

Brown, J. H., and A. Kodric-Brown. 1977. Turnover rates in insular biogeography: Effects of immigration on extinction. *Ecology* 58:445–449.

Burgman, M. A., S. Ferson, and H. R. Akcakaya. 1993. *Risk assessment in conservation biology*. London: Chapman & Hall.

Caceres, C. E. 1997. Temporal variation, dormancy, and coexistence: A field test of the storage effect. *Proceedings of the National Academy of Science, USA* 94:9171–9175.

Caswell, H. 2000. Prospective and retrospective perturbation analyses: Their roles in conservation biology. *Ecology* 81:619–627.

Caswell, H. 2001. *Matrix population models: Construction, analysis and interpretation*. Sunderland, MA: Sinauer Associates.

Clewell, A., J. Rieger, and J. Munro. 2000. *Guidelines for developing and managing ecological restoration projects*. Society for Ecological Restoration International (SERI). www.seri.org.

Collingham, Y. C., and B. Huntley. 2000. Impacts of habitat fragmentation and patch size upon migration rates. *Ecological Applications* 10:131–144.

Crowder, L. B., D. T. Crouse, S. S. Heppell, and T. H. Martin. 1994. Predicting the impact of turtle excluder devices on loggerhead sea turtle populations. *Ecological Applications* 4:437–445.

Dennis, B., P. L. Munholland, and J. M. Scott. 1991. Estimation of growth and extinction parameters for endangered species. *Ecological Monographs* 61:115–144.

Doak, D. F., and L. S. Mills. 1994. A useful role for theory in conservation. *Ecology* 75:615–626.

Doak, D. F., D. Thomson, and E. S. Jules. 2002. Population viability analysis for plants: Understanding the demographic consequences of seed banks for population health. In *Population viability analysis*, ed. S. R. Beissinger and D. R. McCullough, 312–337. Chicago: University of Chicago Press.

Drechsler, M., K. Frank, I. Hanski, R. B. O'Hara, and C. Wissel. 2003. Ranking metapopulation extinction risk: From patterns in data to conservation management decisions. *Ecological Applications* 13: 990–998.

Drechsler, M., and M. A. Burgman. 2004. Combining population viability analysis with decision analysis. *Biodiversity and Conservation* 13:115–139.

Dudash, M. R. 1990. Relative fitness of selfed and outcrossed progeny in a self-compatible, protandrous species, *Sabatia angularis* L. (Gentianaceae): A comparison in three environments. *Evolution* 44:1129–1139.

Dupre, C., and J. Ehrlen. 2002. Habitat configuration, species traits and plant distributions. *Journal of Ecology* 90:796–806.

Earn, D. J. D., S. A. Levin, P. Rohani. 2000. Coherence and conservation. *Science* 290:1360–1364.

Ecological Society of America (ESA), ad hoc committee on endangered species. 1996. Strengthening the use of science in achieving the goals of the Endangered Species Act: An assessment by the Ecological Society of America. *Ecological Applications* 6:1–11.

Ehrlen, J., and O. Eriksson. 2000. Dispersal imitation and patch occupancy in forest herbs. *Ecology* 81:1667–1674.

Ellner, S. P., J. Fieberg, D. Ludwig, and C. Wilcox. 2002. Precision of population viability analysis. *Conservation Biology* 16:258–261.

Eriksson, O., and K. Kiviniemi. 1999. Site occupancy, recruitment and extinction thresholds in grassland plants: An experimental study. *Biological Conservation* 87:319–325.

Fagan, W. F., P. J. Unmack, C. Burgess, W. L. Minckley. 2002. Rarity, fragmentation, and extinction risk in desert fishes. *Ecology* 83:3250–3257.

Falk, D. A., C. I. Millar, and M. Olwell, editors. 1996. *Restoring diversity: Strategies for reintroduction of endangered plants.* Washington, DC: Island Press.

Fiedler, P. L., and R. D. Laven. 1996. Selecting reintroduction sites. In *Restoring diversity: Strategies for reintroduction of endangered plants,* ed. D. A. Falk, C. I. Millar, and M. Olwell, 157–170. Washington, DC: Island Press.

Frank, K. 2004. Ecologically differentiated rules of thumb for habitat network design—Lessons from a formula. *Biodiversity and Conservation* 13:189–206.

Frank, K., and C. Wissel. 1998. Spatial aspects of metapopulation survival—From model results to rules of thumb for landscape management. *Landscape Ecology* 13:363–379.

Franklin, I. R. 1980. Evolutionary change in small populations. In *Conservation biology: An evolutionary-ecological perspective,* ed. M. E. Soulé and B. A. Wilcox, 135–150. Sunderland, MA: Sinauer Associates.

Freckleton, R. P., and A. R. Watkinson. 2002. Large-scale spatial dynamics of plants: Metapopulations, regional ensembles and patchy populations. *Journal of Ecology* 90:419–434.

Gilpin, M. E., and M. E. Soulé. 1986. Minimum viable populations: Processes of species extinction. In *Conservation biology: The science of scarcity and diversity,* ed. M. E. Soulé, 19–34. Sunderland, MA: Sinauer Associates.

Gotelli, N. J. 1991. Metapopulation models: The rescue effect, the propagule rain, and the core-satellite hypothesis. *American Naturalist* 138:768–776.

Grant, A., and T. G. Benton. 2000. Elasticity analysis for density-dependent populations in stochastic environments. *Ecology* 81:680–693.

Groom, M. J. 1998. Allee effects limit population viability of an annual plant. *American Naturalist* 151:487–496.

Groom, M. J. 2001. Consequences of subpopulation isolation for pollination, herbivory, and population growth in *Clarkia concinna concinna* (Onagraceae). *Biological Conservation* 100:55–63.

Guerrant Jr., E. O. 1996. Designing populations: Demographic, genetic and horticultural dimensions. In *Restoring diversity: Strategies for reintroduction of endangered plants,* ed. D. A. Falk, C. I. Millar, and M. Olwell, 171–208. Washington, DC: Island Press.

Hanski, I. A. 1991. Single-species metapopulation dynamics: Concepts, models, and observations. In *Metapopulation dynamics,* ed. M. E. Gilpin and I. Hanski, 17–38. London: Academic Press.

Hanski, I. 1994. A practical model of metapopulation dynamics. *Journal of Animal Ecology* 63:151–162.

Hanski, I. 1999. *Metapopulation ecology.* New York: Oxford University Press.

Hanski, I. 2001. Spatially realistic theory of metapopulation ecology. *Naturwissenschaften* 88:372–381.

Hanski, I., and M. Gyllenberg. 1993. Two general metapopulation models and the core-satellite hypothesis. *American Naturalist* 142:17–41.

Hanski, I., A. Moilanen, and M. Gyllenberg. 1996. Minimum viable metapopulation size. *American Naturalist* 147:527–541.

Hanski, I., and O. Ovaskainen. 2003. Metapopulation theory for fragmented landscapes. *Theoretical Population Biology* 64:119–127.

Harrison, S. 1991. Local extinction in a metapopulation context: An empirical evaluation. In *Metapopulation dynamics*, ed. M. E. Gilpin and I. Hanski, 73–88. London: Academic Press.

Harrison, S., J. Maron, and G. Huxel. 2000. Regional turnover and fluctuation in populations of five plants confined to serpentine seeps. *Conservation Biology* 14:769–779.

Hess, G. R. 1996a. Linking extinction to connectivity and habitat destruction in metapopulation models. *American Naturalist* 148:226–236.

Hess, G. R. 1996b. Disease in metapopulation models: Implications for conservation. *Ecology* 77:1617–1632.

Holmes, E. E. 2004. Beyond theory to application and evaluation: Diffusion approximations for population viability analysis. *Ecological Applications* 14:1272–1293.

Holsinger, K. E. 2000. Demography and extinction in small populations. In *Genetics, demography, and viability of fragmented populations*, ed. A. G. Young and G. M. Clarke, 55–74. Cambridge, UK: Cambridge University Press.

Horvitz, C., D. W. Schemske, H. Caswell. 1997. The relative "importance" of life-history stages to population growth: Prospective and retrospective analyses. In *Structured population models in marine, terrestrial and freshwater systems*, ed. S. Tuljapurkar and H. Caswell, 247–271. New York: Chapman & Hall.

Huntley, B. 1991a. Historical lessons for the future. In *The scientific management of temperate communities for conservation*, ed. I. E. Spellerberg, F. B. Goldsmith, and M. G. Morris. Oxford, UK: Blackwell Scientific Publications.

Huntley, B. 1991b. How plants respond to climate change, migration rates, individualism and the consequences of plant communities. *Annals of Botany* 67:15–22.

International Union for Conservation of Nature and Natural Resources (IUCN). 1998. Guidelines for Reintroductions. Prepared by IUCN/SSC Re-introduction Specialist Group. Gland, Switzerland and Cambridge, UK.

Jacquemyn, H., J. Butaye, and M. Hermy. 2003. Impacts of restored patch density and distance from natural forests on colonization success. *Restoration Ecology*11:417–423.

Kareiva, P. 1990. Population dynamics in spatially complex environments: Theory and data. *Philosophical Transactions of the Royal Society London: Biological Sciences* 330:175–190.

Kareiva, P., and U. Wennergren. 1995. Connecting landscape patterns to ecosystem and population process. *Nature* 373:299–302.

Kauffman, M. J., W. F. Frick, and J. Linthicum. 2003. Estimation of habitat-specific demography and population growth for peregrine falcons in California. *Ecological Applications* 13:1802–1816.

Kendall, B. E., and G. A. Fox. 2002. Variation among individuals and reduced demographic stochasticity. *Conservation Biology* 16:109–116.

Kephart, S. R. 2004. Inbreeding and reintroduction: Progeny success in rare *Silene* populations of varied density. *Conservation Genetics* 5:49–61.

Lande, R. 1988. Demographic models of the northern spotted owl. *Oecologia* 75:601–607.

Lande, R. 1993. Risks of population extinction from demographic and environmental stochasticity and random catastrophes. *American Naturalist* 142:911–927.

Lande, R. 1995. Mutation and conservation. *Conservation Biology* 9:782–791.

Lande, R. 2002. Incorporating stochasticity in population viability analysis. In *Population viability analysis*, ed. S. R. Beissinger and D. R. McCullough, 18–40. Chicago: University of Chicago Press.

Lande, R., S. Engen, and B. E. Saether. 2003. *Stochastic population dynamics in ecology and conservation.* Oxford, UK: Oxford University Press.

Levin, D. A. 1990. The seed bank as a source of genetic novelty in plants. *American Naturalist* 135:563–572.

Levins, R. 1969. Some demographic and genetic consequences of environmental heterogeneity for biological control. *Bulletin of the Entomological Society of America* 15:237–240.

Levins, R. 1970. Extinction. In *Lectures on mathematics in the life sciences*, ed. M. Gerstenhaber, 77–107. Volume 2. Providence: American Mathematical Society.

MacArthur, R. H., and E. O. Wilson. 1967. *The theory of island biogeography*. Princeton: Princeton University Press.

Madden, L. V., and F. Van den Bosch. 2002. A population-dynamics approach to assess the threat of plant pathogens as biological weapons against annual crops. *BioScience* 52:65–74.

Marshall, K., and G. Edwards-Jones. 1998. Reintroducing capercaillie (*Tetrao urogallus*) into southern Scotland: Identification of minimum viable populations at potential release sites. *Biodiversity and Conservation* 7:275–296.

Maschinski, J. 2001. Extinction risk of *Ipomopsis sancti-spiritus* in Holy Ghost Canyon with and without management intervention. In *Southwestern Rare and Endangered Plant Conference: Proceedings of the Third Conference*; 25–38 September 2000; Flagstaff, AZ., ed. J. Maschinski and L. Holter, 206–212. RMRS-P-23. Rocky Mountain Research Station, Fort Collins, CO: USDA, Forest Service. 250 pp.

Maschinski, J., J. E. Baggs, and S. Murray. 2002. Annual report on the long-term research on *Purshia subintegra* in the Verde Valley for 2001. Report to the AZ Department of Transportation. Phoenix, AZ.

Maschinski, J., J. E. Baggs, P. F. Quintana-Ascencio, and E. S. Menges. 2006. Using PVA to predict the effects of climate change on the extinction risk of an endangered limestone endemic shrub, Arizona cliffrose. *Conservation Biology*. 20:218–228.

Maschinski, J., J. E. Baggs, and C. F. Sacchi. 2004. Seedling recruitment and survival of an endangered limestone endemic in its natural habitat and experimental reintroduction sites. *American Journal of Botany* 91:689–698.

McCue, K. A., and T. P. Holtsford. 1998. Seed bank influences on genetic diversity in the rare annual *Clarkia springvillensis* (Onagraceae). *American Journal of Botany* 85:30–36.

Meir, E., and W. J. Fagan. 2000. Will observation error and biases ruin the use of simple extinction models? *Conservation Biology* 14:148–154.

Melbourne, B. A., K. F. Davies, C. R. Margules, D. B. Lindemayer, D. A. Saunders, C. Wissel, and K. Henle. 2004. Species survival in fragmented landscapes: Where to from here? *Biodiversity and Conservation* 13:275–284.

Menges, E. S. 1991. The application of minimum viable population theory to plants. In *Genetics and conservation of rare plants*, ed. D. A. Falk and K. E. Holsinger, 45–61. Oxford, UK: Oxford University Press.

Menges, E. S. 1998. Evaluating extinction risks in plant populations. In *Conservation biology for the coming decade*, ed. P. L. Fieldler and P. M. Kareiva, 49–65. 2nd Edition. New York: Chapman & Hall.

Menges, E. S. 2000. Population viability analyses in plants: Challenges and opportunities. *Trends in Ecology & Evolution* 15:51–56.

Menges, E. S., and P. F. Quintana-Ascencio. 2003. Modeling the effects of disturbance, spatial variation, and environmental heterogeneity on population viability of plants. In *Population viability in plants*, ed. C. A. Brigham and M. W. Schwartz, 289–312. Berlin: Springer-Verlag.

Menges, E. S., and P. F. Quintana-Ascencio. 2004. Evaluating population viability analysis with fire in *Eryringium cuneifolium:* Deciphering a decade of demographic data. *Ecological Monographs* 74:79–99.

Menges, E. S., E. O. Guerrant, and S. Hamze. 2004. Effects of seed collection on extinction risk of perennial plants. In *Ex situ plant conservation: Supporting species survival in the wild*, ed. E. O. Guerrant Jr., K. Havens, and M. Maunder, 305–324. Washington, DC: Island Press.

Noss, R. 1987. Corridors in real landscapes: A reply to Simberloff and Cox. *Conservation Biology* 1:159–164.

Opdam, P., and D. Wascher. 2004. Climate change meets habitat fragmentation: Linking landscape and biogeographical scale levels in research and conservation. *Biological Conservation* 117:285–297.

Ovaskainen, O., and I. Hanski. 2003. How much does an individual habitat fragment contribute to metapopulation dynamics and persistence? *Theoretical Population Biology* 64:481–495.

Pake, C. E., and D. L. Venable. 1996. Seed banks in desert annuals: Implications for persistence and coexistence in variable environments. *Ecology* 77:1427–1435.

Pfister, C. A. 1998. Patterns of variance in stage-structure populations: Evolutionary predictions and ecological implications. *Proceedings of the National Academy of Science, USA* 95:213–218.

Pico, F. X., P. F. Quintana-Ascencio, E. S. Menges, and F. Lopez-Barrera. 2003. Recruitment rates exhibit high elasticities and high temporal variation in population of a short-lived perennial herb. *Oikos* 103:69–74.

Piessens, K., O. Honnay, and M. Hermy. 2005. The role of fragment area and isolation in the conservation of heathland species. *Biological Conservation* 122:61–69.

Platt, W. J., G. W. Evans, and S. L. Rathbun. 1988. The population dynamics of a long-lived conifer (*Pinus palustris*). *American Naturalist* 131:491–525.

Possingham, H. P., I. R. Ball, and S. Andelman. 2000. Mathematical models for reserve design. In *Quantitative methods for conservation biology*, ed. S. Ferson and M. Burgman, 291–306. New York: Springer-Verlag.

Pulliam, H. R. 1988. Sources, sinks and population regulation. *American Naturalist* 132:652–661.

Pulliam, H. R., and B. J. Danielson. 1991. Sources, sinks, and habitat selection: A landscape perspective on population dynamics. *American Naturalist* 137:S50–S66.

Quintana-Ascencio, P. F., R. W. Dolan, and E. S. Menges. 1998. *Hypericum cumulicola* demography in unoccupied and occupied Florida scrub patches with different time-since-fire. *Journal of Ecology* 86:640–651.

Quintana-Ascencio, P. F., E. S. Menges, and C. Weekley. 2003. A fire-explicit population viability analysis of *Hypericum cumulicola* in Florida rosemary scrub. *Conservation Biology* 17:433–449.

Reed, J. M., L. S. Mills, J. B. Dunning Jr., E. S. Menges, K. S. McKelvey, R. Frye., S. R. Beissinger, M. C. Anstett, and P. Miller. 2002. Emerging issues in population viability analysis. *Conservation Biology* 16:7–19.

Ricketts, T. H., and W. F. Morris. 2001. The matrix matters: Effective isolation in fragmented landscapes. *American Naturalist* 158:87–99.

Robinson, G. R., and S. N. Handel. 2000. Directing spatial patterns of recruitment during an experimental urban woodland reclamation. *Ecological Applications* 10:174–188.

Ruthrof, K. X., W. A. Loneragan, and C. J. Yates. 2003. Comparative population dynamics of *Eucalyptus cladocalyx* in its native habitat and as an invasive species in an urban bushland in south-western Australia. *Diversity and Distributions* 9:469–484.

Satterthwaite, W. S., P. F. Quintana-Ascencio, and E. S. Menges. 2002. Assessing scrub buckwheat population viability in relation to fire using multiple modeling techniques. *Ecological Applications* 12:1672–1687.

Schemske, D. W., B. C. Husband, M. H. Ruckelshaus, C. Goodwillie, I. M. Parker, and J. G. Bishop. 1994. Evaluating approaches to the conservation of rare and endangered plants. *Ecology* 75:584–606.

Schoener, T. W., and D. A. Spiller. 1987. High population persistence in a system with high turnover. *Nature* (London) 330:474–477.

Schwartz, M. W. 2003. Assessing population viability in long-lived plants. In *Population viability in plants*, ed. C. A. Brigham and M. W. Schwartz, 239–266. Berlin: Springer Verlag.

Shaffer, M. 1987. Minimum viable populations: Coping with uncertainty. In *Viable populations for conservation*, ed. Michael Soulé, 69–86. Cambridge, UK: Cambridge University Press.

Silvertown, J., M. Franco, and E. Menges. 1996. Interpretation of elasticity matrices as an aid to the management of plant population for conservation. *Conservation Biology* 10:591–597.

Simberloff, D. 1976. Experimental zoogeography of islands: Effects of island size. *Ecology* 57:629–648.

Simberloff, D., and J. Cox. 1987. Consequences and costs of conservation corridors. *Conservation Biology* 2:40–56.

Society for Ecological Restoration International (SERI), Science and Policy Working Group. 2002. *The SERI primer on ecological restoration*. www.seri.org.

Sork V. L., R. J. Dyer, F. W. Davis, and P. E. Smouse. 2002. Mating system in California Valley oak, *Quercus lobata* Neé. In *Oaks in California's changing landscape: Fifth symposium on oak woodland savannas*, ed. R. Standiford and D. McCreary, 427–440. San Diego, CA.

Soulé, M. E. 1980. Thresholds for survival: Maintaining fitness and evolutionary potential. In *Conservation biology: An evolutionary-ecological perspective*, ed. M. E. Soulé and B. A. Wilcox, 151–170. Sunderland, MA: Sinauer Associates.

Soulé, M. E. 1987. *Viable populations for plant conservation*. Cambridge, UK: Cambridge University Press.

Szacki, J. 1999. Spatially structured populations: How much do they match the classic metapopulation concept? *Landscape Ecology* 14:369–379.

Tallmon, D. A., E. S. Jules, N. J. Radke, and L. S. Mills. 2003. Of mice and men and trillium: Cascading effects of forest fragmentation. *Ecological Applications* 12:1193–1203.

Thornton, H. 2003a. Genetic structure and conservation of *Jacquemontia reclinata*, an endangered coastal species of southern Florida. MS thesis, Florida International University, Miami.

Thornton, H. 2003b. Genetic variation within and among *Jacquemontia reclinata* populations studied. In *Restoration of* Jacquemontia reclinata *to the south Florida ecosystem*, ed. J. Maschinski, S. J. Wright, and H. Thornton. Final Report to the U.S. Fish and Wildlife Service for Grant Agreement 1448-40181-99-G-173.

Thrall, P. H., J. J. Burdon, and B. R. Murray. 2000. The metapopulation paradigm: A fragmented view of conservation biology. In *Genetics, demography, and viability of fragmented populations*, ed. A. G. Young and G. M. Clarke, 75–96. Cambridge, UK: Cambridge University Press.

Turner, I. M., and R. T. Corlett. 1996. The conservation value of small, isolated fragments of lowland tropical rain forest. *Trends in Ecology & Evolution* 11:330–333.

U.S. Fish and Wildlife Service (USFWS). 1999. *South Florida multi-species recovery plan*. Atlanta: Southeast Region.

van der Meijden, E. P., G. L. Klinkhamer, T. J. de Jong, and C. A. M. van Wijk. 1992. Metapopulation dynamics of biennial plants: How to exploit temporary habitats. *Acta Botanica Neerlandica* 41:249–270.

Watkinson, A. R., R. P. Freckleton, and L. Forrester. 2000. Population dynamics of *Vulpia ciliata*: Regional, metapopulation and local dynamics. *Journal of Ecology* 88:1012–1029.

Wendelberger, K., K. S. Griffin, and J. Maschinski. 2003. *Tephrosia angustissima* var. *corallicola* microhabitat experiment/outplanting. In *Conservation of south Florida endangered and threatened flora*, ed. J. Maschinski, S. J. Wright, K. Wendelberger, H. Thornton, and A. Muir. Final report to Florida Department of Agriculture and Consumer Services: Gainesville. Contract #007182.

Wilcove, D. S., D. Rothstein, J. Dubow, A. Phillips, and E. Losos. 1998. Quantifying threats to imperiled species in the United States. *Bioscience* 48:607–615.

Wisdom, M. J., L. S. Mills, and D. F. Doak. 2000. Life stage simulation analysis: Estimating vital-rate effects on population growth for conservation. *Ecology* 81:628–641.

Wolf, A. T., and S. P. Harrison. 2001. Effects of habitat size and patch isolation on reproductive success of the serpentine morning glory. *Conservation Biology* 15:111–121.

Wright, S. J., and H. Thornton. 2003. Identification of restoration sites for *Jacquemontia reclinata*. In *Restoration of* Jacquemontia reclinata *to the south Florida ecosystem*, ed. J. Maschinski, S. J. Wright, and H. Thornton. Final Report to the US Fish and Wildlife Service for Grant Agreement 1448-40181-99-G-173.

Chapter 5

Restoring Ecological Communities: From Theory to Practice

HOLLY L. MENNINGER AND MARGARET A. PALMER

In 1999, ecologist Simon Levin wrote that "the central challenge of our time is embodied in the staggering losses, both recent and projected, of biological diversity at all levels, from the smallest organisms to charismatic large animals and towering trees. Largely through the actions of humans, populations of animals and plants are declining and disappearing at unprecedented rates; these losses endanger our way of life and, indeed, our very existence." As we write this chapter five years later, the loss of species continues to escalate such that species extinctions may have ramifications for entire ecological communities and the ecological processes they support (Loreau et al. 2001; Kareiva and Levin 2003).

Evidence has also accumulated that diverse communities may be more resistant and resilient to perturbations (McGrady-Steed et al. 1997; Cottingham et al. 2001; Ives and Cardinale 2004). Recognition that the goods and services provided by many ecological communities needed to be recovered and protected from collapse is growing and efforts to restore entire communities or the ecosystems that encompass them is growing dramatically (Perrow and Davy 2002; Allen 2003). Thus, while many early restoration efforts targeted populations, efforts to restore entire ecological communities or ecosystems are becoming more common.

The shift in focus from restoration of single species or populations (Box 5.1) to entire assemblages of plants and animals has not been without controversy. First, the very definition of an ecological community has been debated almost as much as ecologists have attempted to identify the factors responsible for shaping communities (Morin 1999). Most ecologists agree that a community is a collection of associated populations, and they often consider just the plant or just the animal assemblages. Yet there is considerable disagreement whether this collection is defined exclusively by spatial boundaries or by the interactions among populations (Ricklefs 1990). In this chapter we are inclusive, working with communities defined either within a spatial frame or by their functional interactions.

At the heart of community ecology is a body of theory relating to the formation and maintenance of species diversity; it is the restoration of this diversity that is the desired endpoint for many projects, from grasslands (Smith et al. 2003) to rivers (Theiling et al. 1999) to tropical forests (Lamb 1998), and thus we have chosen to focus on these theories explaining diversity.

Box 5.1
Spillover Effects from Single Species Restoration Efforts

Community restoration is often intimately linked to the recovery of an endangered species. Moreover, there is evidence that efforts to restore single species often have spillover effects on other species in the associated community. For example, to create nesting habitat for the endangered Kirtland's warbler (*Dendroica kirtlandii*), land managers in jack pine plantations in Michigan are using prescribed burns to open the canopy. In addition to creating important habitat for the warblers, these openings provide refugia for the unique jack pine-barrens understory plant community (Houseman and Anderson 2002). In southeastern Australia, efforts to restore the golden sun moth (*Synemon plana*) have included the restoration of its critical habitat, the endangered temperate native grasslands (O'Dwyer and Attiwill 2000). Similarly, the upland prairie community in the Willamette Valley in Oregon is a focus of restoration for the endangered Fender's blue butterfly (*Icaricia icarioides fenderi*) and its host plant, Kincaid's lupine (*Lupinus sulphureus kincaidii*) (Severns 2003).

Spillover effects of single species restoration are not necessarily limited to the flora of the target species' habitat. In some cases, other members of the associated animal assemblage benefit. In the southeastern United States, efforts to restore habitat of the endangered Red-cockaded woodpecker (*Picoides borealis*) include thinning midstory hardwood vegetation and performing prescribed burns, thus promoting loblolly-shortleaf and longleaf pine forest with a more open herbaceous plant understory. In these sites specifically managed for the Red-cockaded woodpecker, Conner et al. (2002) found significantly more diverse and numerous breeding bird communities than in unmanaged mature forest control sites. Physical habitat improvement to a mountain stream in Quebec resulted in increases in target trout biomass as well as benefits to nontarget animals, including crayfish, mink, and raccoons, which increased in biomass and activity (Armantrout 1991).

A Brief History of Community Ecology

Community ecology is rooted in a rich tradition of the observation and description of patterns of community structure. Community structure includes species composition and diversity as well as the relative abundance of given species. Early plant ecologists like Cowles and Clements described plant species associations and discrete stages of succession (Cowles 1899; Clements 1936) while Gleason (1926) emphasized the role of the environment and individual plant characteristics in determining plant community structure. Early animal ecologists examined the feeding relations within communities, pioneering food-web ecology (Forbes 1887; Elton 1966). The study of structure was considered the first step in understanding how communities were put together (MacArthur and MacArthur 1961; Brown 1975; Pianka 1975).

The development of ecological theory and an emphasis on experimentation led community ecologists from the study of pattern to the underlying processes responsible for community structure. Ecologists began to ask how communities form, and what specifically enables the species in a community to coexist. Much work focused on the role of competition and understanding how communities became saturated with species—that is, how communities reached an equilibrium state (MacArthur and Levins 1967; May 1973a, 1973b; Ricklefs

1990). The study of other deterministic biotic factors, specifically predation, led to some elegant field studies and a theoretical understanding of the role of keystone predators in structuring communities (Paine 1969, 1974). Niche theory suggested that species occur in a multidimensional space, the axes of which are both biotic and abiotic environmental factors along which resources are partitioned (Hutchinson 1957). MacArthur and MacArthur (1961) introduced the role of habitat heterogeneity and environmental complexity in structuring communities, whereas island-biogeography theory (MacArthur and Wilson 1967; Simberloff and Wilson 1969) introduced the role of population isolation and species dispersal ability as important determinants of species number. In the 1980s and 1990s, emphasis moved away from the prevalent deterministic perspective of communities to a nonequilibrium view defined by disturbance and stochastic processes (Pickett 1980; Pickett and McDonnell 1989, 1990; Sprugel 1991; Cornell 1999).

As mechanisms became elucidated, community ecologists expanded their search for patterns of species richness beyond local scales to regional and global scales. Ecologists examined relationships between biodiversity and habitat area, latitude; and productivity at large scales (see recent reports and reviews by Stevens and Willig 2002; Hawkins et al. 2003; Ricklefs 2004; Hillebrand 2004; Pimm and Brown 2004) and considered the consequences of diversity to ecosystem function (Loreau et al. 2001; Tilman et al. 2001; Naeem 2002). The most famous relationship, largely considered a law in ecology, is the species-area relationship, where the number of species (S) scales with local habitat area (A) in the form of a power function

$$(S = cA^z)$$

(MacArthur and Wilson 1967; Connor and McCoy 1979). Most recently, the species-area relationship has been used in restoration efforts to estimate local species richness (Plotkin et al. 2000), predict effects of habitat fragmentation and loss (Kinzig and Harte 2000; Peintinger et al. 2003), and design conservation reserves (Neigel 2003). A focus on regional patterns in species richness has resulted in a reemergence of scaling laws whereby physical processes and thermodynamics provide the bases for species diversity and the community organization (Allen et al. 2002; Brown et al. 2002).

Although we will spend the majority of this chapter describing how the rich body of community ecological theory can inform and guide restoration, it is important to note that restoration ecology correspondingly has tremendous potential to contribute to the further development of community ecological theory. Indeed, restoration projects can be viewed as the ultimate way to test and refine theories (Bradshaw 1987). Community restoration projects may provide excellent new opportunities for ecologists to examine specific theoretical constructs in a more applied and biologically relevant context than closed model or experimental systems.

Making the Connection Between Community Ecology Theory and Restoration Ecology

A central issue in community ecology is understanding the factors that govern the composition and abundance of species in ecological communities. This question has been tackled by many theoretical ecologists and is central to the work of restoration practitioners. We will review the answers to this question in the context of restoration using a hierarchical "filter" framework, first introduced by T. R. E. Southwood (1977), modified by Tonn et al. (1990) and Poff (1997), and applied toward the restoration of communities by Temperton et al. (2004) (Figure 5.1). This framework recognizes that the species diversity of a community is a function of many factors, biotic and abiotic, working at different temporal and spatial scales. A restoration practitioner who wishes to restore an ecological community will need to ask the following questions:

1. How do *regional processes* determine species composition?
2. What *environmental conditions* and *habitat characteristics* favor species survival and influence community structure?
3. How do *biotic interactions* shape community structure?

We will use these three questions to demonstrate the important role that ecological theory can play in restoration. It should be noted that while one can make distinctions between the terms *theory*, *model*, *concept*, and *hypothesis* (Pickett et al. 1994), in this chapter, we acknowledge all as belonging to the body of theoretical ecology that tries to explain and generalize the patterns and processes we observe in ecological communities.

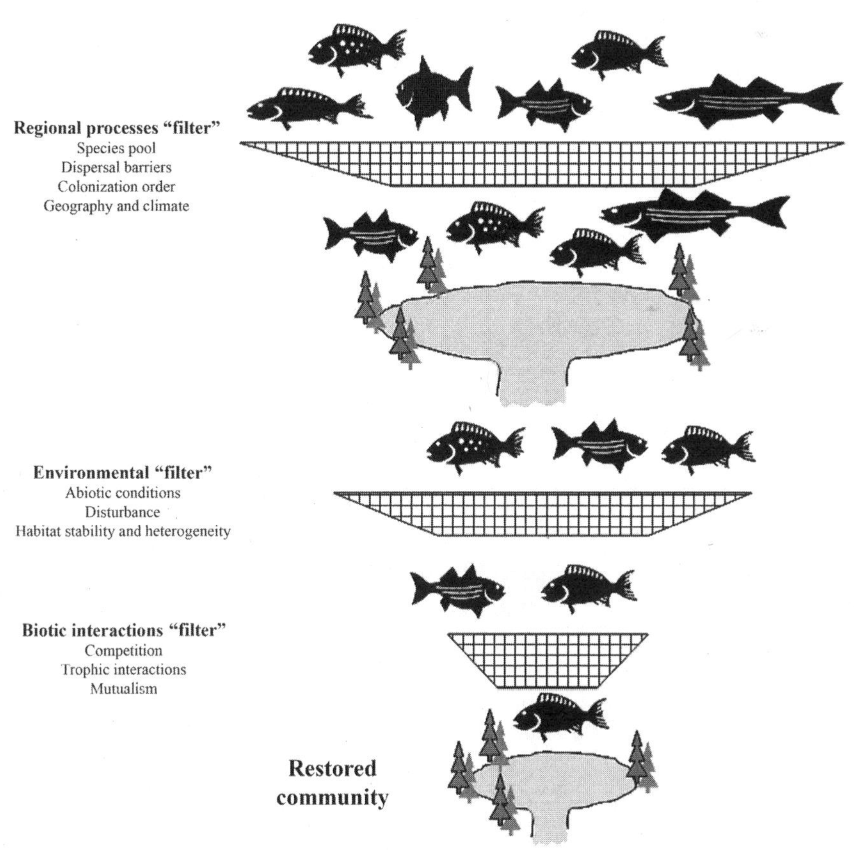

FIGURE 5.1 The species diversity of a restored community, like natural communities, is a function of many biotic and abiotic factors operating at different temporal and spatial scales, as represented by this hierarchical filter framework. Modified from Tonn et al. (1990).

The Influence of Regional Processes

Regional processes, operating at large temporal and spatial scales, strongly affect and even constrain local community structure. Processes affecting the composition of the species pool, dispersal, and recruitment all bear on the number and species composition at a restored site. In addition, historical legacies—the temporal equivalent of regional processes—may constrain community structure. Below, we consider a number of ecological theories that attempt to explain the roles of these larger scale processes in the context of community restoration.

Species Pool

The composition and abundance of species in a local community is largely a function of the species diversity in the regional species pool (Figure 5.2). Local species diversity is strongly and positively correlated with regional species distributions in many taxa and systems (Ricklefs 1987; Caley and Schluter 1997; Ricklefs 2005). Several models attempt to describe the

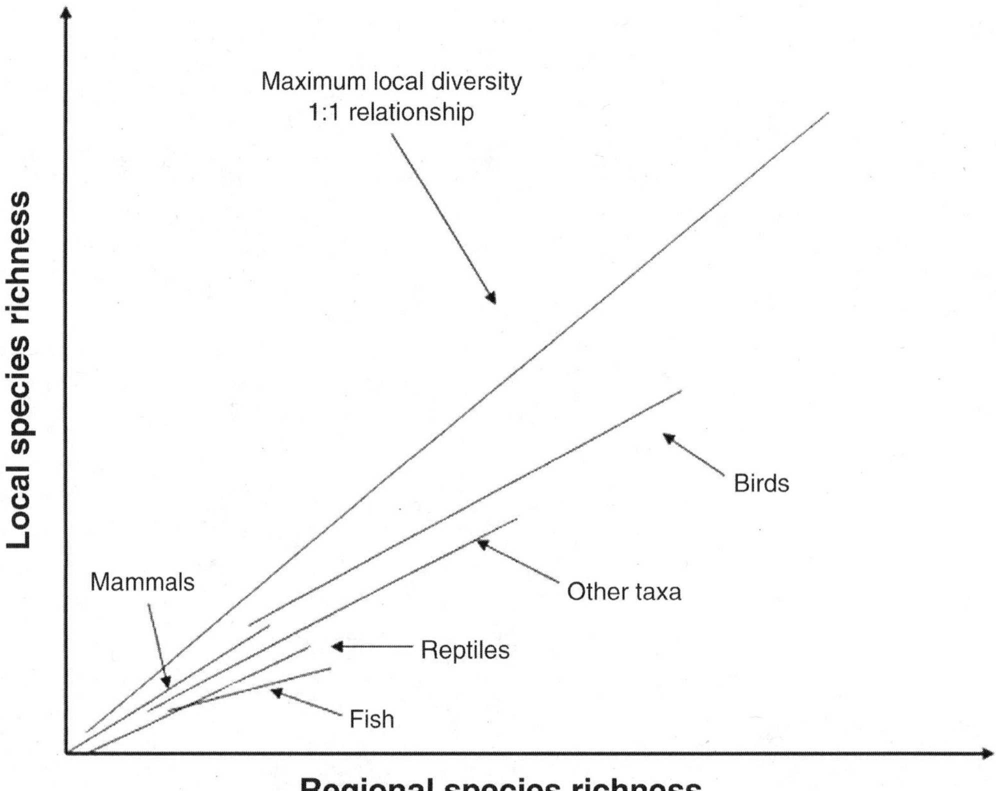

FIGURE 5.2 Local species diversity (measured as 10% of regional area) is a strong linear function of regional species diversity across many taxa (e.g., amphibians, birds, mammals, reptiles, fish, butterflies, corals, odonates, eucalyptus, trees) as well as spatial scale (from local to intercontinental). Note, however, that all slopes are less than 1, suggesting variation in environmental conditions among habitats that may lead to local species turnover and variability, causing rare species to be inadequately represented in local biotas. Adapted from Caley and Schluter (1997).

mechanism by which regional processes maintain community diversity. For example, lottery models suggest that openings in communities, like coral reefs, are filled at random by recruits from a large pool of potential colonists, where more abundant species are more likely to fill empty slots (Sale 1977; Chesson and Warner 1981). The core-satellite hypothesis predicts that regional species distributions are bimodal (Hanski 1982). "Core" species are the most abundant species found at almost all sites and "satellite" species are much less frequent and abundant. While core species may frequently colonize other patches, satellite species are less likely to colonize patches and are more likely to go extinct in individual patches. This theory has been suggested to explain regional distributions for mangrove-island insects (Hanski 1982) and prairie grasses (Gotelli and Simberloff 1987). With respect to restoration, being able to identify core versus satellite taxa may be critical for prioritizing what species to protect in order to repopulate restored habitats, as suggested by modeling of Unionid mussels by Lee et al. (1998).

Dispersal

While the presence of a diverse regional species pool is critical, dispersal processes may govern if regional species colonists are candidates for establishment at a restoration site. Island-biogeography theory would suggest that distance from the reference (intact) site and size of the recipient site (area to be restored) play an important role in determining the number of species colonizing and inhabiting (MacArthur and Wilson 1967). These ideas are particularly relevant when we think about restoring fragmented habitats, like forests for example, where fragment distance from the natural habitat may influence dispersal probability (Jacquemyn et al. 2003). In habitats or for biota where dispersal may be limited, such as the immigration of non-aerial invertebrates to restored wetlands, facilitation of recruitment via inoculation from natural habitats may be important for restoration of the invertebrate community (Brady et al. 2002). In a series of grassland restoration experiments, an unreliable seed bank and slow seed rain limited the assembly of plant communities in fields (Pywell et al. 2002). Similarly, Galatowitsch and vanderValk (1996) found that the dispersal of sedge and other wet-meadow species propagules to restored prairie pothole wetlands was much slower than those of submerged aquatic plant species and may hinder the reestablishment of the full plant community. Studies of recruitment limitation in marine and freshwater benthic habitats have demonstrated that regional, physical processes like fluid and bed load transport may govern dispersal and settlement rates of aquatic biota, as well as trophic dynamics in these communities (Palmer et al. 1996). For example, understanding the physical process of transport, including wind and currents, are essential for the restoration of seagrass beds in coastal regions like the Chesapeake Bay, where submerged aquatic vegetation has been declining as a result of anthropogenic disturbance (Harwell and Orth 2002). In sum, theory and empirical research suggest that restoration efforts will often fail unless there are both sources of colonists and a means for those colonists to reach the site.

Colonization Sequence

The sequence in which colonists arrive to a habitat can play an important role in the further development of some communities. This idea dates back to the 1950s when Egler (1954) suggested that initial floristic composition was a critical factor in determining the further development of plant communities in old fields. More recently, Walker and Vitousek (1991) found that if *Myrica faya*, an introduced, nitrogen-fixing tree, first invades and establishes on young, disturbed, volcanic soils on Hawaii, the native *Metrosideros polymorpha* does not regenerate. In primary succession of newly exposed habitats (e.g., glacial moraine, lava flows, newly emerged islands), inputs of heterotrophic invertebrates and allochthonous organic materials facilitate the success of subsequently colonizing primary producers in the system (Hodkinson et al. 2002). In some ephemeral aquatic habitats, the first colonizing species may inhibit the following species' establishment (termed *priority effects*) via competition and predation (Blaustein and Margalit 1996). In a successional field, leaf litter from a first-year dominant grass hindered growth and dominance of other plants in the community the following year (Facelli and Facelli 1993). In habitats that are frequently disturbed, colonization by pioneering species may promote community development, as, for example, net-building caddisflies (Trichoptera: Hydropsychidae) may enhance stream invertebrate recruitment following floods (Cardinale et al. 2001). The sequence in which species are added (or with current patterns of environmental degradation, the sequence in which species are lost) may have impacts beyond community biodiversity and directly on ecosystem functioning (Ostfeld and LoGiudice 2003). Thus, the order in which colonists arrive to a habitat may result in either negative or positive interactions among members of the developing community. Restoration practitioners have specifically relied on these positive interactions to shorten the time to a diverse community, particularly in disturbed habitats, by introducing fast-growing shade trees to accelerate forest regeneration and employing nurse plants to assist recovering vegetation (Pywell et al. 1995; Temperton and Kirr 2004).

Temporal Scale of Processes

While most modeling of ecological communities takes place at short "ecological time scales," we must not forget that large and long "evolutionary" scale events, those changing geography and climate, may also shape local community structure (Ricklefs 2004). For example, glaciation during the Pleistocene and the differential availability of refugia in parts of Europe and North America dramatically changed the regional species pool of freshwater fishes for Europe and North America (Tonn et al. 1990). Speciation and extinction rates in regions that varied with degree of climate stability during the Pleistocene are hypothesized to have determined regional plant diversity patterns in the Cape Floristic Region in southwestern Africa (Cowling and Lombard 2002). While there has been a call in the recent community ecology literature to consider the phylogenetic history of species assemblages in the development of community structure (Losos 1996; Webb et al. 2002), it is imperative that we also include the effects of current global anthropogenic change in community restoration (Millar and Brubaker, this volume). Restoration practitioners must use lessons from both a community's evolutionary history and its current response to environmental change to guide community restoration. For example, understanding the paleoecology of peatlands, as well as peatland response to increased nitrogen and elevated atmospheric carbon dioxide, will be critical for successful restoration in the coming decades, given the expected high rates of nitrogen deposition and increases in carbon dioxide (Gorham 1994; Vitousek et al. 1997;

IPCC 2001b; Mitchell et al. 2002). It is possible that historical legacies over recent or evolutionary time (Harding et al. 1998) will limit restoration progress at a particular site.

Environmental Conditions and Habitat Characteristics

Following colonization and dispersal, organisms from the regional species pool must pass through an "environmental filter" if they are to survive and successfully reproduce in a given habitat. Only those species that have traits matching the habitat characteristics of the filter can pass through (Tonn et al. 1990; Poff 1997). Habitat, therefore, is the template on which a species' ecological strategy and, ultimately, community structure is built (Southwood 1977). We know that both biotic interactions and abiotic factors structure ecological communities, yet the roles played by environmental conditions and habitat are particularly relevant, for these are often the targets of restoration. A popularly held mantra in restoration is, "if we build it, the community will come" (Palmer et al. 1997). In stream systems, for example, the manipulation of physical habitat structures (e.g., substrate material, woody debris, channel geometry) is often the first step toward restoring the biological community (Shields et al. 2003). Note, however, that restoration of physical habitat structure will not lead to ecological restoration unless the links to the regional species pool of colonists are intact and physiological or historical (e.g., legacies of past land use) barriers are not present.

Abiotic Filters

Abiotic factors, including light, chemical, hydrological and substrate characteristics, may certainly limit the ability of a species to become successfully established at a site. Sunlight, temperature, water availability and movement, soil texture, and nutrient and salt concentrations are critical in determining an organism's tolerance, performance, and productivity in a given habitat. The harsh-benign hypothesis suggests that ecosystems often fall along a gradient of environmental conditions from physically harsh (i.e., those producing physiological stress to organisms) to benign (i.e., supporting a highly productive consumer community). Biotic interactions, namely competition and predation, play a stronger role in shaping communities with more benign physical conditions (Peckarsky 1983). Predator-stress models suggest that predators are more susceptible to environmental stress (conditions taxing the organism beyond its physiological tolerance) than their prey (Menge and Sutherland 1976). Thus, by affecting the distribution of individual species, abiotic factors have direct consequences on community and trophic structure.

In practice, abiotic factors have often been directly manipulated for community restoration; for example, buffering was added to acid-rain-damaged fens in the Netherlands (Bootsma et al. 2002) and controlled floods have been employed in restoration efforts of the Colorado River in the Grand Canyon (Patten et al. 2001). We know organisms are adapted to abiotic factors within a range of natural variation. Restoring this variation rather than a singular standard level of a variable set by a regulatory agency, particularly for factors like water quality in stream networks, may be a more effective and attainable restoration strategy (Poole et al. 2004). Changes in abiotic factors beyond the scope of natural variation may result in a shift in the distribution and abundance of organisms, thereby affecting community structure.

Indeed, systems that are chronically disturbed (e.g., by anthropogenic factors such as pollution) may be constantly adjusting, and restoration to a particular "reference" state may be impossible (Zedler 2000). Restoration ecologists may need to turn to the theory behind alternative stable states to guide their efforts (Suding et al. 2004).

Natural Disturbance Regimes

Natural, physical disturbances such as fire and floods play a significant role in shaping the structure of many communities, particularly conifer forests, prairies, and rivers (Sousa 1984). Disturbance not only incurs mortality or displacement of organisms but also creates environmental heterogeneity and influences major ecosystem processes, such as primary production and nutrient cycling, further affecting community structure. The magnitude and frequency of disturbance to a habitat and its consequences on community structure has been the subject of theoretical discussion. For example, the intermediate disturbance hypothesis suggests that habitat disturbance plays a role in managing biotic interactions, particularly dominance by a given competitor or predator (Connell 1978). Connell described this hypothesis in the context of tropical forests and coral reefs, whereby diversity may be maximized when disturbances intermediate in frequency or intensity are common. These might include physical processes like storms, floods, and lightning strikes as well as biotic events such as outbreaks of predators and herbivores. Researchers have found further evidence for the intermediate disturbance hypothesis acting on many communities (Shea et al. 2004), including benthic organisms living both in the rocky intertidal zone where the most diverse community occurs on medium-sized boulders that move only periodically (Sousa 1979) and stream bottoms with intermediate levels of bed mobility (Townsend et al. 1997).

Given that disturbance appears to be critical in shaping the diversity of many habitats, the absence and loss of disturbance has led to the decline of a number of communities, particularly fire-adapted conifer forests in North America, through loss of diversity, altered community structure, and susceptibility to alien invasion (Savage 1997; Williams 1998; Heuberger and Putz 2003). The restoration of natural disturbance regimes to disturbance-adapted communities has been critical to their recovery as seen by the reintroduction of fire and large herbivore grazing to prairie grass communities (Howe 1995; Collins et al. 1998; Knapp et al. 1999; Copeland et al. 2002). In fact, a rich body of literature involving empirical and theoretical work, as well as a number of "how to" guides, support the use of controlled burning as a widespread restoration tool (Packard and Mutel 1997).

In aquatic systems, human modification of natural flow regimes (including the magnitude, frequency, duration, predictability, and rate of change of hydrologic conditions) has resulted in the decline of plants and animals in running-water communities (Poff et al. 1997). For example, dams reduce the frequency and magnitude of high flows in rivers, effectively removing the floods to which many stream and riparian biota are adapted. Below dams where high flow releases are used to generate hydroelectric power, native species are often replaced by species that can tolerate more thermally and hydrologically stable flows (e.g., introduced trout in the Colorado River below dams versus native warm-water fish such as the humpback chub). River restoration practitioners are beginning to incorporate the importance of the natural flow regime into their recent efforts; for example, regulated flows in many western U.S. rivers now mimic the seasonal magnitude and timing of natural flows, resulting in the recovery of a number of fish species (Poff et al. 1997). Reestablishing the hydrological dynamics between rivers and their floodplains via the flood pulse has also been critical to wetland restoration efforts (Middleton 1998, 2002).

Habitat Heterogeneity and Biotic Modifications of Habitat

Species diversity typically increases with habitat complexity. MacArthur and MacArthur (1962) documented the relationship between the number of bird species and forest foliage height diversity, and this pattern has been recognized in a number of other communities including mammals (Rosenzweig and Winakur 1969), lizards (Pianka 1967), insects (Murdoch et al. 1972), and demersal fish (Kaiser et al. 1999). Habitat complexity may provide more physical space, refuge, resource availability, and open niches for members of a community, thus promoting diversity. Increasing physical complexity is an often-used tactic by practitioners of habitat restoration with the implied understanding that a diverse suite of species will respond positively (Larkin et al., this volume). For example, oyster reef restoration in a Chesapeake Bay estuary created structurally complex habitat that increased abundance and size of the transient fish community (Harding and Mann 2001). Interestingly, in addition to known effects on species diversity (Downes et al. 1998), habitat heterogeneity in aquatic systems may also enhance the restoration of ecosystem function (e.g., primary production) (Cardinale et al. 2002). Similarly, researchers in central U.S. prairies found that soil texture and nutrient heterogeneity influenced restoration of grass community structure and productivity (Baer et al. 2003, 2004).

The biota themselves often have important reciprocal impacts on the environment. A simplistic reading of the filter framework might suggest that the relationship between an organism and its environment is one-directional (environment constrains organisms). But in many cases, there is a more dynamic feedback relationship: organisms respond to the environment, and the environment is significantly modified by organisms. For example, following glaciation, plant succession processes in newly exposed habitats led to the development of soils via the accumulation of litter (increases in carbon and nitrogen) and pH changes accompanying vegetation establishment (Chapin et al. 1994). Ecosystem engineers are capable of changing habitat in such a way that they can completely restructure the community (Lawton and Jones 1995). For example, the activity of beavers can transform lotic habitats to lentic, flooded wetlands, and, as a consequence of increasing habitat heterogeneity in the landscape, they can dramatically increase herbaceous plant species richness (Wright et al. 2002). In fact, restorationists are now taking advantage of this natural engineering by beavers in the restoration of wetlands and riparian habitats (McKinstry et al. 2001).

Biotic Interactions

Organisms do not live in isolation in their environment. To restore whole communities one must also consider the biotic interactions among organisms within the community (D'Antonio and Chambers, this volume). Indeed, species interactions may determine restoration outcomes. This is particularly true as we enter a century in which a potentially huge impediment to restoration looms: the invasion of alien species that rapidly garner resources and exclude native species.

Competitive Interactions

Competition theory predicts that interactions within and among species may act to exclude species from a community, yet identifying the key players and their interactions may be difficult. For many years, studies of what factors control community structure emphasized interactions among organisms within a single guild or trophic level (Denno et al. 1995). Interference or exploitative competition for limiting resources—including nutrients, primary producers, prey, space—was considered the dominant mechanism responsible for structuring communities and the driver for segregating species into niches. Gause (1934) described the competitive exclusion principle, concluding that no two species can occupy the exact same niche. Classic descriptive (Lotka 1925; Volterra 1926) and mechanistic models that include consumer competition and resource dynamics (MacArthur 1972; Levine 1976; Tilman 1982) suggest that the outcomes of competition (survival, coexistence, extinction) depend on consumer differences in resource use. Many empirical studies have provided evidence for competition in communities (Gurevitch et al. 1992), and more recent work has attempted to tease apart the role competition plays in structuring communities, particularly given environmental variation (Chesson and Huntly 1997) and the interaction between competition and predation (Chase et al. 2002).

Resource limitation theory has been extended beyond the single trophic level to food webs in the form of "bottom-up" (versus "top-down") theory where resources control trophic interactions (Vander Zanden et al., this volume). Evidence for bottom-up control of food webs has been well documented in streams where the manipulation of basal food resources (e.g., algae, leaf detritus) resulted in significant changes to stream trophic structure (Wootton and Power 1993; Wallace et al. 1997). While there has been much debate regarding the strength of top-down versus bottom-up forces in structuring communities, there is consensus that primary productivity ultimately controls whether or not a community may even exist (Hunter and Price 1992), that is, plants have primacy in food webs (Power 1992). This provides a compelling argument for the restoration of basal food resources (e.g., reforesting riparian zones) as an important step toward community restoration.

While we discuss predator-prey interactions extensively below, it is worth discussing the interaction between predation and competition here. In classic top-down predation theory, if a predator depletes its prey the expectation is that the next trophic level down will increase in abundance and these effects will cascade down the food web. However, the scenario can be much more complicated when both predation and competition are involved in a community. For example, if a top predator feeds on an intermediate predator and both feed on a shared prey (intraguild predation) (Polis et al. 1989), then the effects of predation will depend on the relative strengths of exploitation competition for prey versus direct predation of one predator on another. This means that if the desire is to restore a particular species, its recovery may depend on complex community level interactions that must be carefully teased apart.

Competition between native and invasive species in communities targeted for restoration is of particular relevance to current restoration practice. Experimental tests of competition theory may elucidate mechanisms for dominance by an invasive species, including competitive superiority, and they could aid practitioners in choosing appropriate management strategies for restoring natives. An example provided by Seabloom et al. (2003) in the California grasslands suggests that invasive grasses are in fact not superior competitors to natives but dominate due to native recruitment limitation and the effects of prior disturbance on native species abundance. Good examples of competition can also be found in the tallgrass prairie, where big bluestem grass (*Andropogon gerardii*) often dominates and out-competes smaller, subdominant flowering species for space and light. Copeland et al. (2002) found that late-growing-season fires, as opposed to the traditionally used early-spring fires, improved subdominant species frequency and richness. Identifying competition as *the* factor limiting restoration may be difficult, however, because two species that may not directly compete for resources may share a predator, thereby negatively influencing each other (apparent competition) (Holt 1977). For example, one strategy for eradicating exotic pest insects is to release a natural enemy to reduce pest numbers; yet, if this natural enemy is a generalist, it may have negative, nontarget effects on native species (Hoddle 2004).

Trophic Interactions

With the incorporation of multiple trophic levels into community studies, greater emphasis was placed on the role of the predator in determining community structure (Hairston et al. 1960). Herbivory is probably the best-known example of trophic control over community structure. It may slow restoration progress (Opperman and Merenlender 2000) or it may be an integral component of restoring a self-sustaining community.

Predators may play an extremely important role in restoration efforts. Keystone predators, such as starfish in the marine intertidal zone, may suppress competitively dominant species, thereby promoting a more diverse and abundant species assemblage (Paine 1969). In communities where keystone predators have been lost or threatened, restoration of predators may be an effective strategy for the restoration of community structure. In East Africa, the decline of red-lined triggerfish (*Balistapus undulatus*), a keystone predator of sea urchins, has led to coral reef decline. Recovery of the triggerfish in marine-protected areas may not be an efficient strategy given a long (30-year) time-lag for recovery of the reef community (McClanahan 2000). The top-down model suggests that predators not only affect the trophic level immediately below them, but also have effects that cascade down the food chain all the way to primary producers (Hairston et al. 1960; Power 1990; Schmitz et al. 2000; Schmitz 2003). Limnologists have employed trophic cascades for lake restorations. However, the manipulation of fish abundance to reduce phytoplankton biomass and improve water quality has been met with mixed success (Sondergaard et al. 1997; Harig and Bain 1998; Drenner et al. 2002; Vander Zanden et al., this volume).

Current food-web syntheses incorporate both top-down and bottom-up controls on community structure where food webs are viewed as a series of feedback loops among trophic levels (see above) that overlay a template of biotic and abiotic heterogeneity (Carpenter and Kitchell 1988; Hunter and Price 1992). In these complex systems, it is highly unlikely that we can attribute community structure to a single factor. While restoration efforts may involve the manipulation of consumer species to affect other trophic levels, the chain of reactions that follows may not be so predictable given environmental heterogeneity.

There is also emerging research to support the theory that higher levels of biodiversity enhance the ability of a community to maintain high levels of ecological functioning (e.g., primary production, resource acquisition) (Kinzig et al. 2002; Loreau et al. 2002; Naeem, this volume). Recent theoretical work suggests that depending on the order of species loss, more diverse communities may be better able to tolerate environmental degradation (Ives and Cardinale 2004). This result is due in part to the ability of some species to compensate (increase in abundance) when other species are lost. Since species compensation is difficult to predict, Ives and Cardinale (2004) argue for ecosystem-based management (e.g., restoration) since the species that may appear "unimportant" (e.g., rare species) may play an important role in community resistance when other species are lost (or are not restored). The important point for restoration is that high species diversity may provide some "insurance" for communities in terms of their function.

Mutualistic Interactions

Positive interactions among species and trophic levels may be just as important for structuring communities as competition and predation (Stachowicz 2001; Bruno et al. 2003). Facilitation by early successional plants, for example, may be critical in mitigating harsh abiotic conditions for other associated species (Crocker and Major 1955). Other examples of facilitation with community-level consequences include the dispersal of fruits and seeds by animals that increase the distribution of plants and overcome dispersal limitation (Chambers and Macmahon 1994) and the stabilization of wave-disturbed, cobble beach habitat for plant and invertebrates by *Spartina alterniflora* (Bruno 2000). Arbuscular mycorrhizal (AM) fungi form mutualistic symbiotic relationships with the roots of many plants. The plants provide carbon for fungal growth and, in exchange, the plants are able to uptake more nutrients, such as phosphorus, and certain plant stresses may be ameliorated. AM fungi are important to plant growth in ecosystems as diverse as grasslands, forests, and urban areas (Allen 1991). Manipulating the species composition or diversity of soil AM fungi, as well as establishing a hyphal network, may promote plant species diversity and ecosystem functioning and therefore may enhance plant restoration efforts (van der Heijden et al. 1998; Cousins et al. 2003). The fact that mutualism has been identified as one of the mechanisms by which a more diverse community may be more productive or process materials at higher rates (Cardinale et al. 2002; Loreau et al. 2002) suggests that reestablishing these relationships (e.g., by ensuring both species are "seeded" at a site) may be very important to the outcome of a restoration project.

Current Thinking and New Directions

Communities are complex living systems often characterized by nonlinear interactions and unexpected outcomes. While this complexity makes the work of the restorationist difficult, it is not impossible. Restoration of the whole community first requires an understanding of its basic biology (Holl et al. 2000). Recent research has suggested that communities have unique emergent properties, including biodiversity, trophic structure, stability, invasibility, and productivity (Brown 1995; Levin 1998). Particularly with respect to ecosystem function, the properties of a diverse community can be nonadditive and larger than predicted by the sum of its parts (Cardinale and Palmer 2002; Cardinale et al. 2002; Swan and Palmer 2004). Restorationists might ask themselves whether the restoration of particular suites of species (e.g., native species versus non-natives) is important. Certainly, the answer to this depends on the purpose of the project and stakeholder values. Does it matter if we can regain only a portion of the native community? Does theory provide guidance on the proportion that is needed?

How much redundancy is there, and how much is critical to a functional system? Understanding the link between individual species and ecosystem function, particularly with respect to functional redundancy and diversity, has already begun (Duffy et al. 2001; Petchey and Gaston 2002; Chalcraft and Resetarits 2003; Wohl et al. 2004); understanding this link between species and function in the context of restoration is a critical new avenue for research.

Work on the functional roles of species is just one example of several new developments in community ecology that have recently emerged and may have important bearing on restoration. Others will require much more work before we fully understand their utility and relevance to restoration. We explore a few of these new findings and theories and ask critical research questions that link them to restoration. Our focus on new directions is not meant to be exhaustive but to demonstrate how rich the opportunities will be for building or extending community ecological research and restoration.

How Does Ecological Commerce to and from Communities Influence Restoration Outcomes?

The exchange of material that may be useful or detrimental to communities has accelerated due to human activities and has been termed *ecological commerce* (Palumbi 2003). The spatial boundaries of communities have been blurred as the importance and extent of cross-habitat resource subsidies have been realized. Prey, energy, or detritus from one habitat subsidizes the community of another, influencing both population dynamics and productivity (Polis et al. 1997; Menge et al. 2003). The importance of resource subsidies is particularly evident at terrestrial-aquatic interfaces (e.g., stream-riparian and ocean-shore interactions) (Wall et al. 2001). Because resources produced in one community may be critical to members of an adjacent community, restoration efforts must incorporate the dynamics of both. This effect can be seen clearly in stream restoration, where the leaf detritus from reforested riparian zones may be just as important as in-stream channel structures to the recovery of the aquatic community. Humans also impact the flow of species and material subsidies from one ecosystem to another (Palumbi 2003). For example, nutrient subsidies flowing from the Mississippi River watershed are creating anoxic zones in the Gulf of Mexico, contributing to the decline of coral reefs. The lesson is that external ecological subsidies, both useful and detrimental, regulate community structure and may ultimately be important to determining restoration outcomes.

What Is the Role of Hidden Players in Community Restoration?

The roles of historically hidden players in communities, such as cryptic invertebrates, microbes, and disease, have come to light through recent research (Thompson et al. 2001). Technological advances have allowed us to explore these realms using new methods for identifying and collecting biota. Crucial in nutrient cycling, detrital processing, and bioturbation, belowground biota may strongly influence aboveground processes (Wall and Moore 1999; Hooper et al. 2000). Recent research suggests that soil invertebrate fauna may even enhance the succession and diversity of grasslands by suppressing dominant plant species (De Deyn et al. 2003). Microbial community diversity and the critical functional roles of microorganisms are being elucidated with molecular genetic approaches (Tyson et al. 2004; Venter et al. 2004). The urgency of soil-microbe and soil-invertebrate interactions to our understanding of how ecosystems can be conserved and restored was highlighted recently in a special focus in *Science* magazine in 2004 ("Soils—The Final Frontier," 11 June 2004). Research in disease ecology has focused on how aspects of human-induced change, for example, climate change, habitat fragmentation, and biodiversity loss, may affect the occurrence and risk of exposure to diseases like Lyme disease (Ostfeld and Keesing 2000; Harvell et al. 2002; Allan et al. 2003; LoGiudice et al. 2003). With better understanding of their basic biology and role in community processes, more hidden players will be incorporated into the theory and practice of community restoration.

Can Ecological Disturbance Theory Be Merged with Newer Theories of the Linkage Between Community Biodiversity and Ecosystem Function in Order to Guide Restoration?

The degree to which biodiversity influences ecosystem function has been highly debated by ecologists over the last ten years (Loreau et al. 2001; Naeem, this volume). A significant contribution of this debate is a new paradigm suggesting that environmental conditions are largely a function of ecosystem processes that are in turn driven by biodiversity and community interactions (Naeem 2002). Recent work has examined diversity at scales beyond the taxonomic level and its implications on function, including the effects of phenotypic diversity in leaf chemistry on the decomposition of leaf litter (Madritch and Hunter 2002). Ecologists are examining how disturbance may mediate the biodiversity-ecosystem function relationship (Cardinale and Palmer 2002) and the implication of the homogenization of the world's fauna on ecosystem processes (Olden et al. 2004). Response diversity, or the variability in reactions to environmental change of species within the same functional group, in addition to functional diversity, may be critical to the resilience of ecosystems and the maintenance of ecosystem services following environmental change (Elmqvist et al. 2003). Research is urgently needed to determine how we can restore communities that are sufficiently resilient to disturbance, and how we can mimic disturbance events that promote restoration.

Do We Need New Theory to Provide Functional Communities in Urban Ecosystems?

By 2030, more than half of the world's population will be living in urban areas (Sadik 1999). Urbanization, even in places where it is not the dominant land use, has major influences on regional and global environments (Alberti et al. 2003). Restoration efforts must be designed to mitigate the impacts arising from the enhanced flux of people, materials, and energy to and from urban centers (Palmer et al. 2004). Further, areas that are not urban will be largely managed (most already are), for agriculture, forestry, aquaculture, and recreation. In such settings, community structure often bears little resemblance to historical conditions. Many suburban areas in high socioeconomic regions actually have higher diversity of plants than surrounding lands, due largely to the use of non-native plants in landscaping. Lower socioeconomic areas may be species depauperate depending on the amount of impervious surface and the success of non-native weedy species (Hope et al. 2003). Restoration theory must be greatly extended to cope with new factors driving community structure: socioeconomics, landscaping, higher levels of NO_x and CO_2 from automobile emissions, and increased movement rates of people. How do humans and human activities influence community assembly and disassembly, species interactions, and our ability to restore communities of plants and animals? How much intervention is necessary to sustain or restore a site to an acceptable baseline? Can functioning of urban ecosystems be enhanced by the maintenance of artificially high species richness?

Can Communities Be Restored to Provide Multiple Functions?

What role can "proactive" restoration of entire communities play in mitigating the loss of key ecosystem functions? By proactive, we mean restoration projects that are designed to accomplish more than returning a system to some prior state (Palmer et al. 2004). Ecological theories concerning the role of individual species in ecosystem function, functional redundancy, and recruitment limitation must be brought together and coupled with experimental restoration projects to test new hypotheses (Foster et al. 2004). Best management practices like no-till farming and planting cover crops (Dendoncker et al. 2004), coupled with plant community restoration, may help to sequester carbon and thus counteract rising atmospheric CO_2 levels (IPCC 2001a). Robertson and Swinton (2005) argue that agroecosystems of the future must be designed or restored (augmented) to not only enhance productivity but also to do so with fewer pesticides (e.g., using biocontrol) and with plantings that enhance water infiltration into soils and mitigate atmospheric emissions. Thus we may be "restoring" old farmlands to areas that provide different functions. What species mixtures optimize restoration of multiple functions and/or target specific functions? Can theory at the interface of ecology, economics, and sociology be developed to support a science to restore ecosystems services?

These topics are not an exhaustive list of new directions, but they do demonstrate rich opportunities for building or extending community ecological research and applying predictions to future restoration efforts. More quantitative approaches will have to be applied to move community ecological theory toward a more predictive framework for restoration ecology. Finally, given the diverse impacts that global change is having on communities and ecosystems worldwide, there is a pressing need for theory describing how to restore communities that are resilient in a changing world (Carpenter and Cottingham 1997).

Summary

A large body of theory has emerged from the rich history of studying species diversity in ecological communities. Using a hierarchical filter framework, we examined how theories regarding regional processes, local environmental and habitat characteristics, and biotic interactions profoundly impact the structure of communities and, ultimately, community restoration outcomes. The dispersal and subsequent colonization of members from a regional species pool is critical, and restoration may be severely constrained if a site is cut off from this pool. Assuming successful colonization, recruits must pass through an environmental filter. Physiological constraints may prevent successful establishment of recruits if abiotic features of a restoration site cannot be restored (e.g., warm-water fish cannot be restored in the cold tailwaters of a dammed river). Aspects of the abiotic habitat that have been restored sufficiently to bring back desired species are typically those linked to habitat structure ("if we build it, they will come"). Other habitat conditions, including natural disturbance regimes and habitat heterogeneity, have been more difficult to restore in practice and often involve large-scale manipulations. The final filter that determines successful restoration of communities is linked to species interactions. Members of a community do not live in isolation; thus, species interactions like competition, predation, and mutualism must be considered. They have been directly manipulated in a number of restoration efforts already. For each of the three filters discussed, we illustrated how theory can assist and inform restoration with specific, real-world examples from the restoration literature. Within each filter we also explored how community restoration efforts may be further challenged by anthropogenic impacts including global climate change and the rapid spread of invasive species. Finally, we highlighted several of the new research developments that have emerged in community ecology that may have important bearing on restoration. We posed questions to stimulate ecologists and practitioners alike to move beyond traditional theory and consider the roles that ecological commerce, hidden players, and functional diversity may play in community restoration. We asked, in the context of our changing world, if disturbance theory can be incorporated into the biodiversity/ecosystem function paradigm to guide restoration and if degraded communities can be rebuilt in a way to restore multiple ecosystem functions.

Acknowledgments

We thank Don Falk, Joy Zelder, and Gina Wimp for thoughtful comments on drafts of this manuscript. We thank Dan Fergus for his generous assistance with the preparation of figures. Support for chapter preparation was provided by the U.S. Environmental Protection Agency, including the Science to Achieve Results Program (EPA number: R828012) and the Global Climate Change Program (EPA numbers: 1W0594NAEX and R83038701). For M.A.P., this is contribution number 3907 from the University of Maryland Center for Environmental Sciences.

Literature Cited

Alberti, M., J. M. Marzluff, E. Shulenberger, G. Bradley, C. Ryan, and C. Zumbrunnen. 2003. Integrating humans into ecology: Opportunities and challenges for studying urban ecosystems. *Bioscience* 53 (12): 1169–1179.

Allan, B. F., F. Keesing, and R. S. Ostfeld. 2003. Effect of forest fragmentation on Lyme disease risk. *Conservation Biology* 17 (1): 267–272.

Allen, A. P., J. H. Brown, and J. F. Gillooly. 2002. Global biodiversity, biochemical kinetics, and the energetic-equivalence rule. *Science* 297 (5586): 1545–1548.

Allen, E. B. 2003. New directions and growth of restoration ecology. *Restoration Ecology* 11 (1): 1–2.

Allen, M. F. 1991. *The ecology of mycorrhizae*. New York: Cambridge University Press.

Armantrout, N. B. 1991. Restructuring streams for anadromous salmonids. *American Fisheries Society Symposium* 10:136–149.

Aronson, J., and D. A. Falk, editors. 2004. The science and practice of ecological restoration. In *Assembly rules and restoration ecology*, ed. V. M. Temperton, R. J. Hobbs, T. Nuttle, and S. Halle. Washington, DC: Island Press.

Baer, S. G., J. M. Blair, S. L. Collins, and A. K. Knapp. 2003. Soil resources regulate diversity and productivity in newly established prairie. *Ecology* 84:724–735.

Baer, S. G., J. M. Blair, S. L. Collins, and A. K. Knapp. 2004. Plant community responses to resource availability and heterogeneity during restoration. *Oecologia* 139:617–629.

Blaustein, L., and J. Margalit. 1996. Priority effects in temporary pools: Nature and outcome of mosquito larva–toad tadpole interactions depend on order of entrance. *Journal of Animal Ecology* 65 (1): 77–84.

Bootsma, M. C., T. van den Broek, A. Barendregt, and B. Beltman. 2002. Rehabilitation of acidified floating fens by addition of buffered surface water. *Restoration Ecology* 10 (1): 112–121.

Bradshaw, A. D. 1987. Restoration: An acid test for ecology. In *Restoration ecology: A synthetic approach to ecological research*, ed. W. R. Jordan III, M. E. Gilpin, and J. D. Aber, 23–29. New York: Cambridge University Press.

Brady, V. J., B. J. Cardinale, J. P. Gathman, and T. M. Burton. 2002. Does facilitation of faunal recruitment benefit ecosystem restoration? An experimental study of invertebrate assemblages in wetland mesocosms. *Restoration Ecology* 10 (4): 617–626.

Brown, J. H. 1975. Geographical ecology of desert rodents. In *Ecology and evolution of communities*, ed. M. L. Cody and J. M. Diamond, 315–341. Cambridge: Belknap Press of Harvard University Press.

Brown, J. H. 1995. Organisms and species as complex adaptive systems: Linking the biology of populations with the physics of ecosystems. In *Linking species and ecosystems*, ed. C. G. Jones and J. H. Lawton, 16–24. New York: Chapman & Hall.

Brown, J. H., V. K. Gupta, B. L. Li, B. T. Milne, C. Restrepo, and G. B. West. 2002. The fractal nature of nature: Power laws, ecological complexity and biodiversity. *Philosophical Transactions of the Royal Society of London, Series B* 357 (1421): 619–626.

Bruno, J. F. 2000. Facilitation of cobble beach plant communities through habitat modification by *Spartina alterniflora*. *Ecology* 81 (5): 1179–1192.

Bruno, J. F., J. J. Stachowicz, and M. D. Bertness. 2003. Inclusion of facilitation into ecological theory. *Trends in Ecology & Evolution* 18 (3): 119–125.

Caley, M. J., and D. Schluter. 1997. The relationship between local and regional diversity. *Ecology* 78 (1): 70–80.

Cardinale, B. J., and M. A. Palmer. 2002. Disturbance moderates biodiversity-ecosystem function relationships: Experimental evidence from caddisflies in stream mesocosms. *Ecology* 83 (7): 1915–1927.

Cardinale, B. J., M. A. Palmer, and S. L. Collins. 2002. Species diversity enhances ecosystem functioning through interspecific facilitation. *Nature* 415:426–429.

Cardinale, B. J., M. A. Palmer, C. M. Swan, S. Brooks, and N. L. Poff. 2002. The influence of substrate heterogeneity on biofilm metabolism in a stream ecosystem. *Ecology* 83 (2): 412–422.

Cardinale, B. J., C. M. Smith, and M. A. Palmer. 2001. The influence of initial colonization by hydropsychid caddisfly larvae on the development of stream invertebrate assemblages. *Hydrobiologia* 455:19–27.

Carpenter, S. R., and K. L. Cottingham. 1997. Resilience and restoration of lakes. *Conservation Ecology* 1 (1): http://www.consecol.org/vol1/iss1/art2.

Carpenter, S. R., and J. F. Kitchell. 1988. Consumer control of lake productivity. *Bioscience* 38 (11): 764–769.

Chalcraft, D. R., and W. J. Resetarits. 2003. Predator identity and ecological impacts: Functional redundancy or functional diversity? *Ecology* 84 (9): 2407–2418.

Chambers, J. C., and J. A. MacMahon. 1994. A day in the life of a seed—Movements and fates of seeds and their implications for natural and managed systems. *Annual Review of Ecology and Systematics* 25:263–292.

Chapin, F. S., L. R. Walker, C. L. Fastie, and L. C. Sharman. 1994. Mechanisms of primary succession following deglaciation at Glacier Bay, Alaska. *Ecological Monographs* 64 (2): 149–175.

Chase, J. M., P. A. Abrams, J. P. Grover, S. Diehl, P. Chesson, R. D. Holt, S. A. Richards, R. M. Nisbet, and T. J. Case. 2002. The interaction between predation and competition: A review and synthesis. *Ecology Letters* 5 (2): 302–315.

Chesson, P., and N. Huntly. 1997. The roles of harsh and fluctuating conditions in the dynamics of ecological communities. *American Naturalist* 150 (5): 519–553.

Chesson, P. L., and R. R. Warner. 1981. Environmental variability promotes coexistence in lottery competitive-systems. *American Naturalist* 117 (6): 923–943.

Clements, F. E. 1936. Nature and structure of the climax. *Journal of Ecology* 24:252–284.

Collins, S. L., A..K. Knapp, J. M. Briggs, J. M. Blair, and E. M. Steinauer. 1998. Modulation of diversity by grazing and mowing in native tallgrass prairie. *Science* 280 (5364): 745–747.

Connell, J. H. 1978. Diversity in tropical rain forests and coral reefs—High diversity of trees and corals is maintained only in a non-equilibrium state. *Science* 199 (4335): 1302–1310.

Conner, R. N., C. E. Shackelford, R. R. Schaefer, D. Saenz, and D. C. Rudolph. 2002. Avian community response to southern pine ecosystem restoration for Red-cockaded woodpeckers. *Wilson Bulletin* 114 (3): 324–332.

Connor, E. F., and E. D. McCoy. 1979. Statistics and biology of the species-area relationship. *American Naturalist* 113 (6): 791–833.

Copeland, T. E., W. Sluis, and H. F. Howe. 2002. Fire season and dominance in an Illinois tallgrass prairie restoration. *Restoration Ecology* 10 (2): 315–323.

Cornell, H. V. 1999. Unsaturation and regional influences on species richness in ecological communities: A review of the evidence. *Ecoscience* 6 (3): 303–315.

Cottingham, K. L., B. L. Brown, and J. T. Lennon. 2001. Biodiversity may regulate the temporal variability of ecological systems. *Ecology Letters* 4 (1): 72–85.

Cousins, J. R., D. Hope, C. Gries, and J. C. Stutz. 2003. Preliminary assessment of arbuscular mycorrhizal fungal diversity and community structure in an urban ecosystem. *Mycorrhiza* 13 (6): 319–326.

Cowles, H. C. 1899. The ecological relations of the vegetation on the sand dunes of Lake Michigan. *Botanical Gazette* 27:95–391.

Cowling, R. M., and A. T. Lombard. 2002. Heterogeneity, speciation/extinction history and climate: Explaining regional plant diversity patterns in the Cape Floristic region. *Diversity and Distributions* 8 (3): 163–179.

Crocker, R. L., and J. Major. 1955. Soil development in relation to vegetation and surface age at Glacier Bay, Alaska. *Journal of Ecology* 43 (2): 427–448.

de Deyn, G. B., C. E. Raaijmakers, H. R. Zoomer, M. P. Berg, P. C. de Ruiter, H. A. Verhoef, T. M. Bezemer, and W. H. van der Putten. 2003. Soil invertebrate fauna enhances grassland succession and diversity. *Nature* 422 (6933): 711–713.

Dendoncker, N., B. Van Wesemael, M. D. A. Rounsevell, C. Roelandt, and S. Lettens. 2004. Belgium's CO_2 mitigation potential under improved cropland management. *Agriculture Ecosystems & Environment* 103 (1): 101–116.

Denno, R. F., M. S. McClure, and J. R. Ott. 1995. Interspecific interactions in phytophagous insects—Competition reexamined and resurrected. *Annual Review of Entomology* 40:297–331.

Downes, B. J., P. S. Lake, E. S. G. Schreiber, and A. Glaister. 1998. Habitat structure and regulation of local species diversity in a stony, upland stream. *Ecological Monographs* 68 (2): 237–257.

Drenner, R. W., R. M. Baca, J. S. Gilroy, M. R. Ernst, D. J. Jensen, and D. H. Marshall. 2002. Community responses to piscivorous largemouth bass: A biomanipulation experiment. *Lake and Reservoir Management* 18 (1): 44–51.

Duffy, J. E., K. S. Macdonald, J. M. Rhode, and J. D. Parker. 2001. Grazer diversity, functional redundancy, and productivity in seagrass beds: An experimental test. *Ecology* 82 (9): 2417–2434.

Egler, F. E. 1954. Vegetation science concepts. Initial floristic composition—a factor in old-field vegetation development. *Vegetatio* 4:412–417.

Elmqvist, T., C. Folke, M. Nystrom, G. Peterson, J. Bengtsson, B. Walker, and J. Norberg. 2003. Response diversity, ecosystem change, and resilience. *Frontiers in Ecology and the Environment* 1 (9): 488–494.

Elton, C. S. 1966. *The pattern of animal communities*. London: Chapman & Hall.

Facelli, J. M., and E. Facelli. 1993. Interactions after death—Plant litter controls priority effects in a successional plant community. *Oecologia* 95 (2): 277–282.

Forbes, S. A. 1887. The lake as a microcosm. *Bulletin of the Peoria Scientific Association* 77–87.

Foster, B. L., T. L. Dickson, C. A. Murphy, I. S. Karel, and V. H. Smith. 2004. Propagule pools mediate community assembly and diversity—Ecosystem regulation along a grassland productivity gradient. *Journal of Ecology* 92 (3): 435–449.

Galatowitsch, S. M., and A. G. vanderValk. 1996. Vegetation and environmental conditions in recently restored wetlands in the prairie pothole region of the USA. *Vegetatio* 126 (1): 89–99.

Gause, G. F. 1934. *The struggle for existence*. Baltimore: Williams & Wilkins.

Gleason, H. A. 1926. The individualistic concept of plant association. *Bulletin of the Torrey Botanical Club* 53:7–26.

Gorham, E. 1994. The future of research in Canadian peatlands—A brief survey with particular reference to global change. *Wetlands* 14 (3): 206–215.

Gotelli, N. J., and D. Simberloff. 1987. The distribution and abundance of tallgrass prairie plants—A test of the core-satellite hypothesis. *American Naturalist* 130 (1): 18–35.

Gurevitch, J., L. L. Morrow, A. Wallace, and J. S. Walsh. 1992. A meta-analysis of competition in field experiments. *American Naturalist* 140 (4): 539–572.

Hairston, N. G., F. E. Smith, and L. B. Slobodkin. 1960. Community structure, population control, and competition. *American Naturalist* 94 (879): 421–425.

Hanski, I. 1982. Dynamics of regional distribution—The core and satellite species hypothesis. *Oikos* 38 (2): 210–221.

Harding, J. M., and R. Mann. 2001. Oyster reefs as fish habitat: Opportunistic use of restored reefs by transient fishes. *Journal of Shellfish Research* 20 (3): 951–959.

Harding, J. S., E. F. Benfield, P. V. Bolstad, G. S. Helfman, and E. B. D. Jones. 1998. Stream biodiversity: The ghost of land use past. *Proceedings of the National Academy of Sciences, USA* 95 (25): 14843–14847.

Harig, A. L., and M. B. Bain. 1998. Defining and restoring biological integrity in wilderness lakes. *Ecological Applications* 8 (1): 71–87.

Harvell, C. D., C. E. Mitchell, J. R. Ward, S. Altizer, A. P. Dobson, R. S. Ostfeld, and M. D. Samuel. 2002. Ecology—Climate warming and disease risks for terrestrial and marine biota. *Science* 296 (5576): 2158–2162.

Harwell, M. C., and R. J. Orth. 2002. Long-distance dispersal potential in a marine macrophyte. *Ecology* 83 (12): 3319–3330.

Hawkins, B. A., R. Field, H. V. Cornell, D. J. Currie, J. F. Guégan, D.M. Kaufman, J. T. Kerr, G. G. Mittelbach, T. Oberdorff, E. M. O'Brien, E. E. Porter, and J. R. G. Turner. 2003. Energy, water, and broad-scale geographic patterns of species richness. *Ecology* 84 (12): 3105–3117.

Heuberger, K. A., and F. E. Putz. 2003. Fire in the suburbs: Ecological impacts of prescribed fire in small remnants of longleaf pine (*Pinus palustris*) sandhill. *Restoration Ecology* 11 (1): 72–81.

Hillebrand, H. 2004. On the generality of the latitudinal diversity gradient. *American Naturalist* 163 (2): 192–211.

Hoddle, M. S. 2004. Restoring balance: Using exotic species to control invasive exotic species. *Conservation Biology* 18 (1): 38–49.

Hodkinson, I. D., N. R. Webb, and S. J. Coulson. 2002. Primary community assembly on land—the missing stages: Why are the heterotrophic organisms always there first? *Journal of Ecology* 90 (3): 569–577.

Holl, K. D., M. E. Loik, E. H. V. Lin, and I. A. Samuels. 2000. Tropical montane forest restoration in Costa Rica: Overcoming barriers to dispersal and establishment. *Restoration Ecology* 8 (4): 339–349.

Holt, R. D. 1977. Predation, apparent competition, and structure of prey communities. *Theoretical Population Biology* 12 (2): 197–229.

Hooper, D. U., D. E. Bignell, V. K. Brown, L. Brussaard, J. M. Dangerfield, D. H. Wall, D. A. Wardle, D. C. Coleman, K. E. Giller, P. Lavelle, W. H. Van der Putten, P. C. De Ruiter, J. Rusek, W. L. Silver, J. M. Tiedje, and V. Wolters. 2000. Interactions between aboveground and belowground biodiversity in terrestrial ecosystems: Patterns, mechanisms, and feedbacks. *Bioscience* 50 (12): 1049–1061.

Hope, D., C. Gries, W. X. Zhu, W. F. Fagan, C. L. Redman, N. B. Grimm, A. L. Nelson, C. Martin, and A. Kinzig. 2003. Socioeconomics drive urban plant diversity. *Proceedings of the National Academy of Sciences, USA* 100 (15): 8788–8792.

Houseman, G. R., and R. C. Anderson. 2002. Effects of jack pine plantation management on barrens flora and potential Kirtland's warbler nest habitat. *Restoration Ecology* 10 (1): 27–36.

Howe, H. F. 1995. Succession and fire season in experimental prairie plantings. *Ecology* 76 (6): 1917–1925.

Hunter, M. D., and P. W. Price. 1992. Playing chutes and ladders—Heterogeneity and the relative roles of bottom-up and top-down forces in natural communities. *Ecology* 73 (3): 724–732.

Hutchinson, G. E. 1957. Concluding remarks. *Cold Spring Harbor Symposia on Quantitative Biology* 22:415–427.

Intergovernmental Panel on Climate Change (IPCC). 2001a. Climate change 2001: Mitigation. Contribution of the Working Group III to the Third Assessment Report of the Intergovernmental Panel on Climate Change. Cambridge, UK: Cambridge University Press.

Intergovernmental Panel on Climate Change (IPCC). 2001b. Climate change 2001: The scientific basis. Contribution of the Working Group I to the Third Assessment Report of the Intergovernmental Panel on Climate Change. Cambridge, UK: Cambridge University Press.

Ives, A. R., and B. J. Cardinale. 2004. Food-web interactions govern the resistance of communities after non-random extinctions. *Nature* 429 (6988): 174–177.

Jacquemyn, H., J. Butaye, and M. Hermy. 2003. Impacts of restored patch density and distance from natural forests on colonization success. *Restoration Ecology* 11 (4): 417–423.

Kaiser, M. J., S. I. Rogers, and J. R. Ellis. 1999. Importance of benthic habitat complexity for demersal fish assemblages. *Proceedings of the American Fisheries Society* 22:212–223.

Kareiva, P., and S. A. Levin, editors. 2003. *The importance of species: Perspectives on expendability and triage.* Princeton: Princeton University Press.

Kinzig, A. P., and J. Harte. 2000. Implications of endemics-area relationships for estimates of species extinctions. *Ecology* 81 (12): 3305–3311.

Kinzig, A. P., S. W. Pacala, and D. Tilman. 2002. *The functional consequences of biodiversity: Empirical progress and theoretical extension.* Princeton: Princeton University Press.

Knapp, A. K., J. M. Blair, J. M. Briggs, S. L. Collins, D. C. Hartnett, L. C. Johnson, and E. G. Towne. 1999. The keystone role of bison in North American tallgrass prairie—Bison increase habitat heterogeneity and alter a broad array of plant, community, and ecosystem processes. *Bioscience* 49 (1): 39–50.

Lamb, D. 1998. Large-scale ecological restoration of degraded tropical forest lands: The potential role of timber plantations. *Restoration Ecology* 6 (3): 271–279.

Lawton, J. H., and C. G. Jones. 1995. Linking species and ecosystems: Organisms as ecosystem engineers. In *Linking species & ecosystems,* ed. C. G. Jones and J. H. Lawton, 141–150. New York: Chapman & Hall.

Lee, H. L., D. DeAngelis, and H. L. Koh. 1998. Modeling spatial distribution of the unionid mussels and the core-satellite hypothesis. *Water Science and Technology* 38 (7): 73–79.

Levin, S. A. 1998. Ecosystems and the biosphere as complex adaptive systems. *Ecosystems* 1 (5): 431–436.

Levin, S. A. 1999. *Fragile dominion: Complexity and the commons.* Reading, MA: Perseus Books.

Levine, S. H. 1976. Competitive interactions in ecosystems. *American Naturalist* 110 (976): 903–910.

LoGiudice, K., R. S. Ostfeld, K. A. Schmidt, and F. Keesing. 2003. The ecology of infectious disease: Effects of host diversity and community composition on Lyme disease risk. *Proceedings of the National Academy of Sciences, USA* 100 (2): 567–571.

Loreau, M., S. Naeem, and P. Inchausti. 2002. *Biodiversity and ecosystem functioning: Synthesis and perspectives.* Oxford, UK: Oxford University Press.

Loreau, M., S. Naeem, P. Inchausti, J. Bengtsson, J. P. Grime, A. Hector, D. U. Hooper, M. A. Huston, D. Raffaelli, B. Schmid, D. Tilman, and D. A. Wardle. 2001. Biodiversity and ecosystem functioning: Current knowledge and future challenges. *Science* 294 (5543): 804–808.

Losos, J. B. 1996. Phylogenetic perspectives on community ecology. *Ecology* 77 (5): 1344–1354.

Lotka, A. J. 1925. *Elements of physical biology.* Baltimore: Williams & Wilkins.

MacArthur, R. H. 1972. *Geographical ecology.* Princeton: Princeton University Press.

MacArthur, R. H., and R. Levins. 1967. Limiting similarity, convergence, and divergence of coexisting species. *American Naturalist* 101 (921): 377–385.

MacArthur, R. H., and J. W. MacArthur. 1961. On bird species diversity. *Ecology* 42 (3): 594–598.

MacArthur, R. H., J. W. MacArthur, and J. Preer. 1962. On bird species diversity II. Prediction of bird census habitat. *American Naturalist* 96 (888): 167–174.

MacArthur, R. H., and E. O. Wilson. 1967. *The theory of island biogeography.* Princeton: Princeton University Press.

Madritch, M. D., and M. D. Hunter. 2002. Phenotypic diversity influences ecosystem functioning in an oak sandhills community. *Ecology* 83 (8): 2084–2090.

May, R. M. 1973a. Qualitative stability in model ecosystems. *Ecology* 54 (3): 638–641.

May, R. M. 1973b. Stability in randomly fluctuating versus deterministic environments. *American Naturalist* 107 (957): 621–650.

McClanahan, T. R. 2000. Recovery of a coral reef keystone predator, *Balistapus undulatus,* in East African marine parks. *Biological Conservation* 94 (2): 191–198.

McGrady-Steed, J., P. M. Harris, and P. J. Morin. 1997. Biodiversity regulates ecosystem predictability. *Nature* 390 (6656): 162–165.

McKinstry, M. C., P. Caffrey, and S. H. Anderson. 2001. The importance of beaver to wetland habitats and waterfowl in Wyoming. *Journal of the American Water Resources Association* 37 (6): 1571–1577.

Menge, B. A., J. Lubchenco, M. E. S. Bracken, F. Chan, M. M. Foley, T. L. Freidenburg, S. D. Gaines, G. Hudson, C. Krenz, H. Leslie, D. N. L. Menge, R. Russell, and M. S. Webster. 2003. Coastal oceanography sets the pace of rocky intertidal community dynamics. *Proceedings of the National Academy of Sciences, USA* 100 (21): 12229–12234.

Menge, B. A., and J. P. Sutherland. 1976. Species-diversity gradients—Synthesis of roles of predation, competition, and temporal heterogeneity. *American Naturalist* 110 (973): 351–369.

Middleton, B. A. 1998. *Wetland restoration, flood pulsing, and disturbance dynamics.* New York: John Wiley & Sons.

Middleton, B. A., editor. 2002. *Flood pulsing in wetlands: Restoring the natural hydrological balance.* New York: John Wiley & Sons.

Mitchell, E. A. D., A. Buttler, P. Grosvernier, H. Rydin, A. Siegenthaler, and J. M. Gobat. 2002. Contrasted effects of increased N and CO_2 supply on two keystone species in peatland restoration and implications for global change. *Journal of Ecology* 90 (3): 529–533.

Morin, P. J. 1999. *Community ecology.* Malden, MA: Blackwell Science.

Murdoch, W. W., C. H. Peterson, and F. C. Evans. 1972. Diversity and pattern in plants and insects. *Ecology* 53 (5): 819–829.

Naeem, S. 2002. Ecosystem consequences of biodiversity loss: The evolution of a paradigm. *Ecology* 83 (6): 1537–1552.

Neigel, J. E. 2003. Species-area relationships and marine conservation. *Ecological Applications* 13 (1): S138–S145.

O'Dwyer, C., and P. M. Attiwill. 2000. Restoration of a native grassland as habitat for the golden sun moth *Synemon plana* Walker (Lepidoptera; Castniidae) at Mount Piper, Australia. *Restoration Ecology* 8 (2): 170–174

Olden, J. D., N. L. Poff, M. R. Douglas, M. E. Douglas, and K. D. Fausch. 2004. Ecological and evolutionary consequences of biotic homogenization. *Trends in Ecology & Evolution* 19 (1): 18–24.

Opperman, J. J., and A. M. Merenlender. 2000. Deer herbivory as an ecological constraint to restoration of degraded riparian corridors. *Restoration Ecology* 8 (1): 41–47.

Ostfeld, R. S., and F. Keesing. 2000. Biodiversity and disease risk: The case of Lyme disease. *Conservation Biology* 14 (3): 722–728.

Ostfeld, R. S., and K. LoGiudice. 2003. Community disassembly, biodiversity loss, and the erosion of an ecosystem service. *Ecology* 84 (6): 1421–1427.

Packard, S., and C. F. Mutel, editors. 1997. *The tallgrass restoration handbook.* Washington, DC: Island Press.

Paine, R. T. 1969. A note on trophic complexity and community stability. *American Naturalist* 103 (929): 91–93.

Paine, R. T. 1974. Intertidal community structure—Experimental studies on relationship between a dominant competitor and its principal predator. *Oecologia* 15 (2): 93–120.

Palmer, M. A., J. D. Allan, and C. A. Butman. 1996. Dispersal as a regional process affecting the local dynamics of marine and stream benthic invertebrates. *Trends in Ecology & Evolution* 11 (8): 322–326.

Palmer, M. A., R. F. Ambrose, and N. L. Poff. 1997. Ecological theory and community restoration ecology. *Restoration Ecology* 5 (4): 291–300.

Palmer, M., E. Bernhardt, E. Chornesky, S. Collins, A. Dobson, C. Duke, B. Gold, R. Jacobson, S. Kingsland, R. Kranz, M. Mappin, M. L. Martinez, F. Micheli, J. Morse, M. Pace, M. Pascual, S. Palumbi, O. J. Reichman, A. Simons, A. Townsend, and M. Turner. 2004. Ecology for a crowded planet. *Science* 304 (5675): 1251–1252.

Palumbi, S. R. 2003. Ecological subsidies alter the structure of marine communities. *Proceedings of the National Academy of Sciences, USA* 100 (21): 11927–11928.

Patten, D. T., D. A. Harpman, M. I. Voita, and T. J. Randle. 2001. A managed flood on the Colorado River: Background, objectives, design, and implementation. *Ecological Applications* 11 (3): 635–643.

Peckarsky, B. L. 1983. Biotic interactions or abiotic limitations? A model of lotic community structure. In *Dynamics of lotic ecosystems,* ed. T. D. Fontaine III and S. M. Bartell, 303–333. Ann Arbor: Ann Arbor Science Publishers.

Peintinger, M., A. Bergamini, and B. Schmid. 2003. Species-area relationships and nestedness of four taxonomic groups in fragmented wetlands. *Basic and Applied Ecology* 4 (5): 385–394.

Perrow, M. R., and A. J. Davy, editors. 2002. *Handbook of ecological restoration.* Volume 1. *Principles of restoration.* Cambridge, UK: Cambridge University Press.

Petchey, O. L., and K. J. Gaston. 2002. Functional diversity, species richness and community composition. *Ecology Letters* 5 (3): 402–411.

Pianka, E. R. 1967. On lizard species diversity—North American flatland deserts. *Ecology* 48 (3): 333–351.

Pianka, E. R. 1975. Niche relations of desert lizards. In *Ecology and evolution of communities*, ed. M. L. Cody and J. M. Diamond. Cambridge: Belknap Press of Harvard University Press.

Pickett, S. T. A. 1980. Non-equilibrium coexistence of plants. *Bulletin of the Torrey Botanical Club* 107 (2): 238–248.

Pickett, S. T. A., J. Kolasa, and C. G. Jones. 1994. *Ecological understanding: The nature of theory and the theory of nature*. San Diego: Academic Press.

Pickett, S. T. A., and M. J. McDonnell. 1989. Changing perspectives in community dynamics—A theory of successional forces. *Trends in Ecology & Evolution* 4 (8): 241–245.

Pickett, S. T. A., and M. J. McDonnell. 1990. Changing perspectives in community dynamics—A reply to Waters. *Trends in Ecology & Evolution* 5 (4): 123–124.

Pimm, S. L., and J. H. Brown. 2004. Domains of diversity. *Science* 304 (5672): 831–833.

Plotkin, J. B., M. D. Potts, D. W. Yu, S. Bunyavejchewin, R. Condit, R. Foster, S. Hubbell, J. LaFrankie, N. Manokaran, L. H. Seng, R. Sukumar, M. A. Nowak, and P. S. Ashton. 2000. Predicting species diversity in tropical forests. *Proceedings of the National Academy of Sciences, USA* 97 (20): 10850–10854.

Poff, N. L. 1997. Landscape filters and species traits: Towards mechanistic understanding and prediction in stream ecology. *Journal of the North American Benthological Society* 16 (2): 391–409.

Poff, N. L., J. D. Allan, M. B. Bain, J. R. Karr, K. L. Prestegaard, B. D. Richter, R. E. Sparks, and J. C. Stromberg. 1997. The natural flow regime. *Bioscience* 47 (11): 769–784.

Polis, G. A., W. B. Anderson, and R. D. Holt. 1997. Toward an integration of landscape and food web ecology: The dynamics of spatially subsidized food webs. *Annual Review of Ecology and Systematics* 28:289–316.

Polis, G. A., C. A. Myers, and R. D. Holt. 1989. The ecology and evolution of intraguild predation—Potential competitors that eat each other. *Annual Review of Ecology and Systematics* 20:297–330.

Poole, G. C., J. B. Dunham, D. M. Keenan, S. T. Sauter, D. A. McCullough, C. Mebane, J. C. Lockwood, D. A. Essig, M. P. Hicks, D. J. Sturdevant, E. J. Materna, S. A. Spalding, J. Risley, and M. Deppman. 2004. The case for regime-based water quality standards. *Bioscience* 54 (2): 155–161.

Power, M. E. 1990. Effects of fish in river food webs. *Science* 250:811–814.

Power, M. E. 1992. Top-down and bottom-up forces in food webs—Do plants have primacy? *Ecology* 73 (3): 733–746.

Pywell, R. F., J. M. Bullock, A. Hopkins, K. J. Walker, T. H. Sparks, M. J. W. Burke, and S. Peel. 2002. Restoration of species-rich grassland on arable land: Assessing the limiting processes using a multi-site experiment. *Journal of Applied Ecology* 39 (2): 294–309.

Pywell, R. F., N. R. Webb, and P. D. Putwain. 1995. A comparison of techniques for restoring heathland on abandoned farmland. *Journal of Applied Ecology* 32 (2): 400–411.

Ricklefs, R. E. 1987. Community diversity—Relative roles of local and regional processes. *Science* 235 (4785): 167–171.

Ricklefs, R. E. 1990. *Ecology*. 3rd Edition. New York: W. H. Freeman & Co.

Ricklefs, R. E. 2004. A comprehensive framework for global patterns in biodiversity. *Ecology Letters* 7 (1): 1–15.

Robertson, G. P., and S. M. Swinton. 2005. Reconciling agricultural productivity and environmental integrity: A grand challenge for agriculture. *Frontiers in Ecology and the Environment* 3 (1): 38–46.

Rosenzweig, M. L., and J. Winakur. 1969. Population ecology of desert rodent communities—Habitats and environmental complexity. *Ecology* 50 (4): 558–572.

Sadik, N. 1999. *The state of world population 1999—6 billion: A time for choices*. New York: United Nations Population Fund.

Sale, P. F. 1977. Maintenance of high diversity in coral-reef fish communities. *American Naturalist* 111 (978): 337–359.

Savage, M. 1997. The role of anthropogenic influences in a mixed-conifer forest mortality episode. *Journal of Vegetation Science* 8 (1): 95–104.

Schmitz, O. J. 2003. Top predator control of plant biodiversity and productivity in an old-field ecosystem. *Ecology Letters* 6 (2): 156–163.

Schmitz, O. J., P. A. Hamback, and A. P. Beckerman. 2000. Trophic cascades in terrestrial systems: A review of the effects of carnivore removals on plants. *American Naturalist* 155 (2): 141–153.

Seabloom, E. W., W. S. Harpole, O. J. Reichman, and D. Tilman. 2003. Invasion, competitive dominance, and resource use by exotic and native California grassland species. *Proceedings of the National Academy of Sciences, USA* 100 (23): 13384–13389.

Severns, P. M. 2003. Propagation of a long-lived and threatened prairie plant, *Lupinus sulphureus* ssp *kincaidii*. *Restoration Ecology* 11 (3): 334–342.

Shea, K., S. H. Roxburgh, and E. S. J. Rauschert. 2004. Moving from pattern to process: Coexistence mechanisms under intermediate disturbance regimes. *Ecology Letters* 7 (6): 491–508.

Shields, F. D., S. S. Knight, N. Morin, and J. Blank. 2003. Response of fishes and aquatic habitats to sand-bed stream restoration using large woody debris. *Hydrobiologia* 494 (1–3): 251–257.

Simberloff, D. S., and E. O. Wilson. 1969. Experimental zoogeography of islands: Colonization of empty islands. *Ecology* 50 (2): 278–296.

Smith, R. S., R. S. Shiel, R. D. Bardgett, D. Millward, P. Corkhill, G. Rolph, P. J. Hobbs, and S. Peacock. 2003. Soil microbial community, fertility, vegetation and diversity as targets in the restoration management of a meadow grassland. *Journal of Applied Ecology* 40 (1): 51–64.

Sondergaard, M., E. Jeppesen, and S. Berg. 1997. Pike (*Esox lucius* L.) stocking as a biomanipulation tool—Effects on lower trophic levels in Lake Lyng, Denmark. *Hydrobiologia* 342:319–325.

Sousa, W. P. 1979. Disturbance in marine inter-tidal boulder fields—The non-equilibrium maintenance of species-diversity. *Ecology* 60 (6): 1225–1239.

Sousa, W. P . 1984. The role of disturbance in natural communities. *Annual Review of Ecology and Systematics* 15:353–391.

Southwood, T. R. E. 1977. Habitat, the templet for ecological strategies? Presidential address to British Ecological Society, 5 January 1977. *Journal of Animal Ecology* 46 (2): 337–365.

Sprugel, D. G. 1991. Disturbance, equilibrium, and environmental variability—What is natural vegetation in a changing environment. *Biological Conservation* 58 (1): 1–18.

Stachowicz, J. J. 2001. Mutualism, facilitation, and the structure of ecological communities. *Bioscience* 51 (3): 235–246.

Stevens, R. D., and M. R. Willig. 2002. Geographical ecology at the community level: Perspectives on the diversity of new world bats. *Ecology* 83 (2): 545–560.

Suding, K. N., K. L. Gross, and G. R. Houseman. 2004. Alternative states and positive feedbacks in restoration ecology. *Trends in Ecology & Evolution* 19 (1): 46–53.

Swan, C. M., and M. A. Palmer. 2004. Leaf diversity alters litter breakdown in a Piedmont stream. *Journal of the North American Benthological Society* 23 (1): 15–28.

Temperton, V. M., and K. Kirr. 2004. Order of arrival and availability of safe sites: An example of their importance for plant community assembly in stressed ecosystems. In *Assembly rules and restoration ecology*, ed. V. M. Temperton, R. J. Hobbs, T. Nuttle, and S. Halle, 285–304. Washington, DC: Island Press.

Theiling, C. H., J. K. Tucker, and F. A. Cronin. 1999. Flooding and fish diversity in a reclaimed river-wetland. *Journal of Freshwater Ecology* 14 (4): 469–475.

Thompson, J. N., O. J. Reichman, P. J. Morin, G. A. Polis, M. E. Power, R. W. Sterner, C. A. Couch, L. Gough, R. Holt, D. U. Hooper, F. Keesing, C. R. Lovell, B. T. Milne, M. C. Molles, D. W. Roberts, and S. Y. Strauss. 2001. Frontiers of ecology. *Bioscience* 51 (1): 15–24.

Tilman, D. 1982. *Resource competition and community structure*. Princeton: Princeton University Press.

Tilman, D., P. B. Reich, J. Knops, D. Wedin, T. Mielke, and C. Lehman. 2001. Diversity and productivity in a long-term grassland experiment. *Science* 294 (5543): 843–845.

Tonn, W. M., J. J. Magnuson, M. Rask, and J. Toivonen. 1990. Intercontinental comparison of small-lake fish assemblages—The balance between local and regional processes. *American Naturalist* 136 (3): 345–375.

Townsend, C. R., M. R. Scarsbrook, and S. Doledec. 1997. The intermediate disturbance hypothesis, refugia, and biodiversity in streams. *Limnology and Oceanography* 42 (5): 938–949.

Tyson, G. W., J. Chapman, P. Hugenholtz, E. E. Allen, R. J. Ram, P. M. Richardson, V. V. Solovyev, E. M. Rubin, D. S. Rokshar, and J. F. Banfield. 2004. Community structure and metabolism through reconstruction of microbial genomes from the environment. *Nature* 428:37–43.

van der Heijden, M. G. A., J. N. Klironomos, M. Ursic, P. Moutoglis, R. Streitwolf-Engel, T. Boller, A. Wiemken, and I. R. Sanders. 1998. Mycorrhizal fungal diversity determines plant biodiversity, ecosystem variability and productivity. *Nature* 396 (6706): 69–72.

Venter, J. C., K. Remington, J. F. Heidelberg, A. L. Halpern, D. Rusch, J. A. Eisen, D. Y. Wu, I. Paulsen, K. E. Nelson, W. Nelson, D. E. Fouts, S. Levy, A. H. Knap, M. W. Lomas, K. Nealson, O. White, J. Peterson, J. Hoffman, R. Parsons, H. Baden-Tillson, C. Pfannkoch, Y. H. Rogers, and H. O. Smith. 2004. Environmental genome shotgun sequencing of the Sargasso Sea. *Science* 304 (5667): 66–74.

Vitousek, P. M., J. D. Aber, R. W. Howarth, G. E. Likens, P. A. Matson, D. W. Schindler, W. H. Schlesinger, and D. G. Tilman. 1997. Human alteration of the global nitrogen cycle: Sources and consequences. *Ecological Applications* 7 (3): 737–750.

Volterra, V. 1926. Variations and fluctuation of the numbers of individuals in animal species living together. In *Animal ecology*, ed. R. N. Chapman. New York: McGraw-Hill.

Walker, L. R., and P. M. Vitousek. 1991. An invader alters germination and growth of a native dominant tree in Hawaii. *Ecology* 72 (4): 1449–1455.

Wall, D. H., and J. C. Moore. 1999. Interactions underground—Soil biodiversity, mutualism, and ecosystem processes. *Bioscience* 49 (2): 109–117.

Wall, D. H., M. A. Palmer, and P. V. R. Snelgrove. 2001. Biodiversity in critical transition zones between terrestrial, freshwater, and marine soils and sediments: Processes, linkages, and management implications. *Ecosystems* 4 (5): 418–420.

Wallace, J. B., S. L. Eggert, J. L. Meyer, and J. R. Webster. 1997. Multiple trophic levels of a forest stream linked to terrestrial litter inputs. *Science* 277:102–104.

Webb, C. O., D. D. Ackerly, M. A. McPeek, and M. J. Donoghue. 2002. Phylogenies and community ecology. *Annual Review of Ecology and Systematics* 33:475–505.

Williams, C. E. 1998. History and status of table mountain pine pitch pine forests of the southern Appalachian Mountains (USA). *Natural Areas Journal* 18 (1): 81–90.

Wohl, D. L., S. Arora, and J. R. Gladstone. 2004. Functional redundancy supports biodiversity and ecosystem function in a closed and constant environment. *Ecology* 85 (6): 1534–1540.

Wootton, J. T., and M. E. Power. 1993. Productivity, consumers, and the structure of a river food-chain. *Proceedings of the National Academy of Sciences, USA* 90 (4): 1384–1387.

Wright, J. P., C. G. Jones, and A. S. Flecker. 2002. An ecosystem engineer, the beaver, increases species richness at the landscape scale. *Oecologia* 132 (1): 96–101.

Zedler, J. B. 2000. Progress in wetland restoration ecology. *Trends in Ecology & Evolution* 15 (10): 402–407.

Chapter 6

Evolutionary Restoration Ecology

Craig A. Stockwell, Michael T. Kinnison, and Andrew P. Hendry

Restoration Ecology and Evolutionary Process

Restoration activities have increased dramatically in recent years, creating evolutionary challenges and opportunities. Though restoration has favored a strong focus on the role of habitat, concerns surrounding the evolutionary ecology of populations are increasing. In this context, previous researchers have considered the importance of preserving extant diversity and maintaining future evolutionary potential (Montalvo et al. 1997; Lesica and Allendorf 1999), but they have usually ignored the prospect of ongoing evolution in real time. However, such *contemporary evolution* (changes occurring over one to a few hundred generations) appears to be relatively common in nature (Stockwell and Weeks 1999; Bone and Farres 2001; Kinnison and Hendry 2001; Reznick and Ghalambor 2001; Ashley et al. 2003; Stockwell et al. 2003). Moreover, it is often associated with situations that may prevail in restoration projects, namely the presence of introduced populations and other anthropogenic disturbances (Stockwell and Weeks 1999; Bone and Farres 2001; Reznick and Ghalambor 2001) (Table 6.1). Any restoration program may thus entail consideration of evolution in the *past*, *present*, and *future*.

Restoration efforts often involve dramatic and rapid shifts in habitat that may even lead to different ecological states (such as altered fire regimes) (Suding et al. 2003). Genetic variants that evolved within historically different evolutionary contexts (the past) may thus be pitted against novel and mismatched current conditions (the present). The degree of this mismatch should then determine the pattern and strength of selection acting on trait variation in such populations (Box 6.1; Figure 6.1). If trait variation is heritable and selection is sufficiently strong, contemporary evolution is likely to occur and may have dramatic impacts on the adaptive dynamics of restoration scenarios. Adaptation to current conditions (the present) may in turn influence the ability of such populations to subsequently persist and evolve over short or long periods (the future). Thus, the success (or failure) of a restoration effort may often be as much an evolutionary issue as an ecological one.

It is also useful to recognize that contemporary evolution may alter the interactions of species with their environments and each other. Restoration ecologists may thus be faced with a changed cast of players, even if many of the same nominal species are restored. Efforts that assume species and populations are evolutionarily stagnant may face frustrating and

TABLE 6.1

Examples of contemporary evolution in plants and animals in nature.

Context/example	Traits	Evolutionary agent	References
Colonization			
Threespine stickleback (*Gasterosteus aculeatus*)	Lateral plate armor	Freshwater habitat	Bell et al. 2004
Pacific salmon (*Oncorhynchus* spp.)	Development and growth, morphology, reproductive timing, ovarian investment	Breeding environment (temperature, flow, migratory rigor)	Hendry et al. 2000; Hendry et al. 2001; Kinnison et al. 2001; Quinn et al. 2001
European Grayling (*Thymallus thymallus*)	Length at termination, yolk-sac volume, growth rate, survival	Water temperature	Koskinen et al. 2002
Guppies (*Poecilia reticulata*)	Age and size at maturity, offspring size, antipredator behavior	Predator regimes	Endler 1980; Reznick et al. 1997; O'Steen et al. 2002
Mosquitofish (*Gambusia affinis*)	Size at maturity, fat storage	Environmental constancy, thermal environment	Stearns 1983; Stockwell and Weeks 1999
White Sands pupfish (*Cyprinodon tularosa*)	Pgdh (Phosphogluconate dehydrogenase); body shape	Salinity/flow	Stockwell and Mulvey 1998; Collyer et al. 2005
Dark-eyed juncos (*Junco hyemalis*)	Morphology (amount of white in tail)	Sexual selection	Rasner et al. 2004; Yeh 2004
Rabbits (*Oryctolagus cuniculus*)	Morphology	Ecoregional variation (temperature and aridity)	Williams and Moore 1989
Isopod (*Asellus aquaticus*)	Pigmentation	Predation	Hargeby et al. 2004
In situ disturbance			
Numerous plant species, including *Mimulus guttatus*, *Anthoxanthum odoratum*, *Agrostis tenuis*, *Lupinus bicolor*, *Lotus pershianus*	Tolerance to metals (e.g., copper)	Metal contaminated soils (e.g., mine waste piles)	Jain and Bradshaw 1966; Antonovics and Bradshaw 1970; Wu and Kruckeberg 1985; Macnair 1987; Bone and Farres 2001
Anthoxanthum odoratum	pH tolerance	Fertilizer, altered soil pH	Snaydon and Davies 1972; Davies and Snaydon 1976; Bone and Farres 2001
Plantago major	Growth rate	High ozone concentration	Davison and Reiling 1995
Arabidopsis thaliana	Seed production	CO_2 concentration	Ward et al. 2000
Oligochaete (*Limnodrilus hoffmeisteri*)	Loss of cadmium resistance	Removal of cadmium	Levinton et al. 2003

Numerous insect species, Diamond back moths (*Plutella xylostella*)	Pesticide resistance (e.g., resistance to *Bt*)	Selective mortality or sterility, *Bt*-related mortality	Mallet 1989; Tabashnik 1994
Soapberry bugs (*Jadera haematoloma*)	Beak length	Introduced host fruit size	Carroll et al. 2001
Pitcher plant mosquitoes (*Wyeomyia smithii*)	Photoperiodic diapause response	Global warming	Bradshaw and Holzapfel 2001
Water flea (*Daphnia galeata*)	Resistance to poor/toxic diet	Cyanobacteria increase following eutrophication	Hairston et al. 1999
Galapágos finches (*Geospiza fortis*)	Body size, beak shape	Drought effects on food resources	Grant and Grant 2002
Red squirrels (*Tamiasciurus hudsonicus*)	Breeding season	Global change (increased temperatures)	Réale et al. 2003
Selective harvest			
European Grayling (*Thymallus thymallus*)	Age and size at maturity	Selectivity of harvest methods and gear (e.g., mesh size of nets)	Haugen and Vøllestad 2001
Northern cod (*Gadus morhua*)	Size, age at maturity	Harvest of large cod	Olsen et al. 2004
Bighorn sheep (*Ovis canadensis*)	Male body and horn size	Selective harvest of large males	Coltman et al. 2003
Introgression			
Ducks, canids, salmonids, sunflowers, etc.	Morphology (other aspects likely)	Hybridization among wild species and between wild and domestic (sub)species	Rhymer and Simberloff 1996; Allendorf et al. 2001; Grant and Grant 2002

Box 6.1
Evolutionary Change in Quantitative Traits

For a quantitative trait (influenced by multiple genes, often of small effect), a simple equation can be used to predict how adaptation should proceed, at least under a number of simplifying assumptions (Lande and Arnold 1983). Specifically, $\Delta z = G\beta$, where Δz is the change in mean trait value from one generation to the next, G is the additive genetic variance for the trait and β is the selection gradient acting on the trait (slope of the relationship between the trait and fitness). When considering a single trait, this equation is analogous to the traditional "breeder's equation" (evolutionary response = heritability * selection; $R = h^2S$) because $G/P = h^2$ and $S/P = \beta$, where P is the phenotypic variance and S is the selection differential (difference between the mean trait value before and after selection). When considering multiple traits, Δz becomes a vector of changes in mean trait values, G becomes a matrix of additive genetic variances/covariances, and β becomes a vector of selection gradients. That is, $\Delta z = G\beta$ (Lande and Arnold 1983; Schluter 2000; Arnold et al. 2001).
In the case of two traits, the multivariate equation expands to

$$\begin{bmatrix} \Delta z_1 \\ \Delta z_2 \end{bmatrix} = \begin{bmatrix} G_{11} & G_{12} \\ G_{21} & G_{22} \end{bmatrix} \begin{bmatrix} \beta_1 \\ \beta_2 \end{bmatrix}$$

where Δz_i is the evolutionary response for trait i, G_{11} and G_{22} are the additive genetic variances for the two traits, G_{12} and G_{21} are identical and are the additive genetic covariance between the two traits, and β_i is the selection gradient acting on the trait. Selection gradients are commonly estimated as partial regression coefficients from a multiple regression of both traits on fitness. In this case, selection gradients represent the effect of each trait on fitness after controlling for the effect of the other trait (i.e., "direct" selection). This equation shows how the evolutionary response for each trait will be a function of selection acting directly on that trait, the additive genetic variance for that trait, selection acting on the other trait, and the additive genetic covariance between the traits. That is, $\Delta z_1 = G_{11}\beta_1 + G_{12}\beta_2$ and $\Delta z_2 = G_{22}\beta_2 + G_{21}\beta_1$. This formulation illustrates how apparently paradoxical evolutionary changes can be observed in some situations. For example, the first trait can evolve to be smaller even if it is under selection to be larger (e.g., Grant and Grant 1995). This can occur when $G_{12}\beta_2 < 0$ and $|G_{12}\beta_2| > G_{11}\beta_1$; that is, when the negative indirect effect of selection on the first trait is stronger than the positive direct effect of selection. These negative indirect effects should increase as selection on the second trait becomes stronger and as the genetic covariance becomes stronger, with one of these quantities necessarily being negative.

Phenotypes in an undisturbed population should be centered around an optimal value (i.e., the population is well adapted). In a restoration context, however, a disturbance to the environment may shift the phenotypic optimum away from the current phenotypes (Figure 6.1). This shift leads to a mismatch between current phenotypes and optimal phenotypes, leaving the population maladapted and subject to directional selection. Under a number of assumptions, the strength of this selection can be represented as:

$$\beta = \frac{-(z - \theta)}{\omega^2 + P}$$

where z is the mean trait value, θ is the optimal trait value, P is the phenotypic variance, and ω^2 is the strength of stabilizing selection around the optimum (for simplicity, we assume ω^2 is the same around the optimum before and after the disturbance). Smaller values of ω^2 correspond to steeper fitness functions and therefore stronger stabilizing selection around the optimum. When a disturbance shifts the optimum away from the current phenotypes, directional selection on the population increases (larger $|\beta|$), causing evolution toward the new optimum.

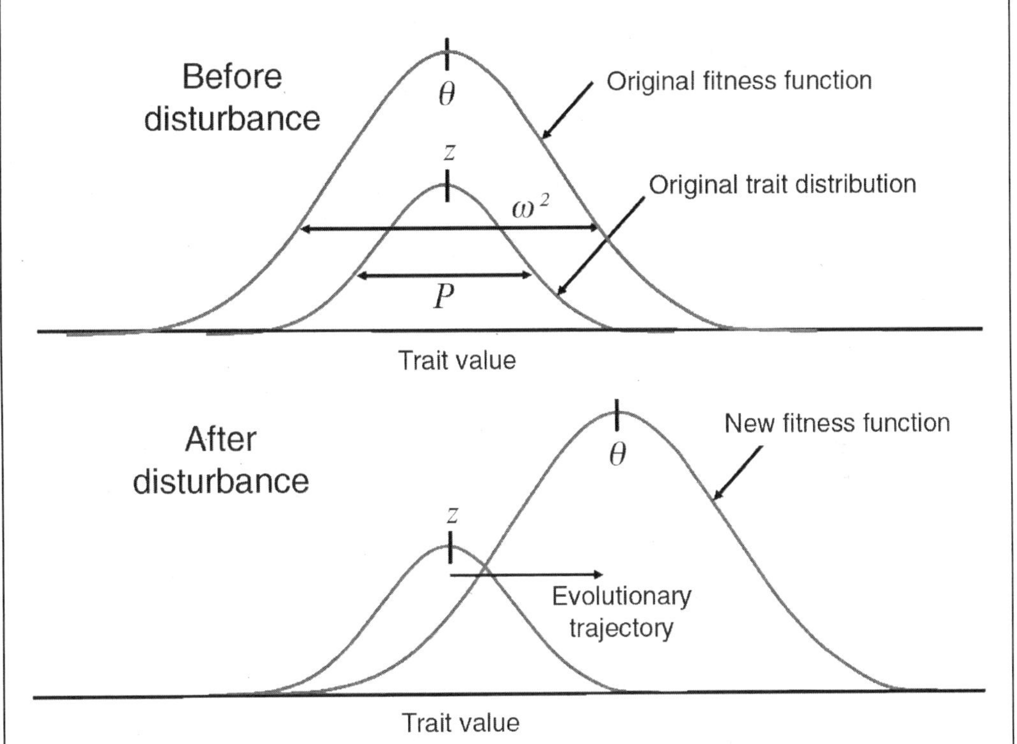

FIGURE 6.1 The distribution of trait values is shown in relation to the fitness function before and after a disturbance (native habitat versus restoration habitat), where z is the mean trait value, θ is the optimal trait value, P is the phenotypic variance, and ω^2 is the strength of stabilizing selection around the optimum. The width of the fitness function reflects the strength of stabilizing selection, which is denoted by ω^2. (Reprinted from *Trends in Ecology and Evolution* 18; Stockwell, Henday and Kinnison, Contemporary evolution meets conservation biology, 96, © 2003 from Elsevier.)

These equations can be used to predict the evolutionary responses of traits following a disturbance and have proven effective in predicting evolutionary responses in natural populations (Grant and Grant 1995, 2002). Figure 6.2 shows evolutionary responses for a population in relation to different additive genetic variances (G = 0.1–0.5) and strengths of stabilizing selection ($\omega = 2 - 5$; i.e., $\omega^2 = 4 - 25$). In each case, we assume the mean trait value is larger than the optimum ($z - \theta = 1$) and the phenotypic variance is $P = 1$. In general, evolutionary responses will increase as genetic variance increases (G increases) and the strength of stabilizing selection increases (ω decreases). Evolutionary responses will also increase with increasing differences between the mean trait value and the new optimal trait value. However, it is important to recognize that several factors can lead to discrepancies between predicted and observed evolutionary responses (Merilä et al. 2001b; see Box 6.2).

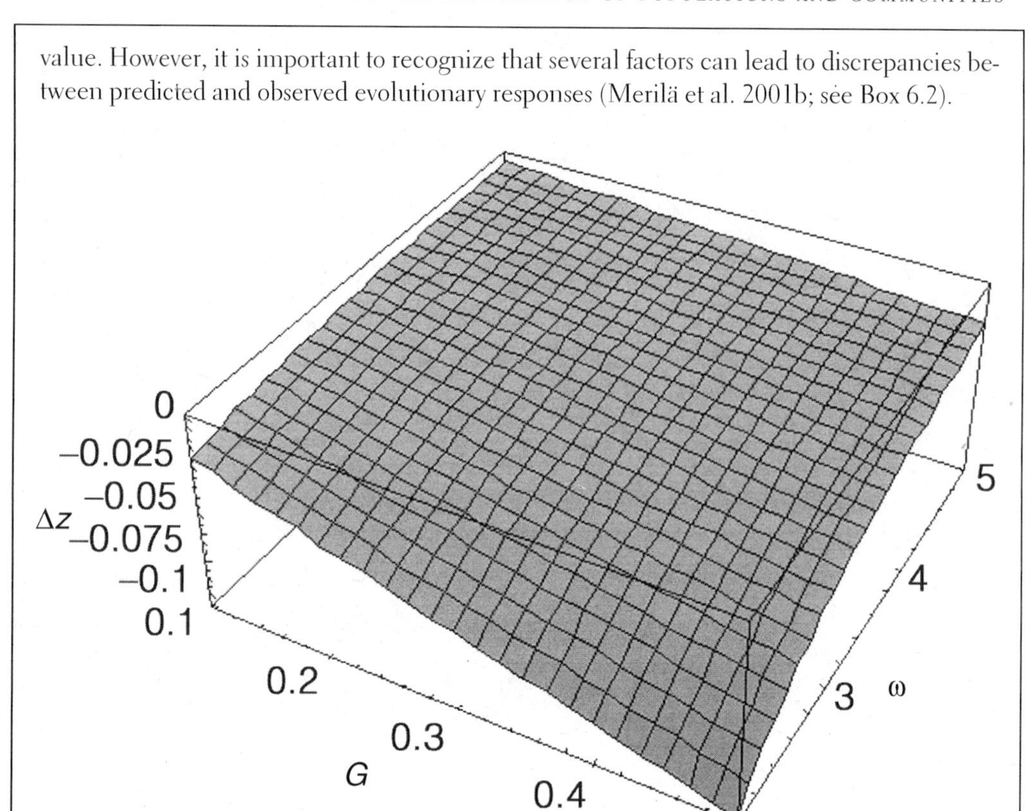

FIGURE 6.2 The evolutionary response (Δz) of a population is shown for different additive genetic variance ($G = 0.1–0.5$) and strengths of stabilizing selection ($\omega = 2 - 5$).

unanticipated outcomes. For instance, mounting evidence suggests that sustained harvest results in rapid life-history evolution toward less preferred phenotypes (e.g., smaller body size) (Haugen and Vøllestad 2001; Coltman et al. 2003; Olsen et al. 2004), potentially influencing both the ecological roles of these species in their communities and our own ecological interactions with them (e.g., rates of harvest).

Simply acknowledging, or even anticipating, that evolution will occur in restoration contexts is, however, probably not an entirely satisfying message for many readers of this book. We wish to go further and suggest that restoration ecology may include evolutionarily enlightened approaches (Ashley et al. 2003) or even constructive management of evolutionary processes. Regardless, it is our hope that the study of evolution may not only inform restoration, but that restoration may also inform the study of evolution. Indeed, the tempo and mode of contemporary evolution is still not well understood (Kinnison and Hendry 2001), and restoration ecology offers an opportunity to study the mechanics of evolution for diverse taxa under a variety of circumstances.

In this chapter, we consider the roles of evolutionary processes in both ecological endangerment and restoration. Because our actions as practitioners of restoration are largely limited to the present, much of our discussion will surround interactions with contemporary evolution. First, we describe the conditions under which contemporary evolution occurs and the factors by which it may be facilitated or constrained. Second, we discuss approaches and tools available for assessing evolutionary mechanics acting in populations of restoration concern. We consider evolutionary dynamics in a landscape context because restoration schemes will generally involve contributions from, and interactions with, the larger metapopulation and metacommunity (Maschinski, this volume; Menninger and Palmer, this volume). Our approach links historically established genetic diversity (the past) with contemporary evolution (the present) and long-term evolutionary potential (the future). Third, we provide an evolutionary perspective on traditional topics within the field of restoration ecology, such as the identification of suitable seed sources. We conclude by discussing research areas ripe for evaluation in the context of evolutionary restoration ecology.

Evolutionary Ecology and Contemporary Evolution

Evolutionary ecology, by analogy to other areas of ecology, is the study of processes that influence the *distribution* and *abundance* of individuals with different genotypes (genetic forms within a species or population). These processes often involve interactions of organisms with their abiotic and biotic environments. The term *evolution* is reserved for *heritable* changes in the relative abundance of trait values among generations. Often, adaptive evolution results from natural selection driven by correlations between heritable trait variation and *fitness* (the likelihood that an individual will contribute to future generations). However, a suite of additional factors, including anthropogenic effects, influence the likelihood and outcome of adaptive evolution and, by association, the performance and sustainability of populations.

In fact, anthropogenic activities are often associated with cases of evolution on contemporary time scales. We use the general term, *contemporary evolution*, in lieu of the other commonly used term, *rapid evolution*, which carries a historic perspective that such evolution is exceptional. In reality, contemporary evolution is now widely documented, and there is no evidence to suspect that it is either uncommon or exceptionally fast, given time-scaling effects (Kinnison and Hendry 2001; Stockwell et al. 2003). In general, contemporary evolution should occur whenever there is sufficient heritable variation for a trait under directional selection (Box 6.1; Figure 6.1). In a restoration program, we might expect a mismatch between the optimal trait value and the actual mean trait value for the given population. This mismatch causes directional selection to act on the trait: that is, selection favoring a shift in the mean trait value toward the optimum (highest net fitness value) (Figure 6.1). If the phenotypic variation has some genetic basis, the population mean should shift toward the optimum in the next generation (i.e., adaptation to the restoration environment). Of course, the population may go extinct, even while it is adapting, if the population is too small or if selection is too strong (Figure 6.3; also see Lynch 1996).

Contemporary evolution should occur in the above situation in general, but adaptation can be influenced by many factors, including population size, gene flow, antagonistic pleiotropy, and life-history constraints (Table 6.2; see also Box 6.2). Small populations are less likely to evolve for two reasons. First, low population size may not provide sufficient phenotypic variation for selection to act (Lande 1995; Lynch 1996). Second, small populations are less likely to persist through the initial reduction in population size during the early stages of

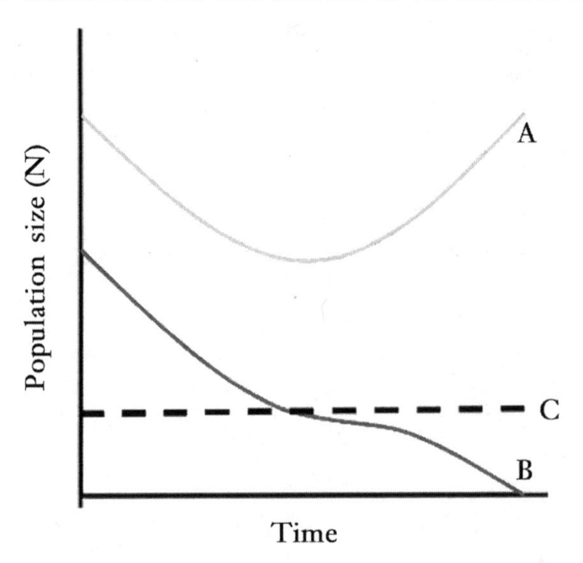

FIGURE 6.3 For large populations (A), novel selection is expected to cause an initial decline in population size (N) until the population adapts and population size increases (A). However, small populations (B) are more vulnerable to extinction because they are more likely to reach sizes where demographic stochasticity becomes overwhelming (C). Figure modified with permission from Gomulkiewicz and Holt (1995).

TABLE 6.2

Factors constraining and facilitating contemporary evolution.

Factors facilitating contemporary evolution	Reference
High genetic variation	Boulding and Hay 2001
Strong selection (directional, stabilizing)	Hendry 2004
Absence of gene flow	Boulding and Hay 2001
Large populations	Gomulkiewicz and Holt 1995; Lande 1995; Lynch 1996; Franklin and Frankham 1998; Lynch and Lande 1998
Factors constraining contemporary evolution (in addition to the converse of those listed above)	**Reference**
Antagonistic pleiotropy	Etterson and Shaw 2001
Life history constraints	Reznick et al. 2004
Environmental deterioration (evolution masked)	Merilä et al. 2001b

adaptive evolution (Figure 6.3). Gene flow has the potential to slow adaptation by introducing maladapted genes from adjacent populations, an effect we discuss at length below. Antagonistic pleiotropy may impede adaptation when two genetically correlated traits (e.g., both partly influenced by the same underlying genes) are under different patterns of selection. Life-history constraints can hinder adaptation by limiting a population's growth rate under selection (Reznick et al. 2004). For instance, populations with delayed maturity were at high extinction risk when introduced to sites with greater predation risk (Reznick et al. 2004).

Box 6.2

The Effects of Gene Flow and Stabilizing Selection on Evolutionary Divergence

One such complication relevant to restoration is gene flow, where populations may be held from separate optima by genetic exchange among populations. The equations from Box 6.1 can be modified quite simply to include the effects of gene flow (Hendry et al. 2001). For example, when selection acts on each population and then some individuals move between populations (e.g., adults but not juveniles move), the difference in mean phenotype between the two populations for a single trait (ΔD) will change according to:

$$\Delta D = -\hat{m}D + [1 - \hat{m}][G_j\beta_j - G_i\beta_i],$$

where $\hat{m}$ is the proportion of individuals exchanged between populations, D is the current difference between the populations, and G and ß are the additive genetic variance and selection gradient for the trait in population i and population j. As above, a multivariate version of the equation would replace ΔD, D, G and ß with their matrix or vector equivalents. One can then predict the equilibrium difference in trait value between the two populations (D^*), by setting $\Delta D = 0$ and solving for D:

$$D^* = \frac{[1 - \hat{m}]}{\hat{m}}[G_j\beta_j - G_i\beta_i].$$

As noted in Box 6.1, it is often useful to express ß as a function of the deviation of the mean trait value from the optimum for that population, the strength of stabilizing selection around the optimum, and the phenotypic variance for the trait. Assuming ω^2, P, and G are the same in both populations, the equilibrium difference when selection takes place before movement is:

$$D^* = D_\theta\left[\frac{G[1 - \hat{m}]}{G[1 - \hat{m}] + [\omega^2 + P]\hat{m}}\right],$$

where D_θ is the difference in the optimum trait value between the two populations. Figure 6.4 uses this last equation to explore how the equilibrium difference between two populations will be a function of varying gene flow ($\hat{m}$ = 0 – 0.05) and stabilizing selection (ω = 2 – 10; i.e., ω^2 = 4 – 100). The analysis assumes a typical heritability ($G = 0.3$, $P = 1$, therefore $h^2 = 0.3$) and an optimal difference in mean trait value of $D_\theta = 1$. Figure 6.4 shows that relative adaptation (deviation from the optimum) decreases with increasing gene flow even when gene flow is quite low and with weaker stabilizing selection around the optimum. The negative effect of gene flow on adaptation is strongest when stabilizing selection is quite weak, as is thought to be the case in nature (Kingsolver et al. 2001).

In the context of restoration, these relationships have implications for how the restored site(s) interacts with surrounding sites. In general, we would expect that increases in gene flow between populations in different environments will reduce their adaptive divergence and therefore fitness (see also Boulding and Hay 2001). This potential negative effect of gene flow should therefore be considered when contemplating artificial manipulations of population density or movements of individuals between environments. These negative effects of gene flow, however, will need to be balanced with the consideration of positive effects that

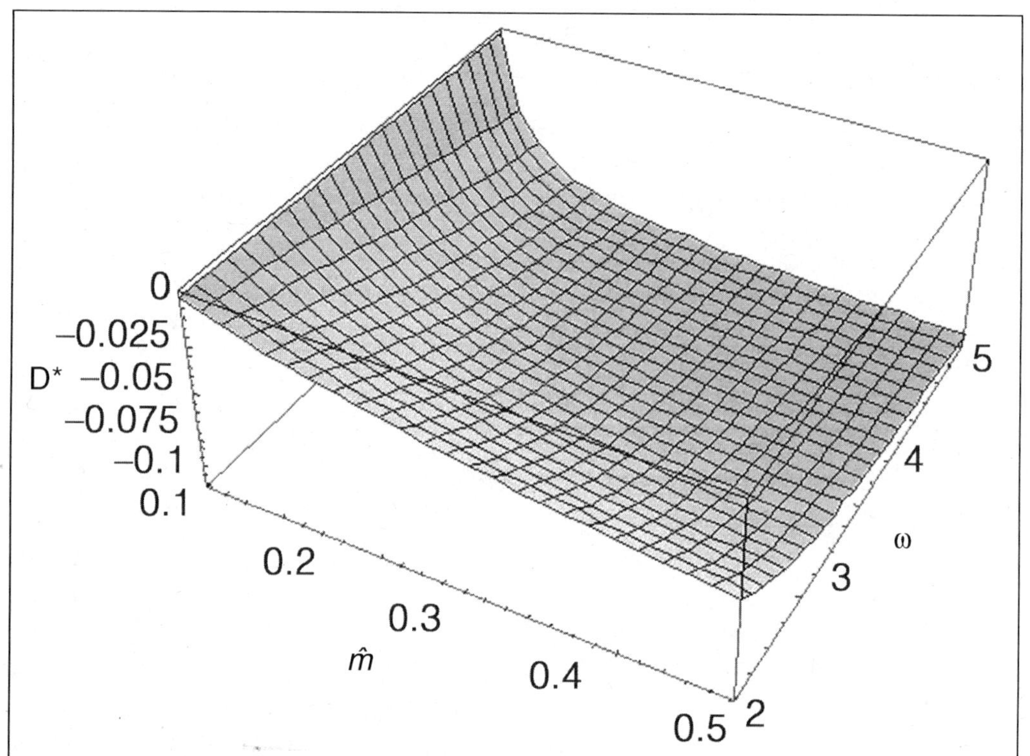

FIGURE 6.4 The effects of gene flow ($\hat{m}$ = 0 – 0.05) and stabilizing selection (ω = 2 – 10) on relative adaptation (D*, divergence from the optimum) are shown. Adaptation decreases with even modest increases in gene flow as long as stabilizing selection is weak, as is thought to be the case in nature.

might arise when populations are inbred (Hedrick 1995), or when gene flow allows a new or marginal population to persist long enough to then adapt to their environment (Holt and Gomulkiewicz 1997; Kawecki and Holt 2002).

Landscape Genetics and Restoration

A profitable way to evaluate current evolutionary processes and future evolutionary potential is through *landscape genetics* (*sensu* Manel et al. 2003; Guillot et al. 2005), which, at its simplest, is the quantification of genetic variation within and among populations (i.e., population genetics) in the context of a spatially and temporally complex landscape. Genetic variation within populations is important because it mediates adaptive evolution in response to changing conditions (Houle 1992; Bürger and Lynch 1995; García-Ramos and Rodríguez 2002; Reed et al. 2003). Genetic variation among populations is important as a reflection of local adaptation and the potential role of gene flow. In some situations such gene flow may be an important source of variation, whereas in others it may prove an impediment to local adaptation (Box 6.2).

Strong adaptation to different ecological environments implies that restoration efforts might benefit from choosing source populations that have phenotypes similar to those at the restoration site or, equivalently, source populations that occupy similar environments. If the restoration site is a novel environment and local adaptation is strong, it may be better to generate a genetically diverse group by mixing source populations. Selection at the restoration site can then weed through the various combinations based on their relative fitness (Lesica and Allendorf 1999; Falk et al., this volume; Maschinski, this volume). Restoration thus has two possible relationships to contemporary evolution. On one hand, the practitioner or researcher should be aware that contemporary evolution is likely in a restoration context. On the other hand, some researchers may choose to modify the conditions of restoration in such a way as to intentionally influence evolutionary change.

One way in which evolution may be managed is by managing gene flow to and from the restoration site. Gene flow is most commonly viewed as a constraining force in adaptive divergence (e.g., Storfer et al. 1999; Riechert et al. 2001; Hendry et al. 2002; Calsbeek and Smith 2003; see also Box 6.2) that contributes to a "migrational load" (Lenormand 2002) limiting the fitness of local populations (Boulding and Hay 2001). Restoration practitioners may consider restricting the rate or pattern of gene flow in cases where migrational load is expected and mediated by anthropogenic influences. For example, translocation rates might be reduced after initial establishment of restoration populations to allow them to adaptively diverge from their sources. However, negative effects of gene flow need to be weighed against potential positive effects, such as reduced inbreeding depression, increased genetic variation, and increased potential for future adaptation (e.g., Hedrick 1995; Newman and Tallmon 2001). Precedent already exists for providing artificial gene flow for the purpose of genetically "rescuing" inbred populations (Hedrick 1995). In either case, information about genetic variation within and among populations, habitat heterogeneity, and dispersal over landscapes would be essential to any attempted management of gene flow.

Landscape genetics, including the assessment of local adaptation, gene flow, and genetic variation, is commonly based on surveying molecular markers (neutral or selected) and quantitative traits (phenotypic traits coded for by multiple loci) (Falk et al., this volume). Variation in neutral markers generally reflects the combined effects of mutation, gene flow, and genetic drift. Variation in markers under selection or in quantitative traits is influenced by these same factors, but also by the strength and direction of selection. In the following section, we consider methods and inferences based on (1) neutral markers, (2) selected markers, and (3) quantitative traits.

Neutral Markers

Much of landscape genetics focuses on variation at presumed neutral loci, including some allozymes, microsatellites, mitochondrial DNA, AFLPs (amplified fragment length polymorphisms), RAPDs (random amplified polymorphic DNA), and SNPs (single nucleotide polymorphisms). Genetic variation at such markers can be summarized with a variety of metrics, the most common being various versions of F_{ST}, the amount of genetic variation among populations (v_a) for a given locus relative to the total variation among (v_a) and within (v_w) populations:

$$F_{ST} = \frac{v_a}{v_a + v_w}$$

For a single locus with two alleles, F_{ST} will equal zero when none of the genetic variation is found among populations (i.e., they have the same allele frequencies), but will equal unity when all of the genetic variation is found among populations (i.e., they are fixed for alternative alleles) (Wright 1969). The situation is more complicated for highly polymorphic markers (Hedrick 1999).

The magnitude of F_{ST} at equilibrium will theoretically reflect three factors: the effective population size (N_e), the rate of mutation (μ), and the rate of gene flow (*m*, proportion of populations that are immigrants). When mutation is low relative to gene flow (and a number of other assumptions are adopted; Whitlock and McCauley 1999), mutation can be ignored, leaving a simple relation between genetic divergence and gene flow: $F_{ST} = 1/(1+4N_em)$. This simplicity has led to the frequent use of F_{ST} to infer the amount of historical (i.e., long-term) gene flow, a quantity typically indexed as the effective number of migrants (N_em). However, it is also useful to estimate the *rate* of gene flow (*m*), because it is this latter quantity that has a direct effect on divergence in selected traits (Hendry et al. 2001). Estimation of *m* from N_em therefore requires the estimation of N_e.

The above estimation procedure has been criticized heavily (Whitlock and McCauley 1999), largely because it rests on a number of dubious assumptions, particularly an "island" model of population structure. Alternatively, gene flow can be estimated using methods such as the rare alleles method or coalescent approaches that are a bit less restrictive to some assumptions (e.g., Slatkin 1985; Beerli and Felsenstein 1999; Pearse and Crandall 2004). Ultimately, however, all of these methods require relatively low rates of gene flow and equilibrium between gene flow and genetic drift, which may take thousands of generations to achieve (Waples 1998). Restoration sites, and the landscapes around them, are often far from equilibrium, and so estimates of historical gene flow must be interpreted with caution. Specifically, gene flow will be overestimated when populations are still diverging from a common ancestor under drift (Kinnison et al. 2001) but underestimated during periods of increased gene flow among previously divergent populations (Whitlock 1992). Fortunately, the problem of non-equilibrium conditions can sometimes (but not always) be circumvented by estimating "current" immigration with genetic assignment methods (Berry et al. 2004; Paetkau et al. 2004). However, practitioners should note that these methods detect immigrants that might not actually contribute to genetic introgression. Further, assignment power should be taken into account before attributing mis-assigned individuals as immigrants.

Gene flow estimates in restoration scenarios can be used to infer evolutionary potential: low gene flow facilitates the independent evolution of different populations, whereas high gene flow constrains it (Hendry et al. 2001; Lenormand 2002). This generalization is largely true for divergence in neutral markers but is heavily nuanced for divergence in selected markers or traits. Large numbers of effective migrants may belie strong selection that weeds out an even larger number of unsuitable immigrants, or their offspring, while favoring more suitable individuals. Selected markers or traits can thus undergo adaptive divergence even in the face of apparently moderate gene flow (Hendry et al. 2001). In short, gene flow estimates can be useful tools in restoration contexts, but are best interpreted along with an understanding of selection acting on traits in the populations of concern.

Selected Markers

Some molecular markers reflect traits that are under direct selection or are physically linked to loci under selection (Mitton 1997). If these markers can be identified, surveying them within and among populations can reveal the amount and structure of adaptive genetic variation. For example, allozyme loci are sometimes under selection (Mitton 1997) and can be easily and quickly assayed for large numbers of individuals. Indeed, variation in allozyme markers has already been considered in a restoration context (Stockwell et al. 1996; Stockwell and Mulvey 1998). For instance, approximately 75‰ of recently established populations have low allelic diversity relative to their ancestral source (Stockwell et al. 1996). Further, contemporary evolution has been reported for several allozymes (Mitton and Koehn 1975; Smith et al. 1983; Stockwell and Mulvey 1998). For example, allele frequencies at the phosphogluconate dehydrogenase (*Pgdh*) locus are correlated with salinity in native and recently introduced populations of the White Sands pupfish (*Cyprinodon tularosa*). This variation likely reflects adaptive variation across space and time (Stockwell and Mulvey 1998).

Another approach to quantifying divergence for selected markers is to infer quantitative trait loci (QTL) based on associations between phenotypic differences and particular alleles at marker loci (e.g., Hawthorne and Via 2001; Peichel et al. 2001). Unfortunately, QTL analyses have a number of limitations. From a practical perspective, they are time consuming and expensive, often requiring a breeding program coupled with extensive controlled rearing. Even if these practical problems can be overcome, however, a serious theoretical limitation is that divergence at QTLs is unlikely to reflect the divergence in the traits influenced by those QTLs (Latta 1998; Pfrender et al. 2000; Le Corre and Kremer 2003). QTL surveys may therefore provide little information about the genetic basis and adaptive significance of phenotypic variation. And so, QTLs may be of limited use in most conservation and recovery programs, which are more concerned with heritable adaptive differences than in the number, nature, and genomic locations of loci influencing those differences.

An emerging approach to the study of adaptation at the molecular level is the use of cDNA (complimentary DNA) microarrays for examining differences in gene expression between populations in different environments (reviews: Gibson 2002; Jackson et al. 2002). At present, microarrays have been developed for only a handful of model species (e.g., *Arabidopsis*) and are very costly and time consuming. For most conservation concerns, microarrays will therefore be prohibitive for the foreseeable future. Ultimately, however, this approach might provide a powerful tool for selecting populations for recovery efforts that best match the gene expression patterns appropriate for a particular restoration environment.

Quantitative Traits

Heritable variation in quantitative traits arguably provides the most direct link to evolutionarily important phenotypic patterns. By far the easiest and most widespread method for estimating such variation is to measure phenotypic traits on wild individuals. The resulting patterns will reflect a combination of genetic and environmental influences, which may interact in complicated ways. For example, phenotypic differences among individuals or populations may not have a genetic basis (e.g., James 1983) if trait expression is directly influenced by environmental factors (i.e., phenotypic plasticity). In this case, the offspring of individuals transplanted between environments may express traits that are expected to suit individuals in those environments. Alternatively, phenotypic similarity among individuals or populations may be maintained by substantial genetic differences that override environmental heterogeneity (e.g., countergradient variation: Conover and Schlutz 1995). In this case, individuals subjected to identical environments may subsequently differ dramatically in trait expression. These two examples illustrate how inferences about heritable variation are tenuous when based on phenotypic variation in the wild alone.

Genetic variation *within* populations is important because it influences the ability of a population to adapt to changing environments (Houle 1992; Bürger and Lynch 1995; García-Ramos and Rodríguez 2002; Reed et al. 2003). The influence of genetic variation on short-term evolutionary potential can be assessed by measuring the narrow-sense heritability (i.e., additive genetic variance divided by total phenotypic variance) of fitness-related traits (Roff 1997), or of fitness itself. With this information, one can predict the evolutionary response of a trait to a given intensity of selection (Box 6.1). The influence of genetic variation on long-term evolutionary potential can be considered by assessing the "evolvability" of the trait (i.e., coefficient of variation of additive genetic variance) (Houle 1992). In some cases features that are strongly correlated with fitness (i.e., under strong selection) may retain less variation for future evolutionary challenges. Hence, current adaptation and future adaptive potential can be at odds with each other.

The most direct, robust, and informative way to infer heritable adaptive differences is through the use of reciprocal transplants (O'Hara Hines et al. 2004). Differences in trait expression and overall performance (e.g., survival, growth, reproductive success) between individuals in "home" versus "foreign" environments can then be used to infer adaptive genetic differences. The optimal design of reciprocal transplant experiments has many important nuances (O'Hara Hines et al. 2004), which we do not discuss further because such experiments are often prohibitively difficult in a restoration context. A more common and tractable approach is to rear/grow individuals from different populations in controlled environments, such as a greenhouse, where conditions can sometimes be set to mimic environmental features at natural sites. Such "common-garden" experiments can reveal whether phenotypic variation within or among populations is heritable and can incorporate breeding designs that reveal the quantitative genetic architecture (e.g., additive, dominance, epistasis) of that variation.

Although such rearing experiments are useful and prevalent, they have important limitations. A seemingly obvious point, often forgotten or ignored, is that genetic variation *within* populations (e.g., significant heritability) does not mean that phenotypic differences *among* populations necessarily have a genetic basis, only that genetic variation exists for evolution of a given population. Common-garden experiments should include all populations about which inferences are to be made. Even when this can be achieved, the phenotypic expression of genetic variation will depend on the specific rearing/growing conditions (Roff 1997; Hoffmann and Merilä 1999). Accordingly, differences observed in a specific common environment may not reflect differences that would be observed under other rearing environments, or in nature. Moreover, common-garden experiments do not directly demonstrate the adaptive significance of phenotypic variation, because they do not expose organisms to the full suite of challenges they would encounter in nature.

Instead, the adaptive significance of heritable phenotypic variation among populations might be inferred by comparing *genetic* divergence in phenotypic traits (i.e., estimated in a common garden) to divergence expected under the absence of natural selection (i.e., based on mutation and genetic drift alone). When additive genetic differences *exceed* these "neutral" expectations, selection is inferred as the basis of phenotypic divergence (review: Turelli et al. 1988). At present, neutral models of trait evolution are limited in that they rarely consider gene flow, variable demographics, or founder effects. Moreover, although the rejection of a neutral model provides evidence that phenotypic patterns are the product of selection, *failure* to reject a neutral model does not provide evidence for or against anything; there is no reason selection cannot produce values consistent with neutral expectations.

A recent extension of such approaches is to estimate the neutral expectation directly from the genomes of the populations under consideration. For example, an analog of F_{ST} has been developed for quantitative traits (termed Q_{ST} by Spitze 1993):

$$Q_{ST} = \frac{\sigma_a^2}{\sigma_a^2 + 2\sigma_w^2}$$

where σ_a^2 is the additive genetic variance among populations and σ_w^2 is the additive genetic variance within populations. Neutral and selected variation within and among population are both subject to the same patterns of demography and immigration. When phenotypic traits are not under selection, Q_{ST} should approximately equal F_{ST} as estimated from neutral markers (Lande 1992; Spitze 1993; Whitlock 1999; but see Hendry 2002). Q_{ST} values greater than F_{ST} thus imply that phenotypic differences are driven by selection and are therefore adaptive.

Q_{ST}/F_{ST} comparisons have attracted considerable recent interest. A point of general agreement is that Q_{ST} often exceeds F_{ST}, suggesting that divergence in quantitative traits is the result of selection (Merilä and Crnokrak 2001; McKay and Latta 2002). Nonetheless, Q_{ST} is sometimes appreciably less than F_{ST}, implying parallel or convergent patterns of selection on some traits in some systems (e.g., Lee and Frost 2002). A point of general disagreement is whether or not F_{ST} and Q_{ST} are correlated. If they are, the easy-to-measure F_{ST} might be used to infer the difficult-to-measure Q_{ST}, which generally requires rearing experiments to estimate genetic variances. Some authors argue that these measures are correlated (Merilä and Crnokrak 2001; Crnokrak and Merilä 2002) and others, that they are not (Pfrender et al. 2000; Latta and McKay 2002; McKay and Latta 2002). Regardless, the correlation is clearly not very strong (see also Reed and Frankham 2001). F_{ST} is therefore unlikely to provide a reliable estimate of relative heritable trait divergence. Despite this limitation, Q_{ST}/F_{ST} comparisons are useful for revealing phenotypic traits that are under strong divergent selection even in the face of some gene flow.

Owing to the difficulty of formally estimating Q_{ST}, several short-cuts have been proposed. The simplest, and hence most appealing, is to assume that phenotypic variation reflects additive genetic variation. If so, Q_{ST} might be estimated simply by measuring the phenotypes of individuals captured from the wild (perhaps with some correction using known heritabilities). Although quantitative trait divergence estimated using additive genetic variation (Q_{ST}) and phenotypic variation (perhaps we can call this P_{ST} by analogy) may sometimes be equivalent, there is no fundamental expectation (see arguments above) or empirical data to support this. Additive genetic variation within populations might be estimated in the wild using phenotypic similarity and genetic relatedness or pedigree (Ritland 2000), but such approaches would not address the problem of estimating heritable variation among populations. Despite all these concerns, evidence that Q_{ST} or P_{ST} exceed F_{ST} would support the importance of environmental heterogeneity in determining trait variation over the landscape.

Application of Contemporary Evolution to Restoration Ecology

Restoration ecology includes a number of strategies that can be considered in the context of contemporary evolution. First, the temporal and spatial context of the restoration site is important. The need for restoration suggests that the site has undergone dramatic transformations. The legacy of previous land uses such as pesticide residue may continue to generate important selective pressure during and after restoration. Alternatively, organisms that persist during a cleanup may actually evolve to post-cleanup conditions (Levinton et al. 2003). As already discussed, the spatial context is also important because the site will interact spatially within its local region in terms of gene flow (or absence thereof). Here, we consider various restoration strategies in an evolutionary context.

Selection of Seed Sources

The restoration of populations to environments from which they have been extirpated requires the selection of suitable source populations (i.e., seed sources). The first order question faced in such cases may be whether to use a captive lineage derived from that site prior to extirpation or to use an exogenous source. If an exogenous source is selected, geographically proximate sources are often chosen under the assumption that these are more likely to have structure and function similar to that required at the restoration site (Jones 2003). However, adaptive divergence and geographic distance are not always well correlated. As a result, more optimal seed sources may be quite distant from the target site. It is also important to remember that populations undergoing restoration may affect surrounding populations through dispersal and gene flow, and this impact will again depend on spatial variation in selection and adaptation (Falk et al., this volume).

Rice and Emery (2003) recommend an approach in which source populations are matched to the general ecological conditions at the restoration site, but the source mixture includes populations from representative microclimates. The idea here is that the general population is sufficiently matched to manage genetic load, but that the mixture from various microclimates maximizes evolutionary potential. We feel that general rules such as these are likely to engender a false sense of security. It is quite possible that increasing the number of sources without respect to understanding their actual performance in the new habitat may simply increase the proportion of individuals that are maladapted and hence the genetic load. Selection of seed sources based on insights into patterns of selection, genetic variation, and local adaptation for the landscape under consideration may have higher odds of success. However, such information is often not readily available. We suggest that the most pragmatic alternative is to start with releases (even small scale) of a suite of likely candidate sources, and then recompose further (and perhaps larger) releases based on *empirical* insights into the best performing sources (e.g., tagging or genetic studies). If a local population takes hold, further empirical evaluations can be made to determine if continued supplementation from outside sources is likely to be beneficial or detrimental.

An additional complication is that preferred donor sites may have limited seed surplus, making it necessary to consider less preferred alternatives. Other options that have been pursued include multiple origin polycrosses, the use of related taxa or interspecific hybrids, or unrelated species that serve a similar function (Jones 2003). Each of these approaches can have important implications in the context of contemporary evolution, and each should again be evaluated *empirically* on a limited scale before risking continued long-term supplementation with the same sources. Literature on exotic and invasive species suggests that some of these options, such as releases of related taxa, carry risks of creating pestiferous invasions. Hence, issues of containment or eradication should be carefully ensured before even performing test releases.

Although captive populations are commonly used in restoration efforts and may retain the traits best suited to a restoration site, they could pose some severe drawbacks. In particular, captive populations may have limited genetic variation, suffer from inbreeding or introgression, and may be adapted to captive conditions ("domestication," see Frankham et al. 2000; Woodworth et al. 2002; Gilligan and Frankham 2003). Such adaptation need not involve intentional selection, or even obvious differences between captive and wild environments. Instead, captive populations can evolve simply due to relaxed selection pressures (Heath et al. 2003). These changes may have negative impacts on the evolution of native populations when captive lines are used for supplementation (Heath et al. 2003). For instance, relaxed selection pressure presumably selected for smaller egg size in a hatchery population of chinook salmon (*Oncorhynchus tshawytscha*) (Heath et al. 2003). Populations supplemented with large numbers of fish from this hatchery showed a reduction in egg size that theoretically could be detrimental to fitness (Heath et al. 2003).

How then might captive populations be managed to prevent unwanted evolutionary changes? Common prescriptions include (1) maintaining limited inputs from any remaining wild source, (2) continual release of captive offspring into the wild to ensure their exposure to natural selection, and (3) manipulations of the captive environment to better match the wild environment. The issue for many captive populations may thus come down to approaches for preventing their domestication and adaptation to captive conditions. Preventing evolution or adaptation is a difficult proposition for the reasons described above. Unfortunately, the evidence that these approaches are effective is limited and considerable debate exists on the role that captive populations should play in conservation and restoration. Captive propagation has been credited with the persistence and reintroduction of some populations and species (e.g., California condor, red wolf). However, supportive releases from captive lineages or even artificial propagation of otherwise wild populations have also been implicated as ineffective or damaging for recovery of remaining wild populations (e.g., salmon) (Myers et al. 2004). Ultimately, fully controlled comparisons simply do not exist to indicate whether already declining wild populations would have fared better or worse in the absence of such measures.

An interesting, but as yet untested, approach may be to mimic natural selection when implementing captive breeding programs. However, accurately determining natural selection is extremely difficult (Hersch and Phillips 2004), and mimicking this selection in captivity is likely to be equally difficult. At present, captive breeding programs usually focus on retaining or increasing genetic variation, reducing inbreeding, and removing individuals with obvious deleterious traits (color variants, deformities, etc.). As noted above, however, supplementation with exogenous sources may actually reduce population fitness if the population is already partly adapted. We suggest that practitioners consider the value of a hybrid approach that favors propagation of individuals having phenotypes/genotypes well suited for the restoration site, while specifically avoiding inbreeding (i.e., not mating with close relatives).

Inoculation, Stocking, and Natural Colonization

One philosophy of restoration is "if you build it, they will come." That is, restoration sites with suitable environmental characteristics will be naturally recolonized by appropriate species. However, the rate of natural colonization can be slow and may vary as a function of species vagility and habitat fragmentation (Maschinski, this volume). Brady et al. (2002) used mesocosms to compare the merits of inoculation versus natural colonization of experimental wetlands. They found that natural colonization resulted in higher species richness but lower species diversity (as measured by richness and evenness) and generally mimicked a natural wetland. Snails dominated inoculated mesocosms, whereas naturally colonized mesocosms were dominated by chironomids. This difference in community structure is likely to influence local selective pressures for a variety of species. For instance, snails act as intermediate hosts for a suite of parasites that also infect fish (Hoffman 1999) and so these different approaches may produce habitats that vary considerably in parasitism risk for various fish species.

The alternative strategies of introduction or colonization may also have important implications for genetic variation. Restoration efforts that proceed by introduction often result in reduced genetic variation at presumably neutral markers (Stockwell et al. 1996; Helenurm and Parsons 1997), which can compromise evolutionary potential. In contrast, Travis et al. (2002) found that naturally colonized sites had high genetic variation. These results suggest that natural colonization is preferable, when possible, but more research should be conducted to confirm these conclusions and to evaluate alternative introduction approaches.

Managing "Refuge" Populations in the Context of Contemporary Evolution

For many actively managed species, refuge populations are sometimes established as a hedge against extinction (Stockwell et al. 1996; Stockwell and Weeks 1999). This approach is especially common for desert fishes, due to the precarious nature of their habitats in the face of anthropogenic disturbance. In most cases, refuge populations are intended as "genetic replicates" of native populations, but this goal can be undermined when refuge populations face different selection pressures to which they adapt. This adaptation may increase the long-term viability of refuge populations, but also decrease their value as genetic replicates of the source population. If adaptive divergence is substantial, refuge populations may no longer possess adaptive variation suited to the original site, although they nonetheless preserve the evolutionary legacy of the lineage. Management actions designed to limit the local adaptation of refuge populations would slow the decay of genetic equivalence but might reduce sustainability of the refuge population.

An example of evolution in refuge populations is provided by White Sands pupfish (*Cyprinodon tularosa*) in New Mexico. Native populations of this species occur at Malpais Spring (brackish spring) and Salt Creek (saline river). These two populations are genetically distinct at microsatellite markers and have nearly fixed differences at an allozyme marker (Stockwell et al. 1998). Circa 1970, two populations of *C. tularosa* were established at Mound Spring (brackish spring) and Lost River (saline river) (Stockwell et al. 1998; Pittenger and Springer 1999). These refuge populations underwent subsequent changes in allele frequencies at an allozyme locus (Stockwell and Mulvey 1998) and in various aspects of body shape (Collyer et al. 2005). Thus, the Mound Spring population has lost some of its value as a genetic replicate of the Salt Creek population. We contend that refuge populations should not necessarily be managed to maintain evolutionary stasis but should instead be managed as reserves of the evolutionary legacy of species (Stockwell et al. 1998, 2003).

Summary and Research Opportunities

Recent work has suggested that contemporary evolution is common, and decades of research support the significance of natural selection, gene flow, and other evolutionary mechanisms in determining the fate of populations. Evolutionary ecology should thus be considered a central element of a more comprehensive restoration science. Indeed, the non-equilibrium conditions associated with restoration efforts are likely to promote evolutionary changes and the success or failure of such efforts is as much an evolutionary problem as an ecological one. Further, *evolutionary restoration ecology* should be considered from the genotype to the landscape scales. After all, it is the heterogeneity of the landscape that determines patterns of selection and gene flow and thus the potential distribution of individuals and their traits.

We have spent the bulk of this chapter describing potential contributions of evolutionary ecology to restoration. However, restoration activities also provide excellent opportunities for evolutionary biologists to study population genetics, natural selection, and contemporary evolution. From an evolutionary biology perspective, exciting opportunities may be afforded by the manipulative experiments in evolution posed by restoration activities. Here we identify a few topics that we think are ripe for collaborative attention.

- *The relationship between contemporary evolution and ecological function:* Seliskar (1995) reported that the genetic background of the founding stock may have profound influence on the ecological function of an introduced population (cord grass, *Spartina alterniflora*). It is also possible that subsequent contemporary evolution may alter the functional role of a species. Collaborations between functional ecologists, evolutionary biologists and restoration practitioners may provide exciting opportunities to explore the cause and effect relationships of contemporary evolution and ecosystem function.
- *Contemporary coevolution:* Species interactions, such as predation and herbivory, are important selective factors associated with contemporary evolution. This creates the potential for contemporary coevolution (e.g., herbivory/defense) or trophic evolutionary cascades (Thompson 1998). For example, selection for smaller size at maturity may lead to a reduction in gap size and a shift in diet, thus releasing (or changing) predation pressure on prey species. This change might lead to evolution in the prey species, which could have cascading effects on additional species.
- *Managed releases:* The traditional approach for reintroduction efforts is to use large numbers of individuals to maintain genetic variation. An alternative is to not only in-

troduce suitable sources, but to select phenotypes within those sources that are best matched to the new environment (Stockwell et al. 2003). This alternative may result in lower initial genetic variation but may facilitate contemporary adaptation. After all, adaptation and population productivity are the product of natural selection associated with reduction of maladaptive variation from the population. Again, the best phenotypes may be gleaned empirically by monitoring preliminary releases. These and other approaches for source selection could be compared in experiments in which multiple sites are targeted for restoration with comparable species and sources.

- *The evolutionary control of exotics:* Some investigators have advocated an evolutionary approach to the control of exotic and weedy species (Palumbi 2001; Stockwell et al. 2003). For instance, simulations by Boulding and Hay (2001) showed that high gene flow among locally adapted populations may cause population extirpation. Perhaps gene flow can be increased in situations where extinction would be desirable (e.g., of an exotic), though the risk of genetically enhancing such populations also exists. Another evolutionary control method may be the simultaneous use of multiple control agents to slow the evolutionary compensation of disease/pest organisms (Palumbi 2001). We agree that evolutionary approaches may be useful for control efforts but caution that the various options must first be evaluated in replicated experiments.
- *Reversing the evolutionary trajectory for managed species:* As indicated above, most populations are likely to evolve when faced with new patterns of selection and such selection is expected to result in some loss of genetic variation. This raises the possibility that they are no longer adapted to their native habitats. A question of critical importance for some actively managed populations may be whether and how rapidly they might evolve back toward their original condition if needed. Restoration ecologists may have unique opportunities to evaluate the reversibility of contemporary evolution by studying cases where exploitation is ceased, for example, when captive lineages are reintroduced to the wild, or where refuge populations are reintroduced to native habitats.
- *The evolution of fitness:* Most studies of contemporary evolution examine changes in just one or a few traits. This atomization of an organism's phenotype will not capture the full implications of contemporary evolution on population productivity and persistence. Much greater insights would be provided by measuring the evolution of fitness itself (Kinnison et al., forthcoming). For example, one might introduce a population to a restoration site, allow that population to adapt for several generations, and then compare its performance (survival and reproductive success) in the introduction site against more individuals from the source population. This sort of comparison would reveal the rate at which fitness evolves in restoration contexts. Such studies would provide a more comprehensive understanding of the implications of contemporary evolution for restoration ecology.

Acknowledgments

We thank D. Falk for commenting on an earlier version of this manuscript. CAS was supported by a Department of Defense Legacy Resource Grant No. DACA87-00-H-0014, administered by M. Hildegard Reiser and Jeanne Dye, 49 CES/CEV, Environmental Flight,

Holloman Air Force Base. MTK was supported by the Maine Agricultural and Forest Experimentation Station. APH was funded by a Natural Sciences and Engineering Research Council of Canada Discovery Grant.

Literature Cited

Allendorf, F., R. Leary, P. Spruell, and J. Wenburg. 2001. The problems with hybrids: Setting conservation guidelines. *Trends in Ecology & Evolution* 17:613–622.

Antonovics, J., and A. D. Bradshaw. 1970. Evolution in closely adjacent plant populations. VII: Clinal patterns of a mine boundary. *Heredity* 25:349–362.

Arnold, S. J., M. E. Pfrender, and A. G. Jones. 2001. The adaptive landscape as a conceptual bridge between micro- and macroevolution. *Genetica* 112–113:9–32.

Ashley, M. V., M. F. Willson, O. R. W. Pergrams, D. J. O'Dowd, S. M. Gende, and J. S. Brown. 2003. Evolutionary enlightened management. *Biological Conservation* 111:115–123.

Beerli, P., and J. Felsenstein. 1999. Maximum-likelihood estimation of migration rates and effective population numbers in two populations using a coalescent approach. *Genetics* 152:763–773.

Bell, M. A., W. E. Aguirre, and N. J. Buck. 2004. Twelve years of contemporary armor evolution in a threespine stickleback population. *Evolution* 58:814–824.

Berry, O., M. D. Tocher, and S. D. Sarre. 2004. Can assignment tests measure dispersal? *Molecular Ecology* 13:551–561.

Bone, E., and A. Farres. 2001. Trends and rates of microevolution in plants. *Genetica* 112–113:165–182.

Boulding, E. G., and T. Hay. 2001. Genetic and demographic parameters determining population persistence. *Heredity* 86:313–324.

Bradshaw, W., and C. Holzapfel. 2001. Genetic shift in photoperiodic response correlated with global warming. *Proceedings National Academy Science, USA* 98:14509–14511.

Brady, V. J., B. J. Cardinale, J. P. Gathman, and T. M. Burton. 2002. Does facilitation of faunal recruitment benefit ecosystem restoration? An experimental study of invertebrate assemblages in wetland mesocosms. *Restoration Ecology* 10:617–626.

Bürger, R., and M. Lynch. 1995. Evolution and extinction in a changing environment: A quantitative-genetic analysis. *Evolution* 49:151–163.

Calsbeek, R., and T. B. Smith. 2003. Ocean currents mediate evolution in island lizards. *Nature* 426:552–555.

Carroll, S. P., H. Dingle, and T. R. Famula. 2001. Genetic architecture of adaptive differentiation in evolving host races of the soapberry bug, *Jadera haematoloma*. *Genetica*. 112–113:257–272.

Collyer, M. L., J. Novak, and C.A. Stockwell. 2005. Morphological divergence in recently established populations of White Sands pupfish (*Cyprinodon tularosa*). *Copeia*. 2005:1–11.

Coltman, D. W., P. O'Donoghue, J. T. Jorgenson, J. T. Hogg, C. Strobeck, and M. Festa-Bianchet. 2003. Undesirable evolutionary consequences of trophy hunting. *Nature* 426:655–658.

Conover, D. O., and E. T. Schlutz. 1995. Phenotypic similarity and the evolutionary significance of countergradient variation. *Trends in Ecology & Evolution* 10:248–252.

Crnokrak, P., and J. Merilä. 2002. Genetic population divergence: Markers and traits. *Trends in Ecology & Evolution* 17:501.

Davies, M. S., and R. W. Snaydon. 1976. Rapid population differentiation in a mosaic environment. III: Measure of selection pressures. *Heredity* 36:59–66.

Davison, A. W., and K. Reiling. 1995. A rapid change in ozone resistance of *Plantago major* after summers with high ozone concentrations. *New Phytologist* 131:337–344.

Endler, J. A. 1980. Natural selection on color patterns in *Poecilia reticulata*. *Evolution* 34:76–91.

Etterson, J. R., and R. G. Shaw. 2001. Constraint to adaptive evolution in response to global warming. *Science* 294:151–154.

Frankham, R., H. Manning, S. H. Margan, and D. A. Briscoe. 2000. Does equalization of family sizes reduce genetic adaptation to captivity? *Animal Conservation* 4:357–363.

Franklin, I. R., and R. Frankham. 1998. How large must populations be to retain evolutionary potential? *Animal Conservation* 1:69–73.

García-Ramos, G., and D. Rodríguez. 2002. Evolutionary speed of species invasions. *Evolution* 56:661–668.

Gibson, G. 2002. Microarrays in ecology and evolution: A preview. *Molecular Ecology* 11:17–24.

Gilligan, D. M., and R. Frankham. 2003. Dynamics of genetic adaptation to captivity. *Conservation Genetics* 4:189–197.

Gomulkiewicz, R., and R. D. Holt. 1995. When does evolution by natural selection prevent extinction? *Evolution* 29:201–207.

Grant, P. R., and B. R. Grant. 1995. Predicting microevolutionary responses to directional selection on heritable variation. *Evolution* 49:241–251.

Grant, P. R., and B. R. Grant. 2002. Unpredictable evolution in a 30-year study of Darwin's finches. *Science* 296:707–711.

Guillot, G., A. Estoup, F. Mortierz, and J. F. Cosson. 2005. A spatial statistical model for landscape genetics. *Genetics* 170:1261–1280.

Hairston, N., W. Lampert, C. Caceres, C. Holmeier, L. Weider, and U. Gaedke. 1999. Rapid evolution revealed by dormant eggs. *Nature* 401:446.

Hargeby, A., J. Johansson, and J. Ahnesjö. 2004. Habitat-specific pigmentation in a freshwater isopod: Adaptive evolution over a small spatiotemporal scale. *Evolution* 58:81–94.

Haugen, T. O., and L. A. Vøllestad. 2001. A century of life-history evolution in grayling. *Genetica* 112–113:475–491.

Hawthorne, D. J., and S. Via. 2001. Genetic linkage of ecological specialization and reproductive isolation in pea aphids. *Nature* 412:904–907.

Heath, D. D., J. W. Heath, C. A. Bryden, R. M. Johnson, and C. W. Fox. 2003. Rapid evolution of egg size in captive salmon. *Science* 299:1738–1740.

Hedrick, P. W. 1995. Gene flow and genetic restoration: The Florida panther as a case study. *Conservation Biology* 9:996–1007.

Hedrick, P. W. 1999. Highly variable loci and their interpretation in evolution and conservation. *Evolution* 53:313–318.

Helenurm, K., and L. S. Parsons. 1997. Genetic variation and the reintroduction of *Cordylanthus maritimus* ssp. *maritimus* to Sweetwater Marsh, California. *Restoration Ecology* 5:236–244.

Hendry, A. P. 2002. $Q_{ST} > = \neq > F_{ST}$? *Trends in Ecology & Evolution* 17:502.

Hendry, A. P. 2004. Selection against migrants contributes to the rapid evolution of ecologically-dependent reproductive isolation. *Evolutionary Ecology Research* 6:1219–1236.

Hendry, A. P., T. Day, and E. B. Taylor. 2001. Population mixing and the adaptive divergence of quantitative traits in discrete populations: A theoretical framework for empirical tests. *Evolution* 55:459–466.

Hendry, A. P., E. B. Taylor, and J. D. McPhail. 2002. Adaptive divergence and the balance between selection and gene flow: Lake and stream stickleback in the misty system. *Evolution* 56:1199–1216.

Hendry, A. P., J. K. Wenburg, P. Bentzen, E. C. Volk, and T. P. Quinn. 2000. Rapid evolution of reproductive isolation in the wild: Evidence from introduced salmon. *Science* 290:516–518.

Hersch, E. I., and P. C. Phillips. 2004. Power and potential bias in field studies of natural selection. *Evolution* 58:479–485.

Hoffman, G. L. 1999. *Parasites of North American freshwater fishes*. Ithaca: Comstock Publishing Associates.

Hoffman, A. A., and J. Merilä. 1999. Heritable variation and evolution under favourable and unfavourable conditions. *Trends in Ecology & Evolution* 14:96–101.

Holt, R. D., and R. Gomulkiewicz. 1997. How does immigration influence local adaptation? A reexamination of a familiar paradigm. *American Naturalist* 149:563–572.

Houle, D. 1992. Comparing evolvability and variability of quantitative traits. *Genetics* 130:195–204.

Jackson, R. B., C. R. Lindar, M. Lynch, M. Purugganan, S. Somerville, and S. S. Thayer. 2002. Linking molecular insight and ecological research. *Trends in Ecology & Evolution* 17:409–414.

Jain, S. K., and A. D. Bradshaw. 1966. Evolutionary divergence among adjacent plant populations. I: Evidence and its theoretical analysis. *Heredity* 21:407–441.

James, F. 1983. Environmental component of morphological differentiation in birds. *Science* 221:184–186.

Jones, T. A. 2003. The restoration gene pool concept: Beyond the native versus non-native debate. *Restoration Ecology* 11:281–290.

Kawecki, T. J., and R. D. Holt. 2002. Evolutionary consequences of asymmetric dispersal rates. *American Naturalist* 160:337–347.

Kingsolver, J. G., H. E. Hoekstra, J. M. Hoekstra, D. Berrigan, S. N. Vignieri, C. E. Hill, A. Hoang, P. Gilbert, and P. Beerli, P. 2001. The strength of phenotypic selection in natural populations. *American Naturalist* 157:245–261.

Kinnison, M. T., and A. P. Hendry. 2001. The pace of modern life. II: From rates to pattern and process. *Genetica* 112–113:145–164.

Kinnison, M. T., M. Unwin, A. P. Hendry, and T. Quinn. 2001. Migratory costs and the evolution of egg size and number in introduced and indigenous salmon populations. *Evolution* 55:1656–1667.

Koskinen, M. T., T. O Haugen, and C. R. Primmer. 2002. Contemporary fisherian life-history evolution in small salmonid populations. *Nature* 419:826–830.

Lande, R. 1992. Neutral theory of quantitative genetic variance in an island model with local extinction and colonization. *Evolution* 46:381–389.

Lande, R. 1995. Mutation and conservation. *Conservation Biology* 9:782–791.

Lande, R., and S. J. Arnold. 1983. The measurement of selection on correlated characters. *Evolution* 37:1210–1226.

Latta, R. G. 1998. Differentiation of allelic frequencies at quantitative trait loci affecting locally adaptive traits. *American Naturalist* 151:283–292.

Latta, R. G., and J. K. McKay. 2002. Genetic population divergence: Markers and traits. *Trends in Ecology & Evolution* 17:501–502.

Le Corre, V., and A. Kremer. 2003. Genetic variability at neutral markers, quantitative trait loci and trait in a subdivided population under selection. *Genetics* 164:1205–1219.

Lee, C. E., and B. W. Frost. 2002. Morphological stasis in the *Eurytemora affinis* species complex (Copepoda: Temoridae). *Hydrobiologia* 480:111–128.

Lenormand, T. 2002. Gene flow and the limits to natural selection. *Trends in Ecology & Evolution* 17:183–189.

Lesica, P., and F. W. Allendorf. 1999. Ecological genetics and the restoration of plant communities: Mix or match? *Restoration Ecology* 7:42–50.

Levinton, J. S., E. Suatoni, W. Wallace, R. Junkins, B. Kelaher, and B. J. Allen. 2003. Rapid loss of genetically based resistance to metals after the cleanup of a Superfund site. *Proceedings of the National Academy of Sciences, USA* 100:9889–9891.

Lynch, M. 1996. A quantitative-genetic perspective on conservation issues. In *Conservation genetics: Case studies from nature,* ed. J. C. Avise and J. L. Hamrick, 471–501. New York: Chapman & Hall.

Lynch, M., and R. Lande. 1998. The critical effective size for a genetically secure population. *Animal Conservation* 1:70–72.

Macnair, M. R. 1987. Heavy metal tolerance in plants: A model evolutionary system. *Trends in Ecology & Evolution* 2:354–359.

Mallet, J. 1989. The evolution of insecticide resistance: Have the insects won? *Trends in Ecology & Evolution* 4:336–340.

Manel, S., M. K. Schwartz, G. Luikart, and P. Taberlet. 2003. Landscape genetics: Combining landscape ecology and population genetics. *Trends in Ecology & Evolution* 18:189–197.

McKay, J. K., and R. G. Latta. 2002. Adaptive population divergence: Markers, QTL and traits. *Trends in Ecology & Evolution* 17:285–291.

Merilä, J., and P. Crnokrak. 2001. Comparison of genetic differentiation at marker loci and quantitative traits. *Journal of Evolutionary Biology* 14:892–903.

Merilä, J., L. E. B. Kruuk, and B. C. Sheldon. 2001a. Cryptic evolution in a wild bird population. *Nature* 412:76–79.

Merilä, J., B. C. Sheldon, and L. E. B. Kruuk. 2001b. Explaining stasis: Microevolutionary studies in natural populations. *Genetica* 112–113:199–122.

Mitton, J. B. 1997. *Selection in natural populations.* New York: Oxford University Press.

Mitton, J. B., and R. K. Koehn. 1975. Genetic organization and adaptive response of allozyme to ecological variables in *Fundulus heteroclitus*. *Genetics* 79:97–111.

Montalvo, A. M., S. L. Williams, K. J. Rice, S. L. Buchmann, C. Cory, S. N. Handel, G. P. Habhan, R. Primack, and R. H. Robichaux. 1997. Restoration biology: A population biology perspective. *Restoration Ecology* 5:277–290.

Myers, R. A., S. A. Levin, R. Lande, F. C. James, W. W. Murdoch, and R. T. Paine. 2004. Hatcheries and endangered salmon. *Science* 303:180.

Newman, D., and D. A. Tallmon. 2001. Experimental evidence for beneficial fitness effects of gene flow in recently isolated populations. *Conservation Biology* 15:1054–1063.

O'Hara Hines, R. J., W. G. S. Hines, and B. W. Robinson. 2004. A new statistical test of fitness set data from reciprocal transplant experiments involving intermediate phenotypes. *American Naturalist* 163:97–104.

Olsen, E. M., M. Helno, G. R. Lilly, M. J. Morgan, J. Brattey, B. Ernande, and U. Dieckmann. 2004. Maturation trends indicative of rapid evolution preceded the collapse of northern cod. *Nature* 428:932–935.

O'Steen, S., A. J. Cullum, and A. F. Bennett. 2002. Rapid evolution of escape ability in Trinidadian guppies (*Poecilia reticulata*). *Evolution* 56:776–784.

Paetkau, D., R. Slade, M. Burden, and A. Estoup. 2004. Genetic assignment methods for the direct, real-time estimation of migration rate: A simulation-based exploration of accuracy and power. *Molecular Ecology* 13:55–65.

Palumbi, S. R. 2001. Humans as the world's greatest evolutionary force. *Science* 293:1786–1790.

Pearse, D. E., and K. A. Crandall. 2004. Beyond F_{ST}: Analysis of population genetic data for conservation. *Conservation Genetics* 5:585–602.

Peichel, C. L., K. S. Nereng, K. A. Ohgi, B. L. E. Cole, P. F. Colosimo, C. A. Buerkle, D. Schluter, and D. M. Kingsley. 2001. The genetic architecture of divergence between threespine stickleback species. *Nature* 414:901–905.

Pfrender, M. E., K. Spitze, J. Hicks, K. Morgan, L. Latta, and M. Lynch. 2000. Lack of concordance between genetic diversity estimates at the molecular and quantitative-trait levels. *Conservation Genetics* 1:263–269.

Pittenger, J. S., and C. L. Springer. 1999. Native range and conservation of the White Sands pupfish (*Cyprinodon tularosa*). *Southwestern Naturalist* 44:157–165.

Quinn, T. P., M. T. Kinnison, and M. Unwin. 2001. Evolution of chinook salmon (*Oncorhynchus tshawytscha*) populations in New Zealand: Pattern, rate, and process. *Genetica* 112–113:493–513.

Rasner, C. A., P. Yeh, L. S. Eggert, K. E. Hunt, D. S. Woodruff, and T. D. Price. 2004. Genetic and morphological evolution following a founder event in the dark-eyed junco, *Junco hyemalis thurberi*. *Molecular Ecology* 13:671–81.

Réale, D., A. G. McAdam, S. Boutin, and D. Berteaux. 2003. Genetic and plastic responses of a northern mammal to climate change. *Proceedings of the Royal Society of London. Series B* 203:591–596.

Reed, D. H., and R. Frankham. 2001. How closely correlated are molecular and quantitative measures of genetic variation? A meta-analysis. *Evolution* 55:1095–1103.

Reed, D. H., E. H. Lowe, D. A. Briscoe, and R. Frankham. 2003. Fitness and adaptation in a novel environment: Effect of inbreeding, prior environment, and lineage. *Evolution* 57:1822–1828.

Reznick, D. N., and C.K. Ghalambor. 2001. The population ecology of rapid evolution. *Genetica* 112–113:183–198.

Reznick, D. N., H. Rodd, and L. Nunney. 2004. Empirical evidence for rapid evolution. In *Evolutionary conservation biology*, ed. R. Ferrière, U. Dieckmann, and D. Couvet, 101–118. New York: Cambridge University Press.

Reznick, D. N., F. H. Shaw, H. Rodd, and R. G. Shaw. 1997. Evaluation of the rate of evolution in natural populations of guppies (*Poecilia reticulata*). *Science* 275:1934–1937.

Rhymer, J., and D. Simberloff. 1996. Extinction by hybridization and introgression. *Annual Review of Ecology Systematics* 27:83–109.

Rice, K. J., and N. C. Emery. 2003. Managing microevolution: Restoration in the face of global change. *Frontiers in Ecology and the Environment* 1:469–478.

Riechert, S. E., F. D. Singer, and T. C. Jones. 2001. High gene flow levels lead to gamete wastage in a desert spider system. *Genetica* 112–113:297–319.

Ritland, K. 2000. Marker-inferred relatedness as a tool for detecting heritability in nature. *Molecular Ecology* 9:1195–1204.

Roff, D. 1997. *Evolutionary quantitative genetics*. Sunderland, MA: Sinauer Associates.

Schluter, D. 2000. *The ecology of adaptive radiation*. New York: Oxford University Press.

Seliskar, D. 1995. Exploiting plant genotypic diversity for coast salt marsh creation and restoration. In *Biology of salt-tolerant plants*, ed. M. A. Khan and I. A. Ungar, 407–417. Karachi, Pakistan: University of Karachi, Department of Botany.

Slatkin, M. 1985. Rare alleles as indicators of gene flow. *Evolution* 39:53–65.

Smith, M. H., M. W. Smith, S. L. Scott, E. H. Liu, and J. C. Jones. 1983. Rapid evolution in a post-thermal environment. *Copeia* 1983:193–197.

Snaydon, R. W., and M. S. Davies. 1972. Rapid population differentiation in a mosaic environment. I: The response of *Anthoxanthum odoratum* to soils. *Evolution* 24:257–269.

Spitze, K. 1993. Population structure in *Daphnia obtusa:* Quantitative genetic and allozyme variation. *Genetics* 135:367–374.

Stearns, S. C. 1983. The genetic basis of differences in life history traits among six populations of mosquitofish (*Gambusia affinis*) that shared ancestors in 1905. *Evolution* 38:618–627.

Stockwell, C. A., A. P. Hendry, and M. T. Kinnison. 2003. Contemporary evolution meets conservation biology. *Trends in Ecology & Evolution* 18:94–101.

Stockwell, C. A., and M. Mulvey. 1998. Phosphogluconate dehydrogenase polymorphism and salinity in the White Sands pupfish. *Evolution* 52:1856–1860.

Stockwell, C. A., M. Mulvey, and A. G. Jones. 1998. Genetic evidence for two evolutionarily significant units of White Sands pupfish. *Animal Conservation* 1:213–225.

Stockwell, C. A., M. Mulvey, and G. L. Vinyard. 1996. Translocations and the preservation of allelic diversity. *Conservation Biology* 10:1133–1141.

Stockwell, C. A., and S. C. Weeks. 1999. Translocations and rapid evolutionary responses in recently established populations of western mosquitofish (*Gambusia affinis*). *Animal Conservation* 2:103–110.

Storfer, A., J. Cross, V. Rush, and J. Caruso. 1999. Adaptive coloration and gene flow as a constraint to local adaptation in the streamside salamander, *Ambystoma barbouri*. *Evolution* 53:889–898.

Suding, K. N., K. L. Gross, and G. R. Houseman. 2003. Alternative states and positive feedbacks in restoration ecology. *Trends in Ecology & Evolution* 19:46–53.

Tabashnik, B. 1994. Evolution of resistance to *Bacillus thuringiensis*. *Annual Review of Entomology* 39:47–79.

Thompson, J. N. 1998. Rapid evolution as an ecological process. *Trends in Ecology & Evolution* 13:329–332.

Travis, S. E., C. E. Proffitt, L.C. Lowenfeld, and T. W. Mitchell. 2002. A comparative assessment of genetic diversity among differently-aged populations of *Spartina alterniflora* on restored versus natural wetlands. *Restoration Ecology* 10:37–42.

Turelli, M., J. H. Gillespie, and R. Lande. 1988. Rate tests for selection on quantitative characters during macroevolution and microevolution. *Evolution* 42:1085–1089.

Waples, R. S. 1998. Separating the wheat from the chaff: Patterns of genetic differentiation in high gene flow species. *Journal of Heredity* 89:438–450.

Ward, J. K., J. Antonovics, R. B. Thomas, and B. R. Strain. 2000. Is atmospheric CO_2 a selective agent on model C_3 annuals? *Oecologia* 123:330–341.

Whitlock, M. C. 1992. Temporal fluctuations in demographic parameters and the genetic variance among populations. *Evolution* 46:608–615.

Whitlock, M. C. 1999. Neutral additive genetic variance in a metapopulation. *Genetical Research* 74:215–221.

Whitlock, M. C., and D. E. McCauley. 1999. Indirect measures of gene flow and migration: $F_{ST} \neq 1/(4Nm + 1)$. *Heredity* 82:117–125.

Williams, C. K., and R. J. Moore. 1989. Phenotypic adaptation and natural selection in the wild rabbit, *Oryctolagus cunniculus*, in Australia. *Journal of Animal Ecology* 58:495–507.

Woodworth, L. M., M. E. Montgomery, D. A. Briscoe, and R. Frankham. 2002. Rapid genetic deterioration in captive populations: Causes and conservation implications. *Conservation Genetics* 3:277–288.

Wright, S. 1969. *Evolution and the genetics of populations. The theory of gene frequencies.* Volume 2. Chicago: University of Chicago Press.

Wu, L., and A. L. Kruckeberg. 1985. Copper tolerance in two legume species from a copper mine habitat. *New Phytologist* 99:565–570.

Yeh, P. J. 2004. Rapid evolution of a sexually selected trait following population establishment in a novel habitat. *Evolution* 58:166–174.

PART TWO

Restoring Ecological Function

The desire to restore species and communities stems both from their intrinsic ecological value as well as the provision of critical ecosystem services (e.g., fisheries, timber, etc.). However, a focus on ecological processes in a restoration context provides a different view of the state and dynamics of ecosystems and the services they may provide. In pragmatic terms, measuring ecological functioning requires appraisal of key ecological processes, such as nutrient processing, productivity, or decomposition. The currency is typically a process *rate*, and it reflects system performance. Because ecosystem function may indicate important elements of system performance, environmental managers are also increasingly interested in the use of functional assessments. Thus, theoretical treatments of the link between structure and function are potentially important for both theoretical and practical reasons.

Historically, many restoration efforts have focused on single species, populations, or the composition of ecological communities. However, it is recognized increasingly that restoration of ecological processes, such as nutrient turnover or hydrological flux, may be critical components of restoration outcomes. This understanding has been paralleled by an upsurge in ecological research on the linkage between ecological structure (e.g., species diversity, habitat complexity) and ecological function (e.g., biogeochemical processes, disturbance regimes). Theoretical and empirical work focused on this linkage has grown exponentially in the last decade alone. Chapters in this section illuminate how that work can inform restoration in practice and, perhaps more important, why understanding this linkage is critical to advancing the science of restoration ecology.

Larkin, Vivian-Smith, and Zedler review the growing evidence that topographic heterogeneity not only plays an important role in influencing community composition, but it can also exert strong influence on ecological processes. The authors summarize a diverse body of theory and directly link this to the practice of restoration by examining how restoration of heterogeneity can enhance or accelerate restoration of ecological processes, constrain invasions, and facilitate recruitment. Using detailed examples and examination of underlying mechanisms, they stress how critical experimentation and additional theoretical work are to understanding the explicit link between restoration and fine-scale ecological heterogeneity.

While food-web manipulations are often not explicit in efforts to restore degraded ecosystems, Vander Zanden, Olden, and Gratton argue that understanding trophic connections may explain outcomes of restoration experiments, and how best to implement or evaluate projects. They review relevant theory including trophic cascades, direct and indirect interactions, keystone predators, biological invasions, and food-web assembly. The concept (and mathematic treatment) of top-down and bottom-up effects is especially important, because many restoration efforts involve the use of fertilizers or are targeted at the recovery of top predators. The authors provide a review of the expanding use of stable isotopes, not only to understand trophic interactions but also to document how food webs have been altered relative to reference conditions and to identify important energy sources for restoration-target organisms.

Linking theoretical models of ecosystem and community change with restoration ecology has the potential to advance both the practice of restoration and our understanding of the dynamics of degraded systems. Suding and Gross summarize theory representing three perspectives on the mechanisms and predictive nature of species turnover and ecosystem development: equilibrium, multiple-equilibrium, and non-equilibrium. Incorporating these perspectives in restoration planning may be critical, yet Suding and Gross also stress the need for experimental work to test which system characteristics indicate the presence or absence of multiple states, how to determine whether thresholds exist, and the relative strengths of different factors affecting resilience in degraded systems. The identification of ecosystem responses that operate at different scales of time and space may be critical to both theoretical and practical advances in restoration ecology.

Intensive theoretical and empirical work on the relationship between biodiversity and ecosystem functioning has taken place in the last decade. Naeem contends that this work is relevant not simply to the field of restoration ecology but is potentially a unifying foundation for the field. While community and ecosystems ecology separately have provided insights into identifying restoration targets, into selecting what ecological processes to monitor, and into designing restoration strategies, these fields typically focus *either* on restoration of populations and communities *or* on nutrient and energetic processes. Naeem points out that, from a biodiversity-ecosystem functioning perspective (BEF), ecological structure and function are inseparable —any change in a community has consequences for ecosystem functioning, and vice versa. Thus, restoration as experiments can both test and inform BEF theory, including the link between diversity and ecological stability.

Ecological modeling offers an opportunity to explore restoration outcomes. Urban makes a strong case for a model-based framework for restoration, including heuristic models similar to those discussed by Naeem or more complicated simulation models. Models can describe how we believe a system behaves and thus guide restoration efforts. Models can also help evaluate where a system is with respect to reference points and can be used to forecast various restoration scenarios. Urban reviews some of the modeling approaches that can be brought to bear on restoration, including schematic diagrams and conceptual models; statistical models, such as regression; and models implemented as numerical algorithms and "solved" by simulation. Throughout, he emphasizes the need for an adaptive framework that evolves from conceptual heuristics to data-driven models. A multivariate framework based on ordination can be appropriate for field data, heuristic models, or forecasts from simulators; consequently, ordination can serve effectively as an integrating framework for restoration ecology.

Ideally, ecological restoration efforts create physical and ecological conditions that promote self-sustainable, resilient systems with the capacity for recovery from rapid change and stress (Holling 1973; Walker et al. 2002). Natural ecological systems are both self-sustaining and dynamic, often with high variability resulting from natural disturbances. But this variability does have limits (Suding et al. 2004), and understanding the relationship between ecological processes and ecological structure in a restoration context should help define those limits. Can a system be restored, or has some threshold already been crossed? What are the mechanistic links between structure and function? Can these be understood using conceptual frameworks, such as those coming out of work on biodiversity-ecosystem functioning, or are more complex statistical and mathematical models required? Such questions about ecological function may constitute the next frontier for restoration ecology research.

Literature Cited

Holling, C. S.1973. Resilience and stability of ecological systems. *Annual Review of Ecology and Systematics* 4:1–23.

Suding, K. N., K. L. Gross, and D. R. Housman. 2004. Alternative states and positive feedbacks in restoration ecology. *Trends in Ecology & Evolution* 19:46–53.

Walker, B., S. Carpenter, J. Anderies, N. Abel, G. S. Cumming, M. Janssen, L. Lebel, J. Norberg, G. D. Peterson, and R. Pritchard. 2002. Resilience management in social-ecological systems: A working hypothesis for a participatory approach. *Conservation Ecology* 6:14.

Chapter 7

Topographic Heterogeneity Theory and Ecological Restoration

DANIEL LARKIN, GABRIELLE VIVIAN-SMITH, AND JOY B. ZEDLER

Natural ecosystems are heterogeneous; their physical, chemical, and biological characteristics display variability in both space and time. In trying to understand vegetation heterogeneity, the earliest ecologists found that species sorted out among habitats according to environmental conditions, as along lakeshore dunes of different size and age (Cowles 1899). Later ecologists acknowledged that some species act as "engineers," by creating variation that in turn affects other species (Jones et al. 1994), as in sedge tussocks (Watt 1947), ant mounds (Vestergaard 1998; Nkem et al. 2000), animal burrows (Inouye et al. 1997; Minchinton 2001), and bison and alligator wallows (Collins and Barber 1985; Gunderson 1997; Coppedge et al. 1999). Abiotic and biotic components of an ecosystem often act together to create variability, making it difficult to separate cause from effect.

Heterogeneous environments are the rule in nature, and the relevant literature is exploding. The ISI Web of Knowledge (Thomson Corporation 2004) shows only nine papers matching the search criteria "heterogene* and ecolog*" from 1970 to1990. The same search for just 2000–2004 returns 1,343 papers. Ecologists urge consideration of heterogeneity in designing reserves (Dobkin et al. 1987; Miller et al. 1987), maintaining threatened species (Fleishman et al. 1997), and preserving ecosystem functions (Ludwig and Tongway 1996). Understanding how habitat heterogeneity regulates the structure and function of biotic communities is one of the most important challenges in modern ecology (Cardinale et al. 2002). Here, we limit our review to topographic heterogeneity (see Box 7.1), which has long been recognized as a key ecological variable (e.g., Watson 1835), and which has important applications in ecological restoration.

A human tendency is to homogenize landscapes, for example, through agriculture (Whisenant et al. 1995; Paz González et al. 2000), forestry (Krummel et al. 1987; Mladenoff et al. 1993), flood control (Koebel 1995), and landscaping. Yet humans can also increase topographic complexity. Restorationists could manipulate sites to create topography ranging from uniform and flat to complicated and spatially variable, although incorporating heterogeneity beyond that in reference systems might not be beneficial, and there are examples of detrimental increases in topographic heterogeneity (e.g., eroded gullies) (Leopold 1999). To manipulate topography for effective restoration, key questions need to be answered:

Box 7.1
Defining and Measuring Topographic Heterogeneity

Topographic heterogeneity is defined as pattern in elevation over a specific area. It is a type of spatial heterogeneity defined as patterns in the distribution of elements in a biotic assemblage (Armesto et al. 1991) or rates of a process (Smith 1972).

The origins of topographic heterogeneity include geologic processes (e.g., hills, cliffs); the movement of water over a landscape (channels, banks, sandbars); wind (dunes, blowouts); waves (beaches of sand, cobble, or rock); and biotic activity (ant mounds, tussocks). The cumulative effect of these processes is a complex three-dimensional landscape that exerts tremendous influence on the composition and function of ecological systems. Microtopographic differences (<1 m) can result in variability in the physical environment and chemical and biological processes over small spatial scales, whereas macrotopography (hills and mountains) can alter the local climate. While macrotopography is generally beyond the scale that can be manipulated for restoration, roughness at the centimeter scale (as in hummocks and hollows of bryophyte bogs) or several-meter scale (as in dunes and swales) is relevant to consider during restoration planning and implementation.

Quantification of topographic heterogeneity must take into account characteristics that are both vertical (e.g., elevation minima, maxima, means, and variance) and horizontal (e.g., frequency and distribution of elevation fluctuations, clustering, density, etc.). Auto-levels, total stations, and high-accuracy global positioning systems facilitate detailed measurements along transects, a grid, or a cyclic sampling design (see Clayton and Hudelson 1995). Even if time and budget constraints limit quantitative analyses, written descriptions (e.g., ridge versus mound, apparent origins, etc.); key measurements (e.g., highest point, lowest point, etc.); and photographs or field sketches are still recommended, as they will inform future restoration efforts.

Literature Cited

Armesto, J. J., S. T. Pickett, and M. J. McDonnell. 1991. Spatial heterogeneity during succession: A cyclic model of invasion and exclusion. In *Ecological heterogeneity*, ed. J. Kolasa and S. T. Pickett, 256–269. New York: Springer-Verlag.

Clayton, M. K., and B. D. Hudelson. 1995. Confidence-intervals for autocorrelations based on cyclic samples. *Journal of the American Statistical Association* 90:753–757.

Smith, F. E. 1972. Spatial heterogeneity, stability and diversity in ecosystems. In *Growth by intussusception; Ecological essays in honor of G. Evelyn Hutchinson*, ed. G. E. Hutchinson and E. S. Deevey, 309–335. New Haven: Connecticut Academy of Arts and Sciences.

- What topographic patterns will facilitate the reestablishment and persistence of desired ecosystem structure and function?
- Will mimicking the topography of a reference ecosystem be sufficient for reestablishing species, or did the current topography develop after organisms modified a different starting condition?
- How much topographic variability needs to be in place for "nature" to provide the rest?
- How much topographic heterogeneity is too much?
- What are the costs (in time, money, and resources) of incorporating topographic heterogeneity into restoration sites?

FIGURE 7.1 A consequence of smoothly mounded islands in a tidal marsh restoration site (dredge spoil excavation in San Diego Bay) was that salts were "wicked" to the surface, and substrate was too saline for plant establishment. This site was graded and made tidal in 1984; the higher elevations of the islands, although within the tidal range, remain unvegetated to date (2004).

- Are there key "ecosystem engineers" (Jones et al. 1994) that need to be introduced first to help structure topography and facilitate establishment of other species?
- Will topographic heterogeneity facilitate sustainability of restoration sites, given their open-ecosystem and non-equilibrium conditions (cf. Pickett and Parker 1994)?

Conversely, one can ask how practices that foster homogeneity affect restoration outcomes (e.g., oversimplified topography) (Figure 7.1).

Current Theory and Testable Hypotheses

Theory about the relationship of topographic heterogeneity to ecosystem structure and function comes largely from theory on habitat diversity. Ecologists tend to agree that habitat diversity enhances species diversity. Brose (2001) viewed area as a "surrogate variable for habitat heterogeneity, which directly enhances vascular plant species diversity." Others hold that heterogeneity contributes to the persistence of entire ecological systems (Wu and Loucks 1995). Levin (1976) proposed that spatial heterogeneity gives rise to a mosaic of locally stable communities, a multiple equilibrium dynamic. Tessier et al. (2002) proposed that heterogeneity can promote diversity by maintaining habitats in a state of non-equilibrium. The apparent contradiction between multiple stable equilibria and non-equilibrium conditions may have more to do with the use of different definitions and scales of observation than with fundamentally different ecological processes (Wu and Loucks 1995). For more about stability and multiple states, see Suding and Gross, this volume.

In community ecology, heterogeneity provides a variety of ecological niches, where a niche is the "N-dimensional hypervolume" that characterizes the spatial distribution and function of a species (Hutchinson 1957). Each "dimension" is composed of the range of an important environmental factor that a species can tolerate. Combined, the total of such dimensions (some number N) constrain a species' distribution (Hutchinson 1957). With more topographic heterogeneity, one expects greater diversity of distinct niche spaces, thus enhancing diversity by facilitating species coexistence (see Jeltsch et al. 1998), as well as greater physical space per unit area.

An interesting feature of many landscapes is that their topographic heterogeneity mediates environmental processes. Sharitz and McCormick (1973) described how slight depressions in barren, granite mountain outcroppings accumulate sediments because of their wind-sheltering capacity, allowing cushion plants to establish. In marine systems, the heterogeneity of coral reefs creates zones of hydrodynamic convergence where debris and organisms are sieved from prevailing currents, increasing meiofaunal density and diversity (Netto et al. 1999). In rangeland restoration, surface soil roughness can increase the capture of water and other limiting resources and slow erosion, accelerating vegetation recovery (Whisenant 1999; Whisenant 2002). These and other examples highlight the importance of viewing heterogeneous topography not just as a static landscape feature but as a "driver" of habitat conditions.

Another perspective on topographic heterogeneity comes from fractal theory (Mandelbrot 1983). Fractals have some degree of self-similarity at multiple scales, though they need not be self-similar at all scales (Williamson and Lawton 1991; Halley et al. 2004). Examples of topographic features that may be fractal are large rivers that fan out into convoluted deltas. A tidal marsh in such a delta might reveal a creek network similar in geometry to the delta itself. Within the marsh, centimeters-wide rivulets mimic this geometry at a still-smaller scale. Topographic convolutions that increase the fractal dimension of habitat may have important ramifications for organisms. Because fractal surfaces disproportionately increase as measurement units decrease, smaller animals experience more absolute and relative space than larger animals (Morse et al. 1985; Williamson and Lawton 1991). Morse et al. (1985) predicted a 560- to 1,780-fold increase in abundance for an order of magnitude decrease in body size from this effect. Palmer (1992) used a simulation model to show that increasing the fractal dimension of a habitat allows more species to coexist. Fractal topographic heterogeneity might enhance diversity by making room for a wider range of organisms.

In this chapter, we propose a theory of topographic heterogeneity: all else being equal, areas with more heterogeneous topography will have greater surface space, environmental variability, and fractal dimensions. The resultant increase in niche space should enhance species diversity and act as a key driver of community structure and ecosystem processes and functions. As such, topographic heterogeneity should be a central component of restoration planning.

Effects of Topographic Heterogeneity

Topographic heterogeneity is known to affect several classes of response variables, including (1) abiotic patterns and ecosystem processes; (2) distributions of organisms; (3) genetic, reproductive, and developmental attributes; and (4) animal habitat use, behavior, and trophic interactions. Following a review of these relationships, we summarize the research of restoration ecologists who have tested the effects of topographic heterogeneity on ecosystem structure and function.

Abiotic Patterns and Ecosystem Processes

Variability in elevation exerts a strong influence on edaphic conditions. In a dune-grassland, Gibson (1988) found that soil under hummocks was moister, more acidic, and higher in conductivity than under hollows, with higher levels of root biomass, organic carbon, and key nutrients. In a deciduous woodlot, low microtopographic positions had greater litter accumulation, lower moisture loss and temperature variability, and three-and-a-half times as many bacteria as level or high sites (Dwyer and Merriam 1981). Cantelmo and Ehrenfeld (1999) found that the intensity of mycorrhizal infection differed on tops, sides, and bottoms of hummocks in Atlantic white cedar swamps. Topographic depressions in a dry tropical forest had more fine roots to support tree growth and maintained fairly high productivity in an area with leached, shallow, and nutrient-poor soils (Roy and Singh 1994). In a variety of herbaceous plant communities, topographic variability induces differential moisture stress, waterlogging, and redox conditions (Pinay et al. 1989; Ehrenfeld 1995a; Li et al. 2001; Werner and Zedler 2002).

Cardinale et al. (2002) tested the effect of fine-scale vertical and horizontal heterogeneity by manipulating variation in the size of stream bed sediments and found immediate impacts on benthic algae and biofilms. Benthic respiration rates were 65% higher in high- versus low-heterogeneity riffles, and benthic biofilms had 39% higher gross productivity. The authors attribute these effects to altered near-bed flow velocity and turbulence due to substrate heterogeneity.

Topographic heterogeneity can also affect gas fluxes, an area of interest in global change studies. Bubier et al. (1993) found that seasonal mean water table position, especially at the microtopographic scale of hummock and hollow, explained most of the variability in methane emissions among wetlands ($R^2 = 0.74$), likely through effects on soil temperature and moisture. Generalizations that do not account for differences at the microtopographic scale inadequately predict methane flux in boreal wetlands and in global methane budget models (Bubier et al. 1993). In a Minnesota peatland, rates of carbon dioxide emission from hummocks were consistently higher than those from hollows, apparently due to peat temperature and water table depth covariates (Kim and Verma 1992). In fertilized grassland and winter wheat plots, highest nitrous oxide emissions were found in areas lying below the average slope, which is attributed to differences in air permeability and nitrate, ammonium, and soil water content (Ball et al. 1997). These examples of effects on methane, carbon dioxide, and nitrous oxide suggest that researchers adjust sampling regimes to characterize small-scale variations adequately.

Distribution of Organisms

Topographic variability can act as an important control on the distribution of organisms over fine to coarse spatial scales. In aquatic communities, this is often through mediation of hydrodynamic processes. For example, Guichard and Bourget (1998) linked the biomass and diversity of intertidal macrobenthos on rocky shores to a spatial cascade that related topographic heterogeneity, hydrodynamics, and community structure. Likewise, topographically controlled hydrodynamic fronts were shown to affect the distribution and diversity of benthic invertebrates in a South Atlantic reef, with hydrodynamic control of particle size distribution and organic matter content leading to increased meiofaunal diversity and complex effects on macrobenthos (Netto et al. 1999). At a fine scale, meiobenthic nematodes were more abundant in the sediment of crests than troughs of 8-cm-wide ripples (Hogue and Miller 1981).

Both terrestrial and wetland communities are subject to topographically controlled spatial structuring. Several examples point to the importance of topographic variability in providing "safe sites" that enhance germination and establishment (Smith and Capelle 1992). Both mounds and hollows have been shown to function as recruitment microhabitats. For example, Collins and Barber (1985) found that hollows created by bison wallowing and prairie dog mounds interacted with fire and grazing to increase habitat heterogeneity and maximize community diversity. In a mangrove forest, more propagules were dispersed to flats than to crab-burrow mounds, yet seedling establishment and sapling abundance were greater on mounds (Minchinton 2001). Likewise, in coastal meadows, anthills and hillocks have been found to be associated with higher plant species richness (Vestergaard 1998). In a controlled field experiment using wetland plants, microtopography promoted establishment of rarer woody species and floristic diversity (Vivian-Smith 1997). *Carex stricta* tussocks present in sedge meadows were shown to increase diversity in three ways: by adding surface area, providing diverse micro-habitats, and supporting progression in species composition during spring warming and late-summer drawdown (Peach 2005).

Patterns of vegetation and elevation are especially strong in intertidal wetlands, where one or a few "low marsh" species occur next to the water's edge and a richer mixture of "high marsh" species occurs further inland in salt marshes worldwide (Adam 1990). Within a meter's rise in elevation, the daily rise and fall of tidewaters creates conditions that range from frequently inundated and relatively buffered from environmental variability at the low end to intermittently wet and highly variable in soil moisture and soil salinity at the high end. In a near-pristine coastal marsh in Baja California, Mexico, 16 halophytes display individual distributions in relation to elevation (Zedler et al. 1999) (see Figure 7.2). One species' distribution, however, responded to both vertical and horizontal heterogeneity. *Spartina foliosa* occurred only in the bayward margin of the wetland; it was absent in areas of appropriate elevation farther inland (Zedler et al. 1999). Similarly, four marsh-plain species of San Diego Bay salt marshes were strongly influenced by both elevation and proximity to tidal creeks, possibly due to better drainage at tidal creek margins (Zedler et al. 1999). Patterns of *Spartina patens* physiology within Gulf of Mexico coastal marshes likewise display interactions between vertical and horizontal topographic variations. Plants in better-drained swale and dune habitats generally respired aerobically while anaerobic metabolism was important in plants in lower, more poorly drained marsh habitat (Burdick and Mendelssohn 1987).

Topographic heterogeneity can enhance diversity by protecting plants from competitive exclusion by dominant species (Tessier et al. 2002) or herbivory (Lubchenco 1983). In Oklahoma mixed-grass prairie, increased spatial heterogeneity associated with trampling by ungulates and bison wallows increased species richness by limiting dominance of the most

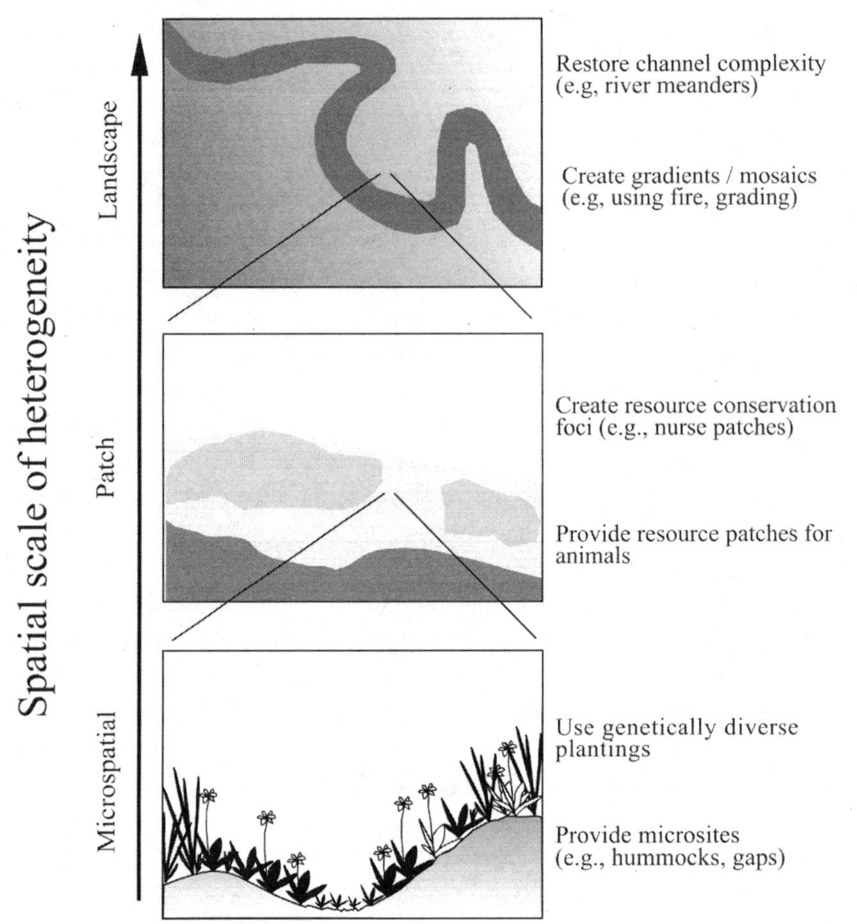

FIGURE 7.2 In this conceptual plan for a large restoration site, topography would be manipulated at three spatial scales. At the landscape scale, a channel would be excavated with a meander that mimics a natural river. Within the meander, patches of different elevations would support mesoscale vegetation patterns. At the microscale, hummocks and gaps favor different plant species. (From Zedler 2001, with permission from CRC Press.)

common "core" species (Collins and Glenn 1990). In a drained Wisconsin marsh, invading prairie species took over higher microtopographic positions, but some wetland species were able to persist in depressions (Zedler and Zedler 1969). In southern Wisconsin sedge meadows, species richness was positively correlated with surface area of *Carex stricta* tussocks (Werner and Zedler 2002). In areas where the tussock microtopography was reduced by sediment accumulation, monotypic stands of reed canary grass (*Phalaris arundinacea*) replaced species-rich mixes (Werner and Zedler 2002). In a rocky intertidal zone, microtopography protected *Fucus* germlings from grazing by periwinkle snails, providing spatial refugia that allowed *Fucus* to escape herbivore control (Lubchenco 1983).

Over very large scales, topography can be an important variable in regional diversity. Topographic heterogeneity has been identified, along with local variation in energy availability, as the best predictor of mammal species richness over most of the continental United States and southern Canada (Kerr and Packer 1997) and second to potential evapotranspiration in predicting richness of North American Papilionidae (swallowtail butterflies) (Kerr et al. 1998). In the Flooding Pampa grasslands of Argentina, latitude played a secondary role to fine-scale topographic features and associated salinity gradients in determining regional species composition (Perelman et al. 2001).

Topographic heterogeneity also influences community composition through interactions with disturbances. In coastal Alaskan wetlands, the spatial variation in flood frequency (SVFF) was explained by microtopographic variation. Sites with high SVFF and intermediate flood frequency were species rich, while sites with frequent, rare, or permanent flooding and low SVFF were species poor (Pollock et al. 1998). In coastal meadows of Queensland, Australia, seed densities were greatest in depressions where vegetation had been removed by disturbance from the foraging of large marine vertebrates and turbulent water currents (Inglis 2000). In New Jersey pinelands, fire reduced microtopographic heterogeneity, but blowdowns increased it (Ehrenfeld 1995b).

Such topography-altering wind disturbances have been shown to influence recruitment. Carlton and Bazzaz (1998) pulled down trees in a southern New England woodland to simulate the effects of a hurricane. They found that seeds accumulated in pits, the majority of seedlings germinated on scarified level areas, and all tree species achieved maximum growth on tip-up mounds. In a southern Appalachian forest, the uprooting of trees by hurricane Opal (1995) created pit and mound microtopography that may have contributed to a greater mix of shade-intolerant and shade-tolerant herbaceous species and higher species richness than were found in undisturbed forest (Elliott et al. 2002).

Genetic, Reproductive, and Developmental Attributes

Several studies have shown that topographic heterogeneity can affect plant and animal reproduction and genetics. Li et al. (2001) sampled allele frequencies and levels of genetic diversity in wild emmer wheat (*Triticum dicoccoides*) in four topographically distinct habitats. They found evidence of adaptively directed allozyme diversity and divergence, effects they attributed to differences in water stress among the different landscape positions. Water stress was also implicated in a study that found populations of the red kangaroo (*Macropus rufus*) to have more restricted genetic populations in areas with greater topographic complexity (Clegg et al. 1998).

In addition to genetic divergence, topography can also influence reproductive success. For example, female plants of *Atriplex canescens* on steep slopes produced lower mass of fruit and lower mass of fruit per gram of leaf tissue than plants on the bases of slopes. In contrast, the efficiency of pollen dispersal was enhanced by growth on steep slopes (Freeman et al. 1993).

Reproductive effects of topographic heterogeneity have also been identified in insects. For larvae of checkerspot butterflies (*Euphydryas editha*), the temporal and spatial pattern of senescence of host plants is determined largely by differences in microclimate resulting from topographic heterogeneity (Dobkin et al. 1987). Such microclimatic differences also affected larval development rates of the federally threatened Bay checkerspot butterfly, *E. editha bayensis* (Fleishman et al. 1997). Reproductive success of the aquatic mayfly *Baetis bicaudcatus* is influenced by streambed heterogeneity, in that the availability of rocks protruding above the water surface appears to limit oviposition and egg survival (Peckarsky et al. 2000).

Animal Habitat Use, Behavior, and Trophic Interactions

Topographic attributes have been shown to affect a variety of animal behaviors and interactions. In a coastal heathland, elevation was a highly significant variable in the habitat selection among two native rodent species (Haering and Fox 1995). In an experiment that manipulated aquatic topographic complexity, Aronson and Harms (1985) found that suspension-feeding ophiuroids (*Ophiothrix oerstedii*) remained in plots with artificially enhanced microtopography while emigrating from unmanipulated control plots, apparently a predator-avoidance behavior. In an experiment that manipulated topographic complexity in temperate-zone shallow water, juvenile reef fish suffered 100% mortality in low-complexity treatments compared to only 13% in high-complexity treatments (Connell and Jones 1991). In another study, prey mortality was intense and constant in a rocky intertidal area with low substratum complexity but spatially and temporally variable where holes and crevices were abundant (Menge et al. 1985).

In salt marshes, small depressions provide habitat for aquatic animals. In New Jersey, loss of intertidal pools due to *Phragmites* invasion is thought to reduce macroinvertebrate biodiversity at small scales (Angradi et al. 2001) and is also implicated in reduced nursery, reproduction, and feeding functions for fishes (Able et al. 2003). At a salt marsh restoration site in San Diego, where floral and faunal establishment have been constrained by hypersalinity and sedimentation (Zedler et al. 2003), microtopographic depressions collect tidal water and appear to act as oases of elevated invertebrate diversity and abundance and as nursery habitats for juvenile fishes (D. Larkin, unpublished data).

Effects of Heterogeneity in Restoration Sites

Restoration projects provide opportunities to alter the topography of landscapes (as in breaching levees and rewetting fields along the Illinois River and within the San Francisco Bay Delta), as well as individual sites (e.g., introducing mounds and depressions and reestablishing dunes) and habitats within sites (e.g., leaving rough instead of smoothly graded surfaces). While experimentation is difficult at large scales, there are many opportunities to establish field experiments that test the effects of topographic heterogeneity within restoration sites. Indeed, restoration ecologists have begun to test various ways that heterogeneity might enhance the stability and resilience of restored, relative to perturbed, sites through diversification of their "ecological portfolios" (Tilman et al. 1998).

Given the variety of cause-effect relationships reviewed above, restorationists could expect many variables to respond to increased topographic heterogeneity. However, rapid assessments and monitoring programs might not detect shifts in community structure until differences between reference and heterogeneity-manipulated treatment sites are extreme. In such cases, ecosystem functions (e.g., production, respiration, decomposition) might be able to measure local-scale changes better than structural attributes (Brooks et al. 2002).

Manipulations of topographic heterogeneity need to be tailored to a restoration site's size, its relationship to the surrounding landscape, and the specific goals of the restoration project (Figure 7.2). The actions needed also depend on how much heterogeneity has been lost. At slightly degraded sites, such as recently drained prairie pothole wetlands, there may be little reason to manipulate topography. In such cases, restoration may simply require removing drainage structures and reestablishing wetland species. At heavily degraded sites, the rationale for manipulating heterogeneity is greater. For example, dredge spoils could be excavated in ways that mimic the topographic complexity of natural marshes, rather than creating oblong pits as in some mitigation projects (Zedler and Callaway 1999). Heterogeneity is best incorporated during the initial construction phase of a project to avoid damaging plants, animals, and processes that might be in place at a later stage.

Here we consider the ramifications of topographic heterogeneity on the restoration of ecological processes across multiple spatial scales (Figure 7.2; Table 7.1). At the macroscale, topography can be manipulated to improve the functioning and diversity of varied habitats connected across a landscape. At the mesoscale, objectives could address species interactions and functional aspects of patch dynamics. At the microscale, soil and aquatic sediment surfaces can be roughened to restore diversity to community composition. We begin by reviewing restoration research at the coarse spatial scale.

Enhancing Landscape Functions

Landscape-level processes, such as movement of organisms, trophic interactions, and fluxes of materials through an ecosystem, are influenced by aspects of large-scale heterogeneity that include variation in the size, shape, edge characteristics, distribution, and connectivity of patches (Turner 1989). For example, forest boundary characteristics (concave versus convex) can affect small-scale seedling establishment patterns on adjacent mines (Hardt and Forman 1989), and riverine sinuosity at a regional scale can affect processes ranging from nutrient cycling to invertebrate abundance (Koebel 1995).

Various strategies aim to restore ecosystem structure and function at the landscape level by restoring natural heterogeneity. Mladenoff et al. (1993) recommended using patterns of landscape-level heterogeneity and spatial complexity in old-growth forest as a model for restoring old-growth forest function in managed landscapes. For tallgrass prairie restoration, Howe (1994) suggested using a variety of management actions to create spatial variation in plant composition, thereby enhancing diversity and creating conditions more similar to those under which tallgrass species evolved.

While benefits from large-scale manipulations of topographic heterogeneity in restoration might be inferred from the literature on landscape heterogeneity, examples of such manipulations are few. Some relevant examples come from river restoration, where management actions that reconnect and reconfigure channels, alter floodplain habitats, and restore meanders (Kern 1992; Toth et al. 1993; Gore and Shields 1995; Stanford et al. 1996) can be viewed as altering topographic heterogeneity over large spatial scales (though the vertical range of these actions may be on the order of only meters or tens of meters). Efforts to reestablish complex tidal creek networks in salt marshes are also relevant to landscape functioning, as they can occur over large spatial scales and enhance connectivity, movement of organisms, and trophic transfer between habitats (see Box 7.2).

TABLE 7.1

Summary of findings and recommendations concerning topographic heterogeneity in a restoration context, organized by spatial scale.

Scale and system	Topographic feature	Finding or recommendation	References
Macro			
Rivers	Restored channels, floodplains, and meanders	Reestablish natural geomorphological features to support biodiversity and enhance river functioning	Kern 1992; Toth et al. 1993; Gore and Shields 1995; Stanford et al. 1996
Salt marsh	Tidal creeks	Creeks conferred bioenergetic advantages to fish by increasing habitat connectivity	Madon et al. 2001; West et al. 2003
Meso			
Salt marsh	Tidal creeks	Creeks provided fish nursery habitat	Desmond 1996
Salt marsh	Tidal creeks	Tidal creeks increased plant diversity	Zedler et al. 1999
Salt marsh	Tidal creeks	Restoring clapper rail nesting habitat requires more complex creek networks with low edges for appropriate *Spartina foliosa* growth	Haltiner et al. 1997
Rangelands	Constructed pits and depressions	Increased the primary productivity and carrying capacity by slowing runoff and increasing infiltration	Slayback and Cable 1970; Hessary and Gifford 1979; Garner and Steinberger 1989
Woodlands	Patches of branches	Elevated soil nutrients, rates of soil accumulation, and water infiltration; buffered soil temperatures; and enhanced perennial grass and ant populations	Ludwig and Tongway 1996
Desert	Constructed pits and mounds	Higher plant species richness, biomass, and density due to better germination and establishment conditions and greater seed-trapping capability	Boeken and Shachak 1994
Semiarid to arid lands	Microcatchments	Depressions or "microcatchment basins" capture water, increase soil nutrients, improve seedling establishment and growth, and accelerate revegetation	Whisenant et al. 1995; Whisenant 2002
Micro			
Abandoned pasture	Small artificial mounds	Higher species richness with mounds that simulated the natural disturbance of burrowing animals	Reader and Buck 1991
Prairie	~20 cm mounds	Effect on growth and survival of plants was positive for three species (neutral or negative for two others)	Ewing 2002

Oak and herbaceous communities	Rabbit and artificially created mounds	Natural mounds increased plant survival, shoot and root biomass, root length, number of tillers, mycorrhizal infection, and nutrient uptake. Artificial mounds more effective in herbaceous vegetation than oaks	Dhillion 1999
Forested wetland	~0.6 m mound-pool topography	Recommend creating mound-pool topography as essential component of forested wetlands restoration	Barry et al. 1996
Former farms restored to wetland	~0.6 m ridge-furrow topography	Higher water tables, reduced outflow and peak outflow rates, and increased duration of outflow	Tweedy and Evans 2001
Freshwater wetland	1–3 cm microtopography	Heterogeneous treatments had higher plant species richness and evenness and higher abundances of most species	Vivian-Smith 1997
Streams	Large woody debris	Recommend patches of large woody debris for organic matter retention and biotic recovery	Hilderbrand et al. 1998
Streams	Woody debris, rock weirs, gravel, etc.	Recommended to improve streambed morphology and increase local heterogeneity	Gore and Shields 1995; Jungwirth et al. 1995; Muhar 1996
Salt marsh	Marsh plain, dune, swale microhabitats	*Spartina patens* genetically differentiated among microhabitats	Silander 1985
Salt marsh	Creek-bank and back-marsh areas	*Spartina alterniflora* from different parts of the marsh genetically distinct	Gallagher et al. 1988

Box 7.2

Restoration Case Study—Southern California Salt Marshes

The need to restore topographic heterogeneity to salt marshes is being tested in Tijuana Estuary, a National Estuarine Research Reserve. The topographic heterogeneity of tidal creeks is considered to confer important ecosystem functions and is associated with increased plant diversity (Zedler et al. 1999) and provision of fish nursery habitat (Desmond 1996). Early salt marsh restoration projects lacked heterogeneity, having smooth marsh plains and highly simplified creeks that are wider, deeper, and less meandering than natural creek networks (Williams and Zedler 1999).

Restoring tidal creek networks should aid salt marsh restoration by providing optimal habitat for tall-growing cordgrass (*Spartina foliosa*). This species is often distributed along tidal creeks, with taller plants found at lower elevations along creek edges (Haltiner et al. 1997). Tall cordgrass provides nesting habitat for an endangered bird, the light-footed clapper rail, *Rallus longirostris levipes* (Zedler 1996). Restoring suitable clapper rail nesting habitat will require improving soil quality and construction of more complex tidal creek networks with low elevation edges (Gibson et al. 1994; Boyer and Zedler 1996; Haltiner et al. 1997).

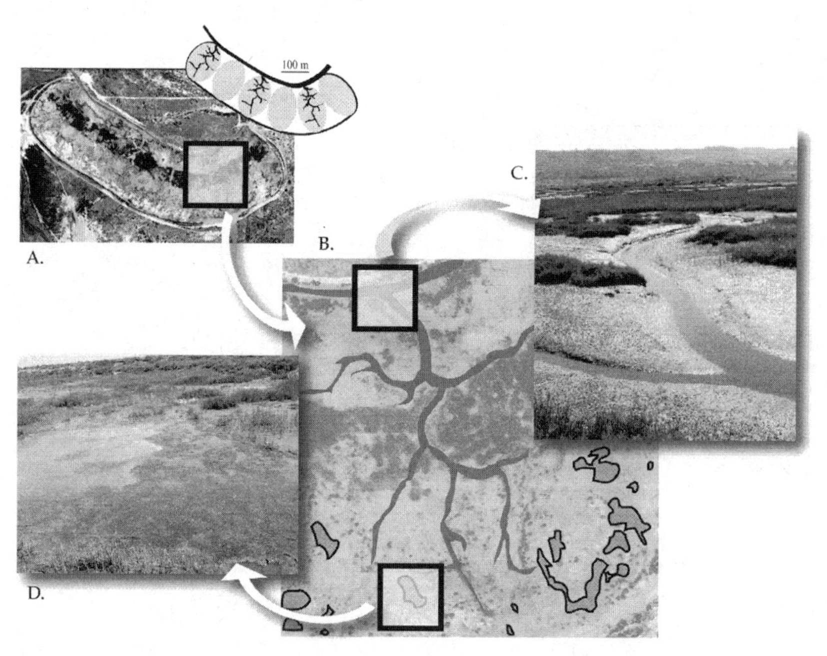

FIGURE 7.3 The Friendship Marsh in Tijuana Estuary, California, is an 8-ha salt marsh restoration site being used to test the ecosystem effects of vertical and horizontal heterogeneity. (A) A conceptual drawing and 2004 aerial photo showing the site's three >1-ha replicates with tidal creek networks and three replicates without tidal creek networks. (B) An experimental tidal creek network and microtopographic pools are highlighted. (C) Lower reach of one of the three tidal creek networks. (D) Naturally formed pool with abundant algae.

Fish habitat mitigation projects along San Diego Bay and Anaheim Bay have lacked the heterogeneity of natural habitats (Zedler 2000). Of particular concern is the lack of connectivity with salt marshes and associated tidal creeks. Where creeks are lacking, fish have less access to food sources on the marsh plain (Kwak and Zedler 1997; Madon et al. 2001; West et al. 2003).

A large-scale experimental test of the role of heterogeneity in salt marshes is being conducted in the "Friendship Marsh," an 8-ha restored site in Tijuana Estuary (Vivian-Smith 2001). Replicate sections of the marsh were constructed with and without tidal creek networks (Figure 7.3). Constructed tidal creeks are modeled on nearby natural creeks in their length, width, orders, and drainage density. Comparing community development and ecosystem functions between homogeneous and heterogeneous areas will provide vital information on the ecological role of tidal creek complexity in salt marshes. Preliminary results on fish feeding indicate that creek-driven increases in marsh access confer important bioenergetic advantages (Madon et al. 2001, unpublished data). Ongoing research is testing the effects of two scales of topographic heterogeneity (tidal creek networks and marsh plain microtopography) on trophic dynamics. These forms of heterogeneity appear to have important effects on patterns of vegetation recruitment, as well as algal, invertebrate, and fish diversity and abundance (O'Brien 2003; D. Larkin, pers. obs.).

Even this mesoscale project required considerable planning and integration of ideas among various stakeholders. The 10-year lag between the idea and its implementation included a lengthy and costly planning procedure, and excavation and sediment disposal alone cost $3.1 million (J. King, California State Coastal Conservancy, pers. comm.).

Literature Cited

Boyer, K. E., and J. B. Zedler. 1996. Damage to cordgrass by scale insects in a constructed salt marsh: Effects of nitrogen additions. *Estuaries* 19:1–12.

Desmond, J. S. 1996. Species composition and size structure of fish assemblages in relation to tidal creek size in southern California coastal wetlands. MS thesis, San Diego State University, San Diego.

Gibson, K. D., J. B. Zedler, and R. Langis. 1994. Limited response of cordgrass (*Spartina foliosa*) to soil amendments in a constructed marsh. *Ecological Applications* 4:757–767.

Haltiner, J., J. B. Zedler, K. E. Boyer, G. D. Williams, and J. C. Callaway. 1997. Influence of physical processes on the design, functioning and evolution of restored tidal wetlands in California (USA). *Wetlands Ecology & Management* 4:73–91.

Kwak, T. J., and J. B. Zedler. 1997. Food web analysis of southern California coastal wetlands using multiple stable isotopes. *Oecologia* 110:262–277.

Madon, S. P., G. D. Williams, J. M. West, and J. B. Zedler. 2001. The importance of marsh access to growth of the California killifish, *Fundulus parvipinnis,* evaluated through bioenergetics modeling. *Ecological Modelling* 136:149–165.

O'Brien, E. 2003. Starting from scratch: A three-factor approach to accelerate vegetation development at a southern California salt marsh restoration site. MS thesis, University of Wisconsin–Madison.

Vivian-Smith, G. 2001. Developing a framework for restoration. In *Handbook for restoring tidal wetlands,* ed. J. B. Zedler, 439. Boca Raton, FL: CRC Press.

West, J. M., G. D. Williams, S. P. Madon, and J. B. Zedler. 2003. Integrating spatial and temporal variability into the analysis of fish food web linkages in Tijuana Estuary. *Environmental Biology of Fishes* 67:297–309.

Williams, G. D., and J. B. Zedler. 1999. Fish assemblage composition in constructed and natural tidal marshes of San Diego Bay: Relative influence of channel morphology and restoration history. *Estuaries* 22:702–716.

Zedler, J. B. 1996. *Tidal wetland restoration: A scientific perspective and southern California focus.* T-038, California Sea Grant College System, University of California, La Jolla.
Zedler, J. B. 2000. Progress in wetland restoration ecology. *Trends in Ecology & Evolution* 15:402–407.
Zedler, J. B., J. C. Callaway, J. S. Desmond, S. G. Vivian, G. D. Williams, G. Sullivan, A. E. Brewster, and B. K. Bradshaw. 1999. Californian salt-marsh vegetation: An improved model of spatial pattern. *Ecosystems* 2:19–35.

Constraining Species Invasions

Topographic heterogeneity might increase plant community diversity at a restoration site by limiting the expansion of competitively dominant species. If competition between species can be moderated, coexistence might be possible (Pacala and Tilman 1994; Hanski 1995). Fugitive species may be better able to persist in heterogeneous environments (see Hanski 1995). However, invasive weeds are often good colonizers, highly plastic in their habitat requirements, and able to exploit patchiness (e.g., Birch and Hutchings 1994). Attempts to restore Wisconsin sedge meadows invaded by *Phalaris arundinacea* (reed-canary grass) should benefit from the topography provided by *Carex stricta* tussocks, which enhance species richness (Werner and Zedler 2002); however, the time required for *C. stricta* to grow tussocks is unknown. Experimentation with artificial tussocks has begun for use in wetland restoration sites (M. Peach, University of Wisconsin–Madison, pers. comm.).

Accelerating Restoration of Extreme Sites and Highly Degraded Habitats

The restoration of severely degraded ecosystems presents special challenges and opportunities with respect to heterogeneity. In degraded rangelands, construction of pits or depressions increased primary productivity and vegetation recovery by slowing runoff and increasing infiltration (Slayback and Cable 1970; Hessary and Gifford 1979; Garner and Steinberger 1989; Whisenant 1999). In semiarid Australian woodland, patches of branches were used to simulate the natural scale of patchiness (Ludwig and Tongway 1996). After three years, soil under branches had more nutrients, higher rates of soil accumulation and water infiltration, and less extreme soil temperatures. In addition, perennial grasses and ant populations had established.

Reintroducing spatial heterogeneity has been a key to restoring productivity and diversity in arid regions. In a desertified area in Israel, constructing pits and adjacent mounds helped to restore natural patch dynamics. The resulting habitat patches developed more species and higher biomass and plant density than areas without this treatment. Differences were attributed to more favorable germination and establishment conditions and a greater seed-trapping capability (Boeken and Shachak 1994). Similar restoration strategies have been used in other arid and semiarid habitats. Workers in Texas rangeland and in Niger used depressions or "microcatchment basins" to capture water, increase soil nutrients, and improve seedling establishment and growth (Whisenant et al. 1995; Whisenant 2002).

Germination and Establishment Microsites

Low plant diversity at many restoration sites may be due to propagule limitation (e.g., Ash et al. 1994), but it can also result from a lack of suitable microhabitats. Fine-scale heterogeneity is thought to provide a variety of germination microsites, or regeneration niches, enabling coexistence of more species (Grubb 1977).

Microhabitat manipulations have been shown to facilitate plant establishment and diversity. In early successional pastures, Reader and Buck (1991) found higher species richness in areas where they created small mounds, simulating the natural disturbance of burrowing animals. In prairie restoration, mounding had a positive effect on growth and survival of three species and only negatively affected one species (Ewing 2002). Natural mounds formed from rabbit-displaced soils in west Texas oak communities were found to increase survival, shoot and root biomass, root length, number of tillers, mycorrhizal infection, and nutrient uptake of *Schizachyrium scoparium*. Artificial mounds created to mimic these mounds were effective in herbaceous vegetation, though less so under oaks (Dhillion 1999). Mounding has also shown promise as a restoration technique in deep-sea benthic communities, deciduous forest, and desert (examples in Ewing 2002).

Restoring microtopography also has value in wetlands. Barry et al. (1996) held that reproducing mound and pool heterogeneity is vital for effective restoration of wetland forests; their design of a 5.3-ha restoration had ~0.6 m mounds created by bulldozers. Tweedy et al. (2001) found that roughly contoured treatments in two formerly agricultural wetland restoration sites had higher water tables, reduced outflow and peak outflow rates, and increased duration of outflow events relative to smooth treatments. Many created wetlands have narrower bands of emergent vegetation than the broad zones or scattered patches of natural wetlands (Confer and Niering 1992). Kentula et al. (1992) recommended using the gradient structure (slope + microtopography) of natural reference marshes as a model for the topography of created marshes. Topography and elevation will influence hydrochory and plant establishment through interactions with flood-pulse dynamics (Middleton 2000). Diversity-support functions would likely be enhanced by shallower surfaces, rougher surfaces, and irregular boundaries between vegetation types.

Wetland plants can be very responsive to subtle changes in topography. Tire ruts were loci of higher plant diversity in a New Jersey freshwater wetland mitigation site (G. Vivian-Smith, pers. obs.) (see Figure 7.4). This was presumably due to greater microhabitat variation in

FIGURE 7.4 A diversity of plants colonized compacted depressions resulting from vehicle traffic over an otherwise homogeneously graded restoration site at the Route 522 mitigation site, Deans Rhode Hall Road, New Jersey.

inundation and soil compaction characteristics. Sensitivity of wetland species to microsite conditions was subsequently demonstrated (Vivian-Smith and Handel 1996; Vivian-Smith 1997). In a mesocosm experiment, Vivian-Smith (1997) found that variation in microtopography on the order of 1–3 cm resulted in greater species richness and evenness and higher abundances of most species than homogeneous treatments.

In stream and river restoration, a return to the small-scale heterogeneity typical of natural systems is a common target (Gore and Shields 1995; Stanford et al. 1996). For example, Hilderbrand et al. (1998) advocated using patches of large woody debris in stream channels to increase retention of organic matter and promote recovery of biota. Woody debris, rock weirs, and gravel are examples of structures used to improve streambed morphology and increase local heterogeneity, with benefits for plants, benthic invertebrates, and fish (Gore and Shields 1995; Jungwirth et al. 1995; Muhar 1996).

Genetic Factors

Topographic heterogeneity may be needed to provide plant communities with appropriate microhabitats for support of genetic diversity. In natural salt marshes, clones of *Spartina patens* are specialized for marsh plain, swale, and dune microhabitats (Silander 1985). A common garden study of *Spartina alterniflora* showed that creek-bank and back-marsh forms of the plant from within the same marsh are genetically distinct (Gallagher et al. 1988). A subsequent restoration experiment with *S. alterniflora* showed that genotypic variation provided differential canopy architecture, above-ground biomass, decomposition rates, algal production, and even fish use (Seliskar et al. 2002). The long-term viability of restored populations of *S. alterniflora* will likely depend on the presence of adequate genetic diversity to provide resistance against disease, invasive species, erosion, storm damage, and other perturbations (Travis et al. 2002).

Future Research Needs

These examples of the role of topographic heterogeneity in natural and restored systems provide strong support for restoring varied elevations where such features have been lost, and to introduce them where they can benefit ecosystem development (e.g., some desert environments). But to ensure that the added effort of increasing topographic heterogeneity is worthwhile, restorationists and researchers need to explore cause-effect relationships further. While the many studies we reviewed provide an emerging theory of topographic heterogeniety, practitioners and researchers are still likely to have system-specific questions:

What Is Topographic Heterogeneity?

- What degree of topographic heterogeneity is typical of relevant reference sites?
- Are topographic conditions in reference sites actually "pristine" or already substantially altered by historic ecological changes?
- What means of quantifying topographic heterogeneity are feasible and useful in this system?

- Are there standardized quantification approaches that could be used for comparison between sites or across habitat types?

What Does It "Do"?

- How does topographic heterogeneity interact with abiotic processes that shape this system (e.g., tides, winds, hydrodynamics, groundwater movement)?
- How does it mediate important ecosystem functions, such as chemical transformations, nutrient cycling, resource pulses, productivity, or decomposition?
- What role does topographic heterogeneity play in shaping community composition in this setting?
- How does topographic variation interact with other forms of spatial heterogeneity and with temporal heterogeneity?

How Should It Be Used?

- How much should be added to a restoration site? At what stage?
- How should it be added? Using what equipment? With what geometry and roughness?
- How should increased topographic complexity be implemented in the face of anthropogenic changes such as increased sedimentation, altered peak flows, or increased erosion rates?
- What design and practice innovations can make use of topographic heterogeneity more effective and efficient?

Answering these and related questions could be of great benefit to both theory and practice. Given well-designed experiments, researchers could advance ecological restoration by elucidating this potent driver of ecosystem dynamics. For instance, future restoration modules in Tijuana Estuary could build on the experience of the Friendship Marsh (Box 7.2) by over-excavating the marsh plain to a mudflat elevation, then adding small islands of higher ground intended for dense plantings, which would reduce the formation of hypersaline surface soil and allow plants to expand vegetatively by accreting sediments from incoming tides or river floodwaters (J. Zedler, pers. obs.). The size, shape, height, location, and frequency of mounds could all be experimental variables in future adaptive restoration efforts. Innovative approaches to site-based restoration in this and other settings should yield useful and informative results.

Summary

A broad review of the ecological literature revealed that topographic heterogeneity acts as an important driver of physical and environmental processes, ecosystem functions, and community composition in settings ranging from terrestrial to aquatic, desert to rain forest, and ocean to bog. Mechanisms to explain these effects come out of habitat heterogeneity, ecological niche, and fractal theory, as well as other ecological literature. Responses have been identified in organisms as different as microbes, cushion plants, and kangaroos and in functions ranging from methane flux to fish nursery support. Findings from restoration settings and experimental manipulations echo effects observed in natural settings. We advocate the use of careful and ambitious experiments to answer remaining questions and develop innovations that can improve restoration practice.

Acknowledgments

The authors are grateful for support from NSF (DEB 0212005) and wish to thank G. Bruland, D. Falk, C. Gratton, M. Palmer, and P. Zedler for insightful reviews that substantially improved this chapter.

Literature Cited

Able, K., S. Hagan, and S. Brown. 2003. Mechanisms of marsh habitat alteration due to *Phragmites:* Response of young-of-the-year Mummichog (*Fundulus heteroclitus*) to treatment for *Phragmites* removal. *Estuaries* 26:484–494.

Adam, P. 1990. *Saltmarsh ecology*. Cambridge, UK: Cambridge University Press.

Angradi, T., S. Hagan, and K. Able. 2001. Vegetation type and the intertidal macroinvertebrate fauna of a brackish marsh: *Phragmites* vs. *Spartina*. *Wetlands* 21:75–92.

Aronson, R. B., and C. A. Harms. 1985. Ophiuroids in a Bahamian saltwater lake: The ecology of a paleozoic-like community. *Ecology* 66:1472–1483.

Ash, H. J., R. P. Gemmell, and A. D. Bradshaw. 1994. The introduction of native plant species on industrial waste heaps: A test of immigration and other factors affecting primary succession. *Journal of Applied Ecology* 31:74–84.

Ball, B. C., G. W. Horgan, H. Clayton, and J. P. Parker. 1997. Spatial variability of nitrous oxide fluxes and controlling soil and topographic properties. *Journal of Environmental Quality* 26:1399–1409.

Barry, W. J., A. S. Garlo, and C. A. Wood. 1996. Duplicating the mound-and-pool microtopography of forested wetlands. *Restoration and Management Notes* 14:15–21.

Birch, C. P. D., and M. J. Hutchings. 1994. Exploitation of patchily distributed soil resources by the clonal herb *Glechoma hederacea*. *Journal of Ecology* 82:653–664.

Boeken, B., and M. Shachak. 1994. Desert plant-communities in human-made patches—Implications for management. *Ecological Applications* 4:702–716.

Brooks, S. S., M. A. Palmer, B. J. Cardinale, C. M. Swan, and S. Ribblett. 2002. Assessing stream ecosystem rehabilitation: Limitations of community structure data. *Restoration Ecology* 10:156–168.

Brose, U. 2001. Relative importance of isolation, area and habitat heterogeneity for vascular plant species richness of temporary wetlands in east-German farmland. *Ecography* 24:722–730.

Bubier, J. L., T. R. Moore, and N. T. Roulet. 1993. Methane emissions from wetlands in the midboreal region of northern Ontario, Canada. *Ecology* 74:2240–2254.

Burdick, D. M., and I. A. Mendelssohn. 1987. Waterlogging responses in dune, swale and marsh populations of *Spartina patens* under field conditions. *Oecologia* 74:321–329.

Cantelmo, A. J., and J. G. Ehrenfeld. 1999. Effects of microtopography on mycorrhizal infection in Atlantic white cedar (*Chamaecyparis thyoides* (L.) Mills). *Mycorrhiza* 8:175–180.

Cardinale, B. J., M. A. Palmer, C. M. Swan, S. Brooks, and N. L. Poff. 2002. The influence of substrate heterogeneity on biofilm metabolism in a stream ecosystem. *Ecology* 83:412–422.

Carlton, G. C., and F. A. Bazzaz. 1998. Regeneration of three sympatric birch species on experimental hurricane blowdown microsites. *Ecological Monographs* 68:99–120.

Clegg, S. M., P. Hale, and C. Moritz. 1998. Molecular population genetics of the red kangaroo (*Macropus rufus*): mtDNA variation. *Molecular Ecology* 7:679–686.

Collins, S. L., and S. C. Barber. 1985. Effects of disturbance on diversity in mixed grass prairie. *Vegetatio* 64:87–94.

Collins, S. L., and S. M. Glenn. 1990. A hierarchical analysis of species' abundance patterns in grassland vegetation. *American Naturalist* 135:633–648.

Confer, S., and W. A. Niering. 1992. Comparison of natural and created freshwater emergent wetlands in Connecticut. *Wetlands Ecology & Management* 2:143–156.

Connell, S. D., and G. P. Jones. 1991. The influence of habitat complexity on postrecruitment processes in a temperate reef fish population. *Journal of Experimental Marine Biology and Ecology* 151:271–294.

Coppedge, B. R., S. D. Fuhlendorf, D. M. Engle, and B. J. Carter. 1999. Grassland soil depressions: Relict bison wallows or inherent landscape heterogeneity? *American Midland Naturalist* 142:382–392.

Cowles, H. C. 1899. The ecological relations of the vegetation on the sand dunes of Lake Michigan. *Botanical Gazette* 27:167–202.

Dhillion, S. S. 1999. Environmental heterogeneity, animal disturbances, microsite characteristics, and seedling establishment in a *Quercus havardii* community. *Restoration Ecology* 7:399–406.

Dobkin, D. S., I. Olivieri, and P. R. Ehrlich. 1987. Rainfall and the interaction of microclimate with larval resources in the population dynamics of checkerspot butterflies (*Euphydryas editha*) inhabiting serpentine grassland. *Oecologia* 71:161–166.

Dwyer, L. M., and G. Merriam. 1981. Influence of topographic heterogeneity on deciduous litter decomposition. *Oikos* 37:228–237.

Ehrenfeld, J. G. 1995a. Microsite differences in surface substrate characteristics in *Chamaecyparis* swamps of the New Jersey pinelands. *Wetlands* 15:183–189.

Ehrenfeld, J. G. 1995b. Microtopography and vegetation in Atlantic white cedar swamps—The effects of natural disturbances. *Canadian Journal of Botany* 73:474–484.

Elliott, K. J., S. L. Hitchcock, and L. Krueger. 2002. Vegetation response to large-scale disturbance in a southern Appalachian forest: Hurricane Opal and salvage logging. *Journal of the Torrey Botanical Society* 129:48–59.

Ewing, K. 2002. Mounding as a technique for restoration of prairie on a capped landfill in the Puget Sound lowlands. *Restoration Ecology* 10:289–296.

Fleishman, E., A. E. Launer, S. B. Weiss, J. M. Reed, C. L. Boggs, D. D. Murphy, and P. R. Ehrlich. 1997. Effects of microclimate and oviposition timing on prediapause larval survival of the Bay checkerspot butterfly, *Euphydryas editha bayensis* (Lepidoptera: Nymphalidae). *Journal of Research on the Lepidoptera* 36:31–44.

Freeman, D. C., E. D. McArthur, S. C. Sanderson, and A. R. Tiedemann. 1993. The influence of topography on male and female fitness components of *Atriplex canescens*. *Oecologia* 93:538–547.

Gallagher, J. L., G. F. Somers, D. M. Grant, and D. M. Seliskar. 1988. Persistent differences in two forms of *Spartina alterniflora*: A common garden experiment. *Ecology* 69:1005–1008.

Garner, W., and Y. Steinberger. 1989. A proposed mechanism for the formation of "fertile islands" in the desert ecosystem. *Journal of Arid Environments* 16:257–262.

Gibson, D. J. 1988. The relationship of sheep grazing and soil heterogeneity to plant spatial patterns in dune grassland. *Journal of Ecology* 76:233–252.

Gore, J. A., and F. D. Shields. 1995. Can large rivers be restored? *Bioscience* 45:142–152.

Grubb, P. J. 1977. The maintenance of species-richness in plant communities: The importance of the regeneration niche. *Biological Review* 52:107–145.

Guichard, F., and E. Bourget. 1998. Topographic heterogeneity, hydrodynamics, and benthic community structure: A scale-dependent cascade. *Marine Ecology Progress Series* 171:59–70.

Gunderson, L. H. 1997. The Everglades: Trials in ecosystem management. In *Principles of conservation biology*, ed. G. K. Meffe and C. R. Carroll, 451–458. Sunderland, MA: Sinauer Associates.

Haering, R., and B. J. Fox. 1995. Habitat utilization patterns of sympatric populations of *Pseudomys gracilicaudatus* and *Rattus lutreolus* in coastal heathland: A multivariate analysis. *Australian Journal of Ecology* 20:427–441.

Halley, J. M., S. Hartley, A. S. Kallimanis, W. E. Kunin, J. J. Lennon, and S. P. Sgardelis. 2004. Uses and abuses of fractal methodology in ecology. *Ecology Letters* 7:254–271.

Hanski, I. 1995. Effects of landscape pattern on competitive interactions. In *Mosaic landscapes and ecological processes*, ed. L. Hansson, L. Fahrig, and G. Merriam, 203–224. London: Chapman & Hall.

Hardt, R. A., and R. T. T. Forman. 1989. Boundary form effects on woody colonization of reclaimed surface mines. *Ecology* 70:1252–1260.

Hessary, I. K., and G. F. Gifford. 1979. Impact of various range improvement practices on watershed protective cover and annual production within the Colorado River basin, USA. *Journal of Range Management* 32:134–140.

Hilderbrand, R. H., A. D. Lemly, C. A. Dolloff, and K. L. Harpster. 1998. Design consideration for large woody debris placement in stream enhancement projects. *North American Journal of Fisheries Management* 18:161–167.

Hogue, E. W., and C. B. Miller. 1981. Effects of sediment microtopography on small-scale spatial distributions of meiobenthic nematodes. *Journal of Experimental Marine Biology and Ecology* 53:181–192.

Howe, H. F. 1994. Managing species diversity in tallgrass prairie: Assumptions and implications. *Conservation Biology* 8:691–704.

Hutchinson, G. E. 1957. Concluding remarks. Cold Spring Harbor symposia. *Quantitative Biology* 22:415–427.

Inglis, G. J. 2000. Disturbance-related heterogeneity in the seed banks of a marine angiosperm. *Journal of Ecology* 88:88–99.

Inouye, R. S., N. Huntly, and G. A. Wasley. 1997. Effects of pocket gophers (*Geomys bursarius*) on microtopographic variation. *Journal of Mammalogy* 78:1144–1148.

Jeltsch, F., S. J. Milton, W. R. J. Dean, N. Van Rooyen, and K. N. Moloney. 1998. Modeling the impact of small-scale heterogeneities on tree-grass coexistence in semi-arid savannas. *Journal of Ecology* 86:780–793.

Jones, C. G., J. H. Lawton, and M. Shachak. 1994. Organisms as ecosystem engineers. *Oikos* 69:373–386.

Jungwirth, M., S. Muhar, and S. Schmutz. 1995. The effects of recreated instream and ecotone structures on the fish fauna of an epipotamal river. *Hydrobiologia* 303:195–206.

Kentula, M. E., R. P. Brooks, S. E. Gwin, C. C. Holland, A. D. Sherman, and J. C. Sifneos. 1992. *An approach to improving decision making in wetland restoration and creation*. Corvallis: U.S. Environmental Protection Agency, Environmental Research Laboratory.

Kern, K. 1992. Restoration of lowland rivers: The German experience. In *Lowland floodplain rivers*, ed. P. A. Carling and G. E. Petts, 279–297. Chichester, UK: John Wiley & Sons.

Kerr, J. T., and L. Packer. 1997. Habitat heterogeneity as a determinant of mammal species richness in high-energy regions. *Nature* 385:252–254.

Kerr, J. T., R. Vincent, and D. J. Currie. 1998. Lepidopteran richness patterns in North America. *Ecoscience* 5:448–453.

Kim, J., and S. B. Verma. 1992. Soil surface carbon dioxide flux in a Minnesota peatland. *Biogeochemistry* 18:37–51.

Koebel Jr., J. W. 1995. An historical perspective on the Kissimmee River restoration project. *Restoration Ecology* 3:149–159.

Krummel, J. R., R. H. Gardner, G. Sugihara, R. V. O'Neill, and P. R. Coleman. 1987. Landscape patterns in a disturbed environment. *Oikos* 48:321–384.

Leopold, A. 1999. Coon Valley: An adventure in cooperative conservation. In *For the health of the land: Previously unpublished essays and other writings*, ed. J. B. Callicott and E. Freyfogle, 47–54. Washington, DC: Island Press.

Levin, S. A. 1976. Population dynamic models in heterogeneous environments. *Annual Review of Ecology and Systematics* 7:287–310.

Li, Y. C., T. Krugman, T. Fahima, A. Beiles, A. B. Korol, and E. Nevo. 2001. Spatiotemporal allozyme divergence caused by aridity stress in a natural population of wild wheat, *Triticum dicoccoides*, at the Ammiad microsite, Israel. *Theoretical and Applied Genetics* 102:853–864.

Lubchenco, J. 1983. *Littorina* and *Fucus:* Effects of herbivores, substratum heterogeneity and plant escapes during succession. *Ecology* 64:1116–1123.

Ludwig, J. A., and D. J. Tongway. 1996. Rehabilitation of semiarid landscapes in Australia. II: Restoring vegetation patches. *Restoration Ecology* 4:398–406.

Mandelbrot, B. B. 1983. *The fractal geometry of nature*. New York: W. H. Freeman & Co.

Menge, B. A., J. Lubchenco, and L. R. Ashkenas. 1985. Diversity, heterogeneity and consumer pressure in a tropical rocky intertidal community. *Oecologia* 65:394–405.

Middleton, B. 2000. Hydrochory, seed banks, and regeneration dynamics along the landscape boundaries of a forested wetland. *Plant Ecology* 146:169–184.

Miller, R. I., S. P. Bratton, and P. S. White. 1987. A regional strategy for reserve design and placement based on an analysis of rare and endangered species' distribution patterns. *Biological Conservation* 39:255–268.

Minchinton, T. E. 2001. Canopy and substratum heterogeneity influence recruitment of the mangrove *Avicennia marina*. *Journal of Ecology* 89:888–902.

Mladenoff, D. J., M. A. White, J. Pastor, and T. R. Crow. 1993. Comparing spatial pattern in unaltered old-growth and disturbed forest landscapes. *Ecological Applications* 3:294–306.

Morse, D. R., J. H. Lawton, M. M. Dodson, and M. H. Williamson. 1985. Fractal dimension of vegetation and the distribution of arthropod body lengths. *Nature* 314:731–733.

Muhar, S. 1996. Habitat improvement of Austrian rivers with regard to different scales. *Regulated Rivers: Research & Management* 12:471–482.

Netto, S. A., M. J. Attrill, and R. M. Warwick. 1999. Sublittoral meiofauna and macrofauna of Rocas Atoll (NE Brazil): Indirect evidence of a topographically controlled front. *Marine Ecology Progress Series* 179:175–186.

Nkem, J. N., L. A. L. de Bruyn, C. D. Grant, and N. R. Hulugalle. 2000. The impact of ant bioturbation and foraging activities on surrounding soil properties. *Pedobiologia* 44:609–621.

Pacala, S. W., and D. Tilman. 1994. Limiting similarity in mechanistic and spatial models of plant competition in heterogeneous environments. *American Naturalist* 143:222–257.

Palmer, M. W. 1992. The coexistence of species in fractal landscapes. *American Naturalist* 139:375–397.

Paz González, A., S. R. Vieira, and M. T. Taboada Castro. 2000. The effect of cultivation on the spatial variability of selected properties of an umbric horizon. *Geoderma* 97:273–292.

Peach, M. 2005. Tussock sedge meadows and topographic heterogeneity: Ecological patterns underscore the need for experimental approaches to wetland restoration despite the social barriers. MS thesis, University of Wisconsin–Madison.

Peckarsky, B. L., B. W. Taylor, and C. C. Caudill. 2000. Hydrologic and behavioral constraints on oviposition of stream insects: Implications for adult dispersal. *Oecologia* 125:186–200.

Perelman, S. B., R. J. C. Leon, and M. Oesterheld. 2001. Cross-scale vegetation patterns of Flooding Pampa grasslands. *Journal of Ecology* 89:562–577.

Pickett, S. T., and V. T. Parker. 1994. Avoiding the old pitfalls: Opportunities in a new discipline. *Restoration Ecology* 2:75–79.

Pinay, G., H. Decamps, C. Arles, and M. Lacassain Seres. 1989. Topographic influence on carbon and nitrogen dynamics in riverine woods. *Archiv fuer Hydrobiologie* 114:401–414.

Pollock, M. M., R. J. Naiman, and T. A. Hanley. 1998. Plant species richness in riparian wetlands. A test of biodiversity theory. *Ecology* 79:94–105.

Reader, R. J., and J. Buck. 1991. Community response to experimental soil disturbance in a midsuccessional, abandoned pasture. *Vegetatio* 92:151–159.

Roy, S., and J. S. Singh. 1994. Consequences of habitat heterogeneity for availability of nutrients in a dry tropical forest. *Journal of Ecology* 82:503–509.

Seliskar, D. M., J. L. Gallagher, D. M. Burdick, and L. A. Mutz. 2002. The regulation of ecosystem functions by ecotypic variation in the dominant plant: A *Spartina alterniflora* salt-marsh case study. *Journal of Ecology* 90:1–11.

Sharitz, R. R., and J. F. McCormick. 1973. Population dynamics of two competing annual plant species. *Ecology* 54:723–740.

Silander, J. A. J. 1985. The genetic basis of the ecological amplitude of *Spartina patens*. II: Variance and correlation analysis. *Evolution* 39:1034–1052.

Slayback, R. D., and D. R. Cable. 1970. Larger pits aid reseeding of semidesert rangeland. *Journal of Range Management* 23:333–335.

Smith, M., and J. Capelle. 1992. Effects of soil surface microtopography and litter cover on germination, growth and biomass production of chicory (Cichorium intybus L.). *American Midland Naturalist* 128:246–253.

Stanford, J. A., J. V. Ward, W. J. Liss, C. A. Frissell, R. N. Williams, J. A. Lichatowich, and C. C. Coutant. 1996. A general protocol for restoration of regulated rivers. *Regulated Rivers: Research & Management* 12:391–413.

Tessier, M., J. C. Gloaguen, and V. Bouchard. 2002. The role of spatio-temporal heterogeneity in the establishment and maintenance of *Suaeda maritima* in salt marshes. *Journal of Vegetation Science* 13:115–122.

Thomson Corportation. 2004. *ISI Web of Knowledge*. http://isi3.isiknowledge.com/portal.cgi (accessed 2004 October 5).

Tilman, D., C. L. Lehman, and C. E. Bristow. 1998. Diversity-stability relationships: Statistical inevitability or ecological consequence? *American Naturalist* 151:277–282.

Toth, L. A., J. T. B. Obeysekera, W. A. Perkins, and M. K. Loftin. 1993. Flow regulation and restoration of Florida's Kissimmee River. *Regulated Rivers: Research & Management* 8:155–166.

Travis, S. E., C. E. Proffitt, R. C. Lowenfeld, and T. W. Mitchell. 2002. A comparative assessment of genetic diversity among differently-aged populations of *Spartina alterniflora* on restored versus natural wetlands. *Restoration Ecology* 10:37–42.

Turner, M. G. 1989. Landscape Ecology—The effect of pattern on process. *Annual Review of Ecology and Systematics* 20:171–197.

Tweedy, K. L., and R. O. Evans. 2001. Hydrologic characterization of two prior converted wetland restoration sites in eastern North Carolina. *Transactions of the American Society of Agricultural Engineers* 44:1135–1142.

Vestergaard, P. 1998. Vegetation ecology of coastal meadows in southeastern Denmark. *Opera Botanica* 134:1–69.

Vivian-Smith, G. 1997. Microtopographic heterogeneity and floristic diversity in experimental wetland communities. *Journal of Ecology* 85:71–82.

Vivian-Smith, G., and S. N. Handel. 1996. Fresh water wetland restoration of an abandoned sand mine: Seed bank recruitment dynamics and plant colonization. *Wetlands* 16:185–196.

Watson, H. C. 1835. *Remarks on the geographical distribution of British plants: Chiefly in connection with latitude, elevation, and climate.* London: Longman et al.

Watt, A. S. 1947. Pattern and process in the plant community. *Journal of Ecology* 35:1–22.

Werner, K. J., and J. B. Zedler. 2002. How sedge meadow soils, microtopography, and vegetation respond to sedimentation. *Wetlands* 22:451–466.

Whisenant, S. G. 1999. *Repairing damaged wildlands: A process-oriented, landscape-scale approach.* Cambridge, UK: Cambridge University Press.

Whisenant, S. G. 2002. Terrestrial systems. In *Handbook of ecological restoration.*Volume 1. *Principles of restoration,* ed. M. R. Perrow and A. J. Davy, 83–105. Cambridge, UK: Cambridge University Press.

Whisenant, S. G., T. L. Thurow, and S. J. Maranz. 1995. Initiating autogenic restoration on shallow semiarid sites. *Restoration Ecology* 3:61–67.

Williamson, M. H., and J. H. Lawton. 1991. Fractal geometry of ecological habitats. In *Habitat structure: The physical arrangement of objects in space,* ed. S. S. Bell, E. D. McCoy, and H. R. Mushinsky, 69–86. New York: Chapman & Hall.

Wu, J. G., and O. L. Loucks. 1995. From balance of nature to hierarchical patch dynamics: A paradigm shift in ecology. *Quarterly Review of Biology* 70:439–466.

Zedler, J. B. 2001. *Handbook for restoring tidal wetlands.* Boca Raton, FL: CRC Press.

Zedler, J. B., and J. C. Callaway. 1999. Tracking wetland restoration: Do mitigation sites follow desired trajectories? *Restoration Ecology* 7:69–73.

Zedler, J. B., J. C. Callaway, J. S. Desmond, S. G. Vivian, G. D. Williams, G. Sullivan, A. E. Brewster, and B. K. Bradshaw. 1999. Californian salt-marsh vegetation: An improved model of spatial pattern. *Ecosystems* 2:19–35.

Zedler, J. B., H. Morzaria-Luna, and K. Ward. 2003. The challenge of restoring vegetation on tidal, hypersaline substrates. *Plant and Soil* 253:259–273.

Zedler, J. B., and P. H. Zedler. 1969. Association of species and their relationship to microtopography within old fields. *Ecology* 50:432–442.

Chapter 8

Food-Web Approaches in Restoration Ecology

M. Jake Vander Zanden, Julian D. Olden, and Claudio Gratton

No species exists in a vacuum. Rather, each species is embedded within a network of predator-prey interactions in what Charles Darwin referred to as an "entangled bank" and is now known in the most general sense as a *food web*. In its most basic form, a food web reveals to us something about the feeding relationships in a system. More broadly, food webs represent a way of thinking about an ecological system that considers trophic (consumer-resource) interactions among species or groups of similar species (trophic guilds or trophic levels). Food-web ecology is an ever-changing subdiscipline of ecology, and it is critical to recognize the diversity of approaches to the study of food webs (Paine 1980; Schoener 1989; Pimm 1991; Polis and Winemiller 1996). The term *food-web structure* can have several meanings to food-web ecologists. Food-web structure can refer simply to the number of trophic levels in a food chain (Figure 8.1a) or, alternatively, can represent the degree of complexity in a food-web network (Figure 8.1b). Food-web diagrams may be used to represent the pathways of energy flow through a system (energetic webs, Figure 8.1c) or, alternatively, the dynamically important linkages for regulating trophic structure (functional webs, Figure 8.1d). An alternative meaning of food-web structure refers to the distribution of biomass across different trophic levels, and ultimately how bottom-up and top-down factors regulate the accumulation of biomass across trophic levels (Figure 8.2a–b). These diverse food-web concepts serve as the basis for our discussion of food-web theory and applications to ecological restoration.

Despite the intuitive importance of explicitly considering trophic connections, food-web approaches have yet to take hold in many applied management endeavors, such as fisheries and wildlife management, conservation biology, and ecological restoration. We argue that food-web ecology has the potential to contribute to ecological restoration by encouraging a more dynamic, interaction-driven view of ecosystems, and it can alert practitioners to the types of trophic interactions that may have bearing on restoration outcomes (Zavaleta et al. 2001). In some situations, adopting a food-web perspective will provide valuable insights into ecological restoration that would not otherwise be attained from a more static, community-based approach. For example, the recent reintroduction of wolves into Yellowstone Park, Wyoming, USA, has precipitated a cascade of food-web changes that has allowed the recovery of riparian vegetation and associated biota from damaging effects of herbivore overgrazing (Berger et al. 2001; Ripple et al. 2001; Ripple and Beschta 2003), an effect that would not

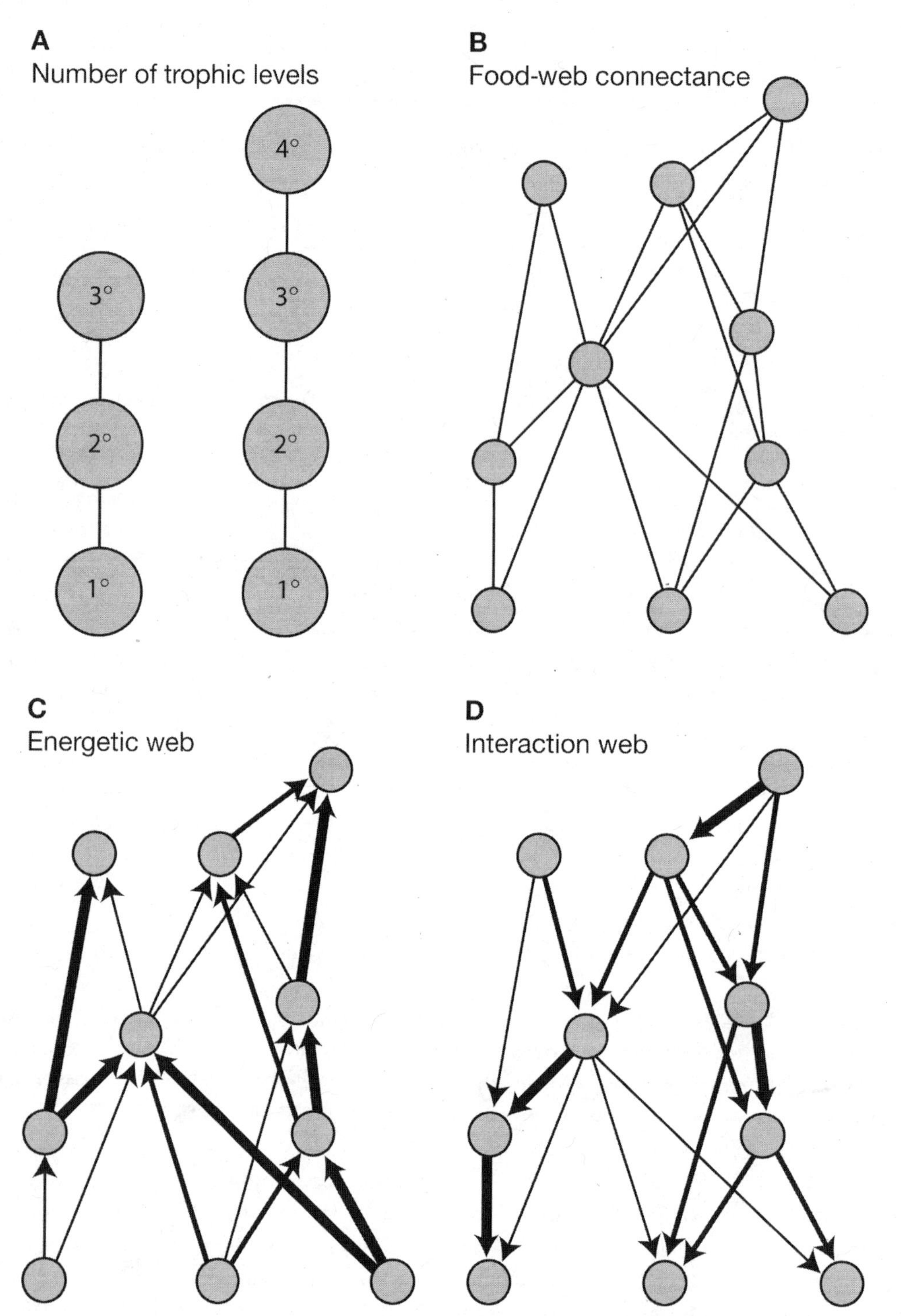

FIGURE 8.1 Different meanings of "trophic structure" used by food-web ecologists: (A) number of trophic levels (three versus four levels); (B) food-web connectance, the pattern of trophic linkages among species in a complex web; (C) energetic web, depicting the pathways of mass or energy flow; and (D) interaction web, showing the dynamically important food-web linkages.

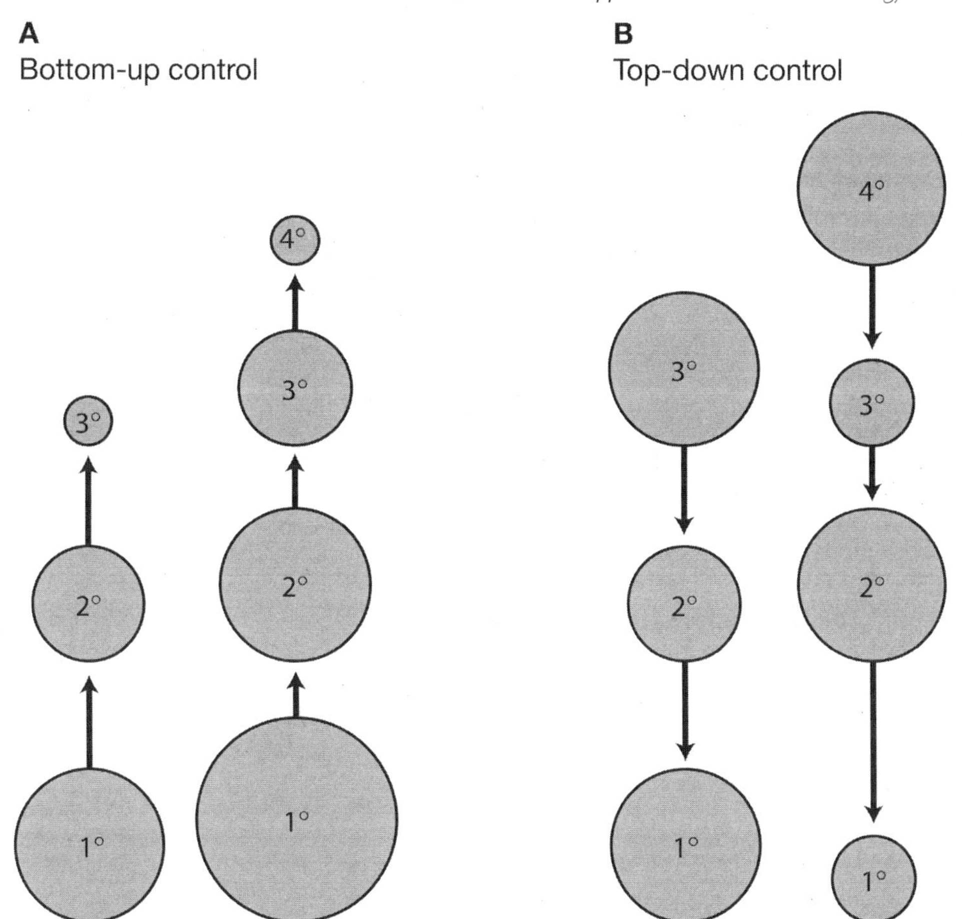

FIGURE 8.2 Bottom-up versus top-down control of the distribution of biomass at different trophic levels. Note that compartment size indicates trophic level biomass: (A) In the case of bottom-up control, primary production is the basis for higher trophic levels. Increasing primary production allows higher biomass at subsequent trophic levels, and possibly the support of additional trophic levels. (B) In the case of top-down control, predation plays a role in determining the distribution of biomass across trophic levels. In a three-level system, herbivores (2°) are suppressed by predators (3°), which allows accumulation of plant (1°) biomass. Addition of 4° controls the biomass of 3°. As a result, herbivore biomass (2°) increases, leading to a reduction in plant biomass (1°).

have been predicted without considering the cascading effects of predator-prey interactions across multiple trophic levels.

Many of the world's ecosystems are highly degraded, and natural recovery processes, particularly in light of the onslaught of biological invasions, are often inadequate to achieve desired goals for ecosystem recovery (Dobson et al. 1997; Hobbs and Harris 2001; D'Antonio and Chambers, this volume). Ecological restoration is undertaken to hasten the recovery of damaged ecosystems, restore ecosystem function, and slow the declines of biodiversity (Jordan III 1987; NRC 1995; Dobson et al. 1997; Young 2000). Restoration in North America has its historical roots in plant community ecology: a perusal of the leading journals in the field such as *Ecological Restoration* and *Restoration Ecology* reveals the botanical nature of the discipline.

As such, succession and community assembly theory have provided the theoretical underpinnings for restoration ecology (Weiher and Keddy 1999; Young 2000; Young et al. 2001). While food-web ecology is often viewed as a subdiscipline of community ecology, community and food-web ecology differ in several significant ways. Community ecologists generally study the factors affecting abundance, species composition, and diversity within a particular trophic guild or group (i.e., the bird community, the plant community) (Drake 1990; Suding and Gross, this volume; Menninger and Palmer, this volume). In contrast, food-web studies consider the energetic and dynamic linkages within a broader spectrum of the ecological community, and the scale of analysis typically spans several trophic levels.

A critical aspect of ecological restoration is the establishment of well-defined restoration targets (Hobbs and Harris 2001). The traditional approach emphasizes *structural* restoration targets, including taxonomic characteristics, such as species richness, or the presence or abundance of indicator species or assemblages. Yet the mere presence of desired taxa or functional groups does not mean that the restored system is functioning as desired, or that the species are performing ecologically relevant roles within the restored system. Thus, a complementary approach considers *functional* targets, which include ecosystem processes, such as primary production, nutrient cycling, and the maintenance of critical food-web linkages (Palmer et al. 1997). Structural and functional approaches are not mutually exclusive, and food-web-based targets may incorporate both components. This chapter examines how food-web theory and, more informally, "food-web thinking" might contribute fruitfully to the planning, implementation, and evaluation of ecological restoration.

Relevant Theory—A Historical Overview

This section provides a brief overview of food-web ecology from a historical perspective. For more in-depth background reading on food-web ecology, we recommend the following sources: Schoener (1989); Pimm (1991); Polis and Winemiller (1996); Persson (1999); Post (2002); and Polis et al. (2004).

Among the first published food web studies was Summerhayes and Elton's description of the food webs of Spitsbergen and Bear Islands (Summerhayes and Elton 1923; Elton 1927). The next major advance in food-web ecology was undoubtedly Lindeman's (1942) trophic-dynamic study of a small Minnesota lake. Lindeman viewed the lake as a chain of energy transformations—solar energy was "fixed" via photosynthesis, a portion was converted to herbivore biomass, and so on up the food chain. Decreasing production was available at successive trophic levels due to metabolic inefficiencies at each trophic step. In this view, primary production limited higher trophic level production, suggesting "bottom-up" control of the distribution of biomass in food webs (Figure 8.2a). This work provided the operational structure for modern food-web research by introducing the concept of trophic levels as well as the use of energy as a currency. One implication of this work was that available energy could limit the number of trophic levels (Pimm 1982), an idea that serves as a basis for assessing whether the energetic needs of higher consumers (often the target of restoration efforts) are likely to be met within a restored ecosystem. Lindeman's ideas also raise the issue of whether variables such as food-chain length could be used as a meaningful restoration endpoint.

Two decades later, a study by Hairston, Smith, and Slobodkin (1960) (now known as HSS) argued that terrestrial food chains have three functional trophic levels—predators keep herbivores in check, thus allowing plant biomass to accumulate. The "top-down" perspective offered in HSS was predicated on the idea that predators control the abundance of their prey, and that these effects can subsequently cascade down food chains, ultimately impacting primary producer biomass (Figure 8.2b). This radical proposition ran counter to the dominant paradigm of the time: that nutrients and/or environmental factors limited plant communities and biomass, which, in turn, constrained higher trophic levels (compare Figure 8.2a and 8.2b). HSS has since inspired major research efforts directed toward the role of predators and resources as determinants of the abundances of organisms at different trophic levels in a variety of ecosystem types (Oksanen et al. 1981; Fretwell 1987; Power 1992; Hairston and Hairston 1993; Polis and Strong 1996; Polis 1999). To illustrate, if top-down factors dominate, removal of predators from a three-level system should produce an increase in herbivore biomass and a decrease in plant biomass. Alternatively, if removal of predators does not cause an increase in herbivore biomass, this indicates bottom-up control, and we might expect that increasing plant productivity would produce an increase in herbivore biomass. In fact, both processes likely operate concurrently and often interact in complex ways (Denno et al. 2003).

Studies predicated upon simple food-chain models have played an important role in ecology. Not only do such models generate easily testable predictions, but many natural systems appear to exhibit dynamics consistent with simple food-chain structures (Oksanen et al. 1981; Carpenter et al. 1985). Interestingly, many descriptive food-web studies offer the paradoxically different view that food webs are immensely complex—with hundreds of species and trophic links, coupled with rampant complications such as ontogenic diet shifts, omnivory, and intraguild predation (Warren 1989; Hall and Raffaelli 1991; Martinez 1991; Polis 1991). In addition, trophic levels themselves are often heterogeneous, such that the addition of grazers to a system may reduce plant biomass or, alternatively, may cause a compensatory shift toward grazer-tolerant plants (Leibold 1989; Hunter and Price 1992). While simple food-chain models undoubtedly overlook certain trophic linkages and interactions, the critical issue to be resolved is whether these complexities are mere details or, alternatively, when and if these trophic linkages are truly important in driving the dynamics of the system (Power 1992).

To understand food-web dynamics, it is critical to distinguish between direct and indirect food-web effects (Abrams et al. 1996). An example of a direct effect is that an increase in species A reduces the density of species B due to predation. An indirect effect implies a change in the density of species B in response to a change in species A, but through interactions with a third species. The three trophic level interaction proposed in HSS (Figure 8.2b) described above is a simple indirect effect (changes in predators affect plant biomass through impacts on herbivores). In the rocky intertidal zone, Robert Paine's (1966) seminal food-web experiments demonstrated that predation by the Ochre sea star (*Pisaster ochraceus*) upon competitively dominant prey reduced competition for space, thereby allowing persistence of inferior competitors. This work highlighted the role of predators in maintaining prey diversity by mediating interspecific competition (Figure 8.3a). Paine labeled *Pisaster* a "keystone species" due to its role in structuring the community. The implication here was that all species are not equally important, and that a few species play central roles in structuring the system (Power et al. 1996c; Lawton 2000). Apparent competition is another type of indirect interaction, whereby two prey species share a common predator, and predation rates on a focal prey species are *increased* due to the presence of the alternative prey (Figure 8.3b) (Holt

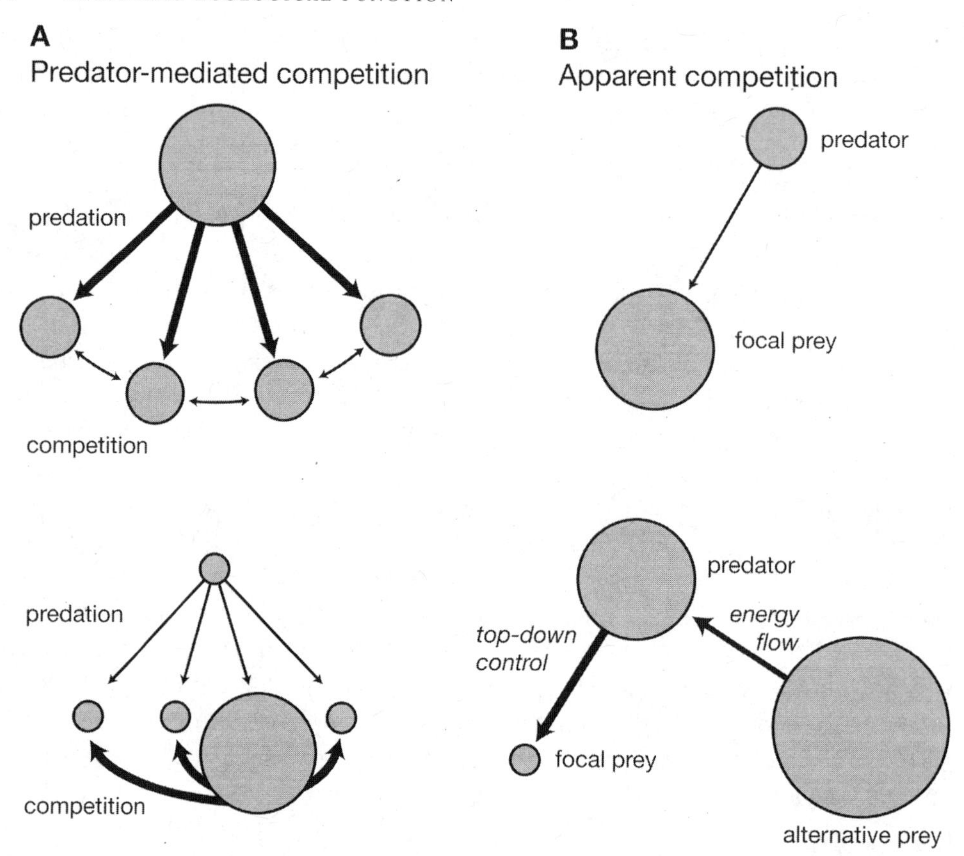

FIGURE 8.3 Examples of complex food-web interactions. Upward arrows represent energy flow pathways, downward arrows represent top-down control. Arrow width represents the strength of the trophic linkage: (A) Predator-mediated competition. High predator biomass suppresses densities of prey taxa, reducing competition among prey. Reduction of predator biomass allows increased prey biomass, thereby increasing competition among prey and domination by the superior competitor taxa. (B) Apparent competition. The predator consumes the focal prey (top panel). Addition of a highly productive alternative prey increases predator biomass, causing greater predation rates on the focal prey than in the absence of alternative prey (bottom panel). The consequence is elevated predator biomass and decreased biomass of focal prey.

1977; Holt and Lawton 1994). If the vulnerability of the two prey differs, consumption of a less vulnerable, abundant prey can have the effect of augmenting the predator population, subsequently increasing predation rates on the more vulnerable prey (Holt 1977; Holt and Lawton 1994).

Application of Food-Web Theory to Restoration Ecology

Restoration ecology has historically been based on a succession-driven, bottom-up view of ecosystems, and current paradigms in ecological restoration do not generally incorporate a food-web perspective. Even if restoration targets do not specifically involve the reestablishment of trophic linkages per se, there may be value in food-web approaches, since the dy-

TABLE 8.1

Areas of food-web research and applications to restoration ecology discussed in this chapter.

Section	Key references
Food-chain approaches	
Ecosystem role of predators	Pace et al. 1999; Soulé and Terborgh 1999
Trophic cascades in lakes	Carpenter et al. 1985
Trophic control in salt marshes	Silliman and Bertness 2002
Mesopredator release	Crooks and Soulé 1999
Trophic control in Yellowstone NP	Beschta 2003
Complex Interactions	
Apparent competition in Guam	Savidge 1987
Complex interactions in Channel Islands	Roemer et al. 2003
Landscape context in Hudson Bay	Jeffries et al. 2004
Invasions and reintroductions	
Trout introductions and restoration	Knapp et al. 2000
Natural flow regimes	Poff et al. 1997
Food-web assembly	
Food-web assembly in recovering lakes	Gunn and Mills 1998
Applications of stable isotopes	
Recovery of salt marsh food webs	Kwak and Zedler 1997
Long-term food-web change in Lake Tahoe	Vander Zanden et al. 2003

namics of any species or community depend critically on interactions among prey and predators (Pimm 1991). For example, identifying keystone species may be of concern in restoration, since they play a critical role in determining community and ecosystem structure (Mills et al. 1993; Power et al. 1996c). Table 8.1 lists the major areas of food-web research examined in this chapter, with potential applications to restoration ecology.

Food-Chain Approaches

There is growing evidence that top predators can have impacts that cascade to lower trophic levels (Pace et al. 1999). In aquatic systems, the decline of top predators (Post et al. 2002) can have cascading effects on lower trophic levels (Carpenter et al. 2001). Similarly, predation is now recognized as a key process in the maintenance of biodiversity and ecosystem function in terrestrial systems (Soulé and Terborgh 1999). Terrestrial conservation biologists and limnologists are recognizing the importance of predation in structuring ecosystems and are focusing efforts on restoring or maintaining predation as a component of ecological restoration efforts. Below, we present several examples where an ecosystem was viewed through the lens of a simple food-chain model, with important implications for ecological restoration.

The importance of simple food-chain interactions in ecosystem restoration has been most thoroughly described for aquatic ecosystems. Human-induced eutrophication caused by excess nutrient loading is a critical environmental problem affecting lakes, resulting in algal blooms, oxygen depletion, loss of aquatic vegetation, and declines in water quality (Carpenter et al. 1998). While nutrient reductions are an obvious approach for improving water quality, food-web manipulations can also play an important role. The trophic cascade hypothesis (Carpenter et al. 1985) was conceived to explain unexplained variance in relationships between nutrient levels and phytoplankton (algae) biomass in lakes, by postulating that changes in predator abundance can "cascade" down the food chain to affect phytoplankton. This recognition of the role of predators in lake ecosystems has led to the use of biomanipulation, particularly the stocking of piscivorous (fish eating) fishes as a lake restoration tool (Shapiro et al. 1975; Jeppesen et al. 1997; Hansson et al. 1998). North-temperate lakes generally function as four trophic level systems comprised of phytoplankton, zooplankton, planktivorous fish, and piscivorous fish (Carpenter et al. 1985). A common goal of biomanipulation is to alter the food web to increase zooplankton grazing, thus reducing the accumulation of algal biomass. Reduction of planktivore biomass "releases" zooplankton from predation, allowing larger and more abundant zooplankton. Reducing planktivore biomass can be achieved by protecting or augmenting populations of piscivorous fishes (Horppila et al. 1998). An important impact of humans on lakes has been the reduction or elimination of piscivore populations due to overfishing and habitat alteration (Post et al. 2002). This decline of piscivores has likely amplified eutrophication effects as a result of changing food-web interactions.

In contrast with lakes, where it is recognized that nutrients and food-web interactions both play a role in determining plant biomass and productivity, predation in Atlantic Coast salt marshes has traditionally been assumed to be unimportant in regulating marsh plant (*Spartina*, cordgrass) productivity. This view has been challenged recently, as work in mid-Atlantic U.S. marshes has demonstrated an important role of periwinkle (*Littoraria*) herbivory in regulating *Spartina* production and biomass (Silliman and Zieman 2001; Silliman and Bertness 2002). This suggests that efforts to restore salt marsh communities may benefit from not only the traditional approach of restoring the system's hydrology and improving abiotic conditions for growth (i.e., nutrient enhancements), but may also be hastened by efforts to manipulate food-web interactions. For example, a temporary reduction of *Littoraria* abundances at restoration sites would benefit *Spartina* growth due to reduced grazing and scarring. Based on simple food-chain concepts, augmenting or protecting populations of blue crabs, a major predator of *Littoraria*, would also be expected to benefit *Spartina* restoration efforts (Silliman and Bertness 2002).

Similar patterns of relatively simple food-chain dynamics have been revealed in terrestrial systems. For example, the rapid suburban development in coastal canyons of southern California has left little remaining habitat, and what remains is highly fragmented. The mesopredator release hypothesis was proposed to explain the dramatic decline of scrub-breeding birds in these fragments. Crooks and Soulé (1999) reported that coyotes (*Canis latrans*), the top predator in the system, have been extirpated from all but the largest habitat patches. Sites lacking coyotes support large numbers of small carnivores (raccoon, grey fox, striped skunk, opossum, domestic cat), which are effective predators on birds and other small vertebrates. An increase in abundance of these mesopredators following the extirpation of coyotes in habitat patches is the likely explanation for the recent avifauna decline in these habitat fragments. Based on this work, efforts to restore the avifauna would not be expected to respond to restoration efforts aimed at improving bird habitat; managing bird predators would perhaps be a more productive approach.

In a similar way, ecologists have recently elucidated the central role of top predators in structuring terrestrial food web in the Rocky Mountains of the western United States. Riparian ecosystems in the Greater Yellowstone Ecosystem (and much of the western United States) have undergone declines over the past century (Ripple et al. 2001; Ripple and Beschta 2003). An important aspect of this decline has been the unexplained recruitment failure of riparian trees such as native cottonwoods and aspens. While a number of possible explanations have been examined, evidence is emerging that food-web interactions play an important role in maintaining riparian vegetation structure (Beschta 2003). More specifically, wolves were extirpated from Yellowstone in the 1920s, which coincided with riparian tree recruitment failure. Reintroduction of wolves in the mid-1990s has altered the foraging behavior of elk (Ripple et al. 2001; Ripple and Beschta 2003); where elk are vulnerable to wolf predation, woody plants are now recovering from past unimpeded browsing by herbivores. Emerging evidence suggests that the regeneration of riparian vegetation associated with wolf reintroductions may have far-reaching positive effects on the broader riparian ecosystems. There is an expectation of long-term benefits to avifaunal communities by improving bird nesting habitat (Berger et al. 2001). Benefits are also expected for aquatic ecosystems, including stabilization of stream banks, strengthened linkages between riparian and riverine habitats, and moderation of water temperatures (Osborne and Kovacic 1993).

Complex Interactions

The above examples illustrate how simple food chains can be useful models for guiding ecological restoration efforts. Yet chain-like interactions may not accurately describe many systems, which are often considerably more complex. Here, we illustrate the value of recognizing food-web complexity, predator-mediated competition, and apparent competition in a restoration context. In addition to food webs being complex, energy and nutrients also move across habitat boundaries and may have important dynamic implications (Polis et al. 2004). Top-down control can be dampened or reinforced by energy "subsidies" from outside the focal habitat, which can cascade to lower trophic levels (Nakano et al. 1999; Polis 1999; Nakano and Murakami 2001). Recognition of landscape context and cross-habitat linkages represents an important conceptual shift in food-web ecology of the past decade (Polis et al. 2004) with potential implications for ecological restoration.

An example of apparent competition in natural systems is the introduction of the exotic brown tree snake (*Boiga irregularis*) to Guam (Savidge 1987). This introduction has caused the near complete elimination of the island avifauna. A simple predator-prey (snake-bird) model would predict snake populations to decline following local extirpation of the avifauna. But *Boiga* are generalist predators, readily consuming alternative prey such as small mammals and lizards. Because of this, *Boiga* has maintained high population densities, even after eliminating bird populations. In effect, the availability of alternative prey sustained high *Boiga* populations, thereby preventing avifaunal recovery.

Studies on islands provide strong evidence for the importance of food-web interactions when conducting ecosystem-level restoration. The eight California Channel Islands off the coast of southern California have been the subject of intensive restoration efforts during recent years. During much of the nineteenth and twentieth centuries, Santa Cruz Island supported exotic populations of cattle, sheep, and pigs, which adversely impacted the native plant community. Restorationists initiated a program to eradicate cattle and sheep. Following the decline of these two exotic herbivores, European fennel (*Foeniculum vulgare*) rapidly became the dominant plant species on the island (Zavaleta et al. 2001). This improved the plant forage base for feral pigs, resulting in an increase in pig numbers. Feral pigs have subsequently devastated native plant communities as a result of their digging and grubbing (Power 2001). These interactions would not have been predicted from a simple herbivore-plant model, as they involve a series of direct and indirect interactions among a mix of native and exotic plants and herbivores.

Food-web interactions involving predators on Santa Cruz Island also have restoration significance (Roemer et al. 2001; Roemer et al. 2002). Santa Cruz Island historically supported two carnivores—the endemic (and endangered) island fox (*Urocyon littoralis*) and the island spotted skunk (*Spilogale gracilis amphiala*). Introduction of feral pigs in the mid-nineteenth century expanded the prey resource base, ultimately allowing the island to be colonized by golden eagles (*Acquila chrysaetos*) from the mainland. Golden eagles have since become significant fox predators, with the result that the endemic island fox has declined dramatically (Roemer et al. 2002). In turn, skunk populations have increased due to competitive release from their main competitor, the island fox. As with the fennel-pig interactions described above, recognizing these more complex food-web interactions will be a central part of developing a restoration strategy for these island ecosystems. These examples demonstrate the potential role of indirect food-web interactions in determining ecosystem response to restoration. Simple food-chain models would not have predicted the observed changes, thus underscoring the importance of being familiar with the other types of food-web interactions.

Because islands are isolated ecosystems, they are free from the heavy influence of landscape context that can complicate restoration at mainland sites. In addition, islands are conducive for whole-ecosystem experimental approaches to restoration, allowing comparisons between experimental and reference ecosystems (Donlan et al. 2002). Yet the majority of restoration projects occur on mainland systems, meaning that restoration sites are nested within a broader landscape context (Ehrenfeld and Toth 1997). For example, while the boundaries of a wetland restoration site may be easily delimited, this target ecosystem is connected in diverse ways to its broader landscape context. Nutrients and consumers may be imported or exported from the wetland via a connecting stream, while mobile consumers (mammals, birds, insects) move across the wetland boundary. Consumers may be dependent on the restoration site to satisfy some needs, and areas outside the restoration site for others (i.e., feeding grounds, reproductive areas, refuge from predators). While restorationists may have some control over what happens within the boundaries of the restoration site, the broader linkages to the surrounding landscape are likely beyond their control. A food-web approach recognizes linkages beyond the boundaries of the restoration site and includes the broader landscape and ecosystem context of ecological restoration (Ehrenfeld and Toth 1997; Nakano and Murakami 2001).

A dramatic example in which the dynamics of distinct habitats are linked by mobile consumers is that of lesser snow geese, which migrate between arctic breeding grounds in Canada and wintering grounds in the central United States (Jefferies 2000; Jefferies et al. 2004). Intensification of agricultural activities and fertilizer use in the central United States during the past century has shifted snow goose wintering grounds from coastal marshes to agricultural areas. In effect, this has subsidized snow goose populations, allowing a 5‰ annual increase in snow goose population size. The effects of this population explosion are readily evident in the coastal breeding habitats around Hudson Bay, Canada, approximately

5,000 km from their winter feeding grounds, producing what has been described as a spatially subsidized trophic cascade (Jefferies et al. 2004). Goose overabundance has intensified grazing and grubbing in breeding grounds. The local impacts of this range from decreased plant cover and productivity, to the transformation of intertidal salt marshes to bare mudflats, a process involving positive feedback mechanisms analogous to that of desertification (Jefferies 2000; Jefferies et al. 2004). Subsequent changes in ecosystem processes, as well as declines in bird and insect communities have also been documented (Jefferies et al. 2004). Restoration of breeding ground habitat would likely necessitate wholesale changes in agricultural management practices in the United States, an unlikely prospect considering the remoteness of the impacted habitat and the vast spatial separation between the two areas. This is a clear example of how the dynamics of spatially separated habitats can be closely linked trophically, and it highlights the need to better understand landscape-level food-web linkages (Polis et al. 2004).

Invasions and Reintroductions

Biological invasions are of global concern because of mounting economic and ecological costs (Lodge and Shrader-Frechette 2003). Exotic species can pose major barriers to achieving restoration goals, which are often focused on native species and communities (D'Antonio and Chambers, this volume). Yet with accumulating numbers of exotics, eradication may not be compatible with restoration goals due to food-web interactions involving native and exotic species (Box 8.1). In addition, exotics are not always considered harmful (Ewel and Putz 2004). In the Laurentian Great Lakes, exotics have adversely affected native biodiversity, though food chains comprising exotic species now support valuable sport fisheries, and the native predators in these systems are now partially reliant on exotic prey (Kitchell et al. 2000). Indeed, non-native species are sometimes used for achieving desired restoration goals and providing ecosystem functions (Ewel and Putz 2004). This does not negate the adverse impacts that exotic species have had on global biodiversity (Wilcove et al. 1998), and reliance on exotics warrants thoughtful consideration of costs, benefits, and other constraints to restoration. Once established, many undesirable invasive species are difficult to control since they tend to be r strategists, with high reproductive rates, broad environmental tolerances, and high dispersal abilities (Elton 1958). In addition, "disturbed" systems, the very sites that require restoration, are more likely to be invasible (Mack et al. 2000), and invasive species may themselves be an agent of disturbance that can promote further invasions, leading to what has been termed an invasion "meltdown" (Ricciardi and MacIsaac 2000). There are also many cases where exotics prevent the reestablishment of desired native species (Simberloff 1990; Vitousek 1990), whose recovery is often a primary goal of ecological restoration (Bowles and Whelan 1994).

Several trout species (brown trout, brook trout, lake trout, rainbow trout) have been widely introduced throughout the world. These species have generally been viewed as "desirable" exotics since they provide valuable recreational fisheries. Yet as the broader ecosystem and food-web consequences of these introductions have been documented, this perspective is shifting toward a more cautious view of trout introductions (Flecker and Townsend 1994; Knapp et al. 2000; Schindler et al. 2000). A notable example of food-web interactions involving exotic trout and native species in a restoration context can be seen in the

Box 8.1

Complex Food Webs and Management of Exotic Species

Conventional wisdom suggests that undesired exotics should be controlled during restoration. But this may not always be the preferred course of action (Ewel and Putz 2004), and the ever-increasing numbers of invaders makes removal decisions more complex. What happens when a desired native animal species comes to depend on an exotic plant for feeding or nesting habitat (Zavaleta et al. 2001)? This was the case in the southwestern United States, where declines in native riparian vegetation forced the endangered Southwestern willow flycatcher (*Empidonax traillii extimus*) to rely on exotic salt cedar *Tamarix* for habitat. Removal of *Tamarix*, in the absence of concurrent efforts to restore native vegetation, would likely adversely affect this endangered species, underscoring the need for thoughtful consideration of secondary effects of invasive species removal.

Another concern is the growing number of systems with several exotics interacting at different trophic levels. Again, food-web interactions may be such that exotic species control may have unexpected consequences for desired native species. Although there are many scenarios, Smith and Quin (1996) reported that declines of Australian island-dwelling mammals were most severe on islands containing both exotic predators (cats, foxes) and prey (rabbits, mice). To explain this pattern, they proposed a "hyperpredation" hypothesis, in which exotic predator populations were maintained at artificially high levels due to consumption of exotic prey, thereby increasing predation rates on native prey species. This process is analogous to apparent competition—in that alternative prey leads to increased predation rates on native species (Courchamp et al. 2000; Courchamp et al. 2003), and highlights the diversity of trophic interactions that can occur where food webs comprise a mix of exotic and native species. In this example, modeling results of Courchamp et al. (1999) indicate that simultaneous control of exotic predators and prey would be the best strategy for conserving native island vertebrate species.

Literature Cited

Courchamp, F., J. L. Chapuis, and M. Pascal. 2003. Mammal invaders on islands: Impact, control and control impact. *Biological Reviews* 78:347–383.

Courchamp, F., M. Langlais, and G. Sugihara. 1999. Control of rabbits to protect island birds from cat predation. *Biological Conservation* 89:219–225.

Courchamp, F., M. Langlais, and G. Sugihara. 2000. Rabbits killing birds: Modelling the hyperpredation process. *Journal of Animal Ecology* 69:154–164.

Ewel, J. J., and F. E. Putz. 2004. A place for alien species in ecosystem restoration. *Frontiers in Ecology and the Environment* 2:354–360.

Smith, A. P., and D. G. Quin. 1996. Patterns and causes of extinction and decline in Australian conilurine rodents. *Biological Conservation* 77:243–267.

Zavaleta, E. S., R. J. Hobbs, and H. A. Mooney. 2001. Viewing invasive species removal in a whole-ecosystem context. *Trends in Ecology & Evolution* 16:454–459.

Colorado River below Glen Canyon Dam (Stevens et al. 2001). The population size of the native humpback chub (*Gila cypha*) has declined precipitously in the last decade (GCMRC 2003). In response, restoration has focused on the removal of rainbow trout (*Oncorhynchus mykiss*) to examine whether trout predation on juvenile chub is limiting their recovery (Marsh and Douglas 1997).

While trout are common exotic species, populations of these same trout species are often extirpated in their native range due to loss of habitat, water quality degradation, exploitation, obstructions to migration, and exotics (Donald and Alger 1993; Gunn et al. 2004). Not only are trout viewed as "sensitive," but theory indicates that extinction risk increases with body size and trophic level and that top predators are vulnerable to habitat fragmentation and degradation (Pimm 1991). This suggests that top predators would be particularly difficult to reestablish (Lawton 2000). Indeed, for a reintroduction to succeed, reintroduced individuals must survive at low population levels and successfully reproduce in spite of predators, competitors, and pathogens. While these are the same challenges faced by invasive species, this highlights the need to better understand food-web interactions involving exotic and native species in the context of ecological restoration.

While biological invasions are an important aspect of global environmental change, and of great importance to ecological restoration (Vitousek et al. 1996), human alteration of physical processes in ecosystems may also have important food-web implications. Restoring or maintaining natural flow regimes is critical for maintaining the integrity of riverine ecosystems (Poff et al. 1997; Richter et al. 1997). In Pacific Northwest rivers, human alteration of stream flow patterns has disrupted food-web interactions (Power et al. 1996a; Wootton et al. 1996). In response, there has been interest in how linkages between flood disturbance and food-chain length in rivers could guide the restoration of riverine food chains (Power et al. 1996b; Marks et al. 2000). In unregulated streams in the southwestern United States, the natural flooding regime has allowed the continued persistence of native fishes despite the presence of exotic predatory fishes (Meffe 1984). Similarly, the occurrence of seminatural flow regimes in dammed rivers during high precipitation years resulted in greater dominance of natives fishes (Probst and Gido 2004). Recognition that natural flow regimes promote the persistence of desired native species has been the basis for experimental flow releases on the Colorado River aimed at rebuilding aquatic habitats that were lost following dam construction (Valdez et al. 2001).

Food-Web Assembly

Ecological communities are not static entities but rather are dynamic in their composition, typically accumulating species through time following disturbances. Community ecologists have examined whether simple rules and the order of species introductions govern the composition of ecological communities. These ideas comprise what are known as the study of ecological assembly rules, which have played a central role in community ecology (Diamond 1975; Weiher and Keddy 1999). It is important to note that "communities" studied by community ecologists are most often a single trophic group (i.e., "the plant community" or "the bird community") (Drake 1990). Few studies have examined ecological assembly involving interacting species across several trophic levels. One approach to examine ecological assembly has been to assemble food webs in small containers or laboratory beakers. These microcosm food-web studies generally find that changing the sequence of species introduction during food-web assembly can produce very different community outcomes (Robinson and Dickerson 1987; Drake 1990). For example, a species that is competitively dominant under one set of circumstances may be unable to establish given a different assembly scenario (Drake 1990, 1991). Simulation models of food-web assembly generally predict that species-rich, complex food webs better resist invaders and their disruptive impacts (Post and Pimm 1983; Robinson and Dickerson 1987; Drake 1991). These studies also indicate that food webs with more links per species are more resistant to invasions (Robinson and Valentine 1979; Post and Pimm 1983). Though the applicability of microcosm studies and simulation models to real ecosystems is uncertain, this work suggests that species diversity, food-web connectivity, and introduction sequence may be important considerations in ecological restoration.

One example of food-web assembly concepts being incorporated into ecological restoration involves lake restoration efforts in the region of Sudbury, Ontario (Gunn 1995). Following the successful control of industrial sulfur emissions in the region, lake pH has improved to levels (pH > 5.5–6.0) capable of supporting top predators such as lake trout (*Salvelinus namaycush*) and smallmouth bass (*Micropterus dolomieu*). The success of predator reintroduction has recently been examined in these acid-recovering Ontario lakes (Gunn and Mills 1998; Snucins and Gunn 2003). Lake trout, the native top predator, recovered (i.e., recruited successfully in the absence of stocking) rapidly in lakes with few fish species, while in species-rich systems lake trout were slow or even unable to reestablish. This suggests that community attributes or reintroduction order (priority effects) may play a role in the recovery of this species. In contrast, reintroduced smallmouth bass established rapidly, regardless of community composition (Snucins and Gunn 2003). Smallmouth bass have well-documented predatory impacts on forage fishes (Whittier and Kincaid 1999; Findlay et al. 2000) and adverse competitive impacts on lake trout (Vander Zanden et al. 1999), though the strength of smallmouth bass–lake trout interactions is mediated by the presence of pelagic forage fishes (Vander Zanden et al. 2004). Restoration of native community assemblages in these lakes will require further attention to priority effects and the order of species reintroductions (Evans and Olver 1995; Gunn and Mills 1998; Snucins and Gunn 2003). In acid-recovering lakes, lake trout should be reintroduced as early as possible in the reassembly process (Snucins and Gunn 2003), although such a strategy may limit the subsequent chance of successfully establishing native prey fishes that are vulnerable to lake trout predation. These lakes should also be protected from unauthorized introductions of rock bass and smallmouth bass, at least until self-sustaining lake trout populations establish. Lake trout are a critical component of shield lake ecosystems—not only do they provide an important fishery, but they are also an important indicator of ecosystem integrity (Gunn et al. 2004). An understanding of food-web interactions in these systems suggests that species introductions should be controlled *during* the restoration process, at least until desired components of the community have established. Yet in other cases, desired native species may come to depend on exotics in various ways (Kitchell et al. 2000; Zavaleta et al. 2001), such that the broader food-web and ecosystem consequences of exotic species removals also need to be carefully considered (Zavaleta et al. 2001).

Application of Stable Isotopes to Restoration

Restoration efforts have traditionally targeted individual species, guilds, or communities, though there is increasing interest in restoration of ecosystem-level processes such as natural flow regimes in rivers (Poff et al. 1997), or fire regimes in terrestrial systems (Baker and Shinneman 2004). Restoration of food-web interactions has also been discussed as a potential restoration goal that incorporates aspects of ecosystem function (Palmer et al. 1997), although the idea has not often been applied in restoration projects. The key reason has been that monitoring food webs is not a trivial task: food webs are complex, and trophic interactions are highly variable in space and time. Stable isotope techniques are used increasingly to infer the movement of energy in food webs (Peterson and Fry 1987; Dawson et al. 2002). Ratios of stable isotopes ($^{13}C/^{12}C$ and $^{15}N/^{14}N$, expressed as δ notation relative to a known standard), vary predictably from resource to consumer tissues. For example, plants with C_4 photosynthetic pathways are enriched in ^{13}C relative to C_3 plants. These differences in plant $\delta^{13}C$ are preserved in consumer tissues, such that $\delta^{13}C$ is an indicator of the ultimate sources of carbon in food webs. In contrast, protein biosynthesis and catabolism tend to excrete the lighter N isotope, resulting in a 3%/4% enrichment of $\delta^{15}N$ from prey to predator. Nitrogen isotopes have therefore been used to infer trophic position of consumers in complex food webs (Vander Zanden and Rasmussen 2001).

Stable isotopes provide a powerful tool for monitoring and evaluating food-web linkages, greatly facilitating the incorporation of food-web approaches into restoration ecology. For example, Gratton and Denno (unpublished) used stable isotopes to monitor arthropod food webs in New Jersey salt marshes that have been restored to *Spartina* following the extirpation of exotic *Phragmites*. The trophic position of most consumers including the top predatory spiders were indistinguishable from those in reference *Spartina* habitats with no history of *Phragmites* invasion (Figure 8.4) indicating that trophic interactions among arthropod consumers had been largely reestablished in restored habitats in less than five years. In the same marshes, Currin et al. (2003) used stable isotopes to show that benthic microalgae and *Spartina*-derived organic matter were a significant component of the diet of mummichogs, *Fundulus heteroclitus*, in *Spartina*-dominated marshes. Reliance on these resources was much lower in *Phragmites*-invaded areas. Energy sources for fish in restored marshes were intermediate between *Phragmites* and *Spartina* marshes. Thus, stable isotopes were useful in delineating resources use by consumers in degraded (invaded), restored, and reference habitats. In the case of arthropods, the isotope data suggests that consumers utilized resources derived primarily from the habitat in which they were collected and as habitats were restored, predators integrated into the local food webs.

Stable isotopes have also been used to assess the restoration of southern California salt marshes (Kwak and Zedler 1997). Recent work indicates that marsh-derived algae and vascular plants, particularly *Spartina*, are important energy sources for invertebrates and fish (Kwak and Zedler 1997; Desmond et al. 2000; West and Zedler 2000; Madon et al. 2001), supporting the idea that these habitats should be managed as a single ecosystem. Mitigation and restoration projects in southern California coastal areas have focused either on the creation of basin or channel habitat for fishes or, alternatively, the creation of coastal salt marshes as habitat for endangered birds (i.e., light-footed clapper rail [*Rallus longirostris levipes*] and Belding's savannah sparrow [*Passerculus sandwichensis beldingi*]). While both are valid restoration targets, restoration of habitat for fishes and endangered birds may have erroneously been viewed as competing objectives (Kwak and Zedler 1997). In light of recent research documenting the importance of linkages between these two habitats (Desmond et al. 2000; West and Zedler 2000; Madon et al. 2001), future restoration efforts should focus on the creation of integrated channel–tidal salt marsh systems, which is expected to simultaneously accomplish both restoration objectives.

Food-web approaches are also valuable for assessing long-term changes and the restoration potential of ecosystems. Lake Tahoe has undergone substantial change during the past

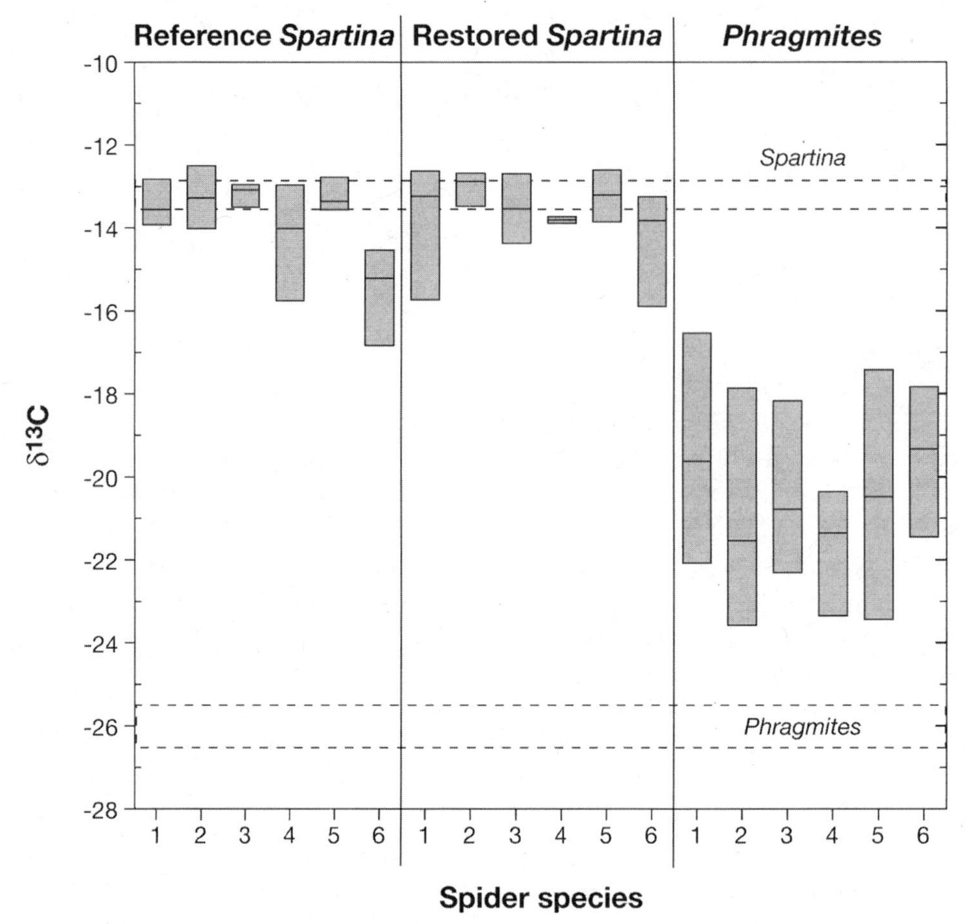

FIGURE 8.4 δ^{13}C stable isotope box-plot (median and interquartile range) of dominant spider predators from reference *Spartina*, restored *Spartina*, and *Phragmites*-dominated habitats within the Alloway Creek Watershed Restoration site (Salem County, New Jersey, USA). Dotted lines indicate the ranges of the basal resources (*Phragmites* or *Spartina*) in each habitat. Spiders in restored habitats are feeding on *Spartina*-based resources (herbivores and other predators) and are indistinguishable from the same species found in reference habitats, while *Phragmites*-collected spiders are feeding on non-*Phragmites*-based resources, likely detritivores. Spider species are (1) *Tetragnatha* sp., (2) *Pachygnatha*, (3) *Grammonota trivittata*, (4) *Hentzia* sp., (5) *Clubiona* sp., (6) *Pardosa* sp. From Gratton and Denno, unpublished.

century, including eutrophication, exotic introductions, and extirpation of the native top predator, Lahontan cutthroat trout (LCT; *Onorchynchus clarki henshawi*) (Jassby et al. 2001). Vander Zanden (2003) used stable isotopes to characterize historical food-web changes in Lake Tahoe based on analysis of contemporary and preserved museum specimens. The introduction of exotic freshwater shrimp (*Mysis relicta*) and lake trout have substantially disrupted the pelagic food-web structure of Lake Tahoe (Figure 8.5). These two exotics are extremely abundant and both have strong impacts on other species in the pelagic zone of Lake Tahoe. For these reasons, it is likely that these food-web alterations may limit the restoration potential of LCT in Lake Tahoe. Interestingly, native food webs in two Tahoe

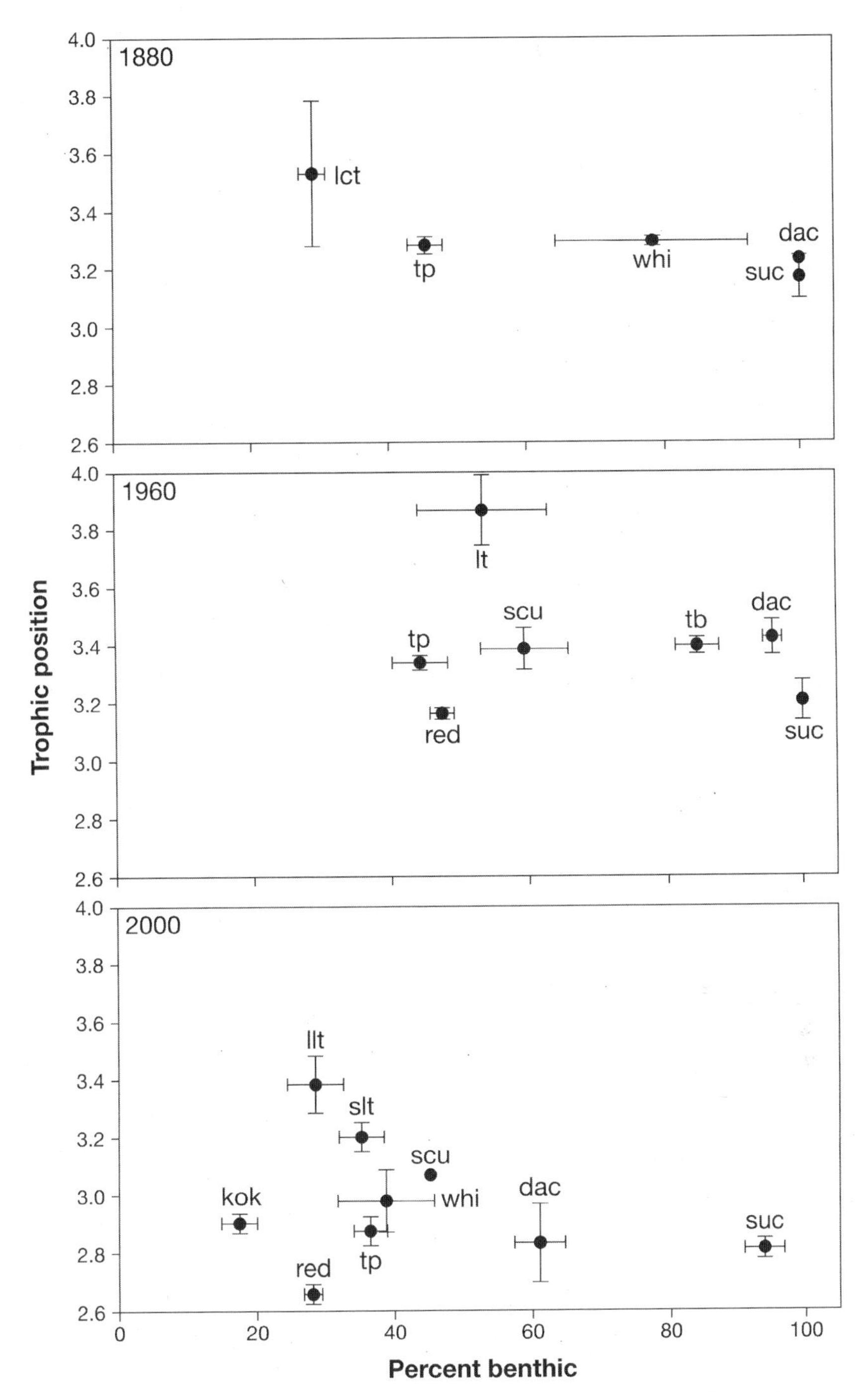

FIGURE 8.5 Food-web structure of Lake Tahoe based on stable isotope analysis of present-day and historical, museum-archived tissue samples. Food webs are presented for several time periods: 1880s, 1960s, 2000s. Species represented are lct = Lahontan cutthroat trout; tp = tui chub (pelagic morph); whi = mountain whitefish; suc = Tahoe sucker; dac = Lahontan speckled dace; scu = Paiute sculpin; red = Lahontan redside shiner; tb = tui chub (benthic morph); kok = kokanee salmon; lt = lake trout (all individuals); llt = large lake trout (>58 cm); slt = small lake trout (< 58 cm). The trend is toward increased pelagic production, and replacement of native Lahontan cutthroat trout with lake trout. Based on Vander Zanden et al. (2003).

basin headwater lakes (Cascade Lake and Fallen Leaf Lake) are still relatively intact despite some non-native introductions, and stable isotopes indicate that food webs in these lakes resemble that of Lake Tahoe prior to exotic introductions (Vander Zanden et al. 2003, unpublished data). These smaller and relatively unaltered systems are ideal candidates for "experimenting" with native LCT reintroductions, and the U.S. Fish and Wildlife Service has been reintroducing LCT into Fallen Leaf Lake since 2002. Ongoing studies are assessing the impact of lake trout predation on LCT in Fallen Leaf Lake during this experimental reintroduction, so that knowledge gained can be meaningfully applied to the restoration of LCT in other systems. The food-web component of this work also provides an opportunity to examine whether historical food-web niches are regained when formerly extirpated species are reestablished (Vander Zanden et al. 2003).

The above studies demonstrate the potential value of stable isotopes as a tool not only for documenting how food webs have been altered relative to reference conditions, but also for identifying important energy sources for restoration target organisms and assessing food-web recovery as systems move along restoration trajectories. Monitoring schemes that only consider presence/absence or abundance of species may overlook important food-web interactions as well as other important functional attributes of recovering ecosystems.

Areas of Research Need and Opportunity

Linkages between basic ecological research and restoration practice are weak, potentially hindering further advancements in both fields (Palmer et al. 1997; Hobbs and Harris 2001). "Bridging the gap," or perhaps "blurring the lines" between basic ecology and on-the-ground restoration represents a major challenge to both researchers and practitioners (Hobbs and Harris 2001). The good news is that restoration ecology has demonstrated that the degradation of ecosystems is often reversible, and there is ample evidence that restoration can be effective in nudging ecosystems toward a desired state (Dobson et al. 1997; Young 2000; Young et al. 2001). As a result, ecological restoration will play a growing role in global efforts to manage ecosystems to maximize ecosystem services and support biodiversity (Dobson et al. 1997). In this section, we have discussed how an understanding of food-web interactions can contribute to ecological restoration. Below, we identify some of the challenges and opportunities likely to be encountered in the application of food-web ecology to ecological restoration.

Food-Web Interactions and Adaptive Management

In some ecosystems, food-web interactions are critical in structuring ecosystems, while in other ecosystems, habitat and bottom-up factors likely drive ecosystem dynamics. How can we identify ecosystems in which predation and top-down forces are important for structuring the food web? Experimental manipulations of consumers and resources can be used to examine this, though in many systems the necessary manipulations are not practical or feasible. Observational studies and a "natural history" understanding of a system can provide some basis for identifying what factors are responsible for structuring a food web, though important food-web interactions may simply not be apparent without experimentation (Carpenter and

Kitchell 1993; Silliman and Zieman 2001; Silliman and Bertness 2002). In the absence of experimentation, there remains a need to understand whether ecosystems are dominated by top-down (predation) or bottom-up (habitat and productivity) forces, how these dual forces interact, and the role of indirect and other complex food-web interactions.

The above issues are difficult to resolve because ecological restoration projects are typically carried out at the whole-ecosystem level, while much of modern ecology is based on small-scale, but highly replicated, experiments. Can we scale up from small-scale experiments to the management and restoration of real ecosystems? Microcosm and small-scale experiments suffer from "cage-effects," whereby the results are simply an artifact of the artificial conditions of the experimental manipulation. Such findings cannot be generalized or "scaled up" to real ecosystems. Small-scale approaches are also likely to fail to capture relevant food-web processes such as cross-habitat linkages (Polis et al. 2004), complex trophic interactions (Carpenter 1996; Roemer et al. 2002), and the role of mobile predators (Soulé and Terborgh 1999). The obvious alternative is to conduct large-scale, whole-ecosystem manipulations (Carpenter et al. 1995; Zedler 2001). Restoration projects provide unique opportunities for whole-ecosystem experiments within an adaptive management, "learning-by-doing," framework (Zedler 2001; Holl et al. 2003). Such experiments speed the accumulation of knowledge about food webs and the response of ecosystems to management actions and hasten the application of ecological knowledge to restoration (Walters 1986; Donlan et al. 2002). In addition, ecological restoration has great potential to improve basic understanding of food webs and inspire new directions in food-web theory with more direct relevance to ecosystem management (Palmer et al. 1997).

The Backdrop of Exotics and Global Change

While the restoration potential of many ecosystems may be high, we have less optimism about restoration in light of accelerating species invasions, which may severely limit prospects for achieving restoration goals (Donlan et al. 2003). Combined with global climate change, it is certain that existing food webs will be torn apart, and new food webs will be reassembled (Root and Schneider 1993). The "rules of engagement" in ecosystems will change, yielding completely new outcomes and interactions (Lawton 2000). Restoring ecosystems within the context of the shifting backdrop of climate change and exotics seriously confounds the task at hand, necessitating a more complete incorporation of food-web, landscape, and ecosystem perspectives (D'Antonio and Chambers, this volume). Restoration ecology will draw increasingly from the field of invasion biology, and it will demand improved methods for controlling undesirable exotics. Perhaps a more critical challenge will be to find ways to manage ecosystems so as to maintain native biodiversity and ecosystem services in the face of invasive exotics (Kitchell et al. 2000; Rosenzweig 2003). In some cases, reliance on non-native species may be crucial for promoting restoration of energy flows and higher trophic levels, and food-web approaches will figure prominently into assessing the value and viability of such efforts (Kitchell et al. 2000; Ewel and Putz 2004). Sustaining native biodiversity will undoubtedly require intensive ecosystem management, which will be carried out by researcher-managers working at the interface of basic and applied ecology (Rosenzweig 2003).

Summary

The study of food webs represents a rapidly expanding subfield of ecology with the goal of understanding and predicting multispecies interactions. Though ecologists have long recognized the interconnectedness of species in ecosystems, new tools and new paradigms are allowing advances in our understanding of food-web interactions, particularly the role of predation and indirect effects in structuring ecosystems. We have suggested that restoration of food-web interactions may not necessarily follow restoration of the plant community or physical habitat features (i.e., the "field of dreams" paradigm—that is, "build it, and they will come") (Palmer et al. 1997). While restoration of habitat is critical, it is not guaranteed that the desired consumer taxa will recolonize and food webs will assemble as expected. The field of dreams approach may be sufficient in some systems, while inadequate in others. In addition, nuisance exotic species can be a barrier to achieving restoration goals. We presented several examples in which food-web interactions affect attributes of the ecosystem in important ways, with important and often poorly appreciated implications for restoration. Viewing restoration at the whole-ecosystem level and incorporating a food-web perspective can contribute in a real way to ecological restoration efforts (Soulé and Terborgh 1999; Roemer et al. 2002; Donlan et al. 2003). We offered that further advances will derive from restorationists incorporating "food-web thinking" into restoration projects and treating their efforts as ecosystem experiments. Food-web ecology has demonstrated the value of more holistic approaches for understanding species and ecosystems, lessons that will undoubtedly contribute toward efforts to restore ecosystems.

Acknowledgments

Thanks to Matt Diebel, Jeff Maxted, Helen Sarakinos, Sudeep Chandra, and Dave Pepin for helpful comments and discussion on the manuscript. Bill Feeny helped prepare the figures. Special thanks to the editors for their useful input and for giving us the opportunity to contribute to this book. M. Jake Vander Zanden received financial support from the Wisconsin Sea Grant Institute, the Great Lakes Fishery Commission, and the Wisconsin Department of Natural Resources.

Literature Cited

Abrams, P. A., B. A. Menge, G. G. Mittelbach, D. A. Spiller, and P. Yodzis. 1996. The role of indirect effects in food webs. In *Food webs: Integration of patterns and dynamics*, ed. G. A. Polis and K. A. Winemiller, 371–395. New York: Chapman & Hall.

Baker, W. L., and D. J. Shinneman. 2004. Fire and restoration of pinon-juniper woodlands in the western United States: A review. *Forest Ecology and Management* 189:1–21.

Berger, J., P. B. Stacey, L. Bellis, and M. P. Johnson. 2001. A mammalian predator-prey imbalance: Grizzly bear and wolf extinction affect avian noetropical migrants. *Ecological Applications* 11:947–960.

Beschta, R. L. 2003. Cottonwoods, elk, and wolves in the Lamar Valley of Yellowstone National Park. *Ecological Applications* 13:1295–1309.

Bowles, M. L., and C. J. Whelan. 1994. *Restoration of endangered species*. Cambridge, UK: Cambridge University Press.

Carpenter, S. R. 1996. Microcosm experiments have limited relevance for community and ecosystem ecology. *Ecology* 77:677–680.

Carpenter, S. R., N. F. Caraco, D. L. Correll, R. W. Howarth, A. N. Sharpley, and V. H. Smith. 1998. Nonpoint pollution of surface waters with phosphorus and nitrogen. *Ecological Applications* 8:559–568.

Carpenter, S. R., S. W. Chisholm, C. J. Krebs, D. W. Schindler, and R. F. Wright. 1995. Ecosystem experiments. *Science* 269:324–327.

Carpenter, S. R., J. J. Cole, J. R. Hodgson, J. F. Kitchell, M. L. Pace, D. Bade, K. L. Cottingham, T. E. Essington, J. N. Houser, and D. E. Schindler. 2001. Trophic cascades, nutrients, and lake productivity: Whole-lake experiments. *Ecological Monographs* 71:163–186.

Carpenter, S. R., and J. F. Kitchell. 1993. *The trophic cascade in lakes*. Cambridge, UK: Cambridge University Press.

Carpenter, S. R., J. F. Kitchell, and J. R. Hodgson. 1985. Cascading trophic interactions and lake productivity. *Bioscience* 35:634–639.

Crooks, K. R., and M. E. Soulé. 1999. Mesopredator release and avifaunal extinctions in a fragmented system. *Nature* 400:563–566.

Currin, C. A., S. C. Waignright, K. W. Able, M. P. Weinstein, and C. M. Fuller 2003. Determination of food web support and trophic position of the mummichog, *Fundulus heteroclitus*, in New Jersey smooth cordgrass (*Spartina alterniflora*), common reed (*Phragmites australis*), and restored salt marshes. *Estuaries* 26:495–510.

Dawson, T. E., S. Mambelli, A. H. Plamboeck, P. H. Templer, and K. P. Tu. 2002. Stable isotopes in plant ecology. *Annual Review of Ecology and Systematics* 33:507–559.

Denno, R. F., C. Gratton, H. Dobel, and D. L. Finke. 2003. Predation risk affects relative strength of top-down and bottom-up impacts on insect herbivores. *Ecology* 84:1032–1044.

Desmond, J., J. B. Zedler, and G. D. Williams. 2000. Fish use of tidal creek habitats in two southern California salt marshes. *Ecological Engineering* 14:233–252.

Diamond, J. M. 1975. Assembly of species communities. In *Ecology and evolution of communities*, ed. M. L. Cody and J. M. Diamond, 342–444. Cambridge: Harvard University Press.

Dobson, A. P., A. D. Bradshaw, and A. J. M. Baker. 1997. Hope for the future: Restoration ecology and conservation biology. *Science* 277:515–522.

Donald, D. B., and D. J. Alger. 1993. Geographic distribution, species displacement, and niche overlap for lake trout and bull trout in mountain lakes. *Canadian Journal of Zoology* 71:238–247.

Donlan, C. J., D. A. Croll, and B. R. Tershy. 2003. Islands, exotic herbivores, and invasive plants: Their roles in coastal California restoration. *Restoration Ecology* 11:524–530.

Donlan, C. J., B. R. Tershy, and D. A. Croll. 2002. Islands and introduced herbivores: Conservation action as ecosystem experimentation. *Journal of Applied Ecology* 39:235–246.

Drake, J. A. 1990. Communities as assembled structures: Do rules govern pattern? *Trends in Ecology & Evolution* 5:159–163.

Drake, J. A. 1991. Community-assembly mechanics and the structure of an experimental species ensemble. *American Naturalist* 137:1–26.

Ehrenfeld, J. G., and L. A. Toth. 1997. Restoration ecology and the ecosystem perspective. *Restoration Ecology* 5:307–317.

Elton, C. 1927. *Animal ecology*. New York: Macmillan.

Elton, C. S. 1958. *The ecology of invasions by animals and plants*. New York: John Wiley & Sons.

Evans, D. O., and C. H. Olver. 1995. Introduction of lake trout (*Salvelinus namaycush*) to inland lakes of Ontario, Canada: Factors contributing to successful colonization. *Journal of Great Lakes Research* 21 (suppl. 1): 30–53.

Ewel, J. J., and F. E. Putz. 2004. A place for alien species in ecosystem restoration. *Frontiers in Ecology and the Environment* 2:354–360.

Findlay, C. S., D. G. Bert, and L. Zheng. 2000. Effect of introduced piscivores on native minnow communities in Adirondack lakes. *Canadian Journal of Fisheries and Aquatic Sciences* 57:570–580.

Flecker, A. S., and C. R. Townsend. 1994. Community-wide consequences of trout introductions in New Zealand streams. *Ecological Applications* 4:798–807.

Fretwell, S. 1987. Food chain dynamics: The central theory of ecology? *Oikos* 50:291–301.

Grand Canyon Monitoring and Research Center (GCMRC). 2003. *An overview of status and trend information for the Grand Canyon population of humpback chub,* Gila cypha. Flagstaff: Grand Canyon Monitoring and Research Center.

Gunn, J. M. 1995. *Restoration and recovery of an industrial region*. New York: Springer-Verlag.

Gunn, J. M., and K. H. Mills. 1998. The potential for restoration of acid-damaged lake trout lakes. *Restoration Ecology* 6:390–397.

Gunn, J. M., R. J. Steedman, and R. A. Ryder. 2004. *Boreal shield watersheds: Lake trout ecosystems in a changing environment.* Boca Raton: Lewis Publishers.

Hairston, N. G. Jr, and N. G. Hairston, Sr. 1993. Cause and effect relationships in energy flow, trophic structure, and interspecific interactions. *American Naturalist* 142:379–411.

Hairston, N. G. Sr, S., F. E. Smith, and L. B. Slobodkin. 1960. Community structure, population control, and competition. *American Naturalist* 94:421–425.

Hall, S. J., and D. Raffaelli. 1991. Food-web patterns: Lessons from a species-rich web. *Journal of Animal Ecology* 60:823–842.

Hansson, L. A., H. Annadotter, E. Bergman, S. F. Hamrin, E. Jeppesen, T. Kairesalo, E. Loukkanen, P. Nilsson, M. Sondergaard, and J. Strand. 1998. Biomanipulation as an application of food-chain theory: Constraints, synthesis, and recommendations for temperate lakes. *Ecosystems* 1:558–574.

Hobbs, R. J., and J. A. Harris. 2001. Restoration ecology: Repairing the Earth's ecosystems in the new millennium. *Restoration Ecology* 9:239–246.

Holl, K. D., E. E. Crone, and C. B. Schultz. 2003. Landscape restoration: Moving from generalities to methodologies. *Bioscience* 53:491–502.

Holt, R. D. 1977. Predation, apparent competition, and the structure of prey communities. *Theoretical Population Biology* 12:197–229.

Holt, R. D., and J. H. Lawton. 1994. The ecological consequences of shared natural enemies. *Annual Review of Ecology and Systematics* 25:495–520.

Horppila, J., H. Peltonen, T. Malinen, E. Luokkanen, and T. Kairesalo. 1998. Top-down or bottom-up effects by fish: Issues of concern in biomanipulation of lakes. *Restoration Ecology* 6:20–28.

Hunter, M. D., and P. W. Price. 1992. Playing chutes and ladders: Heterogeneity and the relative roles of bottom-up and top-down forces in natural communities. *Ecology* 73:724–732.

Jassby, A. D., C. R. Goldman, J. E. Reuter, R. C. Richards, and A. C. Heyvaert. 2001. Lake Tahoe: Diagnosis and rehabilitation of a large mountain lake. In *The great lakes of the world (GLOW): Food-web, health, and integrity,* ed. M. Munawar and R. E. Hecky, 431–454. Leiden, The Netherlands: Backhuys Publishers.

Jefferies, R. L. 2000. Allochthonous inputs: Integration population changes and food web dynamics. *Trends in Ecology & Evolution* 15:19–22.

Jefferies, R. L., H. A. L. Henry, and K. F. Abraham. 2004. Agricultural nutrient subsidies to migratory geese and change in arctic coastal habitats. In *Food webs at the landscape level,* ed. G. A. Polis, M. E. Power, and G. R. Huxel, 268–283. Chicago: University of Chicago Press.

Jeppesen, E., J. P. Jensen, M. Sondergaard, T. Lauridsen, L. J. Pedersen, and L. Jensen. 1997. Top-down control in freshwater lakes: The role of nutrient state, submerged macrophytes and water depth. *Hydrobiologia* 342–343:151–164.

Jordan III, W. R., M. E. Gilpin, and J. D. Aber. 1987. *Restoration ecology: A synthetic approach to ecological research.* Cambridge, U.K.: Cambridge University Press.

Kitchell, J. F., S. P. Cox, C. J. Harvey, T. B. Johnson, D. M. Mason, K. K. Schoen, K. Aydin, C. Bronte, M. Ebener, M. Hansen, M. Hoff, S. Schram, D. Schreiner, and C. J. Walters. 2000. Sustainability of the Lake Superior fish community: Interactions in a food web context. *Ecosystems* 3:545–560.

Knapp, R. A., P. S. Corn, and D. E. Schindler. 2000. The introduction of non-native fish into wilderness lakes: Good intentions, conflicting mandates and unintended consequences. *Ecosystems* 4:275–278.

Kwak, T. J., and J. B. Zedler. 1997. Food web analysis of southern California coastal wetlands using multiple stable isotopes. *Oecologia* 110:262–277.

Lawton, J. H. 2000. *Community ecology in a changing world.* Luhe, Germany: Ecology Institute.

Leibold, M. A. 1989. Resource edibility and the effect of predators and productivity on the outcome of trophic interactions. *American Naturalist* 134:922–949.

Lindeman, R. L. 1942. The trophic-dynamic aspect of ecology. *Ecology* 23:399–418.

Lodge, D. M., and K. Shrader-Frechette. 2003. Nonindigenous species: Ecological explanation, environmental ethics, and public policy. *Conservation Biology* 17:31–37.

Mack, R. N., D. Simberloff, W. M. Lonsdale, H. Evans, M. Clout, and F. A. Bazzaz. 2000. Biotic invasions: Causes, epidemiology, global consequences, and control. *Ecological Applications* 10:689–710.

Madon, S. P., G. D. Williams, J. M. West, and J. B. Zedler. 2001. The importance of marsh access to growth of the California killifish, *Fundulus parvipinnis,* evaluated through bioenergetics modeling. *Ecological Modelling* 136:149–165.

Marks, J. C., M. E. Power, and M. S. Parker. 2000. Flood disturbance, algal productivity, and interannual variation in food chain length. *Oikos* 90:20–27.

Marsh, P. C., and M. E. Douglas. 1997. Predation by introduced fishes on endangered humpback chub and other native species in the Little Colorado River, Arizona. *Transactions of the American Fisheries Society* 126:343–346.

Martinez, N. D. 1991. Artifacts or attributes? Effects of resolution on the Little Rock Lake food web. *Ecological Monographs* 61:367–392.

Meffe, G. K. 1984. Effects of abiotic disturbance on coexistence of predator-prey fish species. *Ecology* 65:1525–1534.

Mills, L., S., M. E. Soulé, and D. F. Doak. 1993. The keystone-species concept in ecology and conservation. *Bioscience* 43:219–224.

Nakano, S., H. Miyasaka, and N. Kuhara. 1999. Terrestrial-aquatic linkages: Riparian arthropod inputs alter trophic cascades in a stream food web. *Ecology* 80:2435–2441.

Nakano, S., and M. Murakami. 2001. Reciprocal subsidies: Dynamic interdependence between terrestrial and aquatic food webs. *Proceedings of the National Academy of Sciences, USA* 98:166–170.

National Research Council. 1995. *Restoration of aquatic ecosystems: Science, technology, and the public.* Washington, D.C.: National Academy Press.

Oksanen, L., S. D. Fretwell, J. Arruda, and P. Liemala. 1981. Exploitation ecosystems in gradients of primary productivity. *American Naturalist* 118:240–261.

Osborne, L. L., and D. A. Kovacic. 1993. Riparian vegetation strips in water-quality restoration and stream management. *Freshwater Biology* 29:243–258.

Pace, M. L., J. J. Cole, S. R. Carpenter, and J. F. Kitchell. 1999. Trophic cascades revealed in diverse ecosystems. *Trends in Ecology & Evolution* 14:483–488.

Paine, R. T. 1966. Food web complexity and species diversity. *American Naturalist* 100:65–75.

Paine, R. T. 1980. Food webs: Linkage, interaction strength and community infrastructure. *Journal of Animal Ecology* 49:667–685.

Palmer, M. A., R. F. Ambrose, and N. L. Poff. 1997. Ecological theory and community restoration ecology. *Restoration Ecology* 5:291–300.

Persson, L. 1999. Trophic cascades: Abiding heterogeneity and the trophic level concept at the end of the road. *Oikos* 85:385–397.

Peterson, B. J., and B. Fry. 1987. Stable isotopes in ecosystem studies. *Annual Review of Ecology and Systematics* 18:293–320.

Pimm, S. L. 1982. *Food webs.* New York: Chapman & Hall.

Pimm, S. L. 1991. *The balance of nature?* Chicago: University of Chicago Press.

Poff, N. L., J. D. Allan, M. B. Bain, J. R. Karr, K. L. Prestegaard, B. D. Richter, R. E. Sparks, and J. C. Stromberg. 1997. The natural flow regime. *Bioscience* 47:769–784.

Polis, G. A. 1991. Complex trophic interactions in deserts: An empirical critique of food web theory. *American Naturalist* 138:123–155.

Polis, G. A. 1999. Why are parts of the world green? Multiple factors control productivity and the distribution of biomass. *Oikos* 86:3–15.

Polis, G. A., M. E. Power, and G. R. Huxel. 2004. *Food webs at the landscape level.* Chicago: University of Chicago Press.

Polis, G. A., and D. R. Strong. 1996. Food web complexity and community dynamics. *American Naturalist* 147:813–846.

Polis, G. A., and K. O. Winemiller. 1996. *Food webs: Integration of patterns and dynamics.* New York: Chapman & Hall.

Post, D. M. 2002. The long and short of food-chain length. *Trends in Ecology & Evolution* 17:269–277.

Post, J. R., M. Sullivan, S. Cox, N. P. Lester, C. J. Walters, E. A. Parkinson, A. J. Paul, L. Jackson, and B. J. Shuter. 2002. Canada's recreational fisheries: The invisible collapse? *Fisheries* 27:6–17.

Post, W. M., and S. L. Pimm. 1983. Community assembly and food web stability. *Mathematical Biosciences* 64:169–192.

Power, M. E. 1992. Top-down and bottom-up forces in food webs: Do plants have primacy? *Ecology* 53:733–746.

Power, M. E. 2001. Field biology, food web models, and management: Challenges of context and scale. *Oikos* 94:118–129.

Power, M. E., W. E. Dietrich, and J. C. Finlay. 1996a. Dams and downstream aquatic biodiversity: Potential food web consequences of hydrologic and geomorphic change. *Environmental Management* 20:887–895.

Power, M. E., M. S. Parker, and J. T. Wootton. 1996b. Disturbance and food chain length in rivers. In *Food webs: Integration of patterns and dynamics*, ed. G. A. Polis and K. O. Winemiller, 286–297. New York: International Thomson Publishing.

Power, M. E., D. Tilman, J. A. Estes, B. A. Menge, W. J. Bond, L. S. Mills, G. Daily, J. C. Castilla, J. Lubchenco, and R. T. Paine. 1996c. Challenges in the quest for keystones. *Bioscience* 46:609–620.

Probst, D. L., and K. B. Gido. 2004. Responses of native and nonnative fishes to natural flow mimicry in the San Juan River. *Transactions of the American Fisheries Society* 133:922–931.

Ricciardi, A., and H. J. MacIsaac. 2000. Recent mass invasion of the North American Great Lakes by Ponto-Caspian species. *Trends in Ecology & Evolution* 15:62–65.

Richter, B. D., J. V. Baumgartner, R. Wigington, and D. P. Braun. 1997. How much water does a river need? *Freshwater Biology* 37:231–249.

Ripple, W. J., and R. L. Beschta. 2003. Wolf reintroduction, predation risk, and cottonwood recovery in Yellowstone National Park. *Forest Ecology and Management* 184:299–313.

Ripple, W. J., E. J. Larsen, R. A. Renkin, and D. W. Smith. 2001. Trophic cascades among wolves, elk and aspen on Yellowstone National Park's northern range. *Biological Conservation* 102:227–234.

Robinson, J. V., and J. E. Dickerson. 1987. Does invasion sequence affect community structure? *Ecology* 68:587–595.

Robinson, J. V., and W. D. Valentine. 1979. Concepts of elasticity, invulnerability and invadability. *Journal of Theoretical Biology* 81:91–104.

Roemer, G. W., T. J. Coonan, D. K. Garcelon, J. Bascompte, and L. Laughrin. 2001. Feral pigs facilitate hyperpredation by golden eagles and indirectly cause the decline of the island fox. *Animal Conservation* 4:307–318.

Roemer, G. W., C. J. Donlan, and F. Courchamp. 2002. Golden eagles, feral pigs, and insular carnivores: How exotic species turn native predators into prey. *Proceedings of the National Academy of Sciences, USA* 99:791–796.

Root, T. L., and S. H. Schneider. 1993. Can large-scale climatic models be linked with multiscale ecological studies? *Conservation Biology* 7:256–270.

Rosenzweig, M. L. 2003. *Win-win ecology*. Oxford, UK: Oxford University Press.

Savidge, J. A. 1987. Extinction of an island forest avifauna by an introduced snake. *Ecology* 68:660–668.

Schindler, D. E., R. A. Knapp, and P. R. Leavitt. 2000. Alteration of nutrient cycles and algal production resulting from fish introduction into mountain lakes. *Ecosystems* 4:308–321.

Schoener, T. W. 1989. Food webs from the small to the large. *Ecology* 70:1559–1589.

Shapiro, J., V. Lamarra, and M. Lynch. 1975. Biomanipulation: An ecosystem approach to lake restoration. In *Symposium on water quality management through biological control*, ed. P. L. Brezonit and J. L. Fox, 85–96. Gainesville: University of Florida.

Silliman, B. R., and M. D. Bertness. 2002. A trophic cascade regulates salt marsh primary production. *Proceedings of the National Academy of Sciences, USA* 99:10500–10505.

Silliman, B. R., and J. C. Zieman. 2001. Top-down control of *Spartina* alterniflora production by periwinkle grazing in a Virginia salt marsh. *Ecology* 82:2830–2845.

Simberloff, D. 1990. Community effects of biological introductions and their implications for restoration. In *Engineered organisms in the environment: Scientific issues*, ed. D. R. Towns, C. H. Daugherty, and I. A. Atkinson, 128–136. Washington, DC: American Society for Microbiology.

Snucins, E. J., and J. M. Gunn. 2003. Use of rehabilitation experiments to understand the recovery dynamics of acid-stressed fish populations. *Ambio* 32:240–243.

Soulé, M. E., and J. Terborgh. 1999. *Continental conservation: Foundations of regional reserve networks*. Washington, DC: Island Press.

Stevens, L. E., T. J. Ayers, J. B. Bennett, K. Christensen, M. J. C. Kearsley, V. J. Meretsky, A. M. Phillips, R. A. Parnell, J. Spence, M. K. Sogge, A. E. Springer, and D. L. Wegner. 2001. Planned flooding and Colorado River riparian trade-offs downstream from Glen Canyon Dam, Arizona. *Ecological Applications* 11:701–710.

Summerhayes, V. S., and C. S. Elton. 1923. Contributions to the ecology of Spitsbergen and Bear Island. *Journal of Ecology* 11:214–286.

Valdez, R. A., T. L. Hoffnagle, C. C. McIvor, T. McKinney, and W. C. Leibfried. 2001. Effects of a test flood on fishes of the Colorado River in Grand Canyon, Arizona. *Ecological Applications* 11:686–700.

Vander Zanden, M. J., J. M. Casselman, and J. B. Rasmussen. 1999. Stable isotope evidence for the food web consequences of species invasions in lakes. *Nature* 401:464–467.

Vander Zanden, M. J., S. Chandra, B. C. Allen, J. E. Reuter, and C. R. Goldman. 2003. Historical food web structure and the restoration of native aquatic communities in the Lake Tahoe (CA-NV) basin. *Ecosystems* 6:274–288.

Vander Zanden, M. J., J. D. Olden, J. H. Thorne, and N. E. Mandrak. 2004. Predicting occurrences and impacts of bass introductions in north temperate lakes. *Ecological Applications* 14:132–148.

Vander Zanden, M. J., and J. B. Rasmussen. 2001. Variation in $d^{15}N$ and $d^{13}C$ trophic fractionation: Implications for aquatic food web studies. *Limnology and Oceanography* 46:2061–2066.

Vitousek, P. M. 1990. Biological invasions and ecosystem processes: Towards an integration of population biology and ecosystem studies. *Oikos* 57:7–13.

Vitousek, P. M., C. M. D'Antonio, L. L. Loope, and R. Westbrooks. 1996. Biological invasions as global environmental change. *American Scientist* 84:468–478.

Walters, C. J. 1986. *Adapative management of renewable resources*. New York: Macmillan.

Warren, P. H. 1989. Spatial and temporal variation in the structure of a freshwater food web. *Oikos* 55:299–311.

Weiher, E., and P. Keddy. 1999. *Ecological assembly rules—Perspectives, advances, retreats*. New York: Cambridge University Press.

West, J. M., and J. B. Zedler. 2000. Marsh-creek connectivity: Fish use of a tidal salt marsh in southern California. *Estuaries* 23:699–710.

Whittier, T. R., and T. M. Kincaid. 1999. Introduced fish in Northeastern USA lakes: Regional extent, dominance, and effects on native species richness. *Transactions of the American Fisheries Society* 128:769–783.

Wilcove, D. S., D. Rothstein, J. Dubow, A. Phillips, and E. Losos. 1998. Quantifying threats to imperiled species in the United States. *Bioscience* 48:607–615.

Wootton, J. T., M. S. Parker, and M. E. Power. 1996. Effects of disturbance on river food webs. *Science* 273:1558–1561.

Young, T. P. 2000. Restoration ecology and conservation biology. *Biological Conservation* 92:73–83.

Young, T. P., J. M. Chase, and R. T. Huddleston. 2001. Community succession and assembly. *Ecological Restoration* 19:5–18.

Zavaleta, E. S., R. J. Hobbs, and H. A. Mooney. 2001. Viewing invasive species removal in a whole-ecosystem context. *Trends in Ecology & Evolution* 16:454–459.

Zedler, J. B. 2001. *Handbook for restoring tidal wetlands*. Boca Raton: CRC Press.

Chapter 9

The Dynamic Nature of Ecological Systems: Multiple States and Restoration Trajectories

KATHARINE N. SUDING AND KATHERINE L. GROSS

One feature of ecological systems is that they are ever-changing and dynamic. As ancient Greek philosopher Heraclitus claimed, "You can never step in the same river twice." Moreover, rates and directions of change in systems are shaped increasingly by human activities. These effects can be intentional or the consequences of engineering of the systems and surrounding landscapes to provide specific services to humans. The dynamics of an ecological system, particularly of a system slated for restoration, is a function of many factors, some deterministic and some stochastic, working at several temporal and spatial scales.

In considering how systems change in restoration, we address several questions:

1. What types of trajectories characterize the recovery of degraded ecosystems? Is the pathway to recovery similar to the pathway to degradation?
2. Can we predict the end states of restoration pathways? Are they similar to states prior to degradation?
3. How will dynamics that occur on very different scales of space and time relate to one another? What should be the scale of focus?
4. How much inherent variability does an ecological system require for adequate recovery and adaptive capacity for change in the future?

In this chapter, we consider ecological theories that help address these questions and may reduce the risk of unpredicted or undesired change in restoration projects. While theory can help guide restoration efforts, it does not provide simple or universal answers for the challenges that confront restoration. Restoration efforts that document species turnover and environmental attributes over time can help test and refine ecological theory related to community dynamics. Links between restoration and community dynamics advance both the practice of restoration and theories of ecological dynamics. We survey the progress and the further potential of this connection.

Major Theories and Connection to Restoration

Over the last one hundred years, extensive work has documented how communities and ecosystems change in response to disturbance. Despite the extensive documentation of patterns (Figure 9.1), a general conceptual framework concerning the controls on species

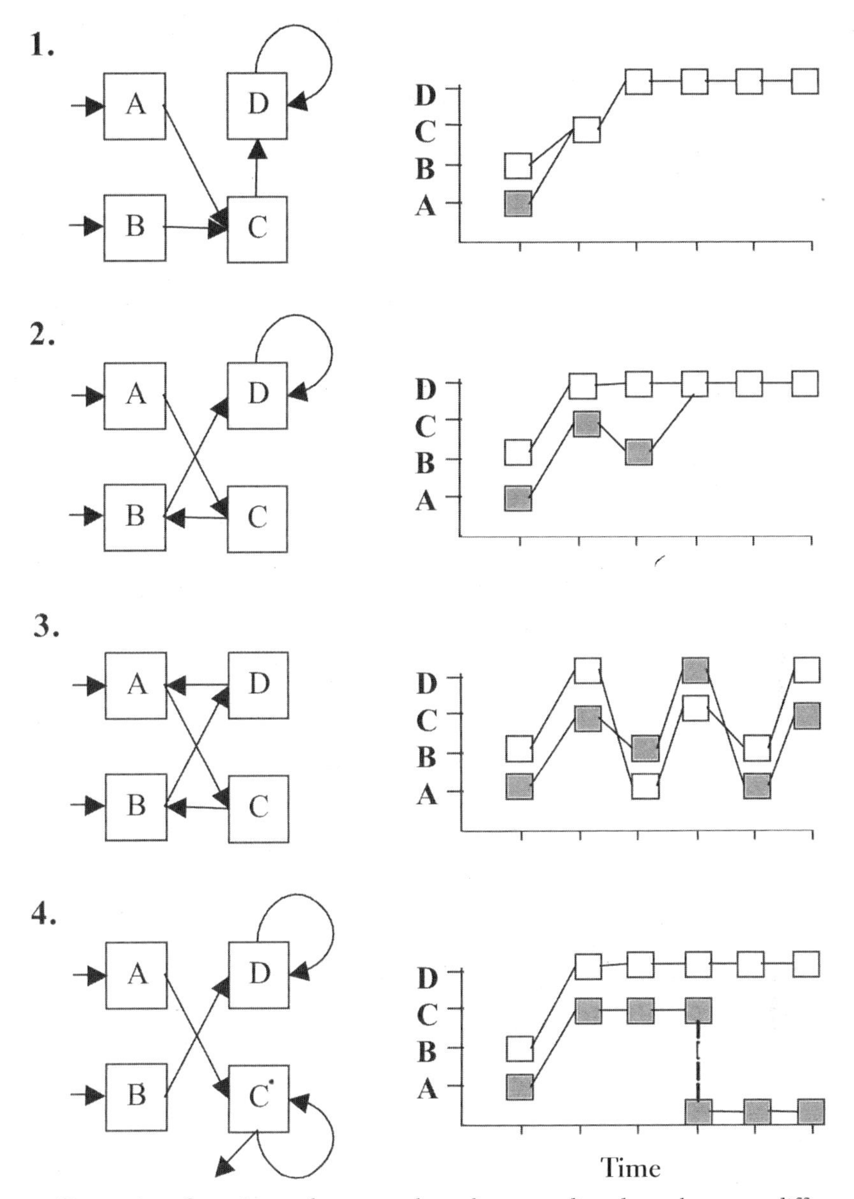

FIGURE 9.1 Dynamics of species replacement have been predicted to take many different forms. Four general patterns of trajectories, each with two starting points (assemblages A and B) are shown here: (1) Convergent trajectories where initial variability eventually converges to similar species composition, often termed the (D) equilibrium "climax" community. (2) Initially divergent trajectories that eventually converge to one equilibrium state. (3) Divergent trajectories that never converge and never reach a permanent state. (4) Divergent trajectories that go to two different stable states (C and D) and, in the case of C, experience an abrupt shift to a third state.

turnover and ecosystem development is still debated. Several contrasting views concerning the mechanisms and predictive nature of these dynamics persist today. In this chapter, we will focus on three views: equilibrium, multiple equilibrium, and non-equilibrium. We discuss each of these and relate them to the concept of fast and slow processes (*sensu* Rinaldi and Scheffer 2000) as a way to evaluate mechanisms of recovery.

Single Equilibrium Endpoint

Equilibrium systems are assumed to return to their predisturbance state or trajectory following disturbance (Table 9.1). This theory predicts a classical successional trajectory: steady, directional change in composition to a single equilibrium point (Clements 1916; Odum 1969) (Figure 9.2a). Recovery in an equilibrium framework is a predictable consequence of interactions among species with different life histories and the development of ecosystem functions. Strong internal regulation occurs through negative feedback mechanisms, including competition and herbivore/predator interactions, as well as climate-ecosystem couplings and life-history tradeoffs. Many of these mechanisms are considered aspects of community assembly rules (Weiher and Keddy 1999; Booth and Swanton 2002), although assembly rules do not necessarily assume single equilibrium dynamics.

In some cases, community development can proceed "spontaneously," with little or no intervention, to reach desirable target states (Prach et al. 2001; Khater et al. 2003; Novak and Prach 2003). Mitsch and Wilson (1996) argue that nature has a "self-design" capacity as species assemble themselves. However, the extent to which this capacity can be expressed in a recovery will depend on how degraded and isolated it has become prior to restoration efforts (Bakker and Berendse 2001). Some restoration efforts are designed to accelerate natural succession so that the ecosystem develops along the same trajectory as it would in the absence of intervention but reaches the goal endpoint sooner. For instance, restoring a severely degraded river back to its more natural flow regime via dam removal can enhance recovery of the surrounding plant communities (Rood et al. 2003; Lytle and Poff 2004). Similarly, prescribed burning of degraded grasslands can promote restoration of native plant assemblages, particularly if the fire management regime is applied according to

TABLE 9.1

General theories that attempt to predict how the composition and function of systems change over time and/or behave following a disturbance.

	Equilibrium	Multiple Equilibrium	Non-equilibrium
Assumptions	Climax equilibrium, unidirectional, continuous	Equilibrium, multidirectional, discontinuous	Persistent non-equilibrium, nondirectional, discontinuous
Permanent states	One (climax)	More than one	None
Trajectories	Convergent	Regime shifts, collapses	Divergent, arrested, cyclic
Predictability	High; based on species attributes	Moderate; possible but difficult	Low; chance and legacies important
Important factors	Species interactions, ecosystem development	Initial conditions, positive feedbacks, landscape position	Chance dispersal, stochastic events

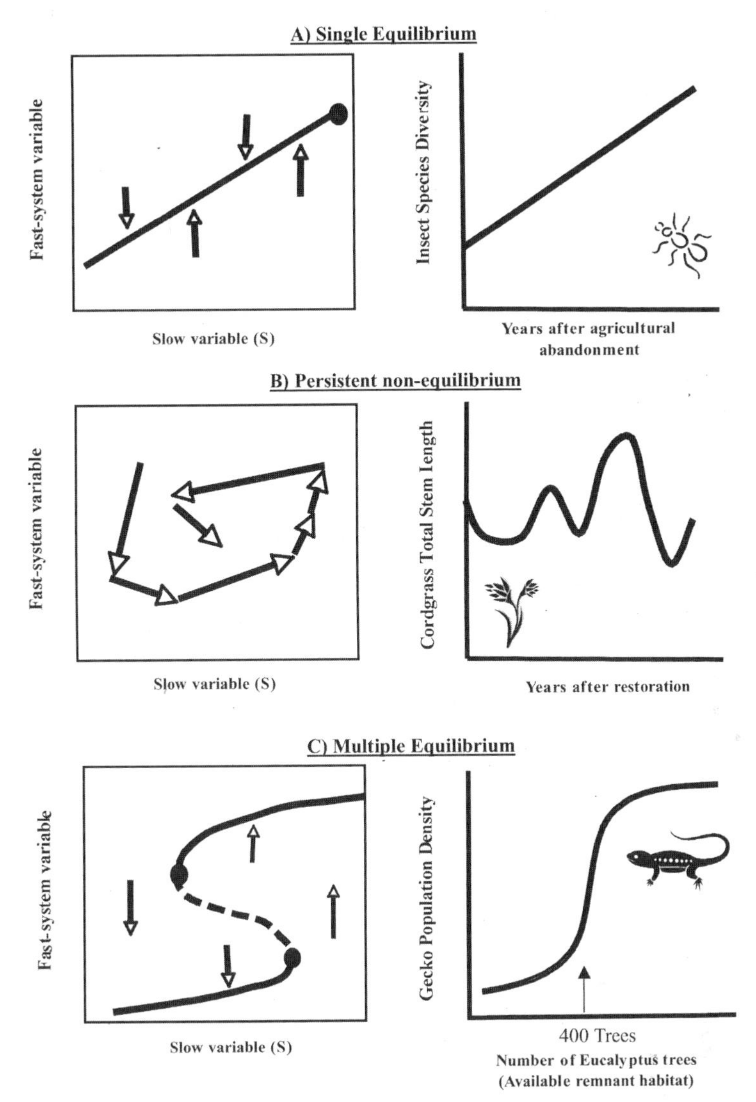

FIGURE 9.2 Examples of dynamics predicted by single equilibrium, persistent non-equilibrium, and multiple equilibium theories (A–C). For each, the left frame shows predicted combinations of "fast" and "slow" variables; arrows indicate direction of change if not at equilibrium. The right frame shows a stylized example from the ecological literature that is consistent with the ecological predictions. In A, changes in the slow and fast variables are linear and unidirectional. Insect species diversity increase linearly in a Minnesota old-field with years since abandonment. Increases in aboveground productivity with time is likely the slow variable that drives the change in insect diversity (Siemann et al. 1999). In B, a persistent non-equilibrium exists with no predictable trajectory. Total stem length of cordgrass (*Spartina foliosa*) shows high interannual variability and no directional trends in time since restoration in San Diego Bay, CA (Zedler and Callaway 1999). In C, at a single level of the slow variable there are two possible equilibrial states. Examples of a pattern predicted by this dynamic, shown in Figure 9.1(4), are strong threshold effects as the slow variable changes. For instance, in a fragmented Eucalyptus forest in Australia, the probability that a gecko species (*Oedura retiulcata*) persists decreases dramatically if the forest remnant contains less than 400 trees (Sarre et al. 1995).

historical patterns (Baer et al. 2002; Copeland et al. 2002). Thus, restoration of some communities can take a single equilibrium approach to spur recovery along a successional trajectory.

Multiple Equilibrium States

Theoretical models predicting multiple stable equilibriums (MSE) (Lewontin 1969; Law and Morton 1993; Rietkerk and vandeKoppel 1997; Pastor et al. 2002; van Nes and Scheffer 2003) may be applicable to restored systems where change among states can be discontinuous, abrupt, and have multiple trajectories. The crash of fishery stocks (Pauly et al. 2002), desertification of arid rangelands (Schlesinger et al. 1990; Foley et al. 2003), and the loss of coral reefs (Nystrom et al. 2000) are examples of discontinuous and abrupt change that suggest MSE.

Evidence that massive, irreversible shifts in species composition and ecosystem processes can occur with little forewarning has increased interest in how to identify thresholds before they are reached (Gunderson 2000; Scheffer et al. 2003). In addition, multiple equilibrium theory provides important insights into system dynamics that can be applied to the recovery of degraded systems (Hobbs and Norton 1996; Prober et al. 2002; Lindig-Cisneros et al. 2003; Suding et al. 2004). The feedbacks that maintain a system in a degraded "alternative" state are likely very different from those in the pristine or target state, and disruption of these feedbacks can be critical to the return to a target state. Consequently, the trajectory to recovery will probably be different from what caused the degradation. Restoration needs to consider positive feedbacks that can make the degraded state resilient to restoration efforts. For example, Prober et al. (2002) found evidence for positive feedbacks between soil nitrogen cycling and persistence of annual exotics in grassy Australian woodlands. They highlight the need for restoration to focus on nitrate-dependent transitions between annual and perennial understory states in these woodlands. Another example comes from wetlands, where secondary salinization may shift systems to different alternative states, and strong, self-reinforcing feedbacks can impede restoration efforts (Davis et al. 2003).

Restoration efforts can sometimes send degraded systems along unintended trajectories, providing evidence that these systems are capable of alternative states. In such cases, management efforts may need to address the issue of thresholds (Hobbs and Norton 1996; Suding et al. 2004). For example, in sand barren prairies in the midwestern United States, areas without frequent burning become dominated by woody vegetation. Reintroduction of fire alone was not effective in restoring these prairies because the woody vegetation (particularly *Salix humulis*) resprouts rapidly following fire (Anderson et al. 2000). Identifying the magnitude and direction of action that is needed in restoration to cross a threshold (e.g., from woody shrubland to prairie) has proven to be difficult (Stylinski and Allen 1999; Nielsen et al. 2003). For instance, adding nitrogen to coastal wetlands increased the height of *Spartina foliosa* and restored the nesting habitat for bird species, but this effect was only temporary. The tall canopies were not self-sustaining after nitrogen amendments ended (Lindig-Cisneros et al. 2003). In lakes, the shift to a turbid state due to nutrient loading (see Box 9.1) often requires more than a nutrient reduction program to bring about the reversal to a clear-water state (Bachmann et al. 1999).

Box 9.1
Two Examples of Application of Theory to Restoration and Management

1. Shallow lakes: In shallow lakes, the interactions among turbidity, nutrients, vegetation, and fish can produce alternative states (Beisner et al. 2003a; Scheffer et al. 1997). Lakes can exist either in a state where the water is clear and rooted plants are abundant or where the water is turbid and phytoplankton are abundant. In clear lakes, rooted plants stabilize the sediment, reducing turbidity, and provide refuges for fish that eat phytoplankton; under these conditions rates of change are relatively slow. However, if the plants are removed or if fishing pressure is high, turbidity blocks light and resuspends sediment for phytoplankton, causing a rapid and dramatic shift. Lakes can switch states for several reasons: increased phosphorus can cause the rooted plants to decline; a decrease in the algae-eating fish can encourage phytoplankton dominance; a disturbance can remove vegetation (Carpenter et al. 1999; Moss et al. 1996). A turbid lake can be restored by manipulating the feedbacks that maintain the system in the turbid state: increasing the population of fish that consume phytoplankton; decreasing the number of predators that eat the phytoplankton-consuming fish; reducing nutrient loading; and installing wave barriers to create refuges for plants (Bachmann et al. 1999; Dent et al. 2002). Feedbacks in this system are strong, and multiple actions may be needed to cause a conversion back to the clear lake state.
2. Grazed semiarid systems: State and transition models of rangeland vegetation dynamics split changes in rangeland systems into discrete states and describe processes that cause transitions between states (Westoby et al. 1989; Friedel 1991; Briske et al. 2003). For example, in an overgrazed system, grass cannot recover quickly and fire suppression enhances the survival of woody plant seedlings. Grazers do not eat the woody plants, and shrubs are able to invade the rangeland. Reduction of grazing intensity is not sufficient to restore the system to a healthy rangeland once this transition occurs (Friedel 1991); burning is needed to remove woody plants (Westoby et al. 1989). State and transition models have altered the general idea of rangeland management, refuting the general dogma that removing grazing from overgrazed rangeland is sufficient for recovery. State and transition models describe states, transitions, and thresholds largely qualitatively, and they base much of their classification on observation of change rather than explicit tests of stability and equilibrium. Such descriptions are accessible and understandable to managers and the general public (Briske et al. 2003; Stringham et al. 2003).

Persistent Non-equilibrium

Non-equilibrium theory assumes external factors play a larger role in the behavior of ecosystems than do internal processes, such as competition and predation. The role of chance and past history in community dynamics gained widespread acceptance in the 1970s and 1980s (Pickett et al. 1987b; Luken 1990). Empirical evidence from a variety of systems has shown that disturbance type, biological legacies, and chance can create multiple trajectories and influence rates of change (Drury and Nisbet 1973; Pickett et al. 2001). This perspective acknowledges the unpredictability of succession (Table 9.1) with no tendencies toward any one permanent state (Zedler and Callaway 1999; Bartha et al. 2003). However, it is often hard to distinguish between stochastic, non-equilibrium, and multiple equilibrium dynamics because it is difficult to demonstrate a stable equilibrium (Box 9.2).

Non-equilibrium theories predict divergent, cyclic, or arrested trajectories that never arrive at an equilibrium state. There are many examples that support this view. For instance, large fluctuations in precipitation are thought to prevent herbivores from regulating primary production, thereby minimizing negative feedbacks and equilibrium behavior in arid rangelands (Ellis and Swift 1988). Following the eruption of Mount St. Helens, Washington, USA, there were variable rates of recovery along several distinct pathways (Franklin and MacMahon 2000). The most rapid rates of recovery were in areas where organisms happened to survive the eruption. In areas completely bare after the eruption, chance colonization determined changes over time: species abundance patterns were predicted extremely well by a stochastic model based on observed frequencies and random accumulation of species (Del Moral 1998, 1999).

Stochastic effects resulting from isolation and dispersal limitations have also been shown to override the deterministic effects of competition that would otherwise lead to convergence (Underwood and Fairweather 1989; Del Moral 1998; Foster et al. 1998). Restorationists

Box 9.2
Testing for Equilibrium, Stability, and Resilience

The idea that communities can develop into alternative or multiple persistent states was first proposed by Lewontin in 1969, and the existence of multiple stable equilibrium (MSE) has met with debate ever since (Sutherland 1974; Connell and Sousa 1983; Grover and Lawton 1994). It has proven difficult to test for the existence of MSE due to several experimental and conceptual requirements (Connell and Sousa 1983; Grover and Lawton 1994; Petraitis and Latham 1999; Sutherland 1974) that are often hard, if not impossible, to meet in natural systems. Connell and Sousa (1983) dismissed most evidence of multiple stable states as insufficient either because researchers used inappropriate scales of space and time (e.g., the study didn't last for one complete turnover of all individuals to establish stability); because they used artificial controls that would not persist naturally (e.g., removal of predators); or because the physical environment differed between the alternate states. While these criteria are becoming more relaxed, rigorous tests of whether a degraded system truly represents an alternative and stable equilibrium are beyond the scope of most restoration efforts.

Experimentation and monitoring of the responses to management can provide evidence as to whether positive feedbacks and biotic constraints will affect the restoration of a degraded system. Given the risk of inappropriate management sending the degraded system in an unintended direction, it might be more costly to assume that a single dimension controls system dynamics rather than that alternative states exist and are determined by interactions among many factors (Gunderson 2000). Adaptive management (monitoring with changes in restoration strategies or implementation procedures) must be used in most restoration contexts—the likelihood of "surprises" should be expected. Perhaps more illuminating is whether the direction and rate of change in a system is uniform and continuous. If so, then past work on the system should allow some degree of linear prediction of future behavior. If not, more complex forecasting might be necessary.

should be particularly aware of divergence in fragmented areas or areas where disturbance has removed seedbank, propagule or larval sources, or where the abiotic environment is highly variable. The role of environmental variability in determining recovery trajectories has been well documented in running-water systems (Poff et al. 1997); less attention has been paid to the influence of variability as it relates to restoration in terrestrial systems. Dispersal limitation is increasingly being identified as a critical constraint in restoration projects in terrestrial and aquatic systems (Underwood and Fairweather 1989; Seabloom et al. 2003).

Restoration in a highly stochastic system may require relatively broad goals, such as restoring functional group presence or particular ecosystem function rather than particular species or community type (Palmer et al. 1997). It may also tactically assume either bounded regions of character states or eventual transition into multiple equilibrium dynamics. For example, dynamics in old-field vegetation may not progress in any one direction or toward any state until, by chance, the community achieves a certain density of woody species. Subsequent dynamics may shift behavior to a more equilibrium pathway (Foster and Gross 1999) (Figure 9.1, 2).

Theories on Mechanism of Change: The Roles of Fast and Slow Processes

Understanding the mechanisms that govern the behavior of change is difficult because it requires relating dynamics that occur on very different scales of space, time, and ecological organization (Pickett et al. 1987a; Rinaldi and Scheffer 2000; Beisner et al. 2003b) (Table 9.2).

TABLE 9.2

Unraveling the mechanisms governing change requires the interfacing of phenomena that occur on very different scales of space, time, and level of ecological organization (Pickett et al. 1989).

		Attributes of Change	
Speed/Scale	Level	Structure	Process
Fast/Small ↓	Individual	Physiology Behavior Size	Mortality Growth Reproduction
	Population	Density Structure (age, size, genetic)	Evolution Extinction
	Community	Diversity Composition Functional groups	Coexistence Competition Mutualism Predation
	Ecosystem	Nutrient Pools/Production	Resistance Resilience Nutrient flux/retention
	Landscape	Exogenous Disturbance Propagule pressure	Connectedness Colonization
Slow/Large	Region	Temperature Precipitation	Pollution inputs Climate change

One way to simplify cross-scale comparisons is to order dynamics along a gradient of responses that are "fast" (occurring at the individual or population level, or measured at small spatial scales) to responses that are "slow" (occurring at the ecosystem or landscape level, or large in scale) (Rinaldi and Scheffer 2000) (Figure 9.2). While we use fast and slow processes to coarsely describe the continuum of time, space, and level of organization, it should be noted that these components are not necessarily correlated (i.e., trait adaptations can be a result of local species interactions but occur over longer time scales). In this section, we discuss dynamics that can influence community and ecosystem behavior based on the distinction between fast and slow process variables (Figure 9.2). In many instances, restoration can be achieved by either fast or slow mechanisms (Figure 9.3).

Fast Processes and Restoration

Understanding the mechanisms for fast processes, such as species replacement early in succession, was the focus of much work starting in the 1970s (Drury and Nisbet 1973; Connell and Slatyer 1977; Tilman 1985). In a classic paper, Connell and Slatyer (1977) propose three mechanisms for species replacement through time: inhibition, facilitation, and tolerance. These models differ primarily in how early colonists affect the establishment of later colonists.

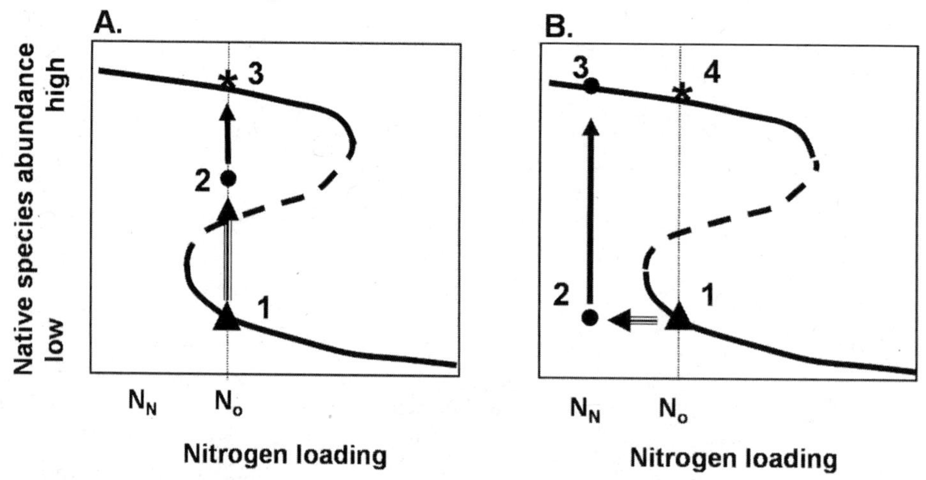

FIGURE 9.3 In a system with multiple equilibriums, restoration could be achieved via "fast" or "slow" mechanisms. In this hypothetical example, there are two states, one dominated by exotic species (the triangle) and one dominated by native species (the asterisk) at a given level of nitrogen input to the system (N_O). (A) Restoration via fast-process mechanisms in a system with multiple equilibriums, one dominated by exotics to one dominated by native species. A restoration action or perturbation could break feedbacks that lead to exotic species abundance (double-lined arrow, from 1 to 2), forcing the system into another basin of attraction and to a state (3) dominated by native species. (B) Restoration via slow-process mechanisms. If nitrogen inputs are decreased from N_O to N_N (double lined arrow to the left, from 1 to 2), perhaps by decreasing nitrogen deposition, native species are predicted to respond as predicted by the trajectory (1–2–3–4).

There have been many tests establishing the potential for all three mechanisms, and although one mechanism may dominate in a system, there is also evidence that multiple mechanisms operate sequentially or in unison in a system (Choi and Wali 1995; Wootton 2002; Franks 2003; Mullineaux et al. 2003; Walker and del Moral 2003). Restoration can use an understanding of positive and negative species interactions to either accelerate rates of change (facilitation) (Luken 1990; Choi and Wali 1995) or to identify times when change is slowed and intervention is needed (inhibition) (Mullineaux et al. 2003; Vander Zanden et al. 2003; Menniger and Palmer, this volume) (Figure 9.3a).

In the resource ratio hypothesis of succession, Tilman (1985) assumes that species interactions are fast and that species turnover along resource gradients is driven by changes in relative competitive and colonization ability. Much of the empirical work was done on abandoned fields in Minnesota, where soil resources, particularly nitrogen, are limiting, and light availability is high. Over time, soil organic matter accumulates in these soils, soil nutrient availability increases, and light becomes limiting. Although this model has not been applied explicitly to restoration projects, it offers promise because it emphasizes tradeoffs that constrain species success (Tilman 1990). Restoration can circumvent tradeoffs such as these by adding propagules, larvae, or individuals of later-successional species, bypassing their colonization limitations. Restoration also can change species interactions by changing the resources for which species compete (Corbin et al. 2004), for example, by adding carbon to reduce nitrogen availability (Blumenthal et al. 2003; Corbin and D'Antonio 2004).

These fast-process models assume a predictable order of species replacement because turnover is driven by species differences in competitive ability, life history, and effects on the environment. However, other less predictable types of species turnover may be crucial elements constraining the recovery of degraded land (Bakker and Berendse 1999; Gunderson 2000). These include priority effects; asymmetric competitive interactions, where one resident species is able to exclude an invading species, and vice versa (D'Antonio et al. 2001; Seabloom et al. 2003); sequential species loss effects (Eriksson and Eriksson 1998; Ostfeld and LoGiudice 2003); and shifts in competitive rankings due to alteration of disturbance regimes (Dudgeon and Petraitis 2001; Suding and Goldberg 2001). Establishment of nonnative species that have distinctive traits that can change rates of resource turnover, nutrient distribution, food-web structure, and disturbance regimes can shift competitive rankings and affect the dynamics and structure of a degraded system (Gordon 1998; Mack and D'Antonio 1998; Ehrenfeld 2003; Vander Zanden et al. 2003; D'Antonio and Chambers, this volume). Once species have changed ecosystem processes, positive feedbacks can increase the resilience of the system in its degraded state and make it resilient to restoration efforts. For example, introduced grasses in woodlands in Hawaii alter nitrogen cycling and promote fire, which further benefits introduced grasses at the expense of native shrub species, creating an internally reenforced state that has proven very difficult to change (Mack et al. 2001). Restoration can break these feedbacks by reducing the abundance of the species with strong impacts or by changing the environment to make the impact less advantageous (Suding et al. 2004).

Changes in trophic-level interactions, involving the removal or additions of predators, pathogens, or prey, can also influence community dynamics by altering internal feedbacks (Paine et al. 1985; McCauley et al. 1999; Chase 2003; Vander Zanden et al., this volume). For instance, herbivory by deer slows recovery of woody species in riparian systems because they feed selectively on regenerating saplings (Opperman and Merenlender 2000). In overgrazed rangelands, herbivores reduce herbaceous plant cover and then move to remaining areas of plant cover. Higher herbivore densities on these remaining areas cause increased losses of the remaining plant cover, and so on. This positive feedback can contribute to the collapse of the rangeland system to desert (vandeKoppel et al. 1997; Van Auken 2000). Knowledge of these interactions can prove useful in restoration to control, accelerate, or bypass particular interactions along a trajectory.

Slow Processes and Restoration

At the landscape level, large-scale processes often drive system dynamics, and these slow processes can have important effects on restoration efforts (Figure 9.3b). There is strong evidence that the lack of landscape connectivity and sufficient native propagule sources can severely limit recovery trajectories (Bakker and Berendse 1999; Holl et al. 2003). For instance, declines of source populations of native species due to habitat destruction and fragmentation limit the regional source of propagules for recolonization. Rivers and streams depend on unidirectional flow of propagules, potentially creating different upstream and downstream dynamics. Loss of source pools, combined with the loss of representation in storage modes, limits the regenerative ability of some species, resulting in patterns of colonization being limited to chance events (Del Moral 1999; Lichter 2000; Amoros and Bornette 2002). Restoration sites that are isolated from sources of propagules can reduce this constraint by adding propagules or enhancing dispersal vectors, such as planting trees to attract birds (Bonet 2004).

Metapopulation dynamics link the landscape and population scales (Maschinski, this volume). A connected set of populations (metapopulation) can exist at either a high-density connected state or a low-density fragmented state. If a population becomes extinct at one site, recolonization is dependent on the combined size of, and distance between, the surrounding populations. A positive feedback exists between regional metapopulation size and the probability that a single population can sustain itself: if a population is not reestablished, the future probability of a population recolonizing another site is lowered, and the regional population could decline rapidly (Hanski and Ovaskainen 2002; Melbourne et al. 2004). The creation of source gardens—not to mimic natural systems but to serve as propagule sources of a wide range of species to a range of environments—could be an important tool in restoration efforts by creating large and diverse species pools for isolated restoration projects.

Changes in external factors, such as pollution, harvesting pressure, and climate, also influence the dynamics of systems (Holmgren and Scheffer 2001; Scheffer et al. 2001). Climatic warming may be a chronic contributing factor to degradation of a system that is accelerated by other human activities (Van Auken 2000). Similarly, increased rates of atmospherically deposited nitrogen can change the effectiveness of natural disturbances and facilitate the invasion of species (Cione et al. 2002; Matson et al. 2002). Degradation due to changes in global or large-scale regional factors will be more difficult, and perhaps impossible, to reverse through local management efforts (Millar and Brubaker, this volume). These processes are considered "slow" and "large" because they do not respond quickly to changes in local population abundance or community structure, and consequently they are often harder to manage within any given reserve or project area. When they do respond, the change is predicted to alter the parameter space and equilibrium points, rather than just push the system along another trajectory (i.e., changes are along the x-axis of Figure 9.3). The possibility of a threshold event is important to consider even if the system does not appear to be tracking the external changes (Carpenter et al. 1999). It is still unclear how to predict a threshold in an unstudied system, or whether there are ways to counteract changes in the "slow" variables with more malleable "fast" variables. In some situations, the goal of restoration may have to be modified if a larger-scale regional approach is not possible.

How Can This Theory Advance Restoration Science and Practice?

In the last section, we highlighted ways in which theory has informed restoration, and vice versa. In this section, we expand upon that work and suggest future avenues for progress. To do this, we revisit the four questions we listed at the beginning of this chapter.

How Do Degraded Ecosystems Recover? Is the Pathway to Recovery Similar to the Pathway to Degradation?

Although the roles of history, chance, and positive feedbacks in affecting trajectories have gained wide appreciation among ecologists, restoration efforts are still too often designed from a perspective of initiating an orderly succession to an equilibrium (Prach 2003; Sheley and Krueger-Mangold 2003). Proponents of this view suggest that, by understanding succession, it is possible to predict, control, and perhaps accelerate community recovery after disturbances. Application of equilibrium theory to restoration has widespread appeal because it implies that we can predict and guide change in a system. It also assumes that there is an ultimate "natural system" that can be identified as the desired endpoint of the restoration effort.

Unfortunately, successional trajectories are still very hard to forecast (Parker 1997; Zedler 2000; Choi 2004; Suding et al. 2004). Assuming any one trajectory could be risky in restoration efforts, particularly if restoration actions have the potential to drive the system in unintended directions. It may be best to not assume directional and continuous change in restoration projects but rather to rely on monitoring programs to document actual trajectories of change. Monitoring the effects of restoration is essential, and the form and magnitude of future interventions must be adjusted, depending on the system response. Nonetheless, surprises could occur even with extensive monitoring programs in place. We need to concurrently focus on improving the ability to forecast trajectories of change using attributes of community structure, indicator species, or generalizations from past monitoring efforts.

The recovery pathways predicted by the different theories are not mutually exclusive. A typical test of stable equilibrium is whether the system is resilient in that state—whether the system returns to that state following a perturbation (Box 9.1). However, equilibrium conditions may depend on the spatial scale being considered. In addition, change can occur episodically, with periods of apparent stasis in species composition punctuated by intervals of rapid species turnover and altered ecosystem function. For instance, interactions between multiple plankton species may give rise to a continuous wax and wane of species within the community. This chaotic behavior implies that plankton dynamics at the species level are not at equilibrium (Rojo and Alvarez-Cobelas 2003; Salmaso 2003). Nonetheless, ecosystem characteristics such as total algal biomass show quite regular and stable patterns. In fact, much of the diversity-stability theory proposes that stable ecosystem functioning depends on less stable population dynamics (Tilman 1996). Thus, systems may express both equilibrium and non-equilibrium dynamics and both continuous and abrupt change, depending on scale.

Incorporating non-equilibrium dynamics into restoration planning suggests that many pathways are likely possible and may depend on restoration actions. While it may be frustrating and difficult to predict the time frame or pathway to recovery, it may allow for more creativity in management and the exploration of the mechanisms that facilitate species replacement or function development (Luken 1990). In many cases, it is likely that we do not know or cannot reach the natural trajectories for a system's recovery. It may be worthwhile to consider taking an unnatural or novel trajectory in restoration to reach desired endpoints. It is critical to focus on establishing natural processes, propagule pressure, and disturbance regimes that maintain the ecosystem in a state where native species can predominate or that can provide desired ecosystem services. This may be more cost-effective and require less human input to achieve and maintain than efforts focused on following a natural trajectory.

Can We Predict the Endpoints of Pathways? Are They Similar to States Prior to Degradation?

One major challenge in guiding the recovery of degraded systems is that degraded communities often do not respond predictably to perturbations, thus producing inconsistent and sometimes unexpected results (Whisenant 1999; Hobbs and Harris 2001). Given the range and variability of possible trajectories (Figure 9.2), it may not be possible to design a specific endpoint for a restoration project. Thus, it may be essential to design and evaluate restoration efforts at large spatial scales and to embrace variable community dynamics as a natural component of the system. Ecological theory also puts into question the assumption of setting goal endpoints that describe particular species composition or function (Parker 1997). If processes are dynamic, the rate of change in a particular ecosystem function or in species diversity might be a more realistic than a static assessment of species composition. Monitoring and experiments that focus on variation in trajectories could help form realistic guidelines about how to incorporate both stochastic and deterministic processes to guide recovery.

How Will Dynamics That Occur on Very Different Scales of Space and Time Relate to One Another? What Should Be the Scale of Focus?

The theoretical framework provides at least two ways to approach the restoration of degraded lands: (1) actions to alter internal feedbacks of the system (fast-process approach), or (2) actions to change external conditions of the system (slow-process approach) (Figure 9.3). Restoration efforts that address fast-process mechanisms would break feedbacks that maintain a degraded state. A unique or novel disturbance followed by sequential species introductions or changes in predator/herbivore communities could change species interactions and priority effects (Figure 9.3a). This type of management could be accomplished on relatively short time scales and small spatial scales. For example, goose herbivory has converted salt marshes to hypersaline mudflats off Hudson Bay. While the reduction of herbivory alone does not facilitate recovery, salt marshes can be restored when seedlings of *Puccinellia phryganodes*, a former dominant grass, are transplanted into the mudflats (Handa and Jefferies 2000). Alternatively, restoration via slow-process mechanisms would address landscape or regional dynamics (Figure 9.3b). This approach may be a viable alternative in systems that have suffered severe overexploitation, such as excessive tree harvesting, overgrazing, overfishing, and agriculture intensification, over large areas of the landscape.

A combination of the two approaches, particularly in cases where unavoidable change in one scale constrains actions at another scale, takes advantage of cross-scale dynamics. For instance, Holmgren and Scheffer (2001) suggest that climatic oscillations such as El Niño Southern Oscillation (ENSO) could be used in combination with grazing reduction to break feedbacks and restore degraded arid rangelands. Although this view has rarely been applied, it should be considered as an option in restoration.

How Much Variability Does an Ecological System Require for Adequate Recovery and Adaptive Capacity for Change in the Future?

Theory and experience suggest that it may be as important to conserve the ability of the system to change compared to restoring specific elements to a system. Ecological systems are, by their nature, dynamic and variable. Management has to be flexible and adaptive, and to leave room for future change to occur. It would be foolhardy to expect that restoration can accelerate change to a given point, and then constrain all subsequent change. Management should consider the importance of both stabilizing and destabilizing forces in a system. Destabilizing forces maintain diversity, resilience, and opportunity, whereas stabilizing forces are important in maintaining productivity and biogeochemical cycles. Maximizing heterogeneity in restoration projects may promote temporally stable and diverse communities and may aid in restoration (Brooks et al. 2002; Levin and Talley 2002; Brown 2003). For instance, Brown (2003) found that temporal variability of invertebrate communities in a northern New Hampshire stream was minimized with high spatial heterogeneity in stream substrates. Levin and Tally (2002) found that natural sources of climate heterogeneity exerted a stronger control on faunal turnover than manipulations of vegetation or soil processes in a southern California marsh restoration project.

Summary

Linking theoretical models of ecosystem and community change with restoration ecology has the potential to advance both the practice of restoration and our understanding of the dynamics of degraded systems. However, tight connections between theory and empirical tests have yet to be developed (for notable exceptions, see Box 9.1). More small-scale experimental studies and large-scale landscape manipulations are needed to test which system characteristics indicate the presence or absence of multiple states, how to determine whether thresholds exist, and the relative strengths of different factors affecting resilience in degraded systems. Addressing these questions involves characterizing internal feedbacks that enforce the resilience of degraded systems (Carpenter et al. 1999), testing the relative effectiveness of different restoration tools across environmental gradients (Pywell et al. 2002), and quantitatively synthesizing results from a range of projects (Palmer et al. 2005) (Box 9.3).

Restoration can inform ecological theory about the controls of dynamics, furthering the investigation of temporal dynamics in communities. Ecological theory could benefit from information on species function and system attributes in restoration projects. Knowledge of

Box 9.3
Critical Research Questions

1. How do we identify ecosystems where we can accelerate succession by skipping stages, while staying on the same trajectory? What are the ecosystems' characteristics? The role of restoration can be to accelerate the recovery of a system to achieve the goal endpoint more quickly. Likely, there are some stages along a trajectory that can be skipped and others that are essential to ensure the success of progressive stages. This question can be likened to identifying critical assembly rules (see Weiher and Keddy 1999) and is based on the appealing but controversial premise that we should be able to know the ecology of a system well enough to assemble it from the ground up. Restoration science has great potential for advancing our understanding of assembly rules.
2. Can we apply multiple stable equilibrium (MSE) theory as we deviate from true stability and equilibrium state assumptions, as is typical in most natural systems? It will be important to develop a tractable approach to distinguish equilibrium and non-equilibrium dynamics (or at least when equilibrium and non-equilibrium assumptions are approximate) in natural systems. Theory could expand to better approximate degraded sites with complex legacies where the distinction between equilibrium and non-equilibrium dynamics may be blurred. While this may be an important distinction in theory, is it important in restoration practice? Further tests are needed to better understand the extent to which theories that assume equilibrium dynamics can be applied in restoration.
3. How can we predict thresholds before a collapse occurs? Studying thresholds presents a conundrum: it is generally very hard to determine a system's proximity to a threshold without actually crossing it. After a system crashes, threshold behavior is relatively clear. Observations from other systems that have crossed thresholds can help predict future collapses, particularly if monitoring is done before and after the collapse. In addition, it would be fruitful to develop experimental approaches to identify threshold behavior. For instance, experimental climate manipulations (warming, precipitation, etc.) can suggest future dynamics.
4. What level of stochastic and destabilizing forces characterizes a self-sustainable, adaptive system? What is the relative importance of stochastic versus deterministic processes in recovery pathways? While variability is an essential feature in any sustainable system, it is much more typical for restoration goals to be oriented around a mean rather than variability. Likewise, ecological research often focuses on deterministic rather than stochastic processes. A better understanding of the balance between stochastic and destabilizing forces that characterize sustainable, healthy systems would help set realistic goals and evaluate recovery trajectories.

species and their interactions are extremely important to understand a restoration trajectory, and yet extensive experimentation would be unrealistic for most restoration projects. It would be fruitful to explore the extent to which generalizations based on functional group or system attributes could advance this front (Pywell et al. 2003).

Because no single ecological theory adequately describes community and ecosystem change, restoration should continue to broaden its perspective to encompass the range and variability of recovery. Knowledge that systems may express both equilibrium and non-equilibrium dynamics and can undergo abrupt change can inform what appears to be insurmountable restoration problems (Hawkins et al. 1999; Zedler 2000; McClanahan et al. 2002; Sutherland 2002). The relevant theory will depend on the temporal and spatial scale, as well as the stage in the recovery process. It will be important to consider the possibilities of persistent non-equilibrium and multiple stable states, along with traditional equilibrium views, particularly in the case of severely degraded or isolated systems that seem resilient to restoration efforts.

Acknowledgments

We thank S. Emery, D. Falk, E. Grman, and M. Palmer for valuable comments on previous versions of this chapter, and S. Collins for suggesting the idea of source gardens. This work was supported by the NSF-LTER Program (Kellogg Biological Station) and the Michigan Department of Military and Veterans Affairs.

Literature Cited

Amorós, C., and G. Bornette. 2002. Connectivity and biocomplexity in waterbodies of riverine floodplains. *Freshwater Biology* 47:761–776.

Anderson, R. C., J. E. Schwegman, and M. R. Anderson. 2000. Micro-scale restoration: A 25-year history of a southern Illinois barrens. *Restoration Ecology* 8:296–306.

Bachmann, R. W., M. V. Hoyer, and D. E. Canfield. 1999. The restoration of Lake Apopka in relation to alternative stable states. *Hydrobiologia* 394:219–232.

Baer, S. G., D. J. Kitchen, J. M. Blair, and C. W. Rice. 2002. Changes in ecosystem structure and function along a chronosequence of restored grasslands. *Ecological Applications* 12:1688–1701.

Bakker, J. P., and F. Berendse. 1999. Constraints in the restoration of ecological diversity in grassland and heathland communities. *Trends in Ecology & Evolution* 14:63–68.

Bartha, S., S. J. Meiners, S. T. A. Pickett, and M. L. Cadenasso. 2003. Plant colonization windows in a mesic old field succession. *Applied Vegetation Science* 6:205–212.

Beisner, B. E., C. L. Dent, and S. R. Carpenter. 2003a. Variability of lakes on the landscape: Roles of phosphorus, food webs, and dissolved organic carbon. *Ecology* 84:1563–1575.

Beisner, B. E., D. T. Haydon, and K. Cuddington. 2003b. Alternative stable states in ecology. *Frontiers in Ecology and the Environment* 1:376–382.

Blumenthal, D. M., N. R. Jordan, and M. P. Russelle. 2003. Soil carbon addition controls weeds and facilitates prairie restoration. *Ecological Applications* 13:605–615.

Bonet, A. 2004. Secondary succession of semi-arid Mediterranean old-fields in south-eastern Spain: Insights for conservation and restoration of degraded lands. *Journal of Arid Environments* 56:213–233.

Booth, B. D., and C. J. Swanton. 2002. Assembly theory applied to weed communities. *Weed Science* 50:2–13.

Briske, D. D., S. D. Fuhlendorf, and F. E. Smeins. 2003. Vegetation dynamics on rangelands: A critique of the current paradigms. *Journal of Applied Ecology* 40:601–614.

Brooks, S. S., M. A. Palmer, B. J. Cardinale, C. M. Swan, and S. Ribblett. 2002. Assessing stream ecosystem rehabilitation: Limitations of community structure data. *Restoration Ecology* 10:156–168.

Brown, B. L. 2003. Spatial heterogeneity reduces temporal variability in stream insect communities. *Ecology Letters* 6:316–325.

Carpenter, S. R., D. Ludwig, and W. A. Brock. 1999. Management of eutrophication for lakes subject to potentially irreversible change. *Ecological Applications* 9:751–771.

Chase, J. M. 2003. Experimental evidence for alternative stable equilibria in a benthic pond food web. *Ecology Letters* 6:733–741.

Choi, Y. D. 2004. Theories for ecological restoration in changing environment: Toward "futuristic" restoration. *Ecological Research* 19:75–81.

Choi, Y. D., and M. K. Wali. 1995. The role of *Panicum virgatum* (switch grass) in the revegetation of iron-mine tailings in northern New York. *Restoration Ecology* 3:123–132.

Cione, N. K., P. E. Padgett, and E. B. Allen. 2002. Restoration of a native shrubland impacted by exotic grasses, frequent fire, and nitrogen deposition in southern California. *Restoration Ecology* 10:376–384.

Clements, F. E. 1916. *Plant succession: An analysis of the development of vegetation*. Washington, DC: Carnagie Institute of Washingon Publication.

Connell, J. H., and R. O. Slatyer. 1977. Mechanisms of succession in natural communities and their role in community stability and organization. *American Naturalist* 111:1119–1144.

Connell, J. H., and W. P. Sousa. 1983. On the evidence needed to judge ecological stability or persistence. *American Naturalist* 121:789–824.

Copeland, T. E., W. Sluis, and H. F. Howe. 2002. Fire season and dominance in an Illinois tallgrass prairie restoration. *Restoration Ecology* 10:315–323.

Corbin, J. D., C. D'Antonio, and S. J. Bainbridge. 2004. Tipping the balance in the restoration of native plants. In *Experimental approaches in conservation biology*, ed. M. S. Gordon and S. M. Bartol, 154–179. Berkeley: University of California Press.

Corbin, J. D., and C. M. D'Antonio. 2004. Can carbon addition increase competitiveness of native grasses? A case study from California. *Restoration Ecology* 12:36–43.

D'Antonio, C. M., R. F. Hughes, and P. M. Vitousek. 2001. Factors influencing dynamics of two invasive C-4 grasses in seasonally dry Hawaiian woodlands. *Ecology* 82:89–104.

Davis, J. A., M. McGuire, S. A. Halse, D. Hamilton, P. Horwitz, A. J. McComb, R. H. Froend, M. Lyons, and L. Sim. 2003. What happens when you add salt: Predicting impacts of secondary salinisation on shallow aquatic ecosystems by using an alternative-states model. *Australian Journal of Botany* 51:715–724.

Del Moral, R. 1998. Early succession on lahars spawned by Mount St Helens. *American Journal of Botany* 85:820–828.

Del Moral, R. 1999. Plant succession on pumice at Mount St. Helens, Washington. *American Midland Naturalist* 141:101–114.

Dent, C. L., G. S. Cumming, and S. R. Carpenter. 2002. Multiple states in river and lake ecosystems. *Philosophical Transactions of the Royal Society of London, Series B* 357:635–645.

Drury, W. H., and I. C. T. Nisbet. 1973. Succession. *Journal of the Arnold Arboretum* 54:331–368.

Dudgeon, S., and P. S. Petraitis. 2001. Scale-dependent recruitment and divergence of intertidal communities. *Ecology* 82:991–1006.

Ehrenfeld, J. G. 2003. Effects of exotic plant invasions on soil nutrient cycling processes. *Ecosystems* 6:503–523.

Ellis, J. E., and D. M. Swift. 1988. Stability of African pastoral ecosystems—Alternate paradigms and implications for development. *Journal of Range Management* 41:450–459.

Eriksson, O., and A. Eriksson. 1998. Effects of arrival order and seed size on germination of grassland plants: Are there assembly rules during recruitment? *Ecological Research* 13:229–239.

Foley, J. A., M. T. Coe, M. Scheffer, and G. L. Wang. 2003. Regime shifts in the Sahara and Sahel: Interactions between ecological and climatic systems in northern Africa. *Ecosystems* 6:524–539.

Foster, B. L., and K. L. Gross. 1999. Temporal and spatial patterns of woody plant establishment in Michigan old fields. *American Midland Naturalist* 142:229–243.

Foster, D. R., D. H. Knight, and J. F. Franklin. 1998. Landscape patterns and legacies resulting from large, infrequent forest disturbances. *Ecosystems* 1:497–510.

Franklin, J. F., and J. A. MacMahon. 2000. Ecology—Messages from a mountain. *Science* 288:1183–1185.

Franks, S. J. 2003. Facilitation in multiple life-history stages: Evidence for nucleated succession in coastal dunes. *Plant Ecology* 168:1–11.

Friedel, M. H. 1991. Range condition assessment and the concept of thresholds—A viewpoint. *Journal of Range Management* 44:422–426.

Gordon, D. R. 1998. Effects of invasive, non-indigenous plant species on ecosystem processes: Lessons from Florida. *Ecological Applications* 8:975–989.

Grover, J. P., and J. H. Lawton. 1994. Experimental studies on community convergence and alternative stable states—Comments. *Journal of Animal Ecology* 63:484–487.

Gunderson, L. H. 2000. Ecological resilience—In theory and application. *Annual Review of Ecology and Systematics* 31:425–439.

Handa, I. T., and R. L. Jefferies. 2000. Assisted revegetation trials in degraded salt-marshes. *Journal of Applied Ecology* 37:944–958.

Hanski, I., and O. Ovaskainen. 2002. Extinction debt at extinction threshold. *Conservation Biology* 16:666–673.

Hawkins, S. J., J. R. Allen, and S. Bray. 1999. Restoration of temperate marine and coastal ecosystems: Nudging nature. *Aquatic Conservation–Marine and Freshwater Ecosystems* 9:23–46.

Hobbs, R. J., and J. A. Harris. 2001. Restoration ecology: Repairing the Earth's ecosystems in the new millennium. *Restoration Ecology* 9:239–246.

Hobbs, R. J., and D. A. Norton. 1996. Towards a conceptual framework for restoration ecology. *Restoration Ecology* 4:93–110.

Holl, K. D., E. E. Crone, and C. B. Schultz. 2003. Landscape restoration: Moving from generalities to methodologies. *Bioscience* 53:491–502.

Holmgren, M., and M. Scheffer. 2001. El Niño as a window of opportunity for the restoration of degraded arid ecosystems. *Ecosystems* 4:151–159.

Khater, C., A. Martin, and J. Maillet. 2003. Spontaneous vegetation dynamics and restoration prospects for limestone quarries in Lebanon. *Applied Vegetation Science* 6:199–204.

Law, R., and R. D. Morton. 1993. Alternative permanent states of ecological communities. *Ecology* 74:1347–1361.

Levin, L. A., and T. S. Talley. 2002. Natural and manipulated sources of heterogeneity controlling early faunal development of a salt marsh. *Ecological Applications* 12:1785–1802.

Lewontin, R. C. 1969. *Meaning of stability*. Upton, NY: Brookhaven Symposia in Biology.

Lichter, J. 2000. Colonization constraints during primary succession on coastal Lake Michigan sand dunes. *Journal of Ecology* 88:825–839.

Lindig-Cisneros, R., J. Desmond, K. E. Boyer, and J. B. Zedler. 2003. Wetland restoration thresholds: Can a degradation transition be reversed with increased effort? *Ecological Applications* 13:193–205.

Luken, J. O. 1990. *Directing ecological succession*. London: Chapman & Hall.

Lytle, D. A., and N. L. Poff. 2004. Adaptation to natural flow regimes. *Trends in Ecology & Evolution* 19:94–100.

Mack, M. C., and C. M. D'Antonio. 1998. Impacts of biological invasions on disturbance regimes. *Trends in Ecology & Evolution* 13:195–198.

Mack, M. C., C. M. D'Antonio, and R. E. Ley. 2001. Alteration of ecosystem nitrogen dynamics by exotic plants: A case study of C-4 grasses in Hawaii. *Ecological Applications* 11:1323–1335.

Matson, P., K. A. Lohse, and S. J. Hall. 2002. The globalization of nitrogen deposition: Consequences for terrestrial ecosystems. *Ambio* 31:113–119.

McCauley, E., R. M. Nisbet, W. W. Murdoch, A. M. de Roos, and W. S. C. Gurney. 1999. Large-amplitude cycles of *Daphnia* and its algal prey in enriched environments. *Nature* 402:653–656.

McClanahan, T., N. Polunin, and T. Done. 2002. Ecological states and the resilience of coral reefs. *Conservation Ecology* 6.

Melbourne, B. A., K. F. Davies, C. R. Margules, D. B. Lindenmayer, D. A. Saunders, C. Wissel, and K. Henle. 2004. Species survival in fragmented landscapes: Where to from here? *Biodiversity and Conservation* 13:275–284.

Mitch, W. J., and R. F. Wilson. 1996. Improving the success of wetland creation and restoration with know-how, time, and self-design. *Ecological Applications* 6:77–83.

Moss, B., J. Stansfield, K. Irvine, M. Perrow, and G. Phillips. 1996. Progressive restoration of a shallow lake: A 12-year experiment in isolation, sediment removal and biomanipulation. *Journal of Applied Ecology* 33:71–86.

Mullineaux, L. S., C. H. Peterson, F. Micheli, and S. W. Mills. 2003. Successional mechanism varies along a gradient in hydrothermal fluid flux at deep-sea vents. *Ecological Monographs* 73:523–542.

Nielsen, S., C. Kirschbaum, and A. Haney. 2003. Restoration of midwest oak barrens: Structural manipulation or process-only? *Conservation Ecology* 7.

Novak, J., and K. Prach. 2003. Vegetation succession in basalt quarries: Pattern on a landscape scale. *Applied Vegetation Science* 6:111–116.

Nystrom, M., C. Folke, and F. Moberg. 2000. Coral reef disturbance and resilience in a human-dominated environment. *Trends in Ecology & Evolution* 15:413–417.

Odum, E. P. 1969. The strategy of ecosystem development. *Science* 164:262–270.

Opperman, J. J., and A. M. Merenlender. 2000. Deer herbivory as an ecological constraint to restoration of degraded riparian corridors. *Restoration Ecology* 8:41–47.

Ostfeld, R. S., and K. LoGiudice. 2003. Community disassembly, biodiversity loss, and the erosion of an ecosystem service. *Ecology* 84:1421–1427.

Paine, R. T., J. C. Castillo, and J. Cancino. 1985. Perturbation and recovery patterns of starfish-dominated intertidal assemblages in Chile, New Zealand, and Washington State. *American Naturalist* 125: 679–691.

Palmer, M. A., R. F. Ambrose, and N. L. Poff. 1997. Ecological theory and community restoration ecology. *Restoration Ecology* 5:291–300.

Palmer, M. A., E. S. Bernhardt, J. D. Allan, P. S. Lake, G. Alexander, S. Brooks, J. Carr, S. Clayton, C. Dahm, et al. 2005. Standards for ecologically successful river restoration. *Journal of Applied Ecology*, forthcoming.

Parker, V. T. 1997. The scale of successional models and restoration objectives. *Restoration Ecology* 5:301–306.

Pastor, J., B. Peckham, S. Bridgham, J. Weltzin, and J. Q. Chen. 2002. Plant community dynamics, nutrient cycling, and alternative stable equilibria in peatlands. *American Naturalist* 160:553–568.

Pauly, D., V. Christensen, S. Guenette, T. J. Pitcher, U. R. Sumaila, C. J. Walters, R. Watson, and D. Zeller. 2002. Towards sustainability in world fisheries. *Nature* 418:689–695.

Petraitis, P. S., and R. E. Latham. 1999. The importance of scale in testing the origins of alternative community states. *Ecology* 80:429–442.

Pickett, S. T. A., M. L. Cadenasso, and S. Bartha. 2001. Implications from the Buell-Small succession study for vegetation restoration. *Applied Vegetation Science* 4:41–52.

Pickett, S. T. A., S. L. Collins, and J. J. Armesto. 1987a. A hierarchical consideration of causes and mechanisms of succession. *Vegetatio* 69:109–114.

Pickett, S. T. A., S. L. Collins, and J. J. Armesto. 1987b. Models, mechanisms and pathways of succession. *Botanical Review* 53:335–371.

Pickett, S. T. A., J. Kolasa, J. J. Armesto, and S. L. Collins. 1989. The ecological concept of disturbance and its expression at various hierarchical levels. *Oikos* 54:129–136.

Poff, N. L., J. D. Allan, M. B. Bain, J. R. Karr, K. L. Prestegaard, B. D. Richter, R. E. Sparks, and J. C. Stromberg. 1997. The natural flow regime. *Bioscience* 47:769–784.

Prach, K. 2003. Spontaneous succession in central-European man-made habitats: What information can be used in restoration practice? *Applied Vegetation Science* 6:125–129.

Prach, K., S. Bartha, C. B. Joyce, P. Pysek, R. van Diggelen, and G. Wiegleb. 2001. The role of spontaneous vegetation succession in ecosystem restoration: A perspective. *Applied Vegetation Science* 4:111–114.

Prober, S. M., K. R. Thiele, and I. D. Lunt. 2002. Identifying ecological barriers to restoration in temperate grassy woodlands: Soil changes associated with different degradation states. *Australian Journal of Botany* 50:699–712.

Pywell, R. F., J. M. Bullock, A. Hopkins, K. J. Walker, T. H. Sparks, M. J. W. Burke, and S. Peel. 2002. Restoration of species-rich grassland on arable land: Assessing the limiting processes using a multi-site experiment. *Journal of Applied Ecology* 39:294–309.

Pywell, R. F., J. M. Bullock, D. B. Roy, L. I. Z. Warman, K. J. Walker, and P. Rothery. 2003. Plant traits as predictors of performance in ecological restoration. *Journal of Applied Ecology* 40:65–77.

Rietkerk, M., and J. vandeKoppel. 1997. Alternate stable states and threshold effects in semi-arid grazing systems. *Oikos* 79:69–76.

Rinaldi, S., and M. Scheffer. 2000. Geometric analysis of ecological models with slow and fast processes. *Ecosystems* 3:507–521.

Rojo, C., and M. Alvarez-Cobelas. 2003. Are there steady-state phytoplankton assemblages in the field? *Hydrobiologia* 502:3–12.

Rood, S. B., C. R. Gourley, E. M. Ammon, L. G. Heki, J. R. Klotz, M. L. Morrison, D. Mosley, G. G. Scoppettone, S. Swanson, and P. L. Wagner. 2003. Flows for floodplain forests: A successful riparian restoration. *Bioscience* 53:647–656.

Salmaso, N. 2003. Life strategies, dominance patterns and mechanisms promoting species coexistence in phytoplankton communities along complex environmental gradients. *Hydrobiologia* 502:13–36.

Sarre, S., G. T. Smith, and J. A. Meyers. 1995. Persistence of 2 species of gecko (*Oedura reticulata* and *Gehyra variegata*) in remnant habitat. *Biological Conservation* 71:25–33.

Scheffer, M., S. Carpenter, J. A. Foley, C. Folke, and B. Walker. 2001. Catastrophic shifts in ecosystems. *Nature* 413:591–596.

Scheffer, M., S. Rinaldi, A. Gragnani, L. R. Mur, and E. H. van Nes. 1997. On the dominance of filamentous cyanobacteria in shallow, turbid lakes. *Ecology* 78:272–282.

Scheffer, M., F. Westley, and W. Brock. 2003. Slow response of societies to new problems: Causes and costs. *Ecosystems* 6:493–502.

Schlesinger, W. H., J. F. Reynolds, G. L. Cunningham, L. F. Huenneke, W. M. Jarrell, R. A. Virginia, and W. G. Whitford. 1990. Biological feedbacks in global desertification. *Science* 247:1043–1048.

Seabloom, E. W., W. S. Harpole, O. J. Reichman, and D. Tilman. 2003. Invasion, competitive dominance, and resource use by exotic and native California grassland species. *Proceedings of the National Academy of Sciences, USA* 100:13384–13389.

Sheley, R. L., and J. Krueger-Mangold. 2003. Principles for restoring invasive plant-infested rangeland. *Weed Science* 51:260–265.

Siemann, E., J. Haarstad, and D. Tilman. 1999. Dynamics of plant and arthropod diversity during old field succession. *Ecography* 22:406–414.

Stringham, T. K., W. C. Krueger, and P. L. Shaver. 2003. State and transition modeling: An ecological process approach. *Journal of Range Management* 56:106–113.

Stylinski, C. D., and E. B. Allen. 1999. Lack of native species recovery following severe exotic disturbance in southern Californian shrublands. *Journal of Applied Ecology* 36:544–554.

Suding, K. N., and D. Goldberg. 2001. Do disturbances alter competitive hierarchies? Mechanisms of change following gap creation. *Ecology* 82:2133–2149.

Suding, K. N., K. L. Gross, and G. Houseman. 2004. Alternative states and positive feedbacks in restoration ecology. *Trends in Ecology & Evolution* 193:46–53.

Sutherland, J. P. 1974. Multiple stable points in natural communities. *American Naturalist* 108:859–873.

Sutherland, W. J. 2002. Restoring a sustainable countryside. *Trends in Ecology & Evolution* 17:148–150.

Tilman, D. 1985. The resource-ratio hypothesis of plant succession. *American Naturalist* 125:827–852.

Tilman, D. 1990. Constraints and tradeoffs—Toward a predictive theory of competition and succession. *Oikos* 58:3–15.

Tilman, D. 1996. Biodiversity: Population versus ecosystem stability. *Ecology* 77:350–363.

Underwood, A. J., and P. G. Fairweather. 1989. Supply-side ecology and benthic marine assemblages. *Trends in Ecology & Evolution* 4:16–20.

Van Auken, O. W. 2000. Shrub invasions of North American semiarid grasslands. *Annual Review of Ecology and Systematics* 31:197–215.

vandeKoppel, J., M. Rietkerk, and F. J. Weissing. 1997. Catastrophic vegetation shifts and soil degradation in terrestrial grazing systems. *Trends in Ecology & Evolution* 12:352–356.

Vander Zanden, M. J., S. Chandra, B. C. Allen, J. E. Reuter, and C. R. Goldman. 2003. Historical food web structure and restoration of native aquatic communities in the Lake Tahoe (California-Nevada) Basin. *Ecosystems* 6:274–288.

van Nes, E. H., and M. Scheffer. 2003. Alternative attractors may boost uncertainty and sensitivity in ecological models. *Ecological Modelling* 159:117–124.

Walker, L. R., and R. del Moral. 2003. *Primary succession and ecosystem rehabilitation*. Cambridge, UK: Cambridge University Press.

Weiher, E., and P. Keddy, editors. 1999. *Ecological assembly rules: Perspectives, advances, retreats*. Cambridge, UK: Cambridge University Press.

Westoby, M., B. Walker, and I. Noymeir. 1989. Range management on the basis of a model which does not seek to establish equilibrium. *Journal of Arid Environments* 17:235–239.

Whisenant, S. G. 1999. *Repairing damaged wildlands: A process-oriented, landscape-scale approach*. Cambridge, UK: Cambridge University Press.

Wootton, J. T. 2002. Mechanisms of successional dynamics: Consumers and the rise and fall of species dominance. *Ecological Research* 17:249–260.

Zedler, J. B. 2000. Progress in wetland restoration ecology. *Trends in Ecology & Evolution* 15:402–407.

Zedler, J. B., and J. C. Callaway. 1999. Tracking wetland restoration: Do mitigation sites follow desired trajectories? *Restoration Ecology* 7:69–73.

Chapter 10

Biodiversity and Ecosystem Functioning in Restored Ecosystems: Extracting Principles for a Synthetic Perspective

Shahid Naeem

The purpose of this chapter is to adapt a synthetic ecological perspective to restoration ecology. I will argue that this perspective provides valuable and interesting insights into the theoretical and empirical foundations of restoration ecology much the way it has for ecology in general. This perspective, which I will refer to as the biodiversity-ecosystem functioning perspective, or BEF perspective, synthesizes the perspectives of community and ecosystem ecology, both of which have been adapted to restoration ecology elsewhere. The *community perspective* focuses on principles of community ecology in which an understanding of the trophic, competitive, and facilitative interactions among populations can provide insights into identifying restoration targets, selecting what ecological processes and properties to monitor, and designing the path or strategies by which one reaches restoration targets. In contrast, the *ecosystem perspective* concerns principles from ecosystem ecology in which energy flow and nutrient cycles in ecosystems provide insights into targets, monitoring, and strategies. The community and ecosystem perspectives, as they relate to restoration ecology, are each nicely summarized by Palmer et al. (1997) and Ehrenfeld and Toth (1997), respectively. This chapter builds upon their contributions.

The BEF perspective differs from the community and ecosystem perspectives because it treats communities and ecosystems as inseparable; any change in a community has its consequences for ecosystem functioning, and vice versa. I will argue that this BEF perspective provides a different approach to understanding the significance of restoration in modern landscapes, landscapes that are increasingly or, inevitably, will be mosaics of unmanaged (e.g., ecological reserves, "wild" or "pristine" lands); managed (e.g., pastures, farms, aquacultural systems, managed lakes and watersheds, or managed wildlife and recreational parks); and degraded ecosystems (e.g., clear-cut old-growth, collapsed fisheries, eutrophied lakes and waterways, or the habitats left at the end of strip-mining and mountain-top removal).

In a nutshell, the BEF perspective considers all ecosystems in the modern landscape as biogeochemical systems on or displaced from a fundamental relationship between biodiversity and ecosystem functioning. From this perspective, restoration is the activity that seeks to restore displaced ecosystems to this fundamental relationship. The three perspectives, community, ecosystem, and BEF perspectives, are compared in Figure 10.1.

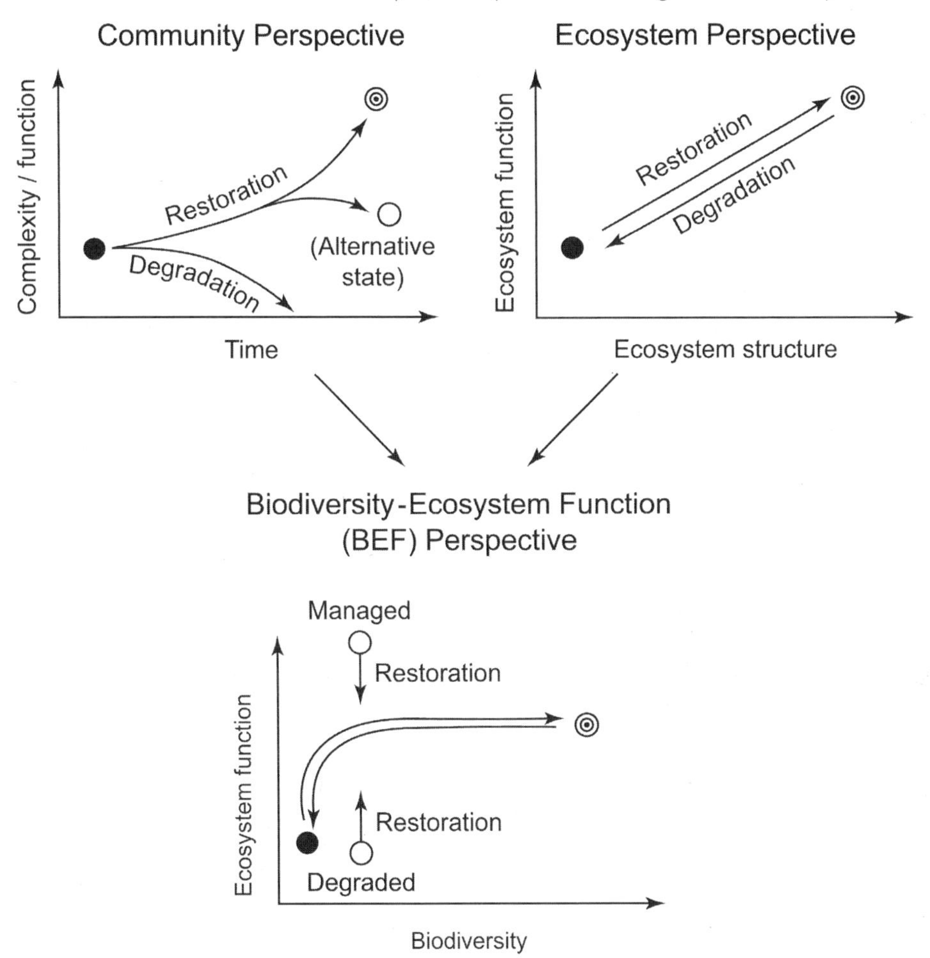

FIGURE 10.1 The community, ecosystem, and biodiversity-ecosystem functioning (BEF) perspectives. This figure compares how each perspective considers the state of an ecosystem designated for ecological restoration (solid black circles) in relationship to the state of the target ecosystem (bull's eye circles) and other states (open circles). The community perspective allows for the possibility that restoration may lead to alternative stable states that complicate efforts to reach designated targets. The BEF perspective considers managed ecosystems to be those whose levels of functioning have been elevated at the expense of biodiversity, while degraded ecosystems are considered ones that have suffered losses in both functioning and biodiversity.

Biodiversity and Ecosystem Functioning (BEF)

Biodiversity and ecosystem functioning is a large and rapidly growing field (Schulze and Mooney 1993; Kinzig et al. 2002), but its significance to restoration ecology has not been explored. While the discipline is new, its foundations are not (Naeem 2002c). The BEF perspective is founded on well-established principles in community ecology, such as resource-competition and niche theory (e.g., Tilman et al. 1997a); Lotka-Volterra theory (e.g., Hughes and Roughgarden 1998, 2000); and theory from ecosystem ecology, such as nutrient cycling and dynamic food webs (e.g., De Mazancourt et al. 1998). Empirical testing of these principles is also extensive and one of the more rapidly growing areas in ecological research (Loreau 1998a, 2001). Its key principles, that of niche complementarity (Hector 1998; Petchey 2000; Loreau and Hector 2001), functional groups (Díaz and Cabido 2001; Lavorel and Garnier 2002; Petchey and Gaston 2002; Naeem and Wright 2003), and species redundancy (Naeem 1998), are all familiar parts of community and ecosystems ecology developed long before the BEF perspective began to emerge.

While its parts are largely familiar, BEF is nevertheless unique in its treatment of biodiversity as a key factor regulating ecosystem functioning within bounds set by climate and geography. For the restorationist, the BEF perspective suggests considering not just targets, whether they be specific community targets or specific ecosystem function targets, but the full realm of possible community configurations and the ecosystem functions associated with those configurations. By realm, I mean the range of possible expressions of ecosystem functioning likely to be obtained for all possible community compositions and structures. This idea is developed further below.

Another important feature of the contemporary BEF perspective is that it is also implicitly multiscale in its construction. It considers the subset of species found in an ecosystem as the product of extrinsic factors (e.g., climate, geography, history, or other abiotic factors) that select for species from the regional or global species pool, while local patterns in distribution, abundance, and dynamics are governed by intrinsic factors (e.g., biotic interactions) (Loreau et al. 2001). In other words, global-scale factors set the stage within which local-scale factors govern the expression of ecosystem functioning.

I will begin by a review of the basic ecological principles that form the foundation for BEF ecology, briefly defining the many terms (Table 10.1) that have come to be associated with BEF research.

Ecosystem Function and Ecosystem Functioning

Though the scientific domain of the BEF perspective is large, biogeochemical principles are at its core. Every ecosystem contains an assemblage of species whose individuals, no matter how evolutionarily or ecologically unique they may be, do one thing: they cycle material back and forth between organic and inorganic forms, generally referred to as ecosystem processes or ecosystem functions. Ecosystem processes are measured by assessing the flux rates of matter among pools of living organic (biomass), dead organic, and inorganic matter with emphasis on C and N, though many other biologically active elements are also measured. The most frequently measured ecosystem processes are presented in Table 10.2. They allow comparisons among ecosystems and the ability to understand ecosystem response to environmental change, whether these changes involve the loss of species, biological invasion, changes in species composition or trophic structure, or habitat modification. These measures are therefore important tools by which ecologists gauge ecosystem response to change, be they attributable to natural causes, management, or restoration. Many contemporary methods are reviewed in Sala et al. (2000) and Chapin III et al. (2002).

In the absence of biological processes, a habitat exhibits magnitudes and rates of geochemical cycling that are governed by abiotic factors, such as climate and geography, and abiotic processes, such as photolysis and chemical weathering. In the presence of organisms, however, the magnitudes and rates of geochemical processes are modified. The field of *biogeochemistry* is the study of biologically modified geochemical processes (Butcher et al. 1992; Schlesinger 1997; Smil 1997, 1999; Ernst 2000). From the BEF perspective, ecosystem functions consist of biogeochemical activities associated with the habitat under investigation. To avoid connotations of "purpose," ecosystem *functioning* is often used, rather than ecosystem function, to emphasize activity and not purpose.

Note that I equate ecosystem function with biogeochemical or ecosystem processes. Ecosystem property, on the other hand, refers to all other ecosystem characteristics (e.g., stability properties). Note also that ecosystem functions are distinct from ecosystem services, the latter being associated with values humans place on ecosystems. These terms, however, take on a variety of meanings within the literature.

Biodiversity: It's Not Just Species

Biodiversity is an all-encompassing term (Table 10.1), but from the standpoint of ecosystem functioning, functional diversity rather than taxonomic diversity is the most relevant component of biodiversity. Functional groups of species are variously defined, but one of the more common classification schemes involves grouping species by traits that relate to ecosystem functioning. There are innumerable ways of classifying species by their contributions to ecosystem functioning, the most common being a hierarchal system in which trophic groups represent the first cut. Within trophic groups, functional groups are more finely divided, such as dividing plants into C_3, C_4, leguminous forbs, and non-leguminous forbs, or dividing phytoplankton into size classes. More recently, ecologists have advocated a dual classification scheme based on organism traits that involve responses to changes in an ecosystem (response-functional) and traits that are directly related to the ecosystem process of interest (effect-functional) (Díaz and Cabido 2001; Lavorel and Garnier 2002; Naeem and Wright 2003).

Classifying species by function is counter to taxonomic classification that is generally based on evolutionary principles. To classify species by function emphasizes similarities in functional or analogous traits, as opposed to homologous or shared, derived traits, these being key to evolutionarily based taxonomic classifications. At very fine scales, the two are likely to be related (Petchey and Gaston 2002), but not at coarse scales.

Classification by function leads immediately to notions of species redundancy or equivalency (Walker 1992; Lawton and Brown 1993; Walker 1995; Naeem 1998). At coarse scales of classification, there are likely to be many species that fall into the same functional group, and such species may be ecologically equivalent to one another even though they are taxonomically different. Such ecological equivalency can be considered species redundancy if the equivalent species can truly replace one another by compensatory growth. However, if two species are identical in their ecology, but the extinction of the dominant species is not compensated for by the rare species (i.e., the species remains rare in the absence of its ecological equivalent), then the rare species does not actually replace the lost species, and the species are equivalent, but not redundant.

Unfortunately, the term *redundancy* often carries a connotation of expendability, as though a redundant species is unnecessary or is redundant across all functions. Indeed, one test for species redundancy would be to remove a species from a functional group and

TABLE 10.1

Terms in biodiversity-ecosystem functioning research and their importance to experiment, theory, or practice in restoration ecology. The selected reference or references listed use or define the terms in greater depth.

Term	Definition	Example	Restoration significance	References
Biodiversity	The genetic, phylogenetic, population, and functional variation of organisms across all temporal and spatial scales within and among ecosystems and their communities.	A landscape of multiple ecosystem types (e.g., a mosaic of grassland and woodland sites) has greater biodiversity than a landscape of a single ecosystem type, even if both landscapes have similar numbers of species. Likewise, a community that has greater functional diversity than another has higher biodiversity.	There is no necessary one-to-one correspondence between taxonomic diversity and biodiversity. Often biodiversity-based restoration targets are purely taxonomic (i.e., species) based when, in fact, biodiversity restoration should seek a balance between restoring genetic, population, taxonomic, and functional diversity.	Harper and Hawksworth 1995
Biological insurance	Having a diversity of species that differ in their responses to changes in environmental conditions lowers the variance of ecosystem functioning in a fluctuating environment.	An ecosystem with many species of herbivorous mammals that each respond differently to climate variability is more likely to show lower variance in primary production over many years than an ecosystem with only a few species of mammalian herbivores or with herbivores that are identical in their response to drought.	Increasing biodiversity by increasing the variety of species that differ in their responses to environmental change may lower the variance of ecosystem functioning in systems where the environment constantly fluctuates.	Perrings 1995; Folke et al. 1996; Yachi and Loreau 1999
Combinatorial experiment	Constructing a series of replicate experimental model ecosystems that are initially identical in all conditions except biodiversity. Biodiversity is explicitly manipulated such that its components (species, functional groups, or any other aspect of diversity) are assigned to replicates at random.	Selecting 16 species of dominant plants in an ecosystem and establishing experimental plots consisting of monocultures, 2-species, 4-species, 8-species, and 16-species polycultures represents a standard combinatorial experiment.	Combinatorial experimental designs provide a means for exploring which combinations of species or other elements of biodiversity are most effective for achieving restoration targets. However, such experiments are costly and time consuming, and one must be careful to disentangle sampling from selection effects in such experiments.	Naeem et al. 1995; Naeem et al. 1996; Garnier et al. 1997; Allison 1999; Hector et al. 1999; Naeem 2002a

Ecosystem process or function	A biogeochemical process.	Examples: primary production, decomposition, biological nitrogen fixation.	Monitoring ecosystem functioning is as important as species composition and structure in terms of outcomes, because restoration efforts are unlikely to be successful without the supporting ecological processes.	de Groot 1992
Ecosystem property	Characteristics of structure, dynamics, or other features of an ecosystem's biogeochemistry, populations, or communities.	Ecosystem properties include stability (e.g., variability, resilience, resistance, or persistence of ecosystem processes or populations), the ability to resist invasion, or the complexity of its food webs.	Monitoring dynamics of both populations and ecosystem functions is important because this may serve as a barometer of restoration progress.	McNaughton 1977
Ecosystem reliability	The probability that an ecosystem will provide a given level of ecosystem functioning over a specified period.	A rare predatory insect may grow during an insect pest outbreak at which point it serves a major role in biocontrol. Thus, the restoration of even rare species can improve reliability.	Restoring rare, seemingly unimportant species, not just dominant species, may prove important for insuring that restoration is retained during future environmental changes.	Naeem and Li 1997; Naeem 1998; Rastetter et al. 1999; Naeem 2003b
Ecosystem service	Goods or services derived from a functioning ecosystem.	Goods: lumber, biofuels, potable water. Services: Greenhouse gas regulation (e.g., C sequestration and storage), soil formation, and recreation.	What ecologists study and what people value are related, but one needs to translate from one to the other (i.e., ecosystem functions to ecosystem goods and services).	Ehrlich and Mooney 1983; Daily et al. 1997
Effect functional trait	Species traits associated with a specific ecosystem function.	The effect functional traits of a drought-tolerant legume associated with soil fertility would be traits associated with symbiotic affiliations with N-fixing microbes.	It may be valuable to consider the effects of species on their environment in order to link species restoration with ecosystem function restoration.	Lavorel and Garnier 2002; Naeem and Wright 2003
Functional diversity	The diversity of species characterized by their functional traits.	If a grassland patch has 100 species of grasses, it is functionally less diverse than a patch with one species of grass and one species of legume. Note that while there is no necessary correlation between functional and taxonomic diversity, in general, it is believed that as one increases, so does the other.	Restoration should maximize functional diversity, though this is often difficult because functional diversity is seldom known. One tends to assume that taxonomic diversity = functional diversity, but there is little evidence for this, and the restorationist should explore species properties in the published literature (see "Functional groups").	Collins and Benning 1996; Díaz and Cabido 2001; Lavorel and Garnier 2002; Petchey and Gaston 2002

TABLE 10.1 *(continued)*

Term	Definition	Example	Restoration significance	References
Functional groups	Species grouped by functional traits.	Legumes, grasses, and non-leguminous forbs. Micro- and macro-arthropods in soil fauna. Denitrifying and N-fixing bacteria.	Functional groups may allow for a more rapid, qualitative assessment of functional diversity.	Körner 1993; Chapin III et al. 1996; Gitay and Noble 1997; Smith et al. 1997
Niche complementarity	When two or more species use resources in complementary ways such that together they more effectively use available resources (e.g., space, light, nutrients) than either can alone.	Deep and shallow rooted plants may collectively make more effective use of soil nutrients than one or the other alone.	Complementary species provide for more effective ecosystem functioning than competitive species. *More effective functioning* refers to greater efficiency or more thorough exploitation of resources in an ecosystem by communities in which niche complementarity is prevalent.	Hector et al. 1999
Portfolio effect	See "Statistical averaging."			
Response functional trait	Species traits that influence/determine response to an environmental change.	The response functional traits of a drought-tolerant legume permit tolerance to drought, such as deep roots, waxy epicutical, or high water-use efficiency.	It may be valuable to consider a species' responses to environmental change, because this may determine the long term success of a restoration effort.	Lavorel and Garnier 2002; Naeem and Wright 2003
Sampling effect	In combinatorial experiments, or in natural habitats where patches are repeatedly recolonized by species from the regional species pool, species with strong positive effects have a higher probability of occurring in higher diversity treatments.	If the species pool consists of 10 species and only one species is a legume, then that species has unique impacts on soil N content. The probability that an experimental plot contains a legume when there are 10 species is 10/10 or 1.0 in the 10-species plots, 5/10 in the 5-species plots, and 1/10 in the monoculture plots. Thus, by simple probabilities, higher diversity will have higher rates of N fixation.	Restoration research that uses combinatorial experiments need to be cognizant of the fact that positive diversity-functioning associations observed in such experiments can be due to selection effects (which include Sampling effects; see below) and complementarity effects.	Huston 1997; Loreau 1998b
Species composition	The particular set of species found in a community.	Communities with similar levels of species and functional diversity and similar levels of ecosystem functioning may nevertheless differ in species composition.	Success in reestablishing a biota will be more likely if composition is adjusted to achieve reliable and sustainable functioning.	Hooper and Vitousek 1997; Tilman 1997; Hooper and Vitousek 1998; Dukes 2002

Selection effect	In combinatorial experiments, or in natural habitats where patches are repeatedly recolonized by species from the regional species pool, species with strong effects, negative or positive, have a higher probability of occurring in higher diversity treatments.	See "Sampling effect" above. Note that sampling effects refer to those species that positively influence ecosystem function, while selection effects refer more generally to any impact a species has on an ecosystem functioning, negative or positive.	Restoration research that uses combinatorial experiments needs to be cognizant of the fact that positive or negative diversity-functioning associations observed in such experiments can be due to both selection and complementarity effects.	Loreau and Hector 2001
Species identity	A species' taxonomic designation.	When biodiversity is altered, the ecosystem consequences of such an alteration will depend on the identity of the species experiencing the change.	The identities of species used in a restoration project may be useful for predicting how ecosystem functions will respond to the restoration activities.	Symstad et al. 1998; Crawley et al. 1999; Vanni et al. 2002
Species redundancy	When two or more species are substitutable with respect to their contributions to a single ecosystem function. This should not be confused with a species having negligible effects on an ecosystem function. If a species has negligible effects, it must be demonstrated that this is due to its replacement by compensatory growth of another species. Note that redundancy for one function does not mean redundancy for other functions.	During a drought, drought tolerant plants that are rare during normal years but increase in abundance during drought years may be important sources of production.	Redundancy is generally a good thing. Some species may play little or no role in an ecosystem's functioning until conditions change, but the role a species plays is context dependent. The more redundancy in a system, the more equitable an ecosystem's functioning in the face of fluctuating environmental conditions.	Walker 1992; Lawton and Brown 1993

TABLE 10.1 *(continued)*

Term	Definition	Example	Restoration significance	References
Species singularity	When a particular ecosystem function is tied to a single species. Note that singularity for one function does not mean singularity for other functions.	A community with only one top predator, or a plant community with only one N-fixing legume, each contain singular species.	If a single species is critical to an ecosystem's function because no other can take its place, then success of a restoration plan may be tied to that species.	Naeem 1998
Statistical averaging	A statistical property of aggregate measures in which the variance of the aggregate measure declines as the diversity of the items in the aggregation increases. Also referred to as the "Portfolio effect."	A landscape made up of many semi-independent ecosystems is, as a whole, more likely to show lower variance over time and space in total ecosystem production than a similar landscape covered by a single, large, homogeneous ecosystem.	Diversify, for the same reasons that corporations and individual investors do. Restored systems with higher biodiversity are likely to provide more predictable and possibly higher levels of desired ecosystem functions.	Doak et al. 1998; Tilman et al. 1998

TABLE 10.2

Common ecosystem processes used in ecology, management, and restoration.

Ecosystem process or function	Biogeochemical activity	How it is measured	Examples
Productivity	Transformation of inorganic C to organic matter.	Biomass of autotroph (plant or algal) produced per unit time.	Whole system sequestration of C per unit time, or simply the difference in autotrophic standing crop over time.
Decomposition	Transformation of dead organic material into inorganic material.	Percent mass loss of dead organic material per unit time.	Loss of litter, woody debris, carrion, or other dead organic mass in bags placed on or in soil, sediments, or water.
N mineralization	Transformation of organic forms of N into inorganic forms.	Accumulation of nitrates, nitrites, and ammonium per gram of soil, sediment, or water, per unit time.	N-free resin bags are buried in soil, retrieved at a later date, and inorganic N that has impregnated the resin is extracted and measured.
Photosynthesis	Acquisition of energy and inorganic C.	Number of moles of C fixed per unit time per mole of photons intercepted, by the area, examined over a unit time.	Measured as oxygen evolution or CO_2 absorption in the presence of light.
Respiration	Loss of energy due to the biochemical transformation of organic C into inorganic C by organisms.	Efflux of CO_2.	Measured as CO_2 evolution in closed or flow-through chambers placed over a portion of the ecosystem of interest.

monitor whether a particular ecosystem function related to that functional group showed a response. If the response to loss was nil, then the species is redundant. That does not mean, however, that it is expendable. Such a test examines only one function, and species redundancy with respect to one function does not mean redundancy with respect to other functions. Additionally, a species that is redundant under one set of conditions does not mean that it is redundant under all conditions.

Another important aspect of biodiversity concerns commonness (or dominance) and rarity. Most communities consist of only a few common or numerically dominant species while the rest are relatively rare. The removal of a dominant species, one whose total biomass (the sum of the biomass of all individuals) is much greater than the average total biomass of other species, is more likely to have a significant impact on ecosystem functioning than the removal of a rare species whose total biomass is small. Thus, the removal of individual species within a trophic group is likely to have little to no impact on ecosystem functioning because most species, on average, are rare (Schwartz et al. 2000). The continuum between redundancy, expendability, and keystone status, or the importance of species, has been reviewed recently by several authors in Kareiva and Levin (2003).

One of the central concepts in the BEF perspective concerns the continuum between species redundancy and its converse, species singularity, where an ecosystem function is governed by a single species. This continuum is key to designating functional groups (as discussed above), designing experiments (discussed below), and constructing the fundamental relationship between biodiversity and ecosystem functioning (also discussed below). Like redundancy, singularity does not mean that a species singular for one function is singular for others, nor does it mean that it is singular under all conditions. Such context dependency of functional classifications, where the redundant or singular status of a species changes if either the community or environmental conditions change, was seen by the late Masahiko Higashi as one of the major challenges for BEF research (Naeem et al. 1998).

Ecosystem Functioning: It's More Than Just Magnitudes

When restoring an ecosystem's functioning, such as its ability to retain nutrients, produce plant and animal biomass, degrade pollutants, or produce clean water, the magnitude of functioning represents only one dimension of restoration; the durability or reliability of the restored system is equally important. Such stability properties of ecosystems as spatial heterogeneity or temporal patterns in biogeochemical processes or the distribution and abundance of organisms have been studied extensively by ecologists. BEF research has contributed significantly to some recent advances in this area of research (McCann 2000; Cottingham et al. 2001; Loreau et al. 2002a). Temporal variability of ecosystem functioning is particularly important in the context of restoration, especially how variable, predictable, or reliable ecosystem functioning is in relation to biodiversity, rather than resilience or resistance. (Note that spatial heterogeneity may be equally important in restoration, but I focus on temporal heterogeneity as it is more relevant to resilience and resistance.)

Variability, predictability, and reliability of ecosystem functioning are all important, closely related properties of ecosystems and constitute an important part of overall restoration goals and practices. Variability concerns how much fluctuation there is around the mean or central tendency of an ecosystem function. Predictability refers to the probability that an ecosystem will exhibit the same level of functioning under the same set of conditions. Reliability is the probability of an ecosystem exhibiting a particular level of functioning over a unit of time, given the norms for environmental variability and the likelihood that some species within the ecosystem are likely to suffer local extinction.

BEF's contributions to these issues are many, but three are particularly important to restoration ecology: statistical averaging, or the portfolio effect; biological insurance; and ecosystem reliability, the latter two having already been discussed a bit in relation to species redundancy. Statistical averaging (Doak et al. 1998) is a simple statistical feature of aggregate measures of multiple populations in which the variance of the aggregate measure declines when the number of populations increases. The portfolio effect, which is closely related (Tilman et al. 1998), nicely captures this statistical property by referring to the well-known business adage that a portfolio of mixed stocks will, over the long term, be more even in its performance than a portfolio dominated by a few high-yielding stocks. Tilman and colleagues noted, however, that the strength of statistical averaging, or the strength of the portfolio effect, is sensitive to deviations from the assumption of neutral covariance in the statistical averaging model. Neutral covariance among interacting species is unlikely to occur in nature, thus one should not automatically expect statistical averaging to lower variance as diversity increases.

Biological insurance is the venerable idea in ecology that having lots of species that are differently adapted to environmental variability provides for more equitable ecosystem functioning than having a few species that are identical in their response to changes in the environment (Perrings 1995; Folke et al. 1996; Yachi and Loreau 1999). Yachi and Loreau (1999) demonstrated theoretically that if the optima for multiple species vary for some factor, then not only is the variance of an ecosystem's function likely to be lower with higher biodiversity, but the magnitude of ecosystem functioning may also increase. (Note that it is not necessarily always desirable to have greater magnitudes of ecosystem functioning, only that Yachi and Loreau suggest that biodiversity loss will simultaneously lower ecosystem predictability and levels of functioning, both of which may need to be restored).

Ecosystem reliability (Naeem 1998; Rastetter et al. 1999; Naeem 2003b) is derived directly from reliability engineering (e.g., Lewis 1987; Süleyman 1996; Ebeling 1997) and draws attention to the fact that the probability that an ecosystem will exhibit a particular level of functioning over time is not likely to be constant unless conditions are constant, as are all of its species and their interactions and dependencies upon one another. This notion of reliability as applied to engineered systems has been applied to aging (e.g., Koltover 1997), genetic robustness (Eagner 2000; Gu et al. 2003), and other properties of biological systems, but rarely applied to ecological systems. The tenet of reliability theory is that reliability declines steadily in any system that is not actively maintained or repaired. The basic reason for this decline is that every species in an ecosystem has a finite probability of local extinction if the system is not actively managed to prevent local extinction and there is no significant immigration. Immigration would ordinarily be the source of recruits in an unmanaged system that would rescue or restore a species from local extinction, but in an increasingly fragmented and depauperate world, immigration may not occur, or it occurs at too low a level to serve this function anymore. Reliability therefore draws attention to the importance of recruitment, immigration, restocking, isolation, and ecological subsidies in restoration.

Community, Ecosystem, and BEF Perspectives

From the above review, it is clear that ecosystem functioning, in terms of localized biogeochemical processes, is governed by the biota within an ecosystem whose boundary conditions are set by abiotic factors such as material fluxes, climate, and geography. Changes in the biota of an ecosystem, such as the loss of a species or widespread declines in diversity caused by an ecological invader or habitat fragmentation, can alter ecosystem functioning, although the magnitude of the change is dependent on which species are affected and how important they are to the ecosystem function under consideration. From the standpoint of ecosystem function, both in terms of response and effect traits, functional classification is more informative than taxonomic classification, though functional and taxonomic diversity are likely to be closely related as one uses increasingly higher resolutions of ecosystem function. Theoretically, one would expect that most changes in biodiversity, in terms of species loss or dramatic reductions in abundance, have little impact on a given ecosystem function since many species are likely to be redundant, but when the change involves nonredundant species, the resulting changes in ecosystem functioning may be dramatic.

These principles can be used to derive a general or fundamental relationship between biodiversity and ecosystem function, which is needed to develop the framework by which we can relate BEF to restoration ecology.

A Fundamental Relationship Between Biodiversity and Ecosystem Functioning

Though still a young field, experimental and theoretical BEF research suggests a basic or fundamental relationship between biodiversity and ecosystem functioning that can be applied to restoration. The fact that all species are singular when they are the only one in a functional group and are likely to be redundant when additional species are added (provided they are capable of compensating for one another in each other's absence) leads to a theoretically asymptotic relationship between biodiversity and ecosystem functioning. The theoretical endpoint in this relationship is the point where there is no ecosystem functioning when no species are present. The addition of species immediately leads to some functioning, but an asymptote or leveling off occurs at the point where all functional groups are represented by at least one species (Figures 10.1 and 10.2). As noted above, one has to first declare what function is under consideration before constructing such a relationship, and the more coarse the function, the less definite the relationship.

As the number of species (S) often correlates with functional diversity, I will use S to represent biodiversity and F to represent the number of functional groups. All functional groups (F) are likely to be present, even at low levels of diversity; thus, the curve begins to saturate beyond the point where every functional group is represented by at least one species ($S/F \geq 1$, Figure 10.2). Debate surrounds just what the actual shape of the curve is (Naeem et al. 2002), but most studies to date have confirmed an asymptotic relationship (e.g., Naeem et al. 1996; Tilman et al. 1996; Hector et al. 1999).

For restoration ecology, BEF suggests that most of an ecosystem's functioning may be achieved with only one or a few species per functional group, but such an ecosystem is less robust than one with many species per functional group. The asymptotic curve and the variability surrounding it provide the key to implementing a strategy based on BEF. (Note that, in reality, there are many species, many ecosystem functions, and many relationships among these, thus these arguments address basic patterns.)

Combinatorial: A View of Nature and a Tool

Envisioning the relationship between biodiversity and ecosystem functioning by imagining an empty ecosystem that is steadily and randomly populated with species (or steadily and randomly depopulated with species) can be translated directly into an experiment, and this is exactly how early BEF studies sought to articulate the BEF perspective. This combinatorial approach also articulates BEF's relevance to restoration ecology.

BEF experimental methods employ combinatorial designs that are immediately applicable to restoration research. Experimental tests involve selecting an ecosystem and a pool of species, and from this pool establishing monocultures of each species. Monocultures (i.e., single-species replicates) and polycultures (i.e., multispecies replicates) are constructed according to a design in which several biodiversity treatment levels are used (Naeem 2002b). This kind of experiment using combinatorial species assembly is referred to as a combinatorial experiment and, for biodiversity experiments that use large numbers of species, it is relatively unique to BEF research. The process is illustrated in Figure 10.3, where an original community is deconstructed into its components, and then reconstructed in a variety of ways, each way involving some random, usually unique, combination (strictly in a combinatorial sense) of species.

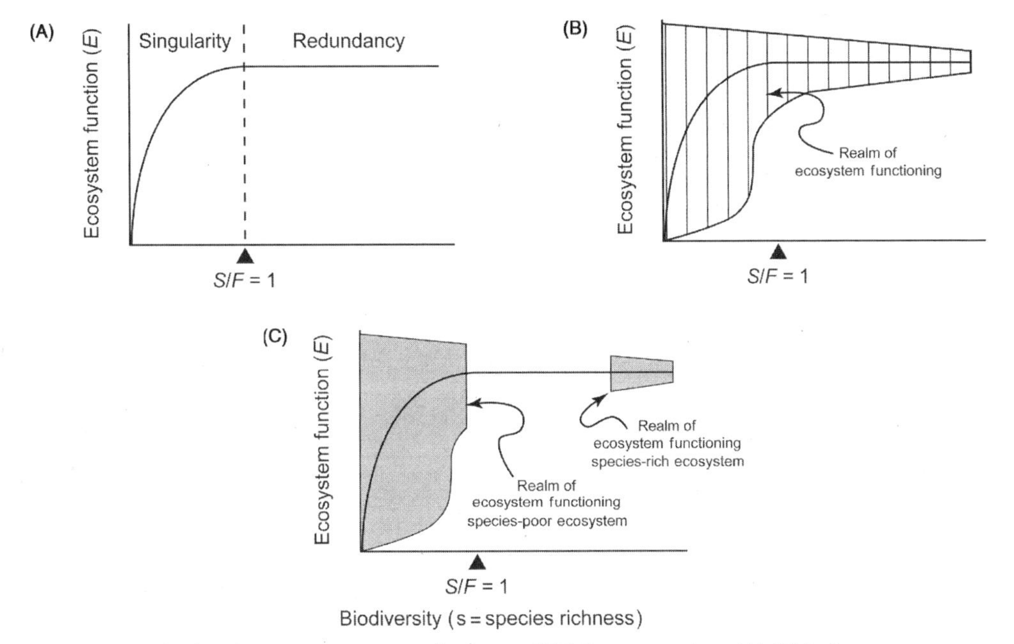

FIGURE 10.2 The biodiversity-ecosystem function (BEF) perspective. (A) This figure captures the basic components of the BEF perspective by illustrating ecosystem function, referring specifically to ecosystem processes (e.g., litter decomposition, NPP, N-mineralization), as a dependent function of biodiversity. From the BEF perspective, most changes in biodiversity result in minor changes in ecosystem function, but at critically low levels, further loss changes ecosystem function. This point is likely to be where there is only one species per functional group such that if the number of species (S) within functional groups (F) drops to an average below one, then any further loss means that an entire functional group is missing, and ecosystem functioning changes significantly. This yields a theoretical asymptotic function that has been widely observed in BEF experiments in which the flat portion results from a prevalence of redundant species, whereas the steep portion to the left results from the predominance of species that are singular in their function. (B) This figure illustrates the theoretical realm of possible expressions of ecosystem functioning based on a couple of assumptions. The first assumption is that it is possible to have higher levels of ecosystem functioning by the restructuring of the biotic composition, though one must be careful to note that if such a change occurs it is not through subsidization. The lower boundary follows a Michalis-Menton function in which it is assumed that small additions of species enhance ecosystem processes, but this enhancement follows a saturating curve. Note that the true realm of ecosystem function is likely to be both system specific and function specific and only qualitatively resemble this figure. (C) This figure compares the realm of ecosystem functioning (shaded regions on left and right that surround the saturating curve) with two versions of the same ecosystem: a species-poor version on the left and a species-rich version on the right. Note that the larger the shaded region, the larger the realm of possible ecosystem functioning and the lower the predictability of an ecosystem's functioning. This comparison illustrates that even though both versions are capable of providing similar levels of ecosystem functioning, in the face of biodiversity loss, the species-poor version is less predictable than the species-rich version.

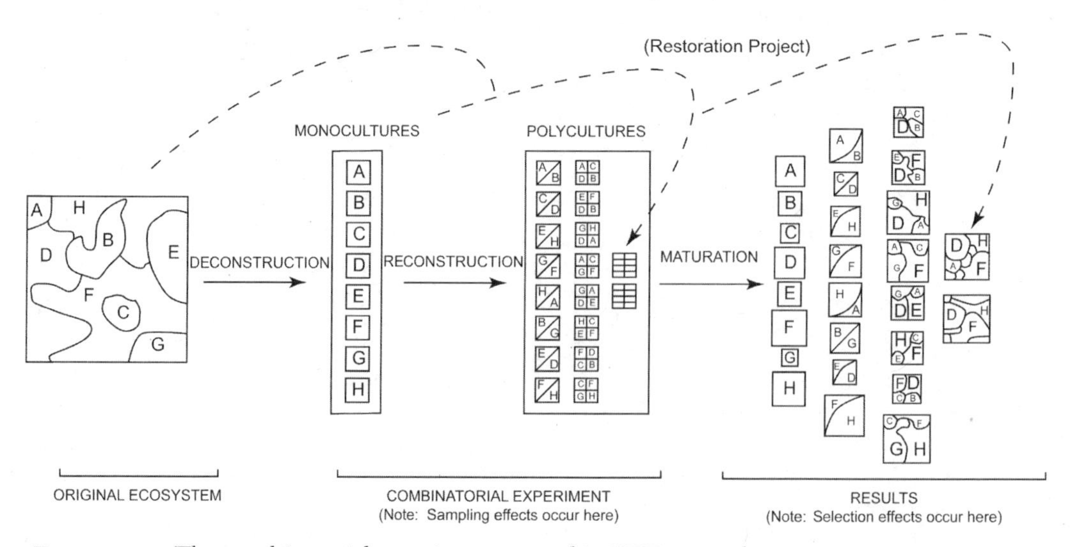

FIGURE 10.3 The combinatorial experiment as used in BEF research in comparison to restoration ecology. Each square represents an ecosystem, the area reflecting the magnitude of functioning. Letters denote species, and the lines surrounding them, their relative contribution to ecosystem function. On the left, the large square represents the functioning of a large, complex ecosystem in its original, unmanaged state. Species D, F, and H are dominant. In the first phase of a combinatorial experiment, replicate experimental ecosystems are established both as monocultures and as polycultures. In polycultures, species are initially established at equal densities. Conceptually, this first phase is equivalent to deconstructing the original system into its species, then reconstructing multiple, replicate ecosystems in which species diversity varies from low to high. The second phase allows for species interactions to take place. The results (farthest right set of squares) are different levels of functioning with different degrees of dominance by species. Which species occur in each replicate reflects sampling effects from the combinatorial design, while the end results are determined by how the species selected for each plot interact with one another (e.g., competition, facilitation, niche complementarity). The dashed arrows indicate how restoration ecology experiments compare with BEF experiments. Typically, restoration involves obtaining species from monocultures or unmanaged systems and using them to construct diverse communities whose maturation is managed to achieve a relative distribution and abundance of species and level of ecosystem functioning that is comparable to the original ecosystem.

Such experiments, and the thinking behind them, require us to revisit Figure 10.2, because the importance of the realm of ecosystem functioning now becomes apparent. At low levels of diversity the range around the central tendency (the asymptotic curve) is enormous. For any set of monocultures made up of species drawn from a community that typically has, as described earlier, only a few numerically dominant species and many rare species, those few numerically dominant species are likely to dominate ecosystem functioning. The rest are likely to show far lower levels of functioning, which is why the realm of possible values or expressions of ecosystem functioning is large when considering monocultures. If one started a monoculture by selecting species at random from a predefined pool, one would have a low probability of selecting the single most productive species at random if the pool of available species is very large. If one starts a polyculture, especially if one selects a large number of species from the pool, then even if selection is entirely random, the probability is much higher that the polyculture will contain the most productive species.

Inherent with such empirical approaches is the need to separate sampling, selection, and niche complementarity effects. Sampling effects concern the simple fact that as one increases the number of species in the replicate, one automatically increases the probability of including those few dominant species with strong positive effects (Huston 1997). Sampling effects are special cases of the more general selection effects in which any species with disproportionate impacts (positive or negative) on an ecosystem occur with higher probabilities in high-diversity replicates. Selection effects mimic niche complementarity effects; the latter occurring when higher diversity exhibits functioning that exceeds what is predicted from monocultures (Loreau and Hector 2001). Combinatorial experiments always yield results that are a mixture of selection and complementarity effects, and Loreau and Hector (2002) devised methods for separating these two effects.

Much of the controversy surrounding experimental BEF research concerns whether sampling or complementarity account for the results, but this issue is not critical for restoration. Whichever the case may be (i.e., sampling, complementarity, facilitation, or some mix of these are responsible for experimental findings), the BEF curve shown in figures 10.1 and 10.2 still stands as a plausible perspective of the general relationship between biodiversity and ecosystem functioning, and it is supported by theoretical, observational, and experimental studies. The asymptotic curve is bounded by two curves or boundary conditions. The top boundary represents the hypothetical maximum functioning achievable for any level of biodiversity (here we assume it declines slightly from the monoculture of the highest performing species simply because some of the resources, even if just a little, are taken up by the other species in the polyculture). The bottom boundary represents the saturation of an ecosystem function using a Michaelis-Menton-like model (Naeem 2002c) where diversity acts like an enzyme—a little enhances ecosystem functioning by facilitating geochemical processes, whereas more and more eventually leads to a saturation where most aspects of the biologically enhanced geochemical process are at their maxima. If one plots the area between these two curves as a function of diversity, one gets plots b and c in Figure 10.2, in which the realm, or the full range of possible levels of ecosystem functioning between the upper and lower bounds and the high and low biodiversity values, is clearly much larger for low diversity systems than for high diversity systems.

The shape of the biodiversity-functioning relationship and its upper and lower boundaries are meant to be heuristic constructs, and so elucidating the actual curves for any system in nature is likely to be difficult. But the fundamental principles underlying each curve—that ecosystem functioning asymptotes with small amounts of diversity; that the upper boundary is likely to be close to what the numerically dominant species can achieve on their own; and that ecosystem functioning is very much like a chemical reaction (in this case a geochemical reaction) that is facilitated by the presence of a biota, much the way a chemical reaction may be facilitated by an enzyme—maps well onto what we know about nature.

Limitations of Current BEF Research

Although its methods represented a radical departure from traditional ecology, BEF research is still founded on rather traditional ecological thinking. Its theory is based on conventional resource-based (e.g., Tilman et al. 1997a; Lehman and Tilman 2000), ecosystem, and Lotka-Volterra models (Loreau 2001). Likewise, the mechanisms of niche complementarity are based on traditional ideas of resource partitioning and ecological niches traceable to MacArthur and Levins (1967) and the intercropping literature (Trenbath 1974; Vandermeer 1989).

While the combinatorial approach is a distinctive feature of BEF research, its experiments have been small scale, short term, focused largely on terrestrial ecosystems, and limited in trophic complexity (Naeem 2001), though these are not uncommon features of experimental ecology. Some of these limitations have been addressed. Some studies (e.g., Mulder et al. 1999; Griffiths et al. 2000; Spehn et al. 2000; Downing and Liebold 2002), especially microbial microcosm studies (reviewed in Petchey et al. 2002), have included some trophic complexity, but this research is relatively limited, given the importance of trophic structure in communities and ecosystems. Further, with rare exceptions (e.g., Cardinale et al. 2002), the mechanisms of sampling, niche complementarity, and facilitation have not been directly tested or demonstrated to be the cause of the patterns observed in BEF experiments.

Another limitation of BEF research might be the debate that surrounds this discipline. After all, if scientists are in disagreement, how does one translate the findings of BEF research into practice? The debates, however, have concerned such issues as experimental design, statistical methods, separating sampling from selection effects, and conflicts between large-scale patterns derived from correlative studies with experimental studies. There is little disagreement on the fundamentals:

1. boundary conditions of ecosystem functioning are set by extrinsic or abiotic factors, while biodiversity regulates functioning within these boundaries;
2. many species may be redundant for single functions;
3. functional diversity is more important than taxonomic diversity;
4. trophic structure is key;
5. small amounts of diversity can achieve the bulk of ecosystem functioning, but over longer terms and large areas, greater diversity may be needed to ensure equitable or consistent functioning.

Both disagreements and agreements are reviewed in a symposium volume edited by Loreau et al. (2002b), in which the initial debates have been clarified and, in most cases, transformed into research objectives. Any ecologist, including restoration ecologists, should review these issues before embarking on research or practice that adopts the BEF perspective.

The BEF Perspective and Current Restoration Research

If we are willing to accept that the fundamental relationship between biodiversity and ecosystem functioning is reasonably approximated by an asymptotic curve, then we can develop a simple model for restoration ecology from the BEF perspective. We divide ecosystems into three useful categories: unmanaged, managed, and degraded. Unmanaged systems are what one might typically consider as wild, pristine, or essentially free of human influences. Since it is almost impossible, however, to find any area in the modern landscape that could be deemed wild or free of human influences, it seems more appropriate to refer to inaccessible ecosystems or areas set aside as reserves as unmanaged, in which, depending on the system, no extraction is allowed and no burning, stocking, culling, or other management activities are occurring. Managed systems would be any that are managed sustainably for ecosystem goods and services. I refer to unsustainably managed systems as degraded systems. Thus, a sustainable plantation or fishery will, at least by definition, never be degraded, while an unsustainably managed system will eventually be exhausted of nutrients or the necessary biota to achieve the levels of ecosystem functioning expected for the region within which it is found. This classification of all ecosystems as managed, unmanaged, or degraded is offered simply to serve the purposes of understanding restoration ecology from the BEF perspective, so it should not be taken as a challenge to any existing schemes.

Managed, unmanaged, and degraded ecosystems are related by their reference to the fundamental, asymptotic BEF relationship. Unmanaged ecosystems would generally have higher native diversity than their managed or degraded counterparts but most likely exhibit lower ecosystem functioning with respect to the ecosystem functions of interest to humans. For example, monoculture pine plantations, croplands, and rangelands each support higher levels of desired autotrophic or heterotrophic production than the unmanaged systems they replaced. These higher levels of functioning are achieved by the use of biocides, irrigation, and the manipulation of species composition or species identity, but if done sustainably, there is little reason to see such managed systems as better or worse than their natural counterparts, with the exception that they require considerable efforts by humans to maintain. By "better or worse" I mean within the context of human well-being. There are, of course, tradeoffs that have to enter into our assessment of whether a sustainable monoculture improves human well-being better than the natural ecosystem it replaced. While maximizing one ecosystem function (e.g., production of edible biomass) one must often minimize another (e.g., storage of nutrient supplies and prevention of erosion). Such tradeoffs need to be part of restoration and management strategies.

Degraded lands are simply unsustainably managed habitats that are lower in both diversity and functioning. Whether the degradation occurred by unsustainable extraction (e.g., overfishing or overhunting) of biotic resources or by physical or chemical intrusions that disrupted the system (e.g., mountaintop-removal mining or eutrophication), such ecosystems contain little of the native biodiversity one would expect to find in the region and no longer provide the ecosystem functions or services once associated with the system.

The business of restoration is primarily that of bringing managed or unmanaged systems, once they have been abandoned and turned over to restorationists, back to the fundamental asymptotic BEF relationship (Figure 10.1). The actual target may be one of much higher biodiversity than what is necessary to achieve basic levels of ecosystem functioning, but the initial steps would be to get ecosystem functioning back on track and then move toward targets of higher biodiversity later.

Comparing Restoration Guidelines

Now that we have explored the foundation of the BEF perspective and related it to the community and ecosystem perspectives, we can compare guidelines each might provide. If general guidelines for restoration were to be constructed, we might envision a slightly different set for community, ecosystem, and BEF perspectives (Table 10.3). Each would recommend a sequence of steps in a different order with different emphases.

These guidelines are not meant to be pitted against one another, but rather to be selected based on the goals of the restoration project. If the goals of a particular restoration project emphasize restoring a community (for example, converting an abandoned habitat to a nature reserve or wildlife park) or some other ecosystem in which species are the primary interest, then one might follow a community-based guideline such as outlined. If goals emphasize restoring ecosystem functioning—for example, an abandoned strip mining field restored to a sustainable rangeland, sustainably harvested forest, or other ecosystem where function is of primary importance—then one would follow the ecosystem-based outline in Table 10.3.

TABLE 10.3

A comparison of possible guidelines for different restoration goals. Each column represents the distinct restoration goals of restoring communities, ecosystems, or the biodiversity-ecosystem functioning (BEF) relationship.

Step	Community	Ecosystem	BEF
1	Restore community structure (basic trophic architecture and dominant species).	Restore ecosystem structure (basic habitat, physical and chemical conditions typical for the ecosystem).	Restore dominant native species of key functional groups (autotrophic groups, consumer groups, and decomposer groups).
2	Verify that linkages among species and community function have been established (e.g., predator regulation of herbivore abundance, stable competitive interactions).	Restore energy input/output.	Verify linkages among species and ascertain that their influences over ecosystem function have been established.
3	Monitor community persistence (community stability).	Restore material input/output.	Restore ecological redundancy by the addition of species that vary in response-functional traits and effect-functional traits.
4	When reasonably persistent, begin restoring rare species.	Restore community function (basic trophic architecture and dominant species).	Monitor community and functional persistence.
5		Verify intrasystem cycling (decomposition, uptake, deposition, turnover, residence times).	Restore to target input and output rates by amendments of combinations of nutrients, energy, or water.
6		Restore to ecosystem targets: efficiency of energy transfer among trophic groups, NPP, standing crop, dynamics (disturbance, resilience, resistance).	Once spatial and temporal variance in ecosystem functioning is within a desired range, manipulate disturbance to create opportunities for rare species.

While at this stage BEF can only provide qualitative guidelines, the BEF perspective (like other perspectives) implies that it is not possible to restore biodiversity without considering ecosystem function, nor is it possible to restore ecosystem function without considering biodiversity. If one wants to recreate "natural ecosystems," the BEF perspective provides useful insights. If one wants to improve a degraded system and is willing to actively manage it (e.g., plantation forests, rangelands), then one may not need to consider the relationships between biodiversity and ecosystem function, although the resulting system is unlikely to be self-sustaining.

Interpreting Current Restoration Research from Three Perspectives

We can apply these three different perspectives and guidelines to current restoration research to get a feel for how our interpretations of findings might differ, depending on which perspective we adopt. In this section, I consider six studies in restoration ecology from a variety of systems to illustrate how we might interpret their findings from the three perspectives of community, ecosystem, and BEF. Each of these studies is excellent and makes important points but, if revisited or expanded using a BEF perspective, they would offer additional insights into developing effective restoration practice.

Example One: Tropical Restoration from a Community Perspective

Jansen (1997) considered the restoration of a tropical rainforest (Lake Barrine National Park in North Queensland, Australia) to be a success because the composition of the decomposer (arthropod) community in planted plots seemed similar to that found in mature forests. This reflects a community perspective in which the goal is to restore species diversity to targets that are based on species diversity in natural or pristine ecosystems that can serve as models for the original systems. From the ecosystem perspective, one might hold off claiming success until decomposition rates were similar between planted and mature sites. From the BEF perspective, one would ask if the functional diversity of trees and functional diversity of the decomposer arthropod community were appropriate (dominated by the major functional groups), and if the rates of decomposition observed in the plots were comparable to the mature forest and robust to perturbation, given litter inputs. Numbers of species are not as key to ecosystem functioning as functional diversity.

Example Two: Desert-Scrub Restoration from a Community/Trophic Perspective

An example of restoration by a community perspective is that of Patten (1997), which examines a desert-scrub restoration in Palm Springs, California, USA. Unlike the first example, this study examines two trophic groups. In this study, vegetation appears to provide little utility for predicting response by mammals to restoration, because vegetation responses were much slower than responses by the mammal community. From an ecosystem perspective, one might ask if the response of mammals outpaces that of vegetation, or if N-cycling or productivity is adversely influenced in such a way that it might hinder restoration. In contrast, the BEF perspective would suggest that the project might aim to restore a set of mammals and plants that represent the key functional groups (dominant plant species, herbivore, granivore, and carnivore) for that system and then ramp up the diversity in each category as persistence is achieved.

Example Three: Montane Forest Restoration from a Community/Succession Perspective

Rhoades et al. (1998) found that planting pasture trees in the lower-montane ecosystems of Ecuador dominated by *Setaria sphacelata*, an exotic foxtail grass that dominates such pasturelands, can help in restoration efforts. The restoration target for this system is montane forests whose woody species are not well adapted to pasture environments. Pasture trees, however, are readily established within pastures, and they modify microclimate and edaphic conditions in ways that favor the establishment of woody montane forest species. (It is unclear if pasture trees would have to be removed or would be naturally replaced during the course of community development.) This is a community perspective again, in which the goal is to restore the plant communities of pasturelands to the original montane forest. The ecosystem perspective would ask how nutrient retention, soil respiration, and other ecosystem functions are affected by these pasture trees before employing them as part of an overall restoration strategy. From the BEF perspective, the goal would be to restore the original functional groups of this ecosystem and manipulate ecosystem functioning back to levels expected for a lower-montane ecosystem in this region. It is possible that further restoration may be hindered by pasture trees, if they adversely alter ecosystem function.

Example Four: Restoration Protocol from a Community and Ecosystem Perspective—Making Use of Functional Groups in Fynbos Restoration in South Africa

Holmes and Richardson (1999) have developed a detailed restoration protocol for the well-studied fynbos of South Africa that is based on a classification of fynbos fragments by their position along a gradient from highly degraded to pristine. At the highly degraded end of the gradient, restoration begins by establishing native cover and reestablishing connections between remnants. At the pristine end of the gradient, management involves maintaining and enhancing biodiversity. The primary goal for all fragment types is to maintain a balanced presence of three guilds of plants that are defined by growth form, regeneration, and nutrient-acquisition traits. This protocol is primarily a community perspective approach, but it includes an ecosystem perspective, as the authors include the importance of monitoring and restoring ecosystem functioning as part of the overall restoration goals. Holmes and Richardson's protocol includes restoring "ecosystem diversity, function, and structure" for fragments that are near pristine. Curiously, however, the monitoring activities all involve monitoring species, not functioning. From the BEF perspective, one would classify the plants in terms of how they respond to environmental variability (response-functional traits) and how they impact ecosystem functioning (effect-functional traits) rather than the guilds used. Growth form (in this study, geophytes, graminoids, annuals/forbs, or shrubs) and regeneration (short-lived, long-lived, or resprouting plants) reflect groupings by response traits. How each species affects nutrient retention, soil organic N and C, erosion resistance, primary production, rates of decomposition and other ecosystem functions would provide a useful means for classifying species by effect traits rather than the guilds used. These may be the authors' nutrient-acquisition guilds, but such guilds are not clearly spelled out in the paper. These functional groups would be used to develop a more integrative protocol for restoring both fynbos diversity and ecosystem functioning.

Example Five: Using Community and Ecosystem Perspectives in Restoring German Fens

In a restoration study of German fens, Richert et al. (2003) found that periodic wetting led to high rates of decomposition that prevented peat buildup (a desired step in fen restoration). The authors also found that the type of vegetation was critical to peat buildup since decomposition, which removes peat, was also affected by plant community composition. In this case, constant, not periodic, wetting was required to retard decomposition, but the appropriate plants were also important. A community perspective would have emphasized only plant community composition, while the ecosystem perspective would have emphasized the response of decomposition to water regime treatments (i.e., continuous or periodic). In some regards, this study reflects the BEF approach in that it considered balancing community (plant composition) and ecosystem factors (decomposition and water inputs and outputs) to achieve the desired result of building up the peat that is a major component of fens. The BEF perspective would suggest also identifying the plant functional traits associated with responses to wetting and effects on decomposition in order to construct a useful functional classification scheme for fen plants in relationship to the ecosystem function of peat accumulation.

Example Six: The Direct Application of the Biodiversity-Ecosystem Function (BEF) Perspective to Wetland Restoration

In a wetland restoration study, Callaway et al. (2003) directly employed the BEF perspective both in experimental design and in interpretation of findings. In this study, the authors examined vegetation biomass and N accumulation in experimental plots of a southern California salt marsh in which plant species richness was experimentally varied between zero, one, three, and six, following a combinatorial design. They found that total N in surface soil was highest in six-species plots at the end of the experiment, with intermediate values for one- and three-species plots, and the lowest values for the zero-species (i.e., unplanted) plots. These differences were not noted at lower soil depths. A similar pattern was found for root biomass and shoot biomass of the planted vegetation. Their data suggests that complementarity among plant species in light utilization may have been the mechanism for the observed biodiversity effects. The implication for wetland restoration is that it would benefit from the use of multiple species both to achieve greater vegetation biomass and N. Had this study been conducted from either the community or ecosystem perspective, this recommendation may not have emerged.

These selected studies demonstrate the gradient of approaches found in current restoration activities. Studies either examined how communities responded to restoration practice (examples One, Two, and Three) or how communities and ecosystems jointly respond to restoration practices (examples Four and Five). While each study provides important insights into restoration ecology, Example Six (Callaway et al. 2003) represents the end of the gradient in which insights in both community and ecosystem processes are obtained. To my knowledge, there are few studies like Callaway et al. (2003) in restoration ecology.

BEF and Restoration Ecology: A Potentially Profitable Union

From the above review of biodiversity and ecosystem functioning research as illuminated by scientific elements of restoration goals, it is clear that even though BEF is fairly young (about a decade, if we date it from the publication of the proceedings of the first symposium on the topic) (i.e., Schulze and Mooney 1993), it has much to offer in terms of developing a foundation for restoration theory.

Perhaps this potentially profitable union is not surprising, because the approach of BEF research is fundamentally similar to that of restoration ecology. BEF explores how nature works, by deconstructing and reconstructing it and, during the process, by extracting principles that inform us about a wide variety of processes and properties of ecosystems and communities (Figure 10.3). Indeed, though there is no way for me to accurately assess this, I will volunteer that objections to restoration ecology as a science are sometimes similar to objections voiced concerning BEF as a science. For instance, both disciplines allow for the possibility that nature can be constructed in short periods of times, in small spaces. This mind set seems contrary to the fact that we generally think of natural systems as being the outcome of an unfathomably large number of complex processes over extraordinary periods of time, in which history, evolution, and long-term ecological processes determine the outcome.

There is, however, no necessary conflict here. For example, we know full well at Cedar Creek, Minnesota, where I and others work on the ecology of grasslands, that it can take 200 years for a field to recover from farming, if it is left to its own devices (e.g., Knops and Tilman 2000). However, that does not mean that we cannot construct grasslands that embody the ecological processes and properties important to understanding how grasslands work using relatively small plots in studies lasting sometimes only a couple of years (consider, for example, the Cedar Creek studies on communities and ecosystems reviewed in Knops et al. 1999). Agreement as to whether such research serves the purpose it was intended to—understanding how grassland communities work—is by no means universal, and it remains the heart of a spirited debate (Tilman et al. 1997b; Wardle et al. 1997; Kaiser 2000; Naeem 2000; Wardle et al. 2000).

The fundamental difference between BEF research and restoration research is, of course, that the former deconstructs and reconstructs nature, while the latter basically reconstructs nature—the deconstruction having happened in an uncontrolled way prior to the arrival of restoration ecologists. It follows, logically, that any finding from restoration ecology is of interest to BEF. For example, a recent BEF study showed that ecosystem responses to variations in biodiversity are sensitive to the sequence of species additions (Fukami and Morin 2003). This research, like many in BEF research, used microcosms, because they provided for the replication and the time necessary for conducting the appropriate experiment—an experiment that would be intractable by conventional field methods. A review of restoration activities in which different sequences of species additions were employed would be incredibly valuable. I imagine that these sorts of experiments go on all the time, and that restoration ecologists exchange, on an informal basis, what sequences work and do not work. If ecosystem processes were part of these restoration experiments, they could readily attest to the efficacy (or lack of it) of assembly theory and the findings from the Fukami and Morin (2003) microcosm study.

One other example of how restoration research can inform BEF research is a study by Pfisterer and Schmid (2002). This study found that whether plant biodiversity will positively influence stability (in this case, recovery from an induced drought) is dependent on whether diversity and productivity are positively related. If diversity increases productivity, it may also decrease stability. This study contradicts other BEF studies (Naeem 2002a), and there is some debate over its findings (Naeem 2003a). Like the Fukami and Morin (2003) study, the scale of this study was relatively small and short term and would need to be repeated before we could make sense of just how robust these results are. Studying existing restoration projects would provide an invaluable, large-scale test of this hypothesis. Given that restoration efforts routinely seed habitats with native plants, and that the diversity of seeds differs among restoration sites, following these sites would be ideal for potentially resolving the debate surrounding the role biodiversity plays in governing ecosystem stability. All it would take is a drought or other perturbation to occur across these sites and to compare vegetation before and after the drought.

These are just two examples of how restoration research can inform BEF research, but there are many others one could imagine. The next few decades represents tremendous potential for these two disciplines to learn from one another. It will, however, require both disciplines to keep track of what the other is doing.

As a proponent of the BEF perspective and an ecological researcher whose activities are structured by this perspective, I am cognizant of the fact that I may be, like the salesperson analogy used before, unwittingly advocating that restoration ecology needs a BMW when, in fact, all it has are the resources for a ten-year-old rust bucket. Restoration projects must allocate resources within the constraints of limited budgets, policy, local and federal legislation, and the desires of the community that has commissioned or will live with the restored ecosystem. Such challenges are seldom deterrents, but they do shape the activities of a restoration ecologist. In advocating the BEF perspective, I do not mean to imply that one must get a BMW, but to use the same decision-making process one would apply when purchasing the ten-year-old old rust bucket as one would apply in purchasing a higher-priced vehicle. Does the system have all its major parts intact and, if not, can they be obtained? Does the system function reliably and, if not, what would be the steps necessary to ensure reliable functioning? If the system has all its major parts and appears to be functioning reliably, what is needed to improve both performance and reliability?

In closing, I suggest that both restoration and BEF research are in the business of understanding the consequences of nature being deconstructed and how best to reconstruct it. Neither group of researchers believes that nature, in its ethereal sense, can be constructed in a short space of time in a small area, but they do believe that the fundamentals of ecology can be studied and employed in small spaces, over short periods, in deconstructed or degraded ecosystems, with the result that the processes and properties of ecosystems can be reconstructed or restored. Given this kinship, they have much to gain from each other.

Acknowledgments

I thank E. Bernhardt, D. Falk, M. Palmer, S. Tjossem, and J. Wright for critically evaluating the manuscript. This work benefited from the support of NSF DEB-0130289.

Literature Cited

Allison, G. W. 1999. The implications of experimental design for biodiversity manipulations. *American Naturalist* 153:26–45.

Butcher, S. S., R. J. Charlson, G. H. Orians, and G. V. Wolfe. 1992. *Global biogeochemical cycles*. London: Academic Press.

Callaway, J. C., G. Sullivan, and J. B. Zedler. 2003. Species-rich plantings increase biomass and nitrogen accumulation in a wetland restoration experiment. *Ecological Applications* 13:1626–1639.

Cardinale, B. J., M. A. Palmer, and S. L. Collins. 2002. Species diversity enhances ecosystem functioning through interspecific facilitation. *Nature* 415:426–429.

Chapin III, F. S., P. A. Matson, and H. A. Mooney, editors. 2002. *Principles of terrestrial ecosystem ecology*. New York: Springer-Verlag.

Collins, S. L., and T. L. Benning. 1996. Spatial and temporal patterns in functional diversity. In *Biodiversity: A biology of numbers and differences*, ed. K. Gaston, 253–282. Oxford, UK: Blackwell Science.

Cottingham, K. L. 2002. Tackling biocomplexity: The role of people, tools, and scale. *BioScience* 52: 793–799.

Crawley, M. J., S. L. Brown, N. S. Heard, and G. R. Edwards. 1999. Invasion-resistance in experimental grassland communities: Species richness or species identity? *Ecology Letters* 2:140–148.

Daily, G. C., S. Alexander, P. R. Ehrlich, L. Gouler, J. Lubchenco, P. A. Matson, H. A. Mooney, S. Postel, S. H. Schneider, D. Tilman, and G. M. Woodwell. 1997. Ecosystem services: Benefits supplied to human societies by natural ecosystems. *Issues in Ecology* 2:1–18.

de Groot, R. S. 1992. *Functions of nature*. Groningen, The Netherlands: Wolters Noordhoff BV.

De Mazancourt, C., M. Loreau, and L. Abbadie. 1998. Grazing optimization and nutrient cycling: When do herbivores enhance plant production? *Ecology* 79:2242–2252.

Díaz, S., and M. Cabido. 2001. Vive la différence: Plant functional diversity matters to ecosystem processes. *Trends in Ecology & Evolution* 16:646–655.

Doak, D. F., D. Bigger, E. Harding-Smith, M. A. Marvier, R. O'Malley, and D. Thomson. 1998. The statistical inevitability of stability-diversity relationships in community ecology. *American Naturalist* 151:264–276.

Downing, A. L., and M. A. Liebold. 2002. Ecosystem consequences of species richness and composition in pond food webs. *Nature* 416:837–841.

Dukes, J. S. 2002. Species composition and diversity affect grassland susceptibility and response to invasion. *Ecological Applications* 12:602–617.

Eagner, A. 2000. Robustness against mutations in genetic networks of yeast. *Nature Genetics* 24:355–361.

Ebeling, C. E. 1997. *An introduction to reliability engineering*. New York: McGraw-Hill.

Ehrenfeld, J. G., and L. A. Toth. 1997. Restoration ecology and the ecosystem perspective. *Restoration Ecology* 5:307–317.

Ehrlich, P. R., and H. A. Mooney. 1983. Extinction, substitution, and ecosystem services. *BioScience* 33:248–253.

Ernst, W. G., editor. 2000. *Earth systems: Processes and issues*. Cambridge, UK: Cambridge University Press.

Folke, C., C. S. Holling, and C. Perrings. 1996. Biological diversity, ecosystems and the human scale. *Ecological Applications* 6:1018–1024.

Fukami, T., and P. J. Morin. 2003. Productivity–biodiversity relationships depend on the history of community assembly. *Nature* 424:423–426.

Garnier, E., M. L. Navas, M. P. Austin, J. M. Lilley, and R. M. Gifford. 1997. A problem for biodiversity-productivity studies: How to compare the productivity of multispecific plant mixtures to that of monocultures? *Acta Oecologica* 18:657–670.

Gitay, H., and I. R. Noble. 1997. What are functional types and how should we seek them? In *Plant functional types*, ed. T. M. Smith, H. H. Shugart, and F. I. Woodward, 20–46. Cambridge, UK: Cambridge University Press.

Griffiths, B. S., K. Ritz, R. D. Bardgett, R. Cok, S. Christensen, F. Ekelund, S. J. Sørenson, E. Bååth, J. Bloem, P. C. De Ruiter, J. Dolfing, and B. Nicolardot. 2000. Ecosystem response of pasture soil communities to fumigation-induced microbial diversity reductions: An examination of the biodiversity-ecosystem function relationship. *Oikos* 90:279–294.

Gu, Z., L. M. Steinmetz, X. Gu, C. Scharfe, R. W. Davis, and W. H. Li. 2003. Role of duplicate genes in genetic robustness against null mutations. *Nature* 421:63–66.

Harper, J. L., and D. L. Hawksworth. 1995. Preface. In *Biodiversity measurement and estimation*, ed. D. L. Hawksworth, 5–23. London: Chapman & Hall.

Hector, A. 1998. The effects of diversity on productivity: Detecting the role of species complementarity. *Oikos* 82:597–599.

Hector, A., B. Schmid, C. Beierkuhnlein, M. C. Caldiera, M. Diemer, P. G. Dimitrakopoulos, J. A. Finn, H. Freitas, P. S. Giller, J. Good, R. Harris, P. Higberg, K. Huss-Danell, J. Joshi, A. Jumpponen, C. Körner, P. W. Leadly, M. Loreau, A. Minns, C. P. H. Mulder, G. O. O'Donovan, S. J. Otway, J. S. Pereira, A. Prinz, D. J. Read, M. Scherer-Lorenzen, E.-D. Schulze, A. S. Siamantziouras, E.-D. Spehn, A. C. Terry, A. Y. Troumbis, F. I. Woodward, S. Yachi, and J. H. Lawton. 1999. Plant diversity and productivity experiments in European grasslands. *Science* 286:1123–1127.

Holmes, P. M., and D. M. Richardson. 1999. Restoration based on recruitment dynamics, community structure, and ecosystem function: Perspectives from South African fynbos. *Restoration Ecology* 7:215–230.

Hooper, D. U., and P. M. Vitousek. 1997. The effects of plant composition and diversity on ecosystem processes. *Science* 277:1302–1305.

Hooper, D. U., and P. M. Vitousek. 1998. Effects of plant composition and diversity on nutrient cycling. *Ecological Monographs* 68:121–149.

Hughes, J. B., and J. Roughgarden. 1998. Aggregate community properties and the strength of species' interactions. *Proceedings of the National Academy of Sciences, USA* 95:6837–6842.

Hughes, J. B., and J. Roughgarden. 2000. Species diversity and biomass stability. *American Naturalist* 155: 618–627.

Huston, M. A. 1997. Hidden treatments in ecological experiments: Re-evaluating the ecosystem function of biodiversity. *Oecologia* 110:449–460.

Jansen, A. 1997. Terrestrial invertebrate community structure as an indicator of the success of a tropical rainforest restoration project. *Restoration Ecology* 5:115–124.

Kaiser, J. 2000. Rift over biodiversity divides ecologists. *Science* 289:1282–1283.

Kareiva, P., and S. A. Levin, editors. 2003. *The importance of species*. Princeton: Princeton University Press.

Knops, J. M. H., and D. Tilman. 2000. Dynamics of soil nitrogen and carbon accumulation for 61 years after agricultural abandonment. *Ecology* 81:88–98.

Knops, J. M. H., D. Tilman, N. M. Haddad, S. Naeem, C. E. Mitchell, J. Haarstad, M. E. Ritchie, K. M. Howe, P. B. Reich, E. Siemann, and J. Groth. 1999. Effects of plant species richness on invasion dynamics, disease outbreaks, insects abundances and diversity. *Ecology Letters* 2:286–293.

Koltover, V. K. 1997. Reliability concept as a trend in biophysics of aging. *Journal of Theoretical Biology* 184:157–163.

Körner, C. 1993. Scaling from species to vegetation: The usefulness of functional groups. In *Biodiversity and ecosystem functioning*, ed. E.-D. Schulze and H. A. Mooney, 117–132. Berlin: Springer-Verlag.

Lavorel, S., and E. Garnier. 2002. Predicting changes in community composition and ecosystem functioning from plant traits: Revisiting the Holy Grail. *Functional Ecology* 16:545–556.

Lawton, J. H., and V. K. Brown. 1993. Redundancy in ecosystems. In *Biodiversity and ecosystem function*, ed. E.-D. Schulze and H. A. Mooney, 255–270. New York: Springer-Verlag.

Lehman, C. L., and D. Tilman. 2000. Biodiversity, stability, and productivity in competitive communities. *American Naturalist* 156:534–552.

Lewis, E. E. 1987. *Introduction to reliability engineering*. New York: John Wiley & Sons.

Loreau, M. 1998a. Biodiversity and ecosystem functioning: A mechanistic model. *Proceedings of the National Academy of Sciences, USA* 95:5632–5636.

Loreau, M. 1998b. Separating sampling and other effects in biodiversity experiments. *Oikos* 82:600–602.

Loreau, M. 2001. Microbial diversity, producer-decomposer interactions and ecosytem processes: A theoretical model. *Proceedings of the Royal Society of London, Series B* 268:303–309.

Loreau, M., and A. Hector. 2001. Partitioning selection and complementarity in biodiversity experiments. *Nature* 412:72–76.

Loreau, M., S. Naeem, P. Inchausti, J. Bengtsson, J. P. Grime, A. Hector, D. U. Hooper, M. A. Huston, D. Raffaelli, B. Schmid, D. Tilman, and D. A. Wardle. 2001. Biodiversity and ecosystem functioning: Current knowledge and future challenges. *Science* 294:806–808.

Loreau, M., A. L. Downing, M. C. Emmerson, A. Gonzalez, J. B. Hughes, P. Inchausti, J. Joshi, J. Norberg, and O. Sala. 2002a. A new look at the relationship between diversity and stability. In *Biodiversity and ecosystem functioning: Synthesis and perspectives*, ed. M. Loreau, S. Naeem, and P. Inchausti, 79–91. Oxford: Oxford University Press.

Loreau, M., S. Naeem, and P. Inchausti. 2002b. *Biodiversity and ecosystem functioning: Synthesis and perspectives*. Oxford: Oxford University Press.

MacArthur, R. H., and R. Levins. 1967. The limiting similarity, convergence, and divergence of coexisting species. *American Naturalist* 101:377–385.

McCann, K. S. 2000. The diversity–stability debate. *Nature* 405:228–233.

McNaughton, S. J. 1977. Diversity and stability of ecological communities: A comment on the role of empiricism in ecology. *American Naturalist* 111:515–525.

Mulder, C. P. H., J. Koricheva, K. Huss-Danell, P. Högberg, and J. Joshi. 1999. Insects affect relationships between plant species richness and ecosystem processes. *Ecology Letters* 2:237–246.

Naeem, S. 1998. Species redundancy and ecosystem reliability. *Conservation Biology* 12:39–45.

Naeem, S. 2000. Reply to Wardle et al. *Bulletin of the Ecological Society of America* 81:241–246.

Naeem, S. 2001. Experimental validity and ecological scale as tools for evaluating research programs. In *Scaling relationships in experimental ecology*, ed. R. H. Gardner, W. M. Kemp, V. S. Kennedy, and J. E. Petersen, 223–250. New York: Columbia University Press.

Naeem, S. 2002a. Biodiversity equals instability? *Nature* 416:23–24.

Naeem, S. 2002b. Disentangling the impacts of diversity on ecosystem functioning in combinatorial experiments. *Ecology* 83:2925–2935.

Naeem, S. 2002c. Ecosystem consequences of biodiversity loss: The evolution of a paradigm. *Ecology* 83:1537–1552.

Naeem, S. 2003a. Continuing debate in the face of biodiversity loss. *Oikos* 100:619.

Naeem, S. 2003b. Models of ecosystem reliability and their implications for species expendability. In *The importance of species: Perspectives on expendability and triage*, ed. P. Kareiva and S. A. Levin, 109–139. Princeton: Princeton University Press.

Naeem, S., K. Haakenson, L. J. Thompson, J. H. Lawton, and M. J. Crawley. 1996. Biodiversity and plant productivity in a model assemblage of plant species. *Oikos* 76:259–264.

Naeem, S., Z. Kawabata, and M. Loreau. 1998. Transcending boundaries in biodiversity research. *Trends in Ecology & Evolution* 13:134–135.

Naeem, S., and S. Li. 1997. Biodiversity enhances ecosystem reliability. *Nature* 390:507–509.

Naeem, S., M. Loreau, and P. Inchausti. 2002. Biodiversity and ecosystem functioning: The emergence of a synthetic ecological framework. In *Biodiversity and ecosystem functioning: Synthesis and perspectives*, ed. M. Loreau, S. Naeem, and P. Inchausti, 3–11. Oxford, UK: Oxford University Press.

Naeem, S., L. J. Thompson, S. P. Lawler, J. H. Lawton, and R. M. Woodfin. 1995. Empirical evidence that declining species diversity may alter the performance of terrestrial ecosystems. *Philosophical Transactions of the Royal Society of London, Series B* 347:249–262.

Naeem, S., and J. P. Wright. 2003. Disentangling biodiversity effects on ecosystem functioning: Deriving solutions to a seemingly insurmountable problem. *Ecology Letters* 6:567–579.

Palmer, M. A., R. F. Ambrose, and N. L. Poff. 1997. Ecological theory and community restoration ecology. *Restoration Ecology* 5:291–300.

Patten, M. A. 1997. Reestablishment of a rodent community in restored desert scrub. *Restoration Ecology* 5:156–161.

Perrings, C. 1995. Biodiversity conservation and insurance. In *The economics and ecology of biodiversity loss*, ed. T. M. Swanson, 69–77. Cambridge, UK: Cambridge University Press.

Petchey, O. L. 2000. Species diversity, species extinction, and ecosystem function. *American Naturalist* 155:696–702.

Petchey, O. L., and K. Gaston. 2002. Functional diversity (FD), species richness and community composition. *Ecology Letters* 5:402–411.

Petchey, O. L., P. J. Morin, F. Hulot, M. Loreau, J. McGrady-Steed, G. Lacroix, and S. Naeem. 2002. Contributions of aquatic model systems to our understanding of biodiversity and ecosystem functioning. In

Biodiversity and ecosystem functioning: Synthesis and perspectives, ed. M. Loreau, S. Naeem, and P. Inchausti, 127–138. Oxford, UK: Oxford University Press.

Pfisterer, A. B., and B. Schmid. 2002. Diversity-dependent productivity can decrease the stability of ecosystem functioning. *Nature* 416:85–86.

Rastetter, E. B., L. Gough, A. E. Hartley, D. A. Herbert, K. J. Nadelhoffer, and M. Williams. 1999. A revised assessment of species redundancy and ecosystem reliability. *Conservation Biology* 13:440–443.

Rhoades, C. C., G. E. Eckert, and D. C. Coleman. 1998. Effect of pasture trees on soil nitrogen and organic matter: Implications for tropical montane forest restoration. *Restoration Ecology* 6:262–270.

Richert, M., O. Dietrich, D. Koppisch, and S. Roth. 2003. The influence of vegetation development and decomposition in a degraded fen. *Restoration Ecology* 8:186–195.

Sala, A. E., R. B. Jackson, H. A. Mooney, and R. W. Howarth, editors. 2000. *Methods in ecosystem science*. New York: Springer-Verlag.

Schlesinger, W. H. 1997. *Biogeochemistry*, 2nd Edition. San Diego: Academic Press.

Schulze, E.-D., and H. A. Mooney, editors. 1993. *Biodiversity and ecosystem function*. New York: Springer-Verlag.

Schwartz, M. W., C. A. Brigham, J. D. Hoeksema, K. G. Lyons, M. H. Mills, and P. J. van Mantgem. 2000. Linking biodiversity to ecosystem function: Implications for conservation ecology. *Oecologia* 122: 297–305.

Smil, V. 1997. *Cycles of life: Civilization and the biosphere*. New York: Scientific American Library.

Smil, V. 1999. *Energies*. Cambridge: MIT Press.

Smith, T. M., H. H. Shugart, and F. I. Woodward, editors. 1997. *Plant functional types*. Cambridge, UK: Cambridge University Press.

Spehn, E. M., J. Joshi, B. Schmid, J. Alphei, and C. Körner. 2000. Plant diversity effects on soil heterotrophic activity in experimental grassland ecosystems. *Plant and Soil* 224:217–230.

Süleyman, Ö. 1996. Reliability and maintenance of complex systems. Berlin: Springer-Verlag.

Symstad, A. J., D. Tilman, J. Wilson, and J. Knops. 1998. Species loss and ecosystem functioning: Effects of species identity and community composition. *Oikos* 81:389–397.

Tilman, D. 1997. Distinguishing the effects of species diversity and species composition. *Oikos* 80:185.

Tilman, D., C. L. Lehman, and C. E. Bristow. 1998. Diversity-stability relationships: Statistical inevitability or ecological consequence? *American Naturalist* 151:277–282.

Tilman, D., C. L. Lehman, and K. T. Thomson. 1997a. Plant diversity and ecosystem productivity: Theoretical considerations. *Proceedings of the National Academy of Science, USA* 94:1857–1861.

Tilman, D., S. Naeem, J. Knops, P. Reich, E. Siemann, D. Wedin, M. Ritchie, and J. Lawton. 1997b. Biodiversity and ecosystem properties. *Science* 278:1866–1867.

Tilman, D., D. Wedin, and J. Knops. 1996. Productivity and sustainability influenced by biodiversity in grassland ecosystems. *Nature* 379:718–720.

Trenbath, B. R. 1974. Biomass productivity of mixtures. *Advances in Agronomy* 26:117–210.

Vandermeer, J. 1989. *The ecology of intercropping*. Cambridge, UK: Cambridge University Press.

Vanni, M. J., A. S. Flecker, J. M. Hood, and J. L. Headworth. 2002. Stoichiometry of nutrient recycling b vertebrates in a tropical stream: Linking species identity and ecosystem proceses. *Ecology Letters* 5:285–293.

Walker, B. 1995. Conserving biological diversity through ecosystem resilience. *Conservation Biology* 9:747–752.

Walker, B. H. 1992. Biological diversity and ecological redundancy. *Conservation Biology* 6:18–23.

Wardle, D. A., M. A. Huston, J. P. Grime, F. Berendse, E. Garnier, W. K. Lauenroth, H. Setälä, and S. D. Wilson. 2000. Biodiversity and ecosystem function: An issue in ecology. *Bulletin of the Ecological Society of America* 81:235–239.

Wardle, D. A., O. Zackrisson, G. Hornberg, and C. Gallet. 1997. Biodiversity and ecosystem properties. *Science* 278:1867–1869.

Yachi, S., and M. Loreau. 1999. Biodiversity and ecosystem functioning in a fluctuating environment: The insurance hypothesis. *Proceedings of the National Academy of Science, USA* 96:1463–1468.

Chapter 11

A Modeling Framework for Restoration Ecology

Dean L. Urban

In this chapter I consider the role of models—ranging from simple heuristics to complicated simulators—in restoration ecology. My intent here is not to provide a perspective on models in ecology more generally (for which, see Canham et al. 2003), nor to promote a particular approach to modeling. Rather, I will advocate a model-based framework for restoration ecology in which various sorts of models can be brought to bear on the diverse tasks of restoration.

If restoration can be defined operationally as the process of returning a degraded system to a healthier, more "normal" state, then there are four fundamental tasks implied by this process. First, we must *pose a model* (conceptual or otherwise) about how we think the system behaves ecologically. That is, what is normal? Next, an *assessment* of the current state of the system should inform us as to how far we are from the target or desired condition. Third, *management experiments* are conducted in an effort to steer the system in the desired direction. Subsequently, *monitoring* is conducted to gauge the relative success of the experimental interventions in achieving their aim. This postmanagement assessment might apply as well to the original model: if our interventions do not have the expected effect, it suggests some inadequacy in the underlying model.

This sequence of steps, of course, outlines the cycle of adaptive management (Holling 1978; Walters 1986; Walters and Holling 1990), and this is familiar to most ecologists in concept, if not in practice (Johnson 1999; Lee 1999). My intention here is to underscore the fundamental role that models play in this adaptive cycle. Indeed, I will argue that models should play *two* separate, but related, roles in this cycle.

Models may be applied to a variety of purposes: (1) to serve as an integrating framework (e.g., Urban et al. 2000); (2) to explore the implications of various management decisions, or to explore alternative scenarios (e.g., Miller and Urban 2000); (3) to design sampling or monitoring schemes (e.g., Urban 2000); (4) to extrapolate understanding across spatial or temporal scales (Peters et al. 2004); or (5) to provide forecasts (predictions) (Clark et al. 2001). Typically, model applications aimed at integration would be developed rather early in a project, while predictive forecasts would be deferred until later, after the model has been well validated. Here I will begin with models posed as our working understanding of how the system behaves and formalize these in as a multivariate framework for assessment. Models aimed at exploration, extrapolation, or forecasting can then be embedded in this framework, which will facilitate model-data comparisons and communication about the models, thus providing a robust framework for assessment and adaptive management.

I should note that I use the term *model* very generally and do so intentionally. This is to recognize that models of various forms might be useful, ranging from simple heuristics (i.e., a hand-drawn graph with postulated relationships on vaguely labeled axes) to more sophisticated, spatially explicit, simulators. In particular, I will argue that the framework I propose can begin with simple heuristic models, which, if used to marshal field studies and as a guide to management experiments, will provide the data needed to implement mechanistically richer and more complicated simulators.

I have three specific objectives in this chapter: (1) to consider some of the various modeling approaches in use in restoration ecology (including some not used, but with clear potential); (2) to construct a general framework in which these various approaches can be reconciled; and (3) to illustrate the potential utility of this framework for restoration applications, especially as an adaptive framework that evolves from conceptual heuristics to data-driven models as its implementation guides data collection.

The examples I choose as illustrations include fire ecology, wetland restoration, and wildlife habitat applications, although there is no reason why this approach could not be applied more generally (Holl et al. 2003). The examples are often projects developed by my own students. My intent is not to provide an exhaustive review of modeling applications in restoration ecology, but to propose a more general framework for their broader application.

Model Applications in Restoration Ecology

While I have no desire to codify a restrictive classification of modeling approaches, it will be convenient here to recognize three broad classes of models:

1. *Heuristic models* include schematic diagrams and conceptual models that illustrate our working understanding of system behavior. These underlie all applications but are not always formalized (i.e., as equations) or implemented as working models (i.e., as computer codes).
2. *Statistical or phenomenological models* include various forms of regressions and are familiar to most ecologists. An example of a phenomenological model would be the theory of island biogeography (MacArthur and Wilson 1967). Although not widely used in restoration ecology, more abstract theoretical models, such as Tilman's (1985) resource ratio hypothesis, might also be included here. Especially, this class also should include multivariate models such as ordinations, which are especially useful in summarizing complicated, multidimensional ecological relationships in a compact, lower-dimensional space (Legendre and Legendre 1998; McCune and Grace 2002). While such models might imply system dynamics, their predictions are typically static or steady-state.
3. *Simulators* are models implemented as numerical algorithms and "solved" by simulation. These include a variety of forest gap models (reviewed by Bugman 2001); fire models (reviewed by Keane et al. 2004); water quality/hydrology models, such as the Better Assessment Science Integrating point and Nonpoint Sources (BASINS)

(USEPA 1998) and Hydrological Simulation Program–Fortran (HSPF) (Bicknell et al. 1996); the soil and water assessment tool SWAT (Arnold et al. 1994) and the regional hydro-ecological simulation system RHESSys (Band et al. 2001); and a wide variety of simulators used to explore wildlife populations in spatially complex habitat mosaics (e.g., McKelvey et al. 1993; Hanski 1994; Dunning et al. 1995, et seq.).

In practice, models often evolve over time (Gardner and Urban 2003): a seminal heuristic model guides data collection to estimate an initial model (probably statistical), which might evolve later into a more complicated simulator (often assembled from statistical model components). Thus, while it might seem that a natural tendency would be to construct a model as a capstone event (e.g., to summarize a mass of data collected from a system), a more useful and strategic approach is to start with an initial model—no matter how crude—and use its formalization, parameterization, and uncertainty to marshal field studies that will build and refine the model most efficiently (Urban 2000, 2002).

This evolutionary approach to model refinement is perhaps nowhere more appropriate than in restoration, where we proceed almost always under conditions of limited data and high uncertainty. This uncertainty invites an adaptive approach that uses management experiments (essentially, all restoration projects!) as the primary tool for resolving uncertainty with crucial data. To the extent that restoration projects can thus discover general insights into ecosystem functioning, this approach also will contribute substantially to ecology in general (Allen et al. 1997).

Illustrations: Models in Restoration

To begin, it will be helpful to review a selection of models to clarify the three types of applications outlined above. These examples are not intended to represent the full range of modeling applications in restoration ecology, but rather to illustrate the sorts of applications that will be especially amenable to the approach I outline below.

Heuristic Models

A conceptual model underlies virtually all management and restoration activities; what varies is the degree to which this model is articulated formally and communicated to our colleagues. For example, Allen et al. (2002) provide an appealing conceptual model for restoration ecology of southwestern fire systems, framed in terms of structural and functional attributes (Figure 11.1). Each dimension (axis) of this conceptual space is scaled arbitrarily from "good" to "poor," generating a simple two-dimensional space in which any site can be located. A similar example could be posed for wetland systems, in which the relevant dimensions might be hydrologic and biogeochemical function (e.g., Weller et al. 1998). In either example, we might ask several questions of the model: How are sites distributed within this space (tightly clustered, dispersed)? How do the sites' locations vary over time (do they remain fixed or vary from year to year)? What management interventions might "move" a degraded site toward the desired "natural" or reference condition? Perhaps most important, what measurable *indicator variables* define these axes? While even a crude conceptual model might provide helpful insights into the first several questions, it is the last issue that translates the conceptual model into a more useful working model.

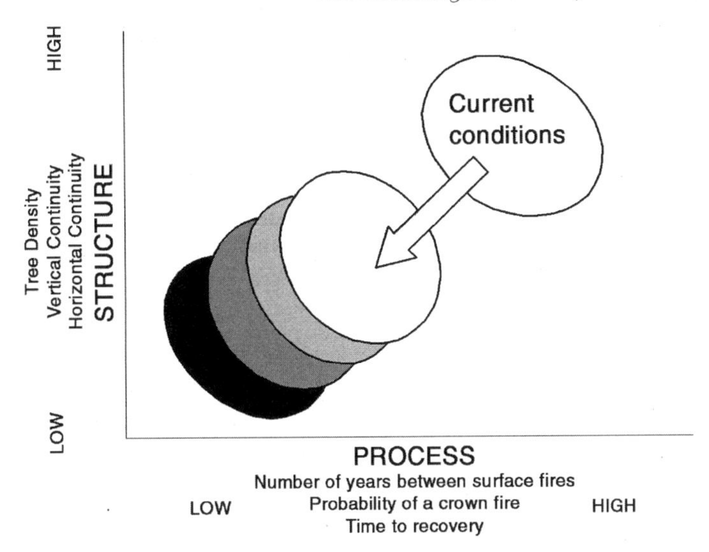

FIGURE 11.1 Schematic diagram of a conceptual, heuristic model for ponderosa pine forest restoration in the southwestern United States (from Allen et al. 2002, reprinted with permission).

Statistical Models

King et al. (2004) developed a statistical model of wetlands of the Everglades, emphasizing vegetation response to increased phosphorus availability resulting from agricultural runoff. They used ordination methods (nonmetric multidimensional scaling) and spatial regressions (partial Mantel tests) to explore species compositional response to a number of correlated environmental variables, over a range of spatial scales (Figure 11.2). While not explicitly a restoration project, the statistical model implies the compositional and structural responses one would expect if phosphorus levels were altered through management interventions. This example nicely illustrates the rich multidimensional structure that characterizes most ecosystem assessment programs and how this complexity can be captured in a compact and intuitive statistical space.

Hierl et al. (forthcoming) developed a conceptually similar assessment framework as a guide to monitoring and restoring artificial wetlands created as wildlife habitat in the Moosehorn National Wildlife Refuge (MNWR) in Maine during the 1950s. These impoundments had been largely unmanaged over several years, and some system for assessing their condition and prioritizing any restoration effort was needed. Hierl identified five dimensions of habitat quality, based on a literature review and the expert opinion of refuge scientists. "Optimal" habitat condition was specified as minimum and maximum values on each of the five axes. Because these five variables were correlated, Hierl used Mahalanobis distance to index how far from "optimal" each impoundment was. Mahalanobis distance is essentially (squared) Euclidean distance, corrected for the correlations among variables (Legendre and Legendre 1998). This was computed for current condition and for their prior condition, based on archival planning maps from the 1980s. This approach provided two useful assessments of the wetlands (Figure 11.3). First, the wetlands could be ranked in terms of their Mahalanobis distances from "optimal" habitat condition. Sites with large distances are farthest from desired condition and most in need of restoration. Second, the distances from two time periods

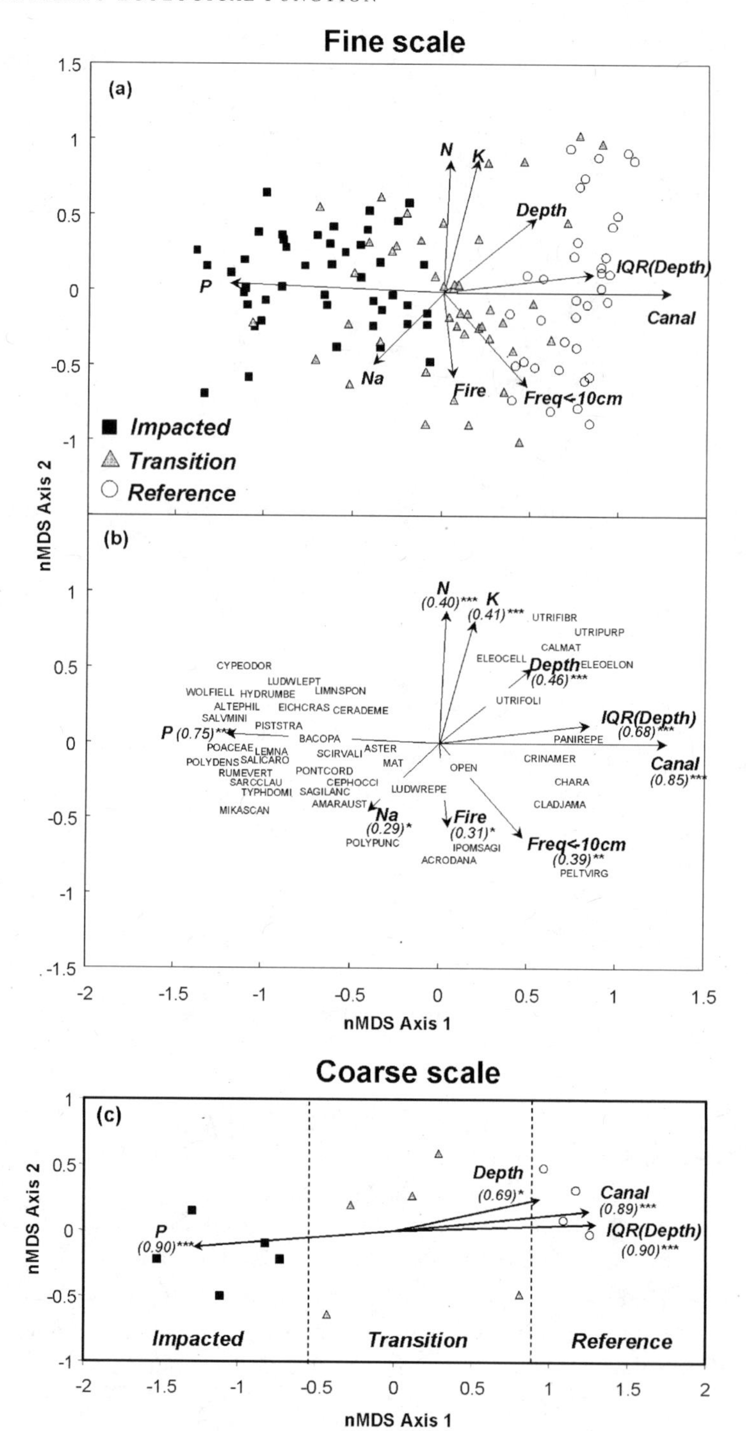

FIGURE 11.2 Distribution of sites impacted by phosphorus-rich effluents in the Everglades, relative to transitional sites and comparatively unperturbed reference conditions, in the framework of a nonmetric multidimensional scaling ordination. The two panels emphasize fine-scale patterns and coarse-scale patterns based on a nested sampling design (coarse-scale samples are aggregated from fine-scale samples) (Figure 11.2 from King et al. 2004, reprinted with kind permission of Springer Science and Business Media).

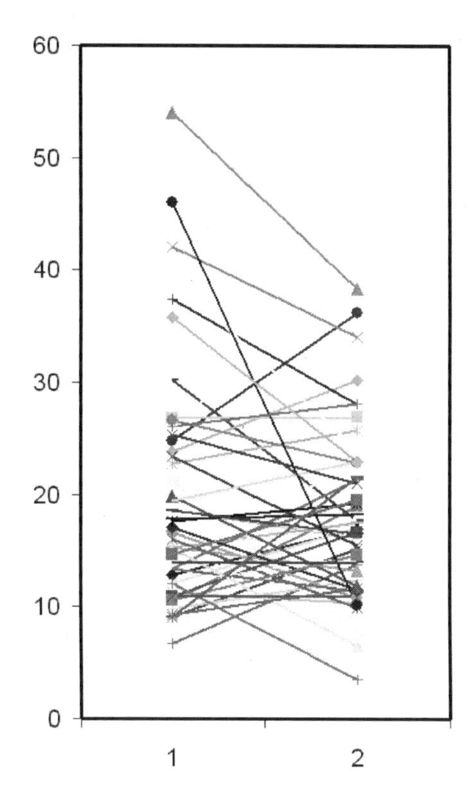

FIGURE 11.3 Statistical model of wildlife habitat assessment based on Mahalanobis distances between impoundment status for waterfowl in 1984–1985 and status in 2002. Each symbol is an impoundment (N = 49). Distances < 20 are in good condition as waterfowl habitat; trajectories emphasize sites that have improved or been degraded over the past two decades (Hierl et al., forthcoming).

indicate which wetlands had improved, degraded, or remained unchanged over the intervening years. The assessment is simple enough to compute that the process can be repeated in the future, and it also provides an easy means of measuring the efficacy of any management interventions aimed at restoring habitat quality. Importantly, the approach can be refined to incorporate more precise information on habitat quality as these data become available. More generally, the same methods can be extended to other wildlife species being managed at MNWR.

Simulation Models

While not developed specifically for ecological restoration, a number of wildlife habitat simulators could be used to illustrate this potential. The general approach is to embed a habitat classification model in a vegetation simulation model, so that the coupled model can simulate the dynamics of habitat availability as a result of natural succession, disturbance, or management (Shugart and Urban 1986). For example, Urban and Smith (1989) used hypothetical bird habitat affinities (random "niches" in an ordination of structural microhabitats) and a forest simulator to illustrate that spatiotemporal dynamics in microhabitat pattern could account for most of the species abundance patterns observed in real communities of forest birds. In this example, microhabitat pattern was summarized statistically in a low-dimensional principal component space. They also used this framework to explore the implications of alternative forestry practices for bird species communities, contrasting selective culling of large trees and stand thinning from below (removing small trees from the understory). Their approach is a useful illustration of the power of the low-dimensional ordination space as a framework for summarizing complicated spatiotemporal dynamics. In similar fashion, Hansen et al. (1995) incorporated habitat classification functions into a forest simulator to explore the consequences of alternative silvicultural practices on forest bird communities in the Pacific Northwest.

Tong and Chen (2002) used BASINS (Arnold et al. 1994) to model relationships between land use and surface-water quality at the regional scale in Ohio. They suggested that their approach could propose guidelines for restoring aquatic ecosystems as well as guidelines for better land-use planning and watershed management. Brun and Band (2000) simulated the impacts of impervious surfaces in urbanizing catchments; their approach, though not aimed at restoration, clearly has implications for restoration.

Miller and Urban (1999) developed a forest simulation model for Sierran mixed-conifer forests. They designed their model to integrate feedbacks among climate, forest process, and fire to explore trends in the Sierran fire regime as a function of elevation (i.e., variability in space) as well as potential implications of anthropogenic climate change (i.e., variability in time). They also used the model to evaluate alternative fire management practices, which would inform efforts to restore fire to these systems after decades of suppression (Miller and Urban 2000). They considered two management scenarios: mechanical intervention to reduce fuels (a structural restoration), and prescribed fire (a restoration of process). They also experimented with varying levels of cutting (as basal area removed) and varying intensities of prescribed fire (i.e., hotter than normal prescriptions). In each scenario, they summarized forest response in terms of the species compositional similarity to pretreatment forest condition (Figure 11.4). Their aim was to assess whether it was feasible to restore species composition to that of presuppression forests, and how long this might take. Similarly, Covington et al. (2001) used the forest succession/fire model (FIRESUM) to evaluate alternative restoration approaches for southwestern Ponderosa pine systems.

A General Framework

Although these examples might seem superficially different, they share several key attributes. First, the model output is framed in terms of attributes that might be measured in real systems. That is, the models speak directly to field data. Second, the applications are inherently multidimensional; the general attributes of interest are expressed in terms of a number of specific measurements. Third, each of these examples compares a managed system (or a degraded system that is a candidate for restoration) in terms of the ecological similarity between the managed system and some reference condition.

These commonalities invite the use of ordination as a reconciling framework. Ordination is a family of well-developed multivariate techniques used to summarize the main trends or patterns in multidimensional data spaces into a concise, low-dimensional ordination space. I suggest that this approach—essentially forcing the restoration and management model to speak the same language as ordination—will provide a powerful integrating framework for restoration ecology.

I propose that we develop these approaches as a general framework for restoration ecology. This approach works well for assessment, monitoring, and evaluating experiments or management treatments (Figure 11.5). What is especially appealing here is that heuristic, statistical, and simulation models all can be reconciled in this common framework; the main difference in these models is the extent to which the relevant dimensions (axes) are defined

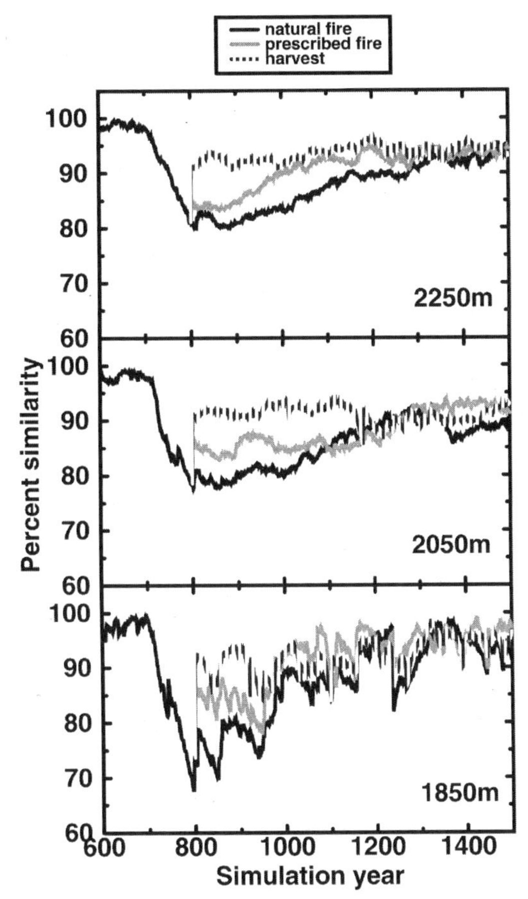

FIGURE 11.4 Simulations exploring alternative management treatments for restoring fire to Sierra Nevada mixed-conifer forests, comparing mechanical thinning and prescribed fire. In each scenario, species compositional change following treatment is summarized as similarity to pretreatment composition using the Bray-Curtis similarity index (from Miller and Urban 1999, reprinted with permission).

quantitatively and, in the case of simulators, the richness of the functional relationships that can be incorporated. In particular, this framework can begin with simple models and evolve toward richer, more complicated simulators without changing the framework itself. The ordination framework is especially powerful as a tool for illustrating complicated information in a compact, graphic form—a key issue in communicating with fellow scientists and other stakeholders.

A Multivariate Framework for Restoration

In this section, I outline in more detail the most promising multivariate frameworks for restoration ecology. I begin with an overview of ordination techniques currently in vogue and close with suggestions for further research in areas that should prove most productive. Along the way, I hope to assuage some apparent misgivings about this general approach (e.g., McCoy and Mishinsky 2002) by clarifying some critical details.

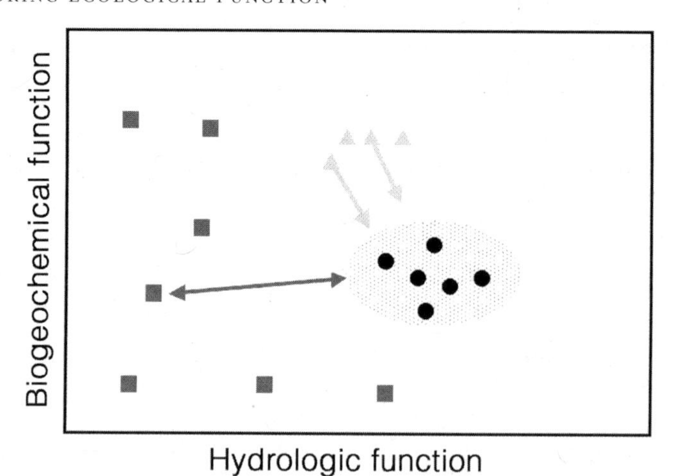

FIGURE 11.5 Schematic of environmental assessment in the framework of ordination space. Round dots are reference sites representing target conditions (as a domain with a centroid implied by the points). Squares are restoration sites being managed actively to move them toward target condition. Change vectors over time monitors their trajectory in terms of direction and rate (vector length). Triangles represent a management experiment posed as a BACI design, with paired control and treatment sites. Paired sites should be close together in ordination space. Successful management implies that the experimental sites should move toward target condition, and at a rate (vector length) that justifies the cost of the intervention.

There are several alternative methods for ecological ordination, but there are only two prevailing lineages. One approach is based on ecological distance or dissimilarity. This lineage began with polar ordination (Beals 1984), which has been largely subsumed by nonmetric multidimensional scaling (NMS) (Minchin 1987). The other common lineage is based on weighted averages (i.e., indicator species or variables), which includes simple weighted averaging as well as more advanced algorithms, such as correspondence analysis (reciprocal averaging) (Hill 1973), and detrended correspondence analysis (Hill and Gauch 1980). All of these techniques are based solely on a single data matrix, typically species composition, and so are *indirect* ordinations (McCune and Grace 2002). In an indirect ordination, relationships between the ordination axes and other variables (typically, environmental factors) must be inferred after the fact. For example, trends in relative species abundances might be used to infer an underlying environmental gradient expressed on the ordination axes. Both ordination lineages also include methods that use a second set of ancillary variables to constrain the ordination of species composition. In a *constrained* ordination, the (compositional) ordination axes are forced to be linear functions of the ancillary (environmental) variables. In the case of distance-based ordinations, the environmentally constrained alternative is distance-based redundancy analysis (Legendre and Anderson 1999). In the case of weighted averaging, canonical correspondence analysis is the technique of choice (ter Braak 1986). Any of these might be appropriate, depending on the application and the nature of the data. All of these techniques, as well as other methods such as principal components analysis (PCA), have been applied in environmental assessment or restoration studies.

Nonmetric Multidimensional Scaling

Polar ordination (Bray and Curtis 1957) is among the earliest ordination techniques available to ecologists. Polar ordination locates each sample, depending on its ecological (compositional) dissimilarity to each of two endpoints (poles) of an ordination axis. While still a viable technique in many applications (Beals 1984), it seems safe to say that polar ordination largely has been subsumed by nonmetric multidimensional scaling (NMS). NMS is essentially a polar ordination done on multiple axes simultaneously, without endpoints. The goal of NMS is to arrange samples (e.g., vegetation quadrats) in ordination space so that distances between samples in the reduced ordination space are as faithful as possible to the compositional dissimilarities between plots. The solution is reached by iterative approximation.

Ecological dissimilarity can be computed using any of a large number of distance metrics (Legendre and Legendre 1998 review dozens). For example, with species compositional data, ecologists often rely on the Bray-Curtis index (Faith et al. 1987), which—when computed from presence/absence data—depends on the number of species shared in common between two sample plots. Computed from species abundance data, the same index depends on the shared species abundances between the two plots. For environmental data (e.g, water chemistry, soils, topographic indices) ecologists often rely on Euclidean distance, the multivariate version of the Pythagorean theorem. (Note that, in this case, the variables must be transformed or relativized to reconcile their disparate measurement units.) In data sets plagued by high levels of correlation among variables, Mahalanobis distance provides a useful measure; this is essentially (squared) Euclidean distance corrected for the correlations among variables (Legendre and Legendre 1998).

Weighted-Averaging Ordination

Perhaps the most popular ordinations have been weighted-averaging methods, including correspondence analysis, or reciprocal averaging (RA) (Hill 1973) and detrended correspondence analysis (DCA) (Hill and Gaugh 1980). In these methods, a sample's ordination score is the simple average of species abundances multiplied by a weight for each species. The original method (Curtis and MacIntosh 1951) assigned weights to tree species based on an inferred gradient from early successional (xeric) to late successional (mesic) species. Similarly, the current approach of rating wetlands based on indicator species (FICWD 1989) amounts to a direct weighted-averaging ordination. The appeal of RA and DCA is that the species weights can be estimated as part of the ordination procedure; they need not be specified in advance. While this flexibility has clearly influenced the popular appeal of RA and DCA, it is worth emphasizing that there are several powerful approaches for identifying indicators or weights, especially including indicator species analysis (ISA) (Dufrêne and Legendre 1997). Thus, weighted averaging might warrant a fresh look by restoration ecologists.

Constrained Ordinations

One of the more widely used ordinations in recent decades has been canonical correspondence analysis (CCA) (ter Braak 1986, 1987, 1988). CCA is a constrained version of RA or DCA, in which the ordination axes are forced to be linear combinations of measured environmental variables provided as ancillary data. More recently, distance-based redundancy analysis (dbRDA) has emerged as an alternative to CCA. Distance-based RDA amounts to a principal components analysis of a distance matrix, constrained by a second (e.g., environmental) data matrix. The use of a distance matrix for the ordination provides considerable flexibility to the analysis, including the potential to sidestep some issues that often plague ordinations based on RA or DCA (Minchin 1987; McCune and Grace 2002; and see below). The two constrained methods differ in that CCA assumes a nonlinear relationship between species and the constraining (environmental) variables, while dbRDA assumes this relationship is linear.

Note that these constrained approaches can be reconciled with indirect ordinations via regression after the fact. For example, any indirect ordination axis can be regressed on ancillary (environmental) variables to derive indicators for that axis. In the ideal case where the ancillary variables are in truth the operative constraints on species composition, this after-the-fact regression analysis should essentially match the constrained ordination. Importantly, in cases where the ancillary variables do *not* explain trends in species composition, a constrained ordination might provide an inadequate or biased depiction of the trends. By contrast, an after-the-fact regression approach would (presumably) faithfully depict the compositional trends while also indicating the inadequacy of the ancillary variables. For example, Urban et al. (2002) found three NMS axes in an ordination of Sierran mixed-conifer forests. One of these axes was clearly characterized by Jeffrey pine (*Pinus jeffreyii*), but this axis was not correlated with any measured environmental variable. A CCA of these data would fail to find the "Jeffrey pine axis" for this reason, essentially hiding this compositional information in the constrained solution. Thus, it is a good idea to corroborate constrained ordinations using other (unconstrained) techniques.

Environmental Assessment in Ordination Space

An appealing aspect of ordination as a framework for assessment is that samples can be connected through time (i.e., monitored) simply by charting their changing positions in the ordination space (Figure 11.5). For example, a time series of sample plots in ordination space delineates a trajectory over time, with each "movement" of a sample summarized by a *change vector* (Halpern 1988; Smith and Urban 1988; McCune and Grace 2002). This opens a wealth of opportunities to examine sample trajectories in terms of the direction and magnitude of these changes. For example, if we apply a management intervention to a sample and monitor it over time, does it move in the *direction* of the target (reference, pristine) condition, as desired? How *fast* does it move (i.e., how long is the vector)? McCune and Grace (2002) describe methods for assessing the directionality of change vectors. Rates of change (vector lengths) might be assessed overall or in terms of ancillary environmental variables. For example, Taverna et al. (2005) examined change vector lengths in relation to environmental gradients in a study of forest understory change over time. This capability to explore trajectories directly from empirical measurements clearly would be a boon to restoration ecology (Zedler and Callaway 1999).

Importantly, inferences about trajectories (including the effects of interventions) require that distance in ordination space be related directly to ecological distance, and this relationship must be consistent throughout the ordination space. These criteria are ensured by NMS to the extent that the numerical algorithm attempts to optimize this relationship (strictly,

NMS optimizes the linear relationship between *rank* distances). This relationship between ordination and ecological distance is not ensured by techniques based on weighted averaging (including RA, DCA, and CCA). Indeed, these weighted-averaging techniques tend to distort this relationship, sometimes quite seriously (McCune and Grace 2002). So while any ordination might provide a workable solution—depending on the nature of the data—it is crucial to verify that the relationship between ordination distances and ecological distances is approximately linear. In practice, this is a straightforward exercise: a Shepard diagram provides the necessary verification. A Shepard diagram is a plot of sample separation in ordination space versus ecological dissimilarity. This relationship must be linear to make robust inferences about sample locations in ordination space.

A second concern in ordination-based assessment is that disjunctions between groups of samples in ordination space actually reflect natural discontinuities in the data (i.e., "natural" communities or habitat types). NMS, by the nature of its algorithm, allows such natural disjunctions to be revealed in ordination space; most other algorithms do not. For example, the algorithm of DCA attempts to array species at equal distances along the ordination axis and so will tend to obscure natural discontinuities in the data.

There have been myriad studies over the past couple of decades that have used various ordination techniques as a framework for assessment or restoration. But because many studies of sample locations or trajectories in ordination space use techniques such as PCA, DCA, or CCA and do not verify the relationship between ordination distance and ecological dissimilarity, inferences about these trajectories (i.e., the implied distances or rates) are highly uncertain. This is not to say these studies are invalid; to be fair, available software and our understanding of this issue have evolved considerably over the past couple of decades. Now, however, it is fair to insist that the underlying assumptions about sample positions within the ordination space must be evaluated before rigorous inferences can be made about these positions or their trajectories over time.

Research Needs and Opportunities

This general modeling framework can be useful for restoration ecology, in part because of the clarity and commonality it can bring to restoration research and, through this, to ecology more generally. In this section, I consider some promising research areas in applying various models in a common framework for restoration ecology.

Model Translation: Converting Models to the Ordination Framework

The model-based illustrations cited above, while consistent with the framework outlined here, also are similar in that none of them is explicitly framed in these terms. For this approach to be useful, model applications must be *translated* into ordination space. In general, this is not a daunting task; it merely requires recasting the data in a slightly different context.

For example, Hierl et al. (forthcoming) based their assessment on Mahalanobis distances computed from five habitat variables. These same variables could be used to ordinate the samples (impoundments) by nonmetric multidimensional scaling. Similarly, Miller and Urban (2000) indexed compositional change under forest management in terms of Bray-Curtis dissimilarity between managed stands and pretreatment conditions. In other applications in the same system, ordinations based on Bray-Curtis dissimilarities have been used to explore compositional trends within these forests (Urban et al. 2002). It would be perfectly natural to recast the management experiments into ordination space, to illustrate how far from current conditions the management interventions might "move" the system. This translation also would have the benefit (partly sociological) of combining research and management in the same analytic framework, helping to dissolve the unnatural distinction between these activities.

More generally, a multivariate framework lends itself naturally to a quantitative and visual representation of the natural range of variability in ecological systems (Landres et al. 1999), an important foundation for restoration ecology (White and Walker 1997). For example, it would be reasonable to draw confidence limits around a set of reference sites in an ordination space (Figure 11.5), delineating the reference condition at any one time, as well as the characteristic dynamics of such sites over time. Alternatively, one might use hierarchical clustering to group samples representing reference conditions (e.g., Harris 1999). The centroid of these reference groups represents the target condition as a synthetic, average point in the ordination space. This approach, in fact, facilitates the conventional application of Mahalanobis distances, in which each sample site is indexed by its ecological distance from the target condition (i.e., as distance to the reference group centroid).

Manipulations to ecosystems—including management experiments, accidents, and natural disturbances—are readily reconciled in an ordination framework. In this, an intervention or perturbation results in a response vector defined by its direction and a magnitude. The direction locates the dynamic relative to the target condition, and the magnitude can indicate the intensity of the effect or impact. For management interventions, this also suggests the efficiency of the intervention in terms of benefit versus cost. Michener (1997) reviews a range of statistical approaches for analyzing experimental and monitoring data in restoration, and Block et al. (2001) provide an in-depth discussion of monitoring needs and approaches for wildlife restoration. Their recommendations are generally consistent with the framework proposed here. Similarly, Philippi et al. (1998) discuss various multivariate methods for analyzing long-term monitoring data on species composition, and the techniques they recommend—based on compositional dissimilarity—are easily embedded in this ordination framework.

Evaluating alternative scenarios (effects forecasting) is a straightforward extension of this framework, requiring only that the simulator (or other model) provide output in terms of the appropriate indicator variables (e.g., in terms of compositional similarity to the reference or initial conditions). Scenarios would be evaluated in much the same way as long-term monitoring data—tempered, of course, by a consideration of uncertainty inherent in model forecasts.

Model Identification and Refinement

Within the perspective espoused here—a multivariate framework that summarizes the principal dimensions of structural, compositional, or functional aspects of the focal system—some research needs are beguilingly simple. One research goal of restoration ecology amounts to identifying the appropriate dimensions (axes) of the reference space, and then finding efficient and robust indicator variables for these axes. The first task in this is to identify the relevant dimensions or axes in a multivariate framework, that is, to "label" these axes in a narrative sense, with labels that participants (academic researchers, managers, regulatory policy agents, community stakeholders) find mutually interpretable. This is essentially an exercise in data analysis and inference. Ordination is specifically designed to find these dominant dimensions, although many regression-based approaches could provide similar results.

The issue of identifying which variables might serve as robust indicators for the relevant dimensions is a somewhat more complicated question. Currently, at least some camps in environmental assessment or restoration are well invested in environmental indicators (e.g., Karr 1991; Tiner 1999; Lenz et al. 2000; McLachlan and Bazely 2001; Ahn et al. 2002; see Dale and Beyeler 2001 for an overview). But the statistical tools for identifying and designing ecological indicators have developed well beyond the current state-of-the-art in restoration ecology or environmental assessment.

There are two general approaches to identifying indicator variables for assessments based on multivariate dimensions of ecological condition. The first is essentially a multiple regression problem, where the task is to identify a variable (or variables) that can capture the large trend represented by a multivariate ordination axis. This approach is especially amenable to after-the-fact regressions of ordination axes on a set of measured variables—either the primary variables (e.g., species composition) or ancillary variables (e.g., environmental factors). For example, what easily measurable variable might best serve as an indicator variable for ecosystem "structure" in a fire model (Figure 11.1)? How might we best index "hydrologic functioning" for a model of wetlands? Within this context, it might be emphasized here that a good indictor variable is not only highly correlated with the ecological dimension of interest, a good indicator is also readily measured (i.e., is logistically affordable) and is highly repeatable (i.e., has low observer bias). Thus, while a high partial r^2 between a measured variable and a complex ecosystem dimension is certainly desirable, this criterion must be weighed against more pragmatic concerns about repeatability and cost.

Although less straightforward than simple regression-based approaches, the use of indicator species is common in ecology in general, and restoration in particular. While many systems of indicators are strongly infused with expert opinion, it is worth emphasizing that there are many other ways to define indicators, some of which have not been explored adequately by ecologists involved with restoration and environmental assessment. Direct methods based on lab experiments (e.g., dosage trials in toxicology, greenhouse trials for environmental tolerances) are one obvious approach to defining indicators. Indirect methods are also readily available. For example, species ordination scores derived from RA, DCA, and CCA are in fact indicator scores that can be used to order new samples by weighted averaging (this presumes, again, that the assumption of linearity in the Shepard diagram is met). A somewhat less familiar but often useful technique, Two-Way Indicator Species Analysis (TWINSPAN) (Hill 1979) provides a simultaneous ordination (based on RA) and classification of samples into discrete types, and also identifies indicator species for each classified group. More recently, Indicator Species Analysis (ISA) (Dufrêne and Legendre 1997) has been developed to identify indicators for groups defined by the user (e.g., as identified by hierarchical clustering, or pristine reference versus restoration sites). These indicators are based on the relative frequency and abundance of species in each group: a good indicator species occurs with high frequency in its group, reaches it greatest abundance in that group, and is uncommon in any other group. Indicators in ISA are tested using Monte

Carlo permutations to verify the most robust indicators. Clearly, these techniques warrant further application in restoration ecology.

A compelling argument for a common modeling framework for restoration ecology is that this commonality invites comparison and synthesis. Restoration activities are necessarily idiosyncratic to the particular application, and this site-specific nature would tend to discourage comparisons among systems. A common assessment framework based on similar analytic techniques would facilitate comparisons, allow researchers to share results, and ultimately yield a more synthetic understanding of regional or cross-system patterns in ecosystem structure and functioning.

Model Evolution: Opportunities in Restoration Ecology

While the scientific benefits of comparison and synthesis in a common framework are obvious, adopting a common modeling framework for environmental assessment offers additional capabilities and benefits. One promising area is to use the parametric summaries explicit in ordination space as a guide to site selection in restoration and sampling design in environmental assessment. The increasingly common use of geographic information systems (GIS) as a framework for research and management makes this especially compelling. Urban (2000, 2002) discusses a variety of ways to use models—from simple heuristics to complicated simulators—to guide sampling design. In this, sites that are especially informative in parameter space (e.g., sites representing an especially sensitive or vulnerable environmental setting) can be selected in parameter space (ordination space) and then mapped into geographic space to select sampling locations. In this way, even a tentative, preliminary model can help marshal the field studies that will refine the model efficiently, so that this model-guided sampling approach is self-improving (Urban and Keitt 2001; Urban 2002).

The natural evolutionary trajectory for many models seems to be from simple to complicated, often culminating in a spatially explicit simulator (Gardner and Urban 2003; Peters et al. 2004). While this complexity does come at some cost in terms of model uncertainty (Peters et al. 2004), the endpoint is nonetheless attractive because spatially articulated simulators provide some capabilities that are crucial to the larger success of restoration ecology. In particular, spatially explicit models provide the capability to evaluate and prioritize restoration projects in a larger context: sites within landscapes, and landscapes within regions. Thus, spatially explicit hydrology models (e.g., RHESSys and similar approaches) could be used to evaluate sites within the context of their larger watersheds, spatially explicit metapopulation simulators could be used to evaluate the conservation value of individual sites, and so on. A powerful method for this evaluation is to use a site-deletion algorithm in model experiments. In a site-deletion algorithm, some index of overall ecological integrity or conservation value is computed at the landscape scale, based on simulations. The simulations are then repeated, with each site withheld in turn. At each iteration, the ecological index is recomputed and that site is ranked in terms of the overall impact resulting from its removal. For example, Keitt et al. (1997) and Urban and Keitt (2001) used two different models and a site-deletion algorithm to evaluate the conservation value of discrete habitat patches to the long-term persistence of the Mexican spotted owl (*Strix occidentalis lucida*). In this instance, the larger context is that of metapopulation dynamics for habitat patches linked by dispersal. But this approach could be generalized to other classes of models operating spatially in a landscape context. Watershed assessment, in particular, invites an analysis in a spatially explicit context. This capability to extrapolate the site-evaluation process to the larger scales of management and policy is an especially compelling argument for the development of robust, spatial models in restoration ecology.

Summary

I have tried to emphasize two themes in this chapter: First, models are an integral part of restoration ecology—whether posed as simple heuristics or implemented as complicated simulators. Second, a common integrating framework based on ordination can be equally appropriate for field data, heuristic models, or forecasts from simulators, and thus it can reconcile these otherwise disparate approaches to restoration ecology. The multivariate framework is a natural construct for evaluating the natural range of variability in ecosystems, management experiments, long-term monitoring data, and model forecasts of the implications of alternative management scenarios. Importantly, the approach I propose also can help to marshal research efforts to facilitate the evolution of initial heuristic models, through statistical models, toward spatially explicit simulators. Central to this effort is the inference of major dimensions of ecosystem variability and the identification of robust indicator variables to characterize these dimensions. I have pointed to several promising techniques for addressing these issues. I also suggest that the natural evolution of ecological models toward spatially explicit simulators is an important trend to foster, as it will provide the means to assess individual sites or restoration projects in a larger landscape context, and thus extrapolate these site-specific analyses to the larger scales of land-use planning and policy.

Literature Cited

Ahn, C., D. C. White, and R. E. Sparks. 2002. Moist-soil plants as ecohydrologic indicators for recovering the flood pulse in the Illinois River. *Restoration Ecology* 12:201–213.

Allen, C. D., M. Savage, D. A. Falk, K. F. Suckling, T. W. Swetnam, T. Schulke, P. Morgan, M. Hoffman, and J.T. Klingel. 2002. Ecological restoration of southwestern Ponderosa pine ecosystems: A broad perspective. *Ecological Applications* 12:1418–1433.

Allen, E. B., W. W. Covington, and D. A. Falk. 1997. Developing the conceptual basis of restoration ecology. *Restoration Ecology* 5:275–276.

Arnold, J. G., J. R. Williams, R. Srinivasan, K. W. King, and R. H. Griggs. 1994. *SWAT—soil and water assessment tool—user manual*. Grassland, Soil, and Water Research Lab: Agricultural Restoration Service, USDA.

Band, L. E., C. L. Tague, P. Groffman, and K. Belt. 2001. Forest ecosystem processes at the watershed scale: Hydrological and ecological controls on nitrogen export. *Hydrological Processes* 15:2013–2028.

Beals, E. W. 1984. Bray-Curtis ordination: An effective strategy for analysis of multivariate ecological data. *Advances in Ecological Research* 14:1–55.

Bicknell, B. R., J. C. Imhoff, J. L. Kittle Jr., A. S. Donigan, and R. C. Johanson. 1996. *Hydrologic simulation program—Fortran user's manual for release 11*. Athens, GA: Office of Research and Development, USEPA.

Block, W. B., A. B. Franklin, J. P. Ward Jr., J. L. Ganey, and G. C. White. 2001. Design and implementation of monitoring studies to evaluate the success of ecological restoration on wildlife. *Restoration Ecology* 9:293–303.

Bray, J. R., and J. T. Curtis. 1957. An ordination of the upland forest communities of southern Wisconsin. *Ecological Monographs* 27:325–349.

Brun, S. E., and L. E. Band. 2000. Simulating runoff behavior in an urbanizing watershed. *Computers, Environment, and Urban Systems* 24:5–22.

Bugmann, H. 2001. A review of forest gap models. *Climatic Change* 51:259–305.

Canham, C. D., J. J. Cole, and W. K. Lauenroth, editors. 2003. *Models in ecosystem science.* Princeton: Princeton University Press.

Clark, J. S., S. R. Carpenter, M. Barber, S. Collins, A. Dobson, J. A. Foley, D. M. Lodge, M. Pascual, R. Pielke Jr., W. Pizer, C. Pringle, W. V. Reid, K. A. Rose, O. Sala, W. H. Schlesinger, D. H. Wall, and D. Wear. 2001. Ecological forecasts: An emerging imperative. *Science* 293:657–661.

Covington, W. W., P. Z. Fule, S. C. Hart, and R. P. Weaver. 2001. Modeling ecological restoration effects on Ponderosa pine forest structure. *Restoration Ecology* 9:421–431.

Curtis, J. T., and R. P. McIntosh. 1951. An upland forest continuum in the prairie-forest border region of Wisconsin. *Ecology* 32:476–496.

Dale, V. H., and S. C. Beyeler. 2001. Challenges in the development and use of ecological indicators. *Ecological Indicators* 1:3–10.

Dufrêne, M., and P. Legendre. 1997. Species assemblages and indicator species: The need for a flexible asymmetrical approach. *Ecological Monographs* 67:345–366.

Dunning Jr., J. B., D. J. Stewart, B. J. Danielson, B. R. Noon, T. L. Root, R. H. Lamberson, and E. E. Stevens. 1995. Spatially explicit population models: Current forms and future uses. *Ecological Applications* 5:3–11.

Faith, D. P., P. R. Minchin, and L. Belbin. 1987. Compositional dissimilarity as a robust measure of ecological distance. *Vegetatio* 69:57–68.

Federal Interagency Committee for Wetlands Delineation (FICWD). 1989. *Federal manual for identifying and delineating jurisdictional wetlands.* Washington, DC: Cooperative Technical Publication, U.S. Army Corps of Engineers, U.S. Environmental Protection Agency, U.S. Fish and Wildlife Service, and U.S. Soil Conservation Service.

Gardner, R. H., and D. L. Urban. Model testing and validation: Past lessons and present challenges. 2003. In *Models in ecosystem science*, ed. C. D. Canham, J. J. Cole, and W. K. Lauenroth, 184–203. Princeton: Princeton University Press.

Halpern, C. B. 1988. Early successional pathways and the resistance and resilience of forest communities. *Ecology* 69:1703–1715.

Hansen, A. J., S. L. Garman, J. F. Weigand, D. L. Urban, W. C. McComb, and M. G. Raphael. 1995. Ecological and economic effects of alternative silvicultural regimes in the Pacific Northwest: A simulation experiment. *Ecological Applications* 5:535–554.

Hanski, I. 1994. A practical model of metapopulation dynamics. *Journal of Animal Ecology* 63:151–162.

Harris, R. R. 1999. Defining reference conditions for restoration of riparian plant communities: Examples from California, USA. *Environmental Management* 24:55–63.

Hierl, L., C. Loftin, J. R. Longcore, D. G. McAuley, and D. Urban. Forthcoming. A multivariate assessment of wetland habitat conditions in Moosehorn National Wildlife Refuge.

Hill, M. O. 1973. Reciprocal averaging: An eigenvector method of ordination. *Journal of Ecology* 61: 237–249.

Hill, M. O. 1979. *TWINSPAN—A FORTRAN program for arranging multivariate data in an ordered two-way table by classification of the individuals and attributes.* Ithaca: Section of Ecology and Systematics, Cornell University.

Hill, M. O., and H. G. Gauch. 1980. Detrended correspondence analysis: An improved ordination technique. *Vegetatio* 42:47–58.

Holl, K. D., E. E. Crone, and C. B. Schultz. 2003. Landscape restoration: Moving from generalities to methodologies. *BioScience* 53:491–502.

Holling, C. S. 1978. *Adaptive environmental assessment and management.* London: John Wiley & Sons.

Johnson, B. L. 1999. The role of adaptive management as an operational approach for resource management agencies. *Conservation Ecology* 3 (2): 8.

Karr, J. R. 1991. Biological integrity: A long-neglected aspect of water resource management. *Ecological Applications* 1:66–84.

Keane, R. E., G. J. Cary, I. D. Davies, M. D. Flannigan, R. H. Gardner, S. Lavorel, J. M. Lenihan, C. Li, and T. S. Rupp. 2004. A classification of landscape fire succession models: Spatial simulations of fire and vegetation dynamics. *Ecological Modelling* 179:3–27.

Keitt, T. H., D. L. Urban, and B. T. Milne. 1997. Detecting critical scales in fragmented landscapes. *Conservation Ecology* 1 (1): 4.

King, R. S., C. J. Richardson, D. L. Urban, and E. A. Romanowicz. 2004. Spatial dependency of vegetation-environment linkages in an anthropogenically influenced wetland ecosystem. *Ecosystems* 7:75–97.

Landres, P. B., P. Morgan, and F. J. Swanson. 1999. Overview of the use of natural variability concepts in managing ecological systems. *Ecological Applications* 9:1179–1188.

Lee, K. N. 1999. Appraising adaptive management. *Conservation Ecology* 3 (2):3.

Legendre, P., and M. J. Anderson. 1999. Distance-based redundancy analysis: Testing multispecies responses in multifactorial ecological experiments. *Ecological Monographs* 69:1–24.

Legendre, P., and L. Legendre. 1998. *Numerical ecology*, 2nd English Edition. Amsterdam: Elsevier.

Lenz, R., I. G. Malkina-Pykh, and Y. Pykh. 2000. Introduction and overview [to special feature on environmental indicators]. *Ecological Modelling* 130:1–11.

MacArthur, R. H., and E. O. Wilson. 1967. *The theory of island biogeography*. Princeton: Princeton University Press.

McCoy, E. D., and H. R. Mushinsky. 2002. Measuring the success of wildlife community restoration. *Ecological Applications* 12:1861–1871.

McCune, B., and J. Grace. 2002. *The analysis of ecological communities*. Gleneden Beach, OR: MjM Software Design.

McKelvey, K., B. R. Noon, and R. H. Lamberson. 1993. Conservation planning for species occupying fragmented landscapes: The case of the Northern Spotted Owl. In *Biotic interactions and global change*, ed. P. Kareiva, J. G. Kingsolver, and R. B. Huey, 424–450. Sunderland, MA: Sinauer Associates.

McLachlan, S. M., and D. R. Bazely. 2001. Recovery patterns of understory herbs and their use as indicators of deciduous forest regeneration. *Conservation Biology* 15:98–110.

Michener, W. K. 1997. Quantitatively evaluating restoration experiments: Research design, statistical analysis, and data management considerations. *Restoration Ecology* 5:324–337.

Miller, C., and D. Urban. 1999. A model of surface fire, climate, and forest pattern in Sierra Nevada, California. *Ecological Modelling* 114:113–135.

Miller, C., and D. Urban. 2000. Modeling the effects of fire management alternatives on Sierra Nevada mixed-conifer forests. *Ecological Applications* 10:85–94.

Minchin, P. R. 1987. An evaluation of the relative robustness of techniques for ecological ordination. *Vegetatio* 69:89–107.

Peters, D. P., J. E. Herrick, D. L. Urban, R. H. Gardner, and D. D. Breshears. 2004. Strategies for ecological extrapolation. *Oikos* 106:627–636.

Philippi, T. E., P. M. Dixon, and B. E. Taylor. 1998. Detecting trends in species composition. *Ecological Applications* 8:300–308.

Shugart, H. H., and D. L. Urban. 1986. Overall summary: A researcher's perspective. In *Modeling habitat relationships of terrestrial vertebrates*, ed. J. Verner, M. L. Morrison, and C. J. Ralph, 425–429. Madison: University of Wisconsin Press.

Smith, T. M., and D. L. Urban. 1988. Scale and resolution of forest structural pattern. *Vegetatio* 74:143–150.

Taverna, K., R. K. Peet, and L. C. Phillips. 2005. Long-term change in ground-layer vegetation of deciduous forests of the North Carolina Piedmont, USA. *Journal of Ecology* 93:202–213.

ter Braak, C. J. F. 1986. Canonical correspondence analysis: A new eigenvector technique for multivariate direct gradient analysis. *Ecology* 67:1167–1179.

ter Braak, C. J. F. 1987. The analysis of vegetation-environment relationships by canonical correspondence analysis. *Vegetatio* 69:69–77.

ter Braak, C. J. F. 1988. *CANOCO*. Wageningen, The Netherlands: Agricultural Mathematics Group Technical Report LWA-88-02.

Tilman, D. 1985. The resource-ratio hypothesis of plant succession. *American Naturalist* 125: 827–852.

Tiner, R. W. 1999. *Wetland indicators: A guide to wetland identification, delineation, classification, and mapping*. Boca Raton: Lewis Publishers.

Tong, S. T., and W. Chen. 2002. Modeling the relationship between land use and surface water quality. *Journal of Environmental Management* 66:377–393.

Urban, D. L. 2000. Using model analysis to design monitoring programs for landscape management and impact assessment. *Ecological Applications* 10:1820–1832.

Urban, D. L. 2002. Tactical monitoring of landscapes. In *Integrating landscape ecology into natural resource management*, ed. J. Liu and W. Taylor, 294–311. Cambridge, UK: Cambridge University Press.

Urban, D. L., and T. H. Keitt. 2001. Landscape connectivity: A graph-theoretic perspective. *Ecology* 82:1205–1218.

Urban, D. L., S. Goslee, K. Pierce, and T. Lookingbill. 2002. Extending community ecology to landscapes. *Ecoscience* 9:200–212.

Urban, D. L., C. Miller, P. N. Halpin, and N. L. Stephenson. 2000. Forest gradient response in Sierran landscapes: The physical template. *Landscape Ecology* 15:603–620.

Urban, D. L., and T. M. Smith. 1989. Microhabitat pattern and the structure of forest bird communities. *American Naturalist* 133:811–829.

U.S. Environmental Protection Agency (USEPA). 1998. Better assessment science integrating point and nonpoint sources. Washington, DC: Office of Water, EPA-823-B-98-006.

Walters, C. J. 1986. *Adaptive managment of renewable resources*. New York: Macmillan.

Walters, C. J., and C. S. Holling. 1990. Large scale management experiments and learning by doing. *Ecology* 71:2060–2068.

Weller, D. E., T. E. Jordan, and D. L. Correll. 1998. Heuristic models for material discharge from landscapes with riparian zones. *Ecological Applications* 8:1156–1169.

White, P. S., and J. L. Walker. 1997. Approximating nature's variation: Selecting and using reference information in restoration ecology. *Restoration Ecology* 5:338–349.

Zedler, J. B., and J. C. Callaway. 1999. Tracking wetland restoration: Do mitigation sites follow desire trajectories? *Restoration Ecology* 7:69–73.

PART THREE

Restoration Ecology in Context

Ecologists promote the need to think about the natural world at many spatial scales because ecosystem structure and functioning are products of processes across the spatial spectrum. From the small-to-large perspective, we acknowledge that microorganisms dictate rates and levels of dead biomass on earth. From the large-to-small view, we recognize that landscape configurations determine where local uplands and wetlands occur. In restoration, likewise, it is not sufficient to consider only the attributes of the site or immediately adjoining lands. A big-picture view is essential for planning restoration projects, as well as for implementing and managing projects. For uplands, the key questions about the ecoregional landscape are what land uses dominate the region; what air-borne contaminants might be carried to the site; and where the nearest habitat blocks and wildlife corridors are. If a migratory songbird is the restoration target, it is necessary to think about wintering grounds in South America, as well as nesting grounds in North America. For wetlands, watersheds are a more appropriate landscape unit for analysis, and initial questions are likely to concern the quantity and quality of water sources and threats of flooding. For a riparian project located in the lower Mississippi River Basin, it is highly relevant to consider how the 40% of the United States that is upstream will affect restoration downstream.

The big picture also demands that we look backward and forward in time. Historical information helps set goals based on former conditions and species that might become targets for reintroduction. Forecasts of changing environmental conditions can help modify historical targets, especially for species that are at the edges of their distribution. Species at their southernmost distributional limits might become more restorable as climate warms; coastal wetlands might become highly vulnerable to river flooding as storms strengthen and increase in frequency.

Space and time considerations can expand the big picture almost indefinitely, leading us to ask of restoration plans: How big is big enough? How long is long enough? Reconstruction of disturbance regimes (for instance, by palynological, geomorphological, or dendroecological studies) can provide valuable information about how systems have operated over long stretches of time. Such information offers an essential perspective on contemporary ecosystem conditions, such as the extent to which current conditions represent departures from a system's states and dynamics.

Strategies for prioritizing restoration efforts over large areas and long time periods are often missing. Much restoration work is site based because so little funding and few incentives are directed toward the big picture. Statisticians have increasingly important tools that allow us to think and plan strategically, but these tools are rarely employed. So, we need to think bigger about spatial, temporal, and strategic issues. With restoration dollars hard to find, efforts need to target the landscapes with the most potential, sites in optimal locations, targets that match the times, and approaches that meet statistical guidelines. In this section, four chapters challenge us to consider the largest contexts within which restoration is being undertaken.

At local to regional scales, invasive species create the need for restoration while threatening the outcomes of restoration efforts. D'Antonio and Chambers discuss the challenges posed by invasives, first, by emphasizing the need to prevent problems in restoration sites and, second, by taking steps to eradicate them where they are already present or when they arrive. In both cases, ecological theory can help design effective restoration strategies. For instance, life-history traits help predict which of the region's invaders might threaten a site, succession theory can help characterize the effect of invaders, population models can help identify when invaders are still vulnerable to control, and competition theory helps explain the effects of invaders on resident species. Theories of resilience and resistance are relevant to the maintenance of native species; theories of top-down and bottom-up control are relevant to the control of invasives.

The evaluation and statistical analysis of restoration outcomes are critical to restoration. Osenberg, Bolker, White, St. Mary, and Shima discuss methods that can assess how well restoration projects achieve their goals. Different statistical approaches can evaluate how closely a site meets a specific endpoint versus determining the effect-size of the restoration activity. The authors argue that the latter, more innovative, approach is more relevant in restoration ecology, where controls may be few and replication difficult. Spatial scale is often a limitation to experimentation in restoration sites, so for ecosystem-level effects to be assessed across large projects, something other than an ANOVA is needed. Using marine protected areas as case studies, they consider how projects achieve their goals with various designs: before-after, control-impact, and the preferred before-after–control-impact paired series. Clearly, big thinking about assessment of restoration challenges us to plan ahead so that more informative evaluations can be accomplished and to apply better statistical approaches in judging outcomes.

How can a site be restored to some historical condition when the historical extent of the ecosystem was much larger? Maurer contributes the macroecology perspective by considering the limitations imposed by restoration sites that are small relative to the lists of species we would like them to support, small relative to the scale of ecological processes that influence them, and often unconnected to other undeveloped lands in modern landscapes. On an immediate level, restoration planners need to acknowledge and address the constraints of small size. The larger, macroecology perspective helps restoration research and practice move toward ecoregional and even continental-scale thinking.

At the global spatial scale and millennium temporal scale, Millar and Brubaker expand our thinking about restoration in relation to climate change. Over long time frames, vegetation has responded to alternating periods of warm/cold and wet/dry climates, with additional patterns of change over shorter time periods, including rapid shifts in the environment.

Given this big picture, Millar and Brubaker consider the implications for restoration ecology. Changes in populations, communities, and other ecological attributes suggest the need to rethink concepts of sustainability and restoration targets. We are then challenged to consider the novel goal of *realignment* in lieu of restoration where future climates will match historical conditions and species' former distributional ranges.

On the whole (and "the whole" is the topic of this final section), we restoration ecologists need to consider just how much can be achieved in sustaining species on this planet and how to maximize the results of our efforts. As demonstrated by Radeloff et al. (2000), restoring entire landscapes requires thinking beyond the norm of "returning a tiny site to a recent condition." We need to be strategic about where we place our efforts in order to accomplish the most with the resources at hand. And "accomplishing the most" needs to take into account the long-term persistence of species as global climate changes.

Literature Cited

Radeloff, V. C., D. J. Mladenoff, and M. S. Boyce. 2000. A historical perspective and future outlook on landscape scale restoration in the northwest Wisconsin pine barrens. *Restoration Ecology* 8:119–126.

Chapter 12

Using Ecological Theory to Manage or Restore Ecosystems Affected by Invasive Plant Species

Carla M. D'Antonio and Jeanne C. Chambers

The widespread degradation of ecosystems around the globe has necessitated implementation of a range of restoration or rehabilitation practices to restore valued ecosystem functions. Invasive non-native plants can contribute directly to the loss of ecosystem services, or they can increase in response to environmental change (Figure 12.1) and thereafter interfere with the achievement of restoration goals. While invasive plants have been the focus of control efforts in agriculture for decades, it is only in the past twenty years that they have been recognized as a significant economic and ecological cost to management and restoration of less managed ecosystems. A relatively small fraction of wildland invaders causes significant ecological or economic damage (Simberloff 1981; Williamson 1996). Those that do can interfere with the maintenance of particular vegetation types by outcompeting more desired species; threatening the persistence of rare species and, at the same time, other trophic levels; co-opting the direction of postdisturbance succession; and maintaining communities in a persistent undesirable state. We refer to these species throughout this chapter as invasive and damaging plant invaders. We do not consider the term *invasive* alone to imply ecological damage. For lists of invasive and damaging wildland invaders in the United States, see the following websites: "related links" on www.bbg.org/gar2/pestalerts/invasives/ and "plant lists" at www.nps.gov/plants/alien/.

Because of the various ways that introduced plant species can interfere with management goals, we discuss different types of management and restoration actions designed to deal with the threat or impact of such invaders. The Society for Ecological Restoration International (SER) defines ecological restoration as the process of assisting the recovery of an ecosystem that has been degraded, damaged, or destroyed (SER 2002). Here we explicitly acknowledge that management actions, or lack thereof, prior to obvious ecosystem degradation, influence ecosystem trajectories and the likelihood of successful plant invasions. The type of management or restoration action employed will depend on the state of degradation of the ecosystem and the causes of degradation (Figure 12.2). If the system is still providing valued ecosystem functions, *preventive management* may be used to reduce the likelihood of invasion by damaging species. If the ecosystem has already been invaded by species that might hinder achievement of management goals and is at risk of further degradation, *removal* of problematic species is an essential step toward restoration. By itself, however, removal may not be enough, and further actions may be necessary to achieve *ecological restoration*. Theories rel-

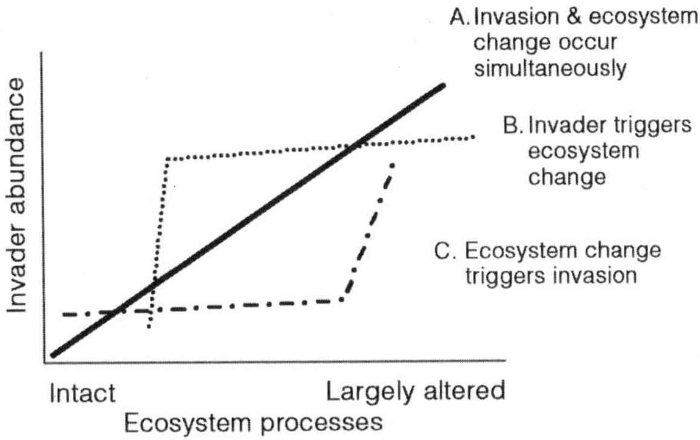

FIGURE 12.1 Hypothetical relationships between invader abundance and ecosystem change. In pathway A, invasion occurs simultaneous with changing conditions. Causal relationships are difficult to determine, as invasion and process change may happen independently. In pathway B, the invader triggers ecosystem change when a threshold of invader abundance is reached. Invaders may also create internal feedbacks that perpetuate the altered state. In pathway C, ecosystem change triggers a large increase in invader abundance. A threshold for environmental change may be required for the invader to become widely abundant.

evant to those tools needed for ecological restoration overlap strongly with those relevant to preventive management because, as detailed below, the goal of management is to maintain or recreate sustainable ecosystems.

We acknowledge that some introduced species can be used to the benefit of land managers or may be important for the achievement of particular restoration goals (D'Antonio and Meyerson 2002; Ewel and Putz 2004). We will not deal with those situations here. Instead, we focus on ecological theories that are relevant to eradication or control of wildland plant invaders and restoration in the face of invasion by species considered to be damaging.

We begin this chapter by discussing differences in two fundamental approaches to restoration of invaded ecosystems: prevention and active restoration. We then examine ecological theory relevant to (1) preventing and controlling invasions, and (2) managing for sustainable conditions during prevention or restoration. We explore ecological concepts that are relevant to preventing the arrival of invaders or controlling them with top-down measures after their establishment and then discuss theories that are relevant to full-scale restoration.

Management-Restoration Approaches for Invasive Species

The ultimate goal of restoration and associated management activities is sustainable ecosystems with a particular composition or series/trajectory of desired states. Sustainable ecosystems, over a cycle of routine disturbance events retain characteristic abiotic and biotic processes including rates and magnitudes of geomorphic activity, hydrologic flux and storage, biogeochemical cycling and nutrient storage, and biological activity and production (modified from Chapin et al. 1996 and Christensen et al. 1996). In their ideal form, sustainable ecosystems are *resilient* in that they return to predisturbance conditions or a trajectory close to that within a reasonable time frame following a disturbance (Holling 1973) without large-

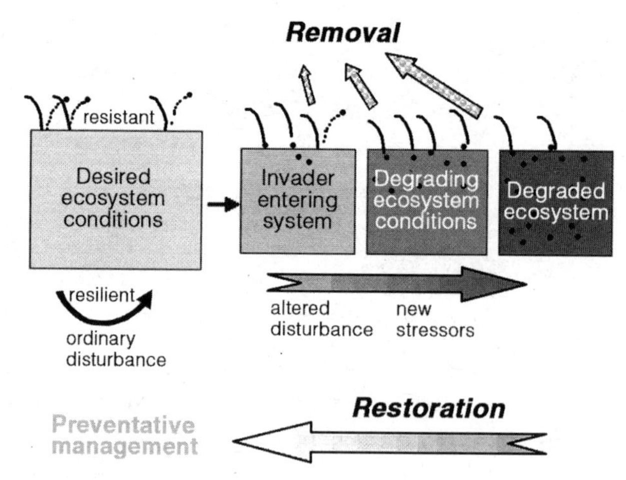

FIGURE 12.2 Conceptual scheme for restoration of systems affected by invasive plant species. Propagule flow of invading species is indicated by black balls. The size of the upper three arrows indicates the intensity of control effort required to stem propagule flow of invaders. The shading intensity of the restoration arrow indicates the likely intensity of effort required to ameliorate stressors or reverse changes in ecosystem processes that affect underlying conditions for plant growth.

scale human intervention (Figure 12.2). Ideally, they are also relatively *resistant* to change following arrival of propagules of potentially damaging species. That is, the ecological processes in resistant systems greatly reduce the likelihood of successful establishment and population growth of invaders.

Ecological theory can provide significant insights into the mechanics and value of different restoration approaches. Because of the threat that invasive non-native species can present to the sustainability of ecosystem services, *preventive management* may be the most valuable approach for keeping intact ecosystems that currently provide valued services free from potentially damaging invaders (Figure 12.2). This approach should be specifically designed to maintain or increase ecosystem resistance prior to or during the early stages of invasion (Masters and Sheley 2001) as well as ecosystem resilience after a disturbance. Resistance and resilience can potentially be enhanced by manipulation or maintenance of structural properties and ecosystem processes known to favor the persistence or recovery of resident or desirable species. Knowledge of controls over resistance and resilience are therefore essential for successful management.

Following the establishment of damaging invaders and subsequent changes in ecosystem properties and processes, active restoration is required and involves two basic elements. *Top-down control* involves removing or eliminating the damaging invader or reducing its abundance and supply of propagules to acceptable levels, while *bottom-up control* emphasizes restoration of properties or processes that contribute to sustainability (*sensu* McEvoy and Coombs 1999). In restoring invaded ecosystems, the top-down approach primarily involves manual removal, herbicides, or biological control but can also involve identifying and controlling the pathways through which propagules of unwanted species are arriving at a site.

Bottom-up control can involve the removal or amelioration of ecosystem stressors that affect the status of more desirable species, manipulation of disturbance regimes, alteration or manipulation of soil conditions to reduce potential growth of undesirable species, and direct seeding to increase the likelihood of competitive dominance being achieved by the desired species. Bottom-up control ultimately includes many elements of preventive management and also encompasses both "designed disturbance" and "controlled species performance" elements of Sheley and Krueger-Mangold's (2003) scheme for how to manage communities toward a desired state (Figure 12.3).

The initial choice of a preventive or restoration approach, and whether either of these should be accomplished using top-down or bottom-up control, or some combination of the two, should be based on knowledge of the known ecological condition and likely trajectory of a given site. Simultaneous top-down and bottom-up control is generally required to achieve restoration of desired conditions in severely infested or highly degraded sites, and bottom-up manipulations can improve the success of top-down measures (for examples see McEvoy et al. 1993; Wilson and Partel 2003).

Preventing and Controlling Invasions

Lonsdale (1999) suggested that the number of invaders with reproducing populations in a site or region (E) is a function of the number of species introduced (I) and the survival and reproduction (S) of those arriving species populations or propagules. Applying this simple concept to damaging invaders and to preventive management, managers should want to control or anticipate I to keep it as low as possible while manipulating systems to reduce the value of S.

Life-History Theory and Species Traits That Predict Arrival (I)

Theory and research relevant to controlling I is based largely on dispersal and life-history traits of populations and species. Restoration actions can be taken in advance of the arrival of invader propagules. Managers can try to protect against species that are likely to be arriving in a site based on their presence in the region and knowledge of their dispersal modes. Indeed, both economists and practitioners have argued that control prior to arrival is the most

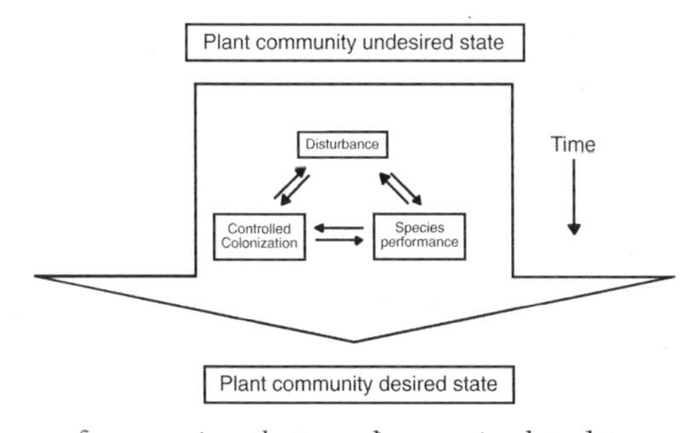

FIGURE 12.3 Three causes of succession that can be manipulated to move communities from an altered condition to a more desired state. Adapted from Sheley and Krueger-Mangold (2003).

cost-effective means of managing damaging invaders (e.g., NISC 2001; Leung et al. 2002). Such arguments are the basis for development of "early detection, rapid response" (EDRR) programs (http://ficmnew.fws.gov/FICMNEW_EDRR_FINAL.pdf). When an invader is widespread in a geographic region or has been planted extensively nearby as an ornamental, its eventual arrival in a reserve can be assumed. The intensity of effort to search for and eradicate incipient populations relates, in part, to dispersal mode. Hence, dispersal and life-history theory can provide insights into predicting which species are most likely to arrive regularly and where they are most likely to establish.

Seeds of invading species represent a wide range of dispersal modes, and thus it has been difficult to find generalizations that characterize damaging plant invaders (Mack et al. 2000; NRC 2002). This is perhaps because most studies looking for traits that predict invasion ability evaluate all established non-native species in a flora or region, including many that have no measurable ecological impact. Studies evaluating traits of invaders versus non-invaders within particular groups (such as pines and a variety of other woody species), have found some general species traits that correlate with a high likelihood of invasion (Rejmanek and Richardson 1996; Reichard and Hamilton 1997). These studies have found that although wind-dispersed seeds tend to be more common among successful pine invaders, both wind- and animal-dispersed species can invade successfully if they produce large seed crops (e.g., Rejmanek 1996). The importance of these traits is likely that high rates of propagule supply, or larger numbers of propagules in an introduction event, lead to a greater likelihood of establishment. This has been shown clearly for birds and insects introduced for biological control (see review in D'Antonio et al. 2001); for woody species introduced into southeastern Australia (Mulvaney 2001); and for eucalypts invading away from plantations (Rejmanek et al. 2005).

Life-History Traits Predicting Reproductive Rate Once Established (S)

Traditional life-history theory divided up species by their reproductive strategies and designated species as falling on a continuum from r- to k-selected species (MacArthur and Wilson 1967; Pianka 1970), referring to species whose reproductive strategy is keyed to maximum growth rate and carrying capacity, respectively. Rapidly colonizing ephemeral species are at the r extreme, while slow-colonizing, slow-growing, long-lived competitors fall near the k end. While r–k theory came under fire during the 1980s, it nonetheless provides a useful framework for thinking about species differences and how they might relate to the types of habitats where species would establish successfully and have subsequent impacts.

Nobel and Slatyer (1980) suggested that species could be divided into groups based on clusters of traits rather than along a simple two-dimensional continuum, and they successfully predicted plant community responses to disturbance based on trait association groups. Invasive non-native plants in any region tend to cover a spectrum of life-history traits, but where particular invaders are likely to show up and where they are likely to persist may be predictable based on both dispersal and life-history trait associations. For example, Rejmanek (1996) found that rapid growth to reproductive stage (short juvenile period) and short intervals between large seed crops are more likely to be traits of invaders than non-invasive species. Ranking potential weeds by their presence in a geographical region and their known traits can help managers and restorationists target and prioritize areas and species for EDRR and preventive management.

Succession Theory as a Means of Prioritizing Species for Removal

The decision to initiate a restoration project that involves the removal of a non-native species is based upon the belief or knowledge that a species interferes with achievement of a desired ecosystem state, successional trajectory, or delivery of ecosystem services. Successional theory can be used as a basis to help make this decision. Prioritization is important since multiple invaders are present in many settings and resources for control are usually limited. Connell and Slatyer (1977) proposed three models to explain the potential influence of a colonizing species on subsequent species compositional change. An invader could (1) facilitate, (2) inhibit, or (3) have no effect on subsequent species colonization in a system. While they did not propose this theory as a framework for evaluating the impact of an invasive species, it nonetheless can be used to provide a baseline for evaluating whether a species is of concern. An invader that can facilitate the establishment of additional potentially undesirable invaders could lead to "invasional meltdown" (Simberloff and Von Holle 1999), a situation in which invaders facilitate one another until little of the original native system remains. For example, nitrogen-fixing invaders have been shown to facilitate the establishment of other undesirable species after their death or removal from a site, potentially with dire consequences for native species (Maron and Connors 1996; Adler et al. 1998; Alexander and D'Antonio 2003). When desired ecosystem services are dependent upon native species, invaders that inhibit their establishment should receive priority for removal.

Evaluating and anticipating the successional effect of invaders is particularly important in the context of restoration projects where a site has been severely degraded or is being converted from one use (e.g., agriculture) to another (e.g., a wetland). Invaders in these contexts may facilitate the establishment of native species, or they may have little effect on them. Non-native species that might facilitate the establishment of more desirable species can be useful in such full-scale restoration, particularly if site conditions are extremely harsh and it is difficult to establish the more desirable species (D'Antonio and Meyerson 2002). In essence, the restoration practitioner might use the facilitation model of traditional primary succession to approach the establishment of desired species under severely altered environmental conditions.

When invaders already dominate a degraded site or rapidly invade after the large-scale disturbance that might be necessary to create or restore a habitat (e.g., unearthing a filled salt marsh), the practitioner can evaluate based on species traits whether invaders are likely to persist. Invaders that respond quickly to disturbance but drop out relatively easily without disrupting succession can be more or less ignored. For example, on severely disturbed sites in semiarid areas of the western United States, *Salsola kali*, an annual Eurasian weed, is often highly abundant the first and second years following restoration but rapidly declines as seeded species become established (Allen 1988). In contrast, invasive species that inhibit succession toward the desirable states should be targeted for early control. An example from more mesic areas of these same semiarid systems is *Lepidium latifolium*, a perennial and highly rhizomatous mustard with high growth rates that rapidly establishes on restored sites and effectively prevents establishment of seeded species (Young et al. 1998). The extent to which disturbance caused by the removal of an invader will reset succession will depend on its abundance and life-history characteristics. Methods that minimize disturbance should be given preference.

Using Population Models to Determine When and Where to Remove a Species

Once a species has been identified as a target for removal, the decision of when and where to remove it can be based on an understanding of population growth curves. Many investigators have noted that introduced species typically show a lag phase between introduction and exponential population growth (Kowarik 1995; Crooks and Soulé 1999). The sources of this lag are debated, and many hypotheses have been proposed to explain it. Regardless of the causes, the existence of lags in many settings is an opportunity for EDRR teams and site managers to implement removal programs. Removing a target at or before exponential population growth occurs is likely to be much more successful and less costly than treating it later (Hobbs and Humphries 1995). Likewise, population growth models at the landscape scale, such as that of Moody and Mack (1988), suggest that targeting outlying small populations (e.g., "nascent foci") is the best way to slow landscape-scale invasion rates once an invader is established in a region.

Targeting small invader populations has several distinct advantages. The goal of EDRR programs is to remove the damaging invader species before it has exerted a significant effect. In this case, the need for full-scale ecological restoration may be avoided. In addition to the practical considerations of costs and access, a focus on small populations should result in minimal disturbance during the removal process, thereby reducing the likelihood of disturbance-stimulated invasion. Also, because the impact of an invader is likely a function of its abundance (Parker et al. 1999), removal of small populations should mean that negative effects have been minimal and can therefore be readily reversed.

Disturbance Theory and Increasing Community Resilience Following Disturbance

Resilient systems are those that return to predisturbance conditions or a trajectory close to that within a reasonable time frame following a disturbance (Holling 1973) without large scale human intervention. Disturbance is a part of virtually all ecosystems. Because it releases resources and opens up space, it can promote invasion if factors promoting resilience of the resident community are absent or impeded. The duration of an invasion window created by a disturbance is influenced by the type, size, and frequency of disturbance events, as well as the tolerance and response of the resident species to those events. For example, severe disturbances can greatly depress resident populations and potentially destroy their seed banks, increasing the length of time that the community is relatively open to invasion by any species. If propagule sources of damaging invaders are limited, the community may eventually recover. Likewise, disturbance frequencies that are higher than what resident species have experienced routinely may select for short-lived species that may or may not be desirable. The trick is to anticipate which species are likely to respond to different types of disturbances.

Most ecosystems have evolved with disturbance, and yet disturbance can also be manipulated to prevent invasion. For example, frequent prescribed burning in tallgrass prairies prevents invasion of prairie stands by Canada thistle because the native regional flora is well adapted to fire (Reever-Morghan et al. 2000). Its role in promoting versus preventing invasion is thus a function of the evolutionary history of both the resident species and invaders (Hobbs and Huenneke 1992; D'Antonio et al. 1999). Fires are essential for the maintenance of native vegetation states in southeastern pine savannas. Yet poorly timed fires (those outside the lightning season) can promote invasive species (especially native and exotic hardwoods) in these species-rich systems. In addition, altering other characteristics, as for example, decreasing fire frequency, appears to favor invasive grasses and shrubs in these sites (e.g., DeCoster et al. 1999; Platt 1999; Platt and Gottshalk 2001; Drewa et al. 2002). The extensive work in this ecosystem emphasizes the complications of manipulating disturbance regimes to obtain desired conditions.

The most widely known theory of species responses to disturbance is the intermediate-disturbance hypothesis, which predicts that species diversity should be highest in landscapes with intermediate sizes, frequencies, or times since disturbance (Connell 1978). To our knowledge, no one has tried to relate the presence and persistence of non-native plant species to the intermediate-disturbance hypothesis in the context of managing against undesirable species. If the regional pool of non-indigenous species is diverse, one would expect invader diversity, and possibly abundance, to increase rapidly after disturbance and then decline over time as the early colonizers drop out but before later colonizers arrive. Invader diversity, however, may not be a good predictor of invader impacts. Many naturalized non-indigenous species are disturbance responsive but may persist for only a relatively short time after disturbance. Others may exist as a component of the community during all subsequent seral stages. It is only those invaders that persist after disturbance and/or that spread into undisturbed areas that are potentially of concern to managers. Such invaders are common on environmental weed lists. In terms of preventive management, the postdisturbance window should be watched closely for incipient invader populations. Planned disturbances should be timed to maximize the likelihood of stimulating the regeneration of native species while depressing undesirable ones.

Competition Theory and Enhancing Community Resistance

Ecological resistance refers to the biotic and abiotic factors in a recipient ecosystem that limit the population of an invading species. The impact of prior residents on invaders is therefore part of the concept of ecological resistance (Elton 1958). Although biotic elements important to resistance can include herbivory and the presence/absence of mutualisms, studies of resistance have largely focused on competition. This reflects the strong focus of plant ecologists on competition as a dominant force structuring plant assemblages (reviewed in Levine and D'Antonio 1999; but see Maron and Vila 2001 for a review of the contribution of herbivores to ecological resistance). In terms of trying to prevent invasion, competition theory predicts that if residents are abundant, they should be able to monopolize available resources, reducing the likelihood that an invader can establish unless the invader has access to unique resources, that is, it is functionally wholly different from the residents. These hypotheses stem from the competitive-exclusion principle and the related limiting-similarity hypothesis, which predict that only invaders that are very different from residents in their use of limiting resources should persist in a site (MacArthur and Levins 1967; May 1973). When a very different invader arrives and establishes successfully, that invader is assumed to have entered an "empty niche" or is utilizing resources distinct enough from residents that interspecific competition is unimportant. Plant invaders that are quite different from residents as adults and are thus invading an "empty, adult-plant niche," nonetheless are still likely to face competition from residents upon arrival. For example, the nitrogen-fixing tree, *Myrica faya,* is invading Hawaiian woodlands previously lacking symbiotic nitrogen fixers (Vitousek et al. 1987; Vitousek and Walker 1989). Thus it appears to be entering a vacant niche for a symbiotic nitrogen fixer. Nonetheless, seedlings of *M. faya* are suppressed by prior residents, slowing the rate of establishment of the invader even though invasion is proceeding (D'Antonio and Mack 2001). The point here is that resistance can slow an early invasion, giving managers more opportunities for controlling a target species even if the species is one whose success appears to be inevitable. Unfortunately, in most sites, we do not know which species contribute most effectively to resistance, but we do know that resistance is real and should be part of restoration planning.

Competition theory is also largely the basis for the hypothesis that diverse (here defined simply as more species rich) plant assemblages utilize available resources more fully than less diverse ones and therefore are less likely to be invaded (reviewed by Levine and D'Antonio 1999). Small plot studies generally confirm the prediction that more diverse assemblages are more resistant to invasion (Tilman 1997; Levine 2000, 2001; Naeem et al. 2000; Kennedy et al. 2002; Dukes 2001, 2002), although there are exceptions (Robinson et al. 1995; Wiser et al. 1999). Studies based on manipulation of functional group richness rather than simply species diversity also generally confirm the prediction that invasion success decreases with increasing functional richness (Pokorny et al., 2005; Hooper et al., forthcoming). Results from these studies suggest that enhancement of diversity within a park, reserve, or restoration plot should reduce the probability of establishment of arriving propagules of many invader species. These hypotheses can thus be guiding principles for practitioners both in preventive management and postinvasion restoration.

The literature debate over the importance of diversity or species richness in limiting invasion (e.g., Stohlgren et al. 1999, 2003) provides interesting lessons for restoration practitioners. In extensive surveys from both their own work and published species lists, Stohlgren et al. (1998, 1999, 2003) evaluated the relationship between native and exotic species diversity across several spatial scales and concluded that hotspots for native biodiversity are also hotspots for introduced species. These tend to be resource-rich sites with high exotic diversity. Lonsdale (1999) also recognized that, at a landscape scale, more nutrient-rich sites have higher diversity of *both* native and non-native species. With regard to restoration, such patterns suggest only that non-native species should be expected to occur in relatively rich sites because they are good places for plants to grow. These data say nothing about whether we would expect to find strong invader impacts in these areas or whether particularly damaging invaders tend to proliferate and have their greatest impacts in these sites. Indeed, Dukes (2001) demonstrated that despite being able to invade and persist in diverse assemblages, the noxious weed *Centaurea solsticialis* showed reduced impact on native species as diversity in experimental plots was increased. The relationship between number of invaders, richness hotspots, and invader impacts needs more careful exploration.

Correlations observed at the landscape scale suggest that factors that covary with diversity similarly affect both native and exotic species. For example, disturbed sites tend to have both more exotics and more natives because disturbance generally promotes diversity. Despite observations like this, resistance may still occur on a local scale as demonstrated experimentally by Levine (2000, 2001). He found that, at the scale of large stretches of river, native and exotic richness were positively correlated, yet he demonstrated experimentally that, at a local scale, native diversity could slow establishment of several invaders. Brown and Peet (2003) have made similar findings.

Such results are directly relevant to restoration, because most restoration projects are concerned with the local scale. Enhancement of local resistance should increase the time available for managers or EDRR teams to find and remove establishing invaders while creating environmental conditions that favor native species and potentially increasing native diversity or desired composition.

Davis et al. (2000) attempted to provide an overarching theory to explain community "invasibility" (susceptibility to invasion) or resistance and referred to it as the *fluctuating resources hypothesis*. This framework predicts that invasibility (or resistance) is a function of the balance between community-level resource uptake and resource renewal or gross resource supply rates (Figure 12.4, adapted from Davis et al. 2000). When supply and uptake are equal, no invasion occurs, presumably because there are no extra resources for invaders to harvest. Disturbances can reduce resource uptake without enhancing supply rates or alter both supply and uptake rates, allowing for invasion. Pulses of resources such as might occur with nitrogen deposition or high rainfall years in an arid system can result in windows of opportunity for invaders to establish. Davis and Pelsor (2001) demonstrated experimentally that even short pulses of high-resource availability can result in new invader populations that persist for many years. For preventive management and restoration, this hypothesis helps to solidify a dynamic understanding of ecosystems, with resistance fluctuating along with climate or other factors. It can also help managers to target invasion windows (e.g., a high rainfall year in an otherwise arid site) as times when an EDRR approach is critical.

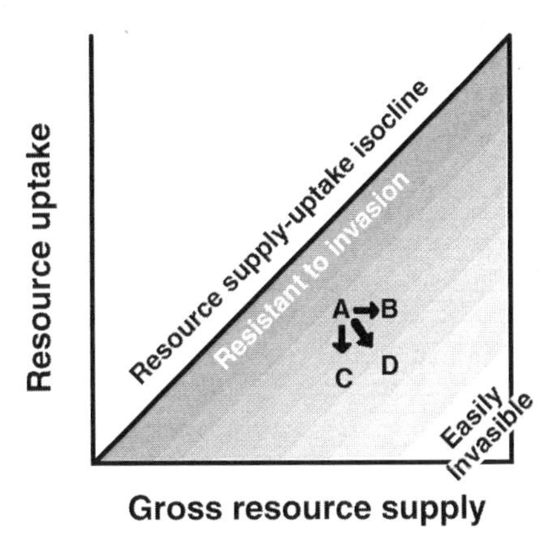

FIGURE 12.4 The theory of fluctuating resource availability holds that a community's susceptibility to invasion increases as resource availability (the difference between gross resource supply and resource uptake) increases. Resource availability can increase due to a pulse in resource supply (A → B), a decline in resource uptake (A → C), or both (A → D). Adapted from Davis et al. 2000, Figure 1.

Integrating Concepts: Disturbance, Fluctuating Resources, and Competition Across Resource Gradients

Ecological restoration may be needed in conjunction with control of damaging invaders in natural ecosystems because the act of removal typically stimulates germination of other often-introduced species and frees up resources for their growth. The fluctuating resources hypothesis of Davis et al. (2000) suggests that removal of invaders reduces resource uptake, thereby moving a site away from the uptake/supply rate isocline, which paradoxically can increase its further invasibility. Without intervention, replacement species may be those that perpetuate the impacts of the original invader or are even worse than the original invader. For example, control efforts to remove *Hakea sericea* in South Africa can lead to invasion of native fynbos sites by *Acacia longiflora*, a species that is much harder to control (Pieterse and Cairns 1986).

Resource conditions at a restoration site or across a management unit have a strong influence on the outcome of management actions because they affect the interactions among both desirable and undesirable species. The relative importance of disturbance, competition, facilitation, and herbivory across abiotic gradients has been a major focus of community ecology for the past several decades and can provide a framework for predicting when and where species interactions are likely to affect the outcome of restoration. Since the classification of species along axes of stress tolerance and competitive ability by Grime (1977), there has been lively debate over whether the intensity of plant competition changes across resource gradients. While the debate currently remains unresolved (for a review, see Gurevitch et al. 2002), it appears that competition for light can be very intense in resource-rich sites. In the context of restoring communities affected by harmful invasive species, site productivity can affect growth rates of establishing species as well as the response rates of resident species to removal of target weeds.

Fertilization has long been a restoration practice because of its positive effects on plant growth in many settings. Yet the now-widespread occurrence of aggressive fast-growing and damaging invaders calls this practice into question. Fertilization studies in many regions of the United States have demonstrated that undesirable invaders can increase with fertilization (e.g., Huenneke et al. 1990; Vinton and Burke 1995; Maron and Jeffries 1999; Green and Galatowitsch 2002; Woo and Zedler 2002) and potentially persist for long time periods after fertilization. By contrast, Chambers (1997) demonstrated that, in a highly infertile alpine soil where almost no invasive species were present, fertilization increased the establishment and growth of desirable species during restoration. In intact organic soils, fertilization was not necessary to restore native plant cover. Thus, careful consideration of soil fertility conditions is an essential first step to restoration in sites where damaging invaders could become important community constituents.

In resource-rich sites, native species with high relative growth rates often respond most strongly to disturbance or resource pulses. If such species are already present or can be readily seeded onto a site, they can quickly take advantage of resources released during disturbance and maintain the resiliency of the system. However, if seeds of fast-growing and damaging invaders are in the area, there is a reasonable likelihood that desirable species will face resource competition. If the invaders suppress the desired species, then active top-down and bottom-up control should be enacted. The application of sucrose or sawdust to soils during or prior to the planting of desirable species has been evaluated as a bottom-up tool to reduce fast-growing weeds (eg., McLendon and Redente 1992; Reever-Morghan and Seastedt 1999;

Alpert and Maron 2000; Paschke et al. 2000; Torok et al. 2000; Corbin and D'Antonio 2004). This approach is based on the assumption that the added carbon will stimulate microbial immobilization of nitrogen and that this in turn will reduce competitive suppression of the desired species. While studies generally confirm that carbon addition immobilizes available nitrogen, they are mixed in support of the prediction that decreased nitrogen will alter the strength of competition, thereby favoring slower growing native species (McLendon and Redente 1992; Reever-Morghan and Seastedt 1999; Alpert and Maron 2000; Paschke et al. 2002; Monaco et al. 2003; Corbin and D'Antonio 2004; Perry et al. 2004).

Other practitioners trying either to restore native species to nitrogen-enriched sites or to increase native species abundances have tried mowing or cutting followed by biomass removal to reduce undesirable species. The ultimate goals appear to be reduction of ecosystem nitrogen while directly reducing the competitive effect of the fast-growing invaders by decreasing their stature and biomass. For example, prior to the last two decades, serpentine grasslands in California were refugia for native species richness, presumably because low soil nutrients limited the growth of fast-growing natives (McNaughton 1968; Harrison 1999). In recent decades, however, nitrogen deposition from automobiles has fertilized these grasslands, which has favored invasion by fast-growing European grasses to the detriment of native annual forbs (Weiss 1999). To ameliorate the effects of nitrogen deposition and promote native richness, Weiss instituted a mowing and biomass removal program. Likewise, at Bodega Marine Reserve in northern California, Alpert and Maron (2000) repeatedly mowed nitrogen-enriched pasture soils and found that both total soil nitrogen and exotic plant cover declined with repeated mowing and biomass removal, although the mechanism of reduction of the exotics was not clearly identified.

The relationship between intensity of competition, disturbance, and availability of water is less well established than competition for soil nutrients. Thus, tools such as sucrose, sawdust, or mowing designed to reduce soil resource levels during restoration are not as appropriate if water is the primary limiting resource at a site. As with nitrogen, too much water can promote invasive weeds (e.g., Kercher and Zedler 2004) but largely because it can be directly detrimental to some native species. Alterations of hydrological regime have been correlated with the invasion of aggressive weeds in several systems (Horton 1977; Everitt 1998; Galatowitsch et al. 2000; Zedler and Kercher 2004), and manipulation/restoration of hydroperiod can be a tool to select against weedy species. However, tolerance of many invaders to a wide range of hydrological conditions or susceptibility of native species to the same conditions that restrict invaders may limit its usefulness (e.g., Maurer and Zedler 2002; Miller and Zedler 2003). For example, it has long been believed that alterations to flooding regimes in western U.S. rivers have contributed to invasion by saltcedar (*Tamarix* spp) (Horton 1977; Everitt 1998; Levine and Stromberg 2001). However, manipulations of flooding regimes to try to reduce now abundant *Tamarix* and restore cottonwoods have proven difficult, because conditions that kill *Tamarix* seedlings also kill cottonwood seedlings (Stevens et al. 2001). Likewise, conditions that favor *Tamarix* establishment also favor cottonwood establishment (Stromberg 1997, 1998; Levine and Stromberg 2001). In addition, flooding severe enough to remove *Tamarix* from established stands can be difficult to simulate (Stevens et al. 2001). Similarly, wetland restoration ecologists do not yet fully understand what aspects of natural hydro-period are critical for sustaining natives, so aiming for the "right" hydro-period is difficult (Joy Zedler, personal communication). Nonetheless, manipulation of hydrological

regime at the same time that other tools are employed for species control may be the key to restoration in riparian and wetland settings.

Alternative States and Breaking the Cycle

While invaders may become abundant because of changing site conditions, they also can be the cause of the changing conditions (Figure 12.1). The persistence and strength of changes created by invaders can affect the type of management activities needed to restore more desired conditions. Indeed, some of the most pernicious invaders are species that create positive feedbacks and enhance their own growth. To adequately restore a site invaded by one of these species requires breaking the positive feedback cycle. While many investigators have suggested that such biologically driven positive feedbacks are important in locking invaded systems into "alternative states," which may be difficult to restore (e.g., D'Antonio and Vitousek 1992; Whisenant 1999; Suding et al. 2004), actual positive feedbacks caused by invasive species are not well documented except in the case of invader impacts on disturbance regimes (Mack and D'Antonio 1998). In these cases the altered disturbance regime typically results in a decline in desirable species and persistence and/or spread of the invader.

Perhaps the most widespread example is the invasion of the semiarid to arid deserts of the western United States by fire-enhancing annual grasses of Eurasian origin at the expense of native species (Figure 12.5). Because of the poor tolerance of fire by the native species as these sites become increasingly arid, breaking the fire cycle or significantly reducing the frequency of the fires is essential to directing the system toward more desired communities. Researchers in the Great Basin and Sonoran deserts are actively exploring potential restoration toward plant communities that will suppress annual grasses. The intent is to reduce the mass and continuity of fine fuels and, thus, decrease the frequency and spread of desert fires. Because of the severely degraded state of the most alien-grass-dominated sites, in the western deserts of the United States, heavy-handed restoration may be needed to achieve desired results. Preventive management is being employed to protect intact areas from fire by planting fire resistant buffers (i.e., green-stripping) and reduction of stresses that might increase the likelihood of cheatgrass (*Bromus tectorus*) conversion (Pellant, http://www.fire.blm.gov/gbri/). A framework for controlling and restoring sites invaded by fire-altering plants has been suggested by Brooks et al. (2004).

Invading species can affect soil nutrient pools and fluxes (Levine et al. 2003; Ehrenfeld 2003), but their long-term legacies are relatively unexplored, and the extent to which they contribute to internally reinforced states is questionable. One significant challenge for restoration appears to be when nutrient distributions or pool sizes are dramatically altered by invasive species, such as nitrogen-fixing shrubs and trees (e.g., Vitousek et al. 1987; Vitousek and Walker 1989; Stock et al. 1995) or halophyte forbs (Vivrette and Muller 1977; Blank and Young 2002), and the growing conditions for potential desired species are impacted. In these cases, soil conditions may need to be manipulated to restore the site.

Summary

Damaging invasive species are among the many significant challenges that land managers and restoration practitioners face now and into the future. Ecological theory, and the body of

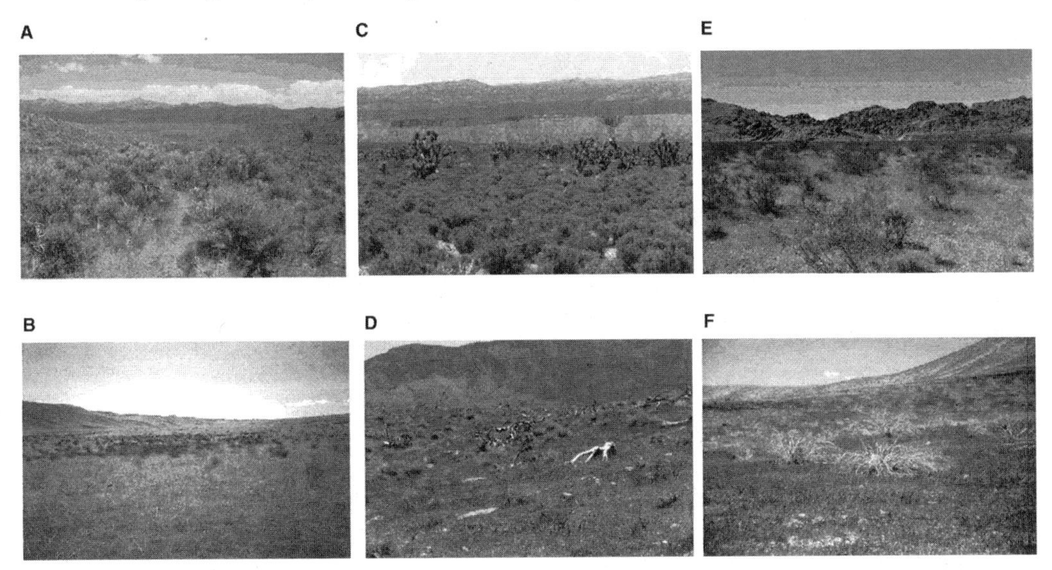

FIGURE 12.5 Examples of desert sites in western North America that have burned in the past 15 years as a result of fires fueled by invasive annual grasses. These grasses create positive feedbacks toward more fire and reduced abundances of desired species, making restoration very difficult. (A) unburned sagebrush (*Artemisia tridentata*) shrubland with cheatgrass (*Bromus tectorum*) in understory. (B) burned sagebrush shrubland now largely dominated by cheatgrass. (C) unburned blackbrush (*Coleogyne ramossisima*)/Joshua tree (*Yucca brevifolia*) shrubland with understory of annual brome grasses. (D) burned blackbrush/Joshua tree shrubland now largely dominated by red brome (*Bromus rubens*). (E) unburned creosote (*Larrea tridentata*) shrubland with scattered annual grasses, *Schismus* spp. and *Bromus rubens*. (F) burned creosote shrubland now largely dominated by *Schismus* and *B. rubens*.

research that supports it, provide a strong foundation for devising management and restoration methods aimed at creating or maintaining sustainable ecosystems. However, our review also indicated that there is still much to be done with regard to integrating relevant theories, collecting the necessary ecosystem and species-specific information, and developing management and restoration approaches that target the appropriate processes at the proper times and scales. Until recently, methods for controlling invaders have often been agronomic in origin and have consequently focused largely on top-down measures. It is increasingly clear that, for wildland ecosystems, a major emphasis needs to be placed on bottom-up approaches that will increase the resistance of the ecosystem to invaders and foster resilience following disturbance. Central to this approach is preventive management that maintains the ecological integrity of ecosystems by facilitating routine disturbance and minimizing degradation caused by human activities and other stressors. Integrating preventive management with top-down control and, when necessary, active restoration is increasingly important, as the pressure from damaging invaders increases.

Regardless of the stage of invasion or the causes of degradation, effective management and restoration will require increased understanding of the properties and processes of ecosystems that convey resistance and resilience. It will also require furthering our ability to recreate communities with those properties and processes. In this chapter we identified and discussed ecological theories relevant to increasing ecosystem resistance and resilience. Using these theories as a basis, we summarize below some of the specific types of information that can be used to improve effectiveness of preventive management and restoration at local to regional scales:

1. Information on the dispersal processes, life-history traits, and the controls on population lags and growth for damaging invaders can be used to minimize invader dispersal to new sites, target sites for monitoring, and facilitate control of both local and source populations.
2. Information on the effects of invaders on community dynamics and ecosystem processes can contribute to our understanding of the conditions that result in communities crossing critical thresholds. Beyond these thresholds, a community may not return to its original state, and restoration may need to focus on feedback processes that must be interrupted for the system to be returned to its original state or to an appropriate alternative state.
3. Information on the pathways through which both natural and anthropogenic disturbances, including management activities, affect individual species, native or otherwise, can help managers to identify ways to use disturbance to their advantage or, conversely, alert them to necessary postdisturbance treatments.
4. Information on the phenology of growth and resource uptake of species in wildland ecosystems could be used to design communities capable of maximizing resource use and minimizing "vacant niches" in the presence and absence of the dominant disturbance(s).
5. An understanding of species life-history characteristics that convey both native and invasive species persistence following the dominant disturbance(s), for example, fire or drought tolerance, will aid in prioritization and development of control methods.
6. An understanding of characteristics of native species that increase their competitive ability with dominant invaders, for example, similar phenologies or resource-use patterns (e.g., Booth et al. 2003), will help in both preventive management and restoration.

The information above, while not all encompassing, provides a basis for using preventive management and restoration to direct successional processes and trajectories. It addresses the primary elements of succession—disturbance, colonization, competitive interactions, and adjustment (MacMahon 1987)—and can thus help move communities at risk of invasion, or already degraded by invasive species, toward more desired states (Figure 12.3; Luken 1990; Sheley and Krueger-Mangold 2003).

The ability to use preventive management or restoration effectively to maintain ecosystems that are both resistant to invaders and resilient to disturbances will depend on the characteristics of the ecosystem of interest, as well as those of the invaders. For many ecosystems, prescribed disturbances coupled with control over propagule supply and manipulation of species performance may be most effective in controlling invaders (e.g., Reever-Morghan et al. 2000). It also is possible that practitioners will have to create "designer ecosystems" (*sensu* MacMahon 1987) that maintain a diverse composition of desired species and that include species tolerant of disturbance and capable of maximizing resource uptake following perturbations. The task for ecologists and practitioners is to obtain and integrate relevant ecological information to test theories and approaches at local and regional scales. Interdisciplinary research and management programs that utilize "restoration experiments" will contribute to valuable advances in this growing and important field.

Acknowledgments

The authors would like to acknowledge financial support from the USDA-ARS, Exotic and Invasive Weeds Research Unit, and the USDA, Forest Service, Rocky Mountain Station. Karen Haubensak provided feedback as the manuscript developed. J. Zedler and D. Falk provided useful suggestions and thorough reviews of an earlier draft.

Literature Cited

Adler, P. B., C. M. D'Antonio, and J. T. Tunison. 1998. Understory succession following a dieback of *Myrica faya* in Hawai'i Volcanoes National Park. *Pacific Science* 52:9–78.

Alexander, J. M., and C. M. D'Antonio. 2003. Grassland species composition after control of broom species in California: Effects of removal method and location. *Ecological Restoration* 21:91–198.

Allen, E. B. 1988. Some trajectories of succession in Wyoming sagebrush grassland: Implications for restoration. In *The reconstruction of disturbed arid lands. An ecological approach*, ed. E. B. Allen, 89–112. AAAS Selected Symposium, 109. Boulder: Westview Press.

Alpert, P., and J. L. Maron. 2000. Carbon addition as a countermeasure against biological invasion by plants. *Biological Invasions* 2:33–40.

Blank, R. R., and J. A. Young. 2002. Influence of the exotic crucifer, *Lepidium latifolium*, on soil properties and elemental cycling. *Soil Science* 167:821–829.

Booth, M. S., M. M. Caldwell, and J. Stark. 2003. Overlapping resource use in three Great Basin species: Implications for community invasibility and vegetation dynamics. *Journal of Ecology* 91:36–48.

Brooks, M. L., C. M. D'Antonio, D. M. Richardson, J. B. Grace, J. E. Keeley, J. M. DiTomaso, R. J. Hobbs, and M. Pellaut. 2004. Effects of invasive alien plants on fire regimes. *Bioscience* 54:677–688.

Brown, R., and R. K. Peet. 2003. Diversity and invasibility of southern Appalachian plant communities. *Ecology* 84:32–39.

Chambers, J. C. 1997. Restoring alpine ecosystems in the western United States: Environmental constraints, disturbance characteristics, and restoration success. In *Restoration ecology and sustainable development*, ed. K. M. Urbanska, N. R. Webb, and P. Edwards, 161–187. Cambridge, UK: Cambridge University Press.

Chapin III, F. S., H. Reynolds, C. M. D'Antonio, and V. Eckhart. 1996. The functional role of species in terrestrial ecosystems. In *Global change in terrestrial ecosystems*, ed. B. Walker and W. Steffen, 403–428. Cambridge, UK: Cambridge University Press.

Christensen, N. L., A. Bartuska, J. H. Brown, S. Carpenter, C. D'Antonio, R. Francis, J. F. Franklin, J. A. MacMahon, R. F. Noss, D. J. Parsons, C. H. Peterson, M. G. Turner, R. G. Woodmansee. 1996. The scientific basis for ecosystem management. *Ecological Applications* 6 (3): 1–34.

Connell, J. H. 1978. Diversity in tropical rainforests and coral reefs. *Science* 199:1302–1310.

Connell, J. H., and R. O. Slatyer. 1977. Mechanisms of succession in natural communities and their role in community stability and organization. *American Naturalist* 3:1119–1144.

Corbin, J., and C. M. D'Antonio. 2004. Can carbon addition increase competitiveness of native grasses? A case study from California. *Restoration Ecology* 12:36–43.

Crooks, J. A., and M. E. Soulé. 1999. Lag times in population explosions of invasive species: Causes and implications. In *Invasive species and biodiversity management*, ed. O. T. Sandlund, P. J. Schei, and A. Viken, 103–125. Kluwer: Dordrecht.

D'Antonio, C. M., T. Dudley, and M. C. Mack. 1999. Disturbance and biological invasions. In *Ecosystems of disturbed ground*, ed. L. Walker, 429–468. New York: Elsevier.

D'Antonio, C. M., and M. C. Mack. 2001. Exotic grasses slow invasion of a nitrogen fixing tree into a Hawaiian woodland. *Biological Invasions* 3:69–73.

D'Antonio, C. M., J. Levine, and M. Thomsen. 2001. Ecosystem resistance to invasion and the role of propagule supply: A California perspective. *Journal of Mediterranean Ecology* 2:233–245.

D'Antonio, C. M., and L. Meyerson. 2002. Exotic species and restoration: Synthesis and research needs. *Restoration Ecology* 10 (4): 703–713.

D'Antonio, C. M., and P. M. Vitousek. 1992. Biological invasions by exotic grasses, the grass/fire cycle, and global change. *Annual Review of Ecology and Systematics* 23:63–87.

Davis, M. A., J. P. Grime, and K. Thompson. 2000. Fluctuating resources in plant communities: A general theory of invasibility. *Journal of Ecology* 88:528–534.

Davis, M. A., and M. Pelsor. 2001. Experimental support for a mechanistic resource-based model of invasibility. *Ecology Letters* 4:421–428.

DeCoster, J., W. J. Platt, and S. A. Riley. 1999. Pine savannas of Everglades National Park: An endangered ecosystem. In *Florida's garden of good and evil, Proceedings of the 1998 joint symposium of the Florida Exotic Pest Plant Council and the Florida Native Plant Society*, ed. D. T. Jones and B.W. Gamble, 81–88. West Palm Beach: South Florida Water Management District.

Drewa, P. B., W. J. Platt, and E. B. Moser. 2002. Fire effects on resprouting of shrubs in headwaters of southeastern longleaf pine savannas. *Ecology* 83:755–767.

Dukes, J. 2001. Biodiversity and invasibility in grassland microcosms. *Oecologia* 126:563–568.

Dukes, J. 2002. Species composition and diversity affect grassland susceptibility and response to invasion. *Ecological Applications* 12:602–617.

Ehrenfeld, J. G. 2003. Effects of exotic plant invasions on soil nutrient cycling processes. *Ecosystems* 6:503–523.

Elton, C. 1958. *The ecology of invasions by animals and plants*. London: Metheun.

Everitt, B. L. 1998. Chronology of the spread of tamarisk in the central Rio Grande. *Wetlands* 18 (4): 658–668.

Ewel, J. J., and F. E. Putz. 2004. A place of alien species in ecosystem restoration. *Frontiers in Ecology and the Environment* 2:354–360.

Galatowitsch, S. M., D. C. Whited, R. Lehtinen, J. Husveth, and K. Schik. 2000. The vegetation of wet meadows in relation to their land use. *Environmental Monitoring Assessment* 60:121–144.

Green, E. K., and S. M. Galatowitsch. 2002. Effects of *Phalaris arundinacea* and nitrate-N addition on the establishment of wetland plant communities. *Journal of Applied Ecology* 39:34–144.

Grime, J. P. 1977. Evidence for the existence of three primary strategies in plants and its relevance to ecological and evolutionary history. *American Naturalist* 11:1169–1194.

Gurevitch, J., S. Scheiner, and G. A. Fox. 2002. *The ecology of plants*. Sunderland, MA: Sinauer Associates.

Harrison, S. 1999. Local and regional diversity in a patchy landscape: Native, alien, and endemic herbs on serpentine. *Ecology* 80:70–80.

Hobbs, R. J., and L. F. Huenneke. 1992. Disturbance, diversity and invasion: Implications for conservation. *Conservation Biology* 6:324–337.

Hobbs, R. J., and S. E. Humphries. 1995. An integrated approach to the ecology and management of plant invasions. *Conservation Biology* 9:761–770.

Holling, C. S. 1973. Resilience and stability of ecological systems. *Annual Review of Ecology and Systematics* 4:1–23.

Horton, J. S. 1977. The development and perpetuation of the permanent tamarisk type in the phreatophyte zone of the Southwest. In *Importance, preservation and management of riparian habitat: A symposium*, coordinated by R. Johnson and D. A. Jones, 124–127. General Technical Report RM-43. Rocky Mountain Forest and Range Experiment Station, Fort Collins, CO: USDA, Forest Service.

Huenneke, L. R., S. P. Hamburg, R. Koide, H. A. Mooney, and P. M. Vitousek. 1990. Effects of soil resources on plant invasion and community structure in California serpentine grassland. *Ecology* 71:478–491.

Kennedy, T. A., S. Naeem, K. M. Howe, J. M. Knops, D. Tilman, and P. Reich. 2002. Biodiversity as a barrier to ecological invasion. *Nature* 417:636–638.

Kercher, S. M., and J. B. Zedler. 2004. Multiple disturbances accelerate invasion of reed canary grass (*Phalaris arundinacea* L.) in a mesocosm study. *Oecologia* 138:455–464.

Kowarik, I. 1995. Time lags in biological invasions with regard to the success and failure of alien species. In *Plant invasions*, ed. P. Pysek, K. Prach, M. Rejmanek, and P. M. Wade, 15–38. The Hague, The Netherlands: SPB Academic Publishing.

Leung B., D. M. Lodge, D. Finnoff, J. F. Shogren, M. A. Lewis, and G. Lamberti. 2002. An ounce of prevention or a pound of cure: Bioeconomic risk analysis of invasive species. *Proceedings of the Royal Society of London, Series B* 269:2407–2413.

Levine, C. M., and J. C. Stromberg. 2001. Effects of flooding on native and exotic plant seedlings: Implications for restoring south-western riparian forests by manipulating water and sediment flows. *Journal of Arid Environments* 49 (1): 111–131.

Levine, J., and C. M. D'Antonio. 1999. Elton revisited: A review of the evidence linking diversity and invasibility. *Oikos* 87:1–12.

Levine, J. M. 2000. Species diversity and biological invasions: Relating local process to community pattern. *Science* 288:852–854.

Levine, J. M. 2001. Local interactions, dispersal and native and exotic plant diversity along a California stream. *Oikos* 95:397–408.

Levine, J. M., M. Vilà, C. M. D'Antonio, J. S. Dukes, K. Grigulis, and S. Lavorel. 2003. Mechanisms underlying the impacts of exotic plant invasions. *Proceedings of the Royal Society of London, Series B* 270:775–781.

Lonsdale, W. M. 1999. Global patterns of plant invasions and the concept of invasibility. *Ecology* 80:1522–1536.

Luken, J. O. 1990. *Directing ecological succession*. London: Chapman & Hall.

MacArthur, R. H., and R. Levins. 1967. The limiting similarity, convergence and divergence of coexisting species. *American Naturalist* 101:377–385.

MacArthur, R .H., and E. O. Wilson. 1967. *The theory of island biogeography*. Princeton: Princeton University Press.

Mack, M. C., and C. M. D'Antonio. 1998. Impacts of biological invasions on disturbance regimes. *Trends in Ecology & Evolution* 13:195–198.

Mack, R. N., D. Simberloff, W. M. Lonsdale, H. Evans, M. Clout, and F. A. Bazzaz. 2000. Biotic invasions: Causes, epidemiology, global consequences, and control. *Ecological Applications* 10:689–710.

MacMahon, J. A. 1987. Disturbed lands and ecological theory: An essay about a mutualistic association. In *Restoration ecology*, ed. W. R. Jordan, M. E. Gilpin, and J. D. Aber, 221–240. Cambridge, UK: Cambridge University Press.

Maron, J. L., and P. G. Connors. 1996. A native nitrogen-fixing shrub facilitates weed invasion. *Oecologia* 105:302–312.

Maron, J. L., and R. L. Jefferies. 2001. Restoring enriched grasslands: Effects of mowing on species richness, productivity and nitrogen retention. *Ecological Applications* 11:1088–1100.

Maron, J. L., and M. Vilà. 2001. When do herbivores affect plant invasion: Evidence for the natural enemies and biotic resistance hypotheses. *Oikos* 95:361–373.

Masters, R., and R. Sheley. 2001. Principles and practices for managing rangeland invasive plants. *Journal of Range Management* 54:502–517.

Maurer, D. A., and J. B. Zedler. 2002. Differential invasion of a wetland grass explained by tests of nutrients and light availability on establishment and clonal growth. *Oecologia* 131:279–288.

May, R. 1973. *Stability and complexity in model ecosystems*. Princeton: Princeton University Press.

McClendon, T., and E. F. Redente. 1992. Effects of nitrogen limitation on species replacement dynamics during early succession on a semiarid sagebrush site. *Oecologia* 91:312–317.

McEvoy, P. B., N. T. Rudd, C. S. Cox, and M. Huso. 1993. Disturbance, competition, and herbivory effects on ragwort, *Senecio jacobaea*, populations. *Ecological Monographs* 63:55–75.

McEvoy, P. B., and E. M. Coombs. 1999. Biological control of plant invaders: Regional patterns, field experiments, and structured population models. *Ecological Applications* 9:387–401.

McNaughton, S. J. 1968. Structure and function in California annual grassland. *Ecology* 49:962–972.

Miller, R. C., and J. B. Zedler. 2003. Responses of native and invasive wetland plants to hydroperiod and water depth. *Plant Ecology* 167:57–69.

Moody, M. E., and R. N. Mack. 1988. Controlling the spread of plant invasions: The importance of nascent foci. *Journal of Applied Ecology* 25:1009–1021.

Mulvaney, M. 2001. The effect of introduction pressure on the naturalization of ornamental woody plants on southeastern Australia. In *Weed risk assessment*, ed. R. H. Groves, F. D. Panetta, and J. G. Virtue, 186–193. Collingwood, Australia: CSIRO.

Naeem, S., J. M. Knops, D. Tilman, K. M. Howe, T. Kennedy, and S. Gale. 2000. Plant diversity increases resistance to invasion in the absence of covarying extrinsic factors. *Oikos* 91 (1): 97–108.

National Invasive Species Council (NISC). 2001. *Meeting the invasive species challenge: National invasive species management plan*. Washington, DC.

National Research Council (NRC). 2002. *Predicting invasions of nonindigenous plants and plant pests*. Washington, DC: National Academy Press.

Nobel, I. R., and R. O. Slatyer. 1980. The use of vital attributes to predict successional changes in plant communities subject to recurrent disturbances. *Vegetatio* 43:5–21.

Parker, I. M., D. Simberloff, W. M. Lonsdale, K. Goodell, M. Wonham, P. M. Kareiva, M. H. Williamson, B. Vonttolle, P. B. Moyle, J. E. Byers, and L. Goldwasser. 1999. Impact: Toward a framework for understanding the effects of invaders. *Biological Invasions* 1:3–19.

Paschke, M. W., T. McLendon, and E. F. Redente. 2000. Nitrogen availability and old-field succession in a short-grass steppe. *Ecosystems* 3:144–158.

Perry, L. G., S. M. Galatowitsch, and C. J. Rosen. 2004. Competitive control of invasive vegetation: A native wetland sedge suppresses *Phalaris arundinacea* in carbon-enriched soil. *Journal Applied Ecology* 41:151–162.

Pianka, E. 1970. On r- and k-selection. *American Naturalist* 104:592–597.

Pieterse, P. J., and A. L. P. Cairns. 1986. The effect of fire on an *Acacia longifolia* seed bank in the southwestern Cape. *South African Journal Botany* 52:233–236.

Platt, W. J. 1999. Southeastern pine savannas. In *The savanna, barren, and rock outcrop communities of North America*, ed. R. C. Anderson, J. S. Fralish, and J. M. Baskin, 23–51. Cambridge: Cambridge University Press.

Platt, W. J., and R. M. Gottschalk. 2001. Effects of exotic grasses on potential fine fuel loads in the groundcover of south Florida slash pine savannas. *International Journal of Wildland Fire* 10:155–159.

Pokorny, M. L., R. L. Sheley, C. A. Zabiuski, R. E. Engel, T. J. Svejcar, and J. J. Borkowski. 2005. Plant functional group diversity as a mechanism for invasion resistance. *Restoration Ecology* 13:448–459.

Reever Morghan, K. J., and T. R. Seastedt. 1999. Effects of soil nitrogen reduction on non-native plants in disturbed grasslands. *Restoration Ecology* 7:51–55.

Reever Morghan, K. J., T. R. Seastedt, and P. J. Sinton. 2000. Frequent fire slows invasion of ungrazed tallgrass prairie by Canada thistle (Colorado). *Ecological Restoration* 18:194–195.

Reichard, S. H., and C. W. Hamilton. 1997. Predicting invasions of woody plants introduced into the United States. *Conservation Biology* 11:193–203.

Rejmanek, M. 1996. A theory of seed plant invasiveness: The first sketch. *Biological Conservation* 78:171–181.

Rejmanek, M., and D. M. Richardson. 1996. What attributes make some plants more invasive? *Ecology* 77:1655–1661.

Rejmanek, M., D. M. Richardson, S. I. Higgins, M. J. Pitcairn, and E. Grotkopp. 2005. Ecology of invasive plants: State of the art. In *Invasive alien species: A New Synthesis*, ed. H. A. Mooney, R. N. Mack, J. A. McNeely, L. Neville, P. J. Schei, and J. K. Waage, 104–161. Washington, DC: Island Press.

Robinson, G. R., J. F. Quinn, and M. L. Stanton. 1995. Invasibility of experimental habitat islands in a California winter annual grassland. *Ecology* 76:786–794.

Sheley, R. L., and J. Krueger-Mangold. 2003. Principles for restoring invasive plant-infested rangeland. *Weed Science* 51:260–265.

Simberloff, D. K. 1981. Community effects of introduced species. In *Biotic crises in ecological and evolutionary time*, ed. T. H. Nitecki, 53–81. New York: Academic Press.

Simberloff, D. K., and M. Von Holle. 1999. Synergistic interactions of nonindigenous species: Invasional meltdown? *Biological Invasions* 1:21–32.

Stevens L. E., T. J. Ayers, M. J. C. Kearsley, J. B. Bennett, R. A. Parnell, A. E. Springer, K. Christensen, V. J. Meretsky, A. M. Phillips, J. Spence, M. K. Sogge, and D. L. Wegner. 2001. Planned flooding and Colorado river riparian trade-offs downstream from Glen Canyon Dam, Arizona. *Ecological Applications* 11 (3): 701–710.

Stock, W. D., K. T. Wienand, and A. C. Baker. 1995. Impacts of invading N_2 fixing *Acacia* species on patterns of nutrient cycling in two Cape ecosystems: Evidence from soil incubation studies and ^{15}N natural abundance values. *Oecologia* 101:375–382.

Stohlgren T. J., D. T. Barnett, and J. T. Kartesz. 2003. The rich get richer: Patterns of plant invasions in the United States. *Frontiers in Ecology and the Environment* 1:1114.

Stohlgren, T. J., D. Binkley, et al. 1999. Exotic plant species invade hot spots of native plant diversity. *Ecological Monographs* 69:25–46.

Stohlgren, T. J., K. A. Bull, G. W. Chong, M. A. Kalkhan, L. D. Schell, K. A. Bull, Y. Otsuki, G. Newman, M. Bashkin, and Y. Son. 1998. Riparian zones as havens for exotic plant species in the central grasslands. *Plant Ecology* 138:113–125.

Stromberg, J. 1998. Dynamics of Fremont cottonwood (*Populus fremontii*) and saltcedar (*Tamarix chinensis*) populations along the San Pedro River, Arizona. *Journal of Arid Environments* 40 (2): 133–155.

Stromberg, J. C. 1997. Growth and survivorship of Fremont cottonwood, Goodding willow, and salt cedar seedlings after large floods in central Arizona. *Great Basin Naturalist* 57 (3): 198–208.

Suding, K. N, K. L. Gross, and G. R. Houseman. 2004. Alternative states and positive feedbacks in restoration ecology. *Trends in Ecology & Evolution* 19:46–53.

Tilman, D. 1997. Community invasibility, recruitment limitation, and grassland biodiversity. *Ecology* 78:81–92.

Torok, K., T. Szili-Kovacs, M. Halassy, T. Toth, Zs. Hayek, M. W. Pascke, and L. J. Wardell. 2000. Immobilization of soil nitrogen as a possible method for the restoration of sandy grassland. *Applied Vegetation Science* 3:7–14.

Vinton, M. A., and I. C. Burke. 1995. Interactions between individual plant species and soil nutrient status in shortgrass steppe. *Ecology* 76:1116–1133.

Vitousek, P. M., L. R. Walker, L. D. Whiteaker, D. Mueller,-Dombuis, and P. A. Matson. 1987. Biological invasion by *Myrica faya* alters ecosystem development in Hawaii. *Science* 238:802–804.

Vitousek, P. M., and L. R. Walker. 1989. Biological invasion by *Myrica faya* in Hawaii: Plant demography, nitrogen fixation and ecosystem effects. *Ecological Monographs* 59:247–265.

Vivrette, N. J., and C. H. Muller. 1977. Mechanism of invasion and dominance of coastal grassland by *Mesembryanthemum crystallinum*. *Ecological Monographs* 47:301–318.

Weiss, S. 1999. Cars, cows and checkerspot butterflies: Nitrogen deposition and management of nutrient-poor grasslands for a threatened species. *Conservation Biology* 13:1–12.

Whisenant, S. G. 1999. *Restoring damaged wildlands: A process-oriented, landscape approach*. Cambridge, UK: Cambridge University Press.

Williamson, M. 1996. *Biological invasions*. London: Chapman & Hall.

Wilson, S. D., and M. Partel. 2003. Extirpation or coexistence? Management of a persistent introduced grass in a prairie restoration. *Restoration Ecology* 11:410–416.

Wiser, S. K., R. B. Allen, P. W. Clinton, and K. H. Platt. 1998. Community structure and forest invasion by an exotic herb cover over 23 years. *Ecology* 79:2071–2081.

Woo, I., and J. B. Zedler. 2002. Can nutrients alone shift a sedge meadow towards invasive *Typha x glauca?* *Wetlands* 22:509–521.

Young, J. A., D. E. Palmquist, and R. R. Blank. 1998. The ecology and control of perennial pepperweed (*Lepidium latifolium* L.). *Weed Technology* 12:402–409.

Zedler, J. B., and S. Kercher. 2004. Causes and consequences of invasive plants in wetlands: Opportunities, opportunists and outcomes. *Critical Reviews in Plant Sciences* 23:431–452.

Chapter 13

Statistical Issues and Study Design in Ecological Restorations: Lessons Learned from Marine Reserves

Craig W. Osenberg, Benjamin M. Bolker, Jada-Simone S. White, Colette M. St. Mary, and Jeffrey S. Shima

Scientists and managers often seek to restore degraded systems to more desirable states. A system might be restored by eliminating a putatively deleterious factor(s) and allowing the system to recover naturally (e.g., by removing a sewage outfall or abolishing pesticide application) or by aggressively managing the system to reduce the time required for natural recovery. Regardless of the approach taken, we need to know if the restoration has fulfilled expectations. Thus, two fundamental questions underlie the scientific assessment of any restoration project: (1) What is the goal (e.g., to what state should the system be restored)? and (2) Did the restoration project achieve this goal (or, more generally, what were the effects of the restoration project)? Both aspects are central to the inferences we draw about restoration efforts and intimately linked to the statistical tools that we use to make these inferences.

Goals of restoration projects fall into two broad categories. The first, which we call *endpoint based,* aims to restore the system to a predefined state. We may define endpoints theoretically (e.g., that the density of an endangered bird species be restored to ≥50 breeding pairs based on a population viability analysis) or empirically, by comparison to a more "pristine" reference site (e.g., that species richness be ≥90% of that found at the reference site). Outcomes can be assessed by sampling the restored system and comparing it with the stated endpoint. To help formulate inferences, we might use a standard statistical null-hypothesis framework in which a single sample is compared with a theoretical expectation, or two samples are directly compared. Statistical power could also be considered in the assessment of restoration effects (low power will reduce our ability to detect the effects of restoration: Mapstone 1995). Although useful in many contexts, these endpoint-based approaches fail to provide an estimate of the *effect* of the restoration activity. In fact, the restoration effort may not have had any effects and yet the site may reach the desired state (e.g., due to natural variation independent of the restoration). This may be satisfactory in many contexts, but such a result would fail to inform future restoration projects.

Thus, we also define *effect-size-based* goals, in which we quantify the effects (and the associated uncertainty) of the restoration activity (e.g., determine the increase in the abundance of a threatened species caused by the restoration project), possibly by comparison with similarly degraded sites (as opposed to pristine sites), so that the response to the restoration can be quantified.

A combination of both approaches is likely ideal—we would like to know how much of an effect we have produced (effect-size-based outcomes) and if that change is "sufficient" (endpoint-based outcomes). In this chapter, however, we focus on effect-size-based goals and the study designs that facilitate this assessment, because endpoint-based approaches can be tackled with well-known statistical tools (e.g., ANOVA). In contrast, the apparently "simple" task of quantifying an effect size requires approaches that often are distinct from the standard quantitative tools we learn in basic statistics or experimental design courses, especially when dealing with large-scale, unreplicated assessments. These solutions are not, therefore, generally appreciated or applied. Fortunately, the complex challenges (and solutions) that are posed are very similar to those shared by assessments of unreplicated human interventions, such as the study of the effects of sewage outfalls, foresting practices, or nuclear power plants. As a result, we borrow heavily from the literature on impact assessment (e.g., Stewart-Oaten et al. 1986; Schmitt and Osenberg 1996). Effect sizes may be either univariate or multivariate, but for simplicity of discussion and presentation, we lay out the framework for univariate measures. Multivariate analogues exist for our univariate examples. For more general discussion of statistical issues in restoration studies, we refer the reader to the useful reviews by Michener (1997) and Schreuder et al. (2004).

To provide context to our discussion of assessment designs, we draw examples from the restoration of marine systems through the establishment of marine protected areas (MPAs) (Allison et al. 1998; Lubchenco et al. 2003; Norse et al. 2003). MPAs share many features with other restoration activities: (1) they are expected to have local effects within the boundaries of the restoration activity; (2) they also may have effects that extend beyond the MPA boundaries and therefore help restore degraded sites that are not actively managed, but may nonetheless benefit from distant restoration activities; and (3) there remains a considerable need for improved tools to document and estimate the local and regional effects of a given restoration effort.

Below, we discuss the central concepts drawn from experimental design and contrast these approaches with those needed in large-scale restorations (and impact assessments in general). We then discuss the major types of assessment designs, including their advantages and limitations, and highlight these issues with a critique of MPA studies. Last, we propose future directions, including more appropriate designs that will address current shortcomings and enhance the practice of restoration ecology.

Central Concepts

The basic question posed in any effect-size-based assessment study is simple to state and hard to solve: how does the state of the system after restoration compare with the state of the system that would have existed had the restoration activity not taken place (Stewart-Oaten et al. 1986; Stewart-Oaten 1996a)? Of course, the latter cannot be observed directly (because the restoration activity *did* take place) and must therefore be estimated. That is the crux of the problem: how do we estimate this unknown state and therefore (i.e., by comparison with the observed state) infer the effect of the restoration activity? The classic approach is experimental and employs null-hypothesis tests. Indeed, experiments are the primary tool of many restoration ecologists, so we begin with a discussion of issues germane to field experiments.

P-values Versus Estimation

Most ecologists use frequentist statistics, epitomized by P-values and tests of null hypotheses. If the observed data are not very unlikely under the null hypothesis (typically, $P > 0.05$), then we tentatively accept the null hypothesis, which is often erroneously interpreted as indicating "no effect" (Yoccoz 1991). Alternatively, if the data are sufficiently unlikely under the null (typically, $P < 0.05$), then we conclude that there was "an effect." The *P*-value itself (or the test statistic), however, gives little indication of the likely effect size or the associated uncertainty; we know only whether the confidence interval on this effect includes or excludes zero.

Consider two studies of the effects of two restoration approaches on the abundance of an endangered species. Approach A leads to an estimated increase in population density of 0.1% per year (± 0.11%), whereas approach B yields an effect of 100% (±101%). Although neither result is "significant," in approach A, we have high confidence that the effect is "small" because of the high precision in the estimate. In B, we do not even know the direction of the effect—the restoration might have very detrimental effects or extremely positive effects. A conclusion of "no effect" cannot be made with any confidence.

Instead of *P*-values we need to estimate the magnitude of effects and their uncertainty (Yoccoz 1991; Stewart-Oaten 1996a; Johnson 1999; Osenberg et al. 1999, 2002; Anderson et al. 2000). This is especially true in assessment studies in which policy makers, the public, and the scientific community should care less about whether there is a demonstrable (but possibly tiny) effect and more about the magnitude (and uncertainty) of the response (Stewart-Oaten 1996a). In this chapter, we emphasize estimation and refer the reader to these other sources for greater detail about the *P*-value culture.

An Experimental Approach: Why Do We Need an Alternative?

A restoration project might be conducted using a standard experimental approach, with multiple treatments (including appropriate controls), replication (multiple independent units that receive a given treatment), and random assignment of units to treatments (Underwood 1997; Scheiner and Gurevitch 2001). Imagine a site in which sea grass was previously present but was severely damaged by an anthropogenic activity (e.g., dredging or an oil spill). An investigator could choose multiple plots within this site and randomly assign them to two or more treatments (e.g., a suitable "control" plus different "restoration" treatments). After some appropriate amount of time the plots could be sampled and compared using standard statistical procedures. In principle, such an approach can be useful, especially to compare different possible restoration techniques. However the extension to a large-scale restoration project requires that (1) the spatial scale of the plots is appropriate to the overall goals of the large-scale restoration project; (2) the plots are independent of one another (e.g., restoration treatments do not affect adjacent control plots); and (3) the analysis focuses on effect sizes and their uncertainty.

To explore this issue, we reviewed all papers from the 2003 volume of *Restoration Ecology*. Of the 68 papers that reported results from studies that could be used to infer effects of a restoration activity (or activities), 41 were experimental, with replication, random assignment, and a control. Of these, the modal scale of manipulation was 10 m^2 (range: ~$0.025–4 \times 10^6$ m^2) with all but 6 occurring on scales <100 m^2.

However, propagules disperse, herbivores colonize, and predators typically forage over scales larger than 100 m^2. Indeed, scaling up small experiments to their larger-scale implications is a continuing challenge for ecologists (Englund and Cooper 2003; Melbourne and Chesson 2005; Schmitz 2005). As a result, small-scale experiments (e.g., conducted on the scale of 10 m^2), although useful for revealing mechanisms and evaluating likely restoration strategies, may be poor predictors of actual effects of a large-scale restoration project or the success of a restoration conducted at a larger scale (e.g., involving hectares or km^2). Of course, we could conduct experiments at larger spatial scales (e.g., using many different sea grass beds as replicates and having half randomly assigned to controls), but this is rarely feasible. For example, in our survey, only 2 of 68 studies were replicated and conducted at a scale $>10,000$ m^2 (also see Michener 1997 and Schindler 1998 for examples of associated constraints).

Even if a replicated large-scale experiment were possible, it would only reveal the *average* effect of a restoration activity on the population of potential sites and not the effect at any particular site. This would be useful to compare among possible restoration approaches; however, we are often most interested in understanding the effect of restoration at a *particular site* (e.g., for mitigation or regulation). Furthermore, if some sites were positively affected by restoration while others were negatively affected, one could conclude "no effect" overall. Instead, we would prefer to know which sites were positively and negatively affected (and, ideally, why). In an experiment, a "positive," site-specific effect cannot be inferred by the deviation of one site from the pool of replicates, because the restoration effect is confounded with other aspects of that site (e.g., initial conditions). This limitation is a generic feature of replicated experiments and standard statistical approaches.

Thus, we propose an approach that departs from our standard experimental training and that (1) can be applied to spatially unreplicated interventions; (2) is site-specific; and (3) yields defensible estimates of the effect of the restoration activity (rather than *P*-values or "yes/no" answers). That approach is the BACIPS (Before-After–Control-Impact Series) assessment design, which is currently used in impact assessments (Stewart-Oaten et al. 1986; Schmitt and Osenberg 1996). Interestingly, none of the studies we reviewed in *Restoration Ecology* used a BACIPS study or presented a cogent description of these assessment issues, suggesting that BACIPS could be a valuable addition to the restoration ecology tool kit.

Local Versus Regional Effects

Local effects, which are the focus of most restoration studies (and all of those we reviewed), arise within the boundaries of the specific restoration activity. However, effects will not be limited to the boundaries of the restoration project. Indeed, we expect that there will be regional effects that arise outside the restored site, for example, due to movement of plant propagules, animals, detritus, or nutrients. Selection of control sites (which need to be independent of the restoration effects) must therefore consider the life history and dispersal capabilities of the interacting species and the transport of materials. Local and regional effects also must be studied with different study designs, because one emphasizes effects that occur within the boundary of the project and the other focuses on effects outside of the boundary. In some cases (as we illustrate below), the regional effects are of equal (if not greater) importance than the local effects, yet they remain understudied because of problems inherent to their assessment.

Assessment Designs and Their Application to Restoration Ecology

For our discussion of effect-size-based approaches, we assume that the restoration effort is unreplicated, that reference and restoration site(s) are not necessarily assigned at random, and that all sites are initially degraded, although these conditions are not required. We refer to the reference site(s) as a "Control" and the site to be restored as the "Impact" site, as in the impact assessment literature (Schmitt and Osenberg 1996). Our goal is to estimate the change in some variable (say population density of a focal species) at the Impact site resulting from the restoration activity. Below we summarize common assessment designs to highlight the differences in their approach and the problems that may arise in drawing conclusions from the resulting data.

Control-Impact (CI) Designs

In this common design, multiple samples are typically taken from plots within an Impact site and at least one Control site. These two sets of samples are compared statistically to determine if the two sites differ. If they do, then we conclude that there was an effect of the restoration activity. Of course, because no two sites are identical (although Control and Impact sites may be *similar*), there will likely be statistically significant differences between the two sites. This will be true even before the restoration project begins. Thus, the Control-Impact design confounds the effect of the restoration project with other processes that produce spatial variation in parameters (e.g., Figure 13.1a).

Before-After (BA) Designs

The Before-After design avoids problems with spatial variation by sampling only the Impact site and comparing its state Before versus After restoration (e.g., see Figure 13.1b). We discuss two variants of this basic design.

BA-Single Time

The Impact site is sampled once Before and once After the restoration activity (with many plots within each site providing "replication"). However, all systems change through time, so any two sets of samples from the same site (but different times) will be different (assuming sufficient sampling). Thus, the BA-single-time design confounds the restoration effect with other processes that produce temporal variation.

BA-Time Series

Multiple sampling times within a period provides a form of replication that allows the investigator to incorporate, and potentially deal with, temporal variation. By using time-series methods that account for serial correlation, BA designs can be used to infer effects. Indeed, one of the most famous of all intervention studies was Box and Tiao's (1975) BA-time-series study of ozone in downtown Los Angeles and its response to two separate interventions: (1) the simultaneous reformulation of gasoline designed to reduce reactive hydrocarbons and

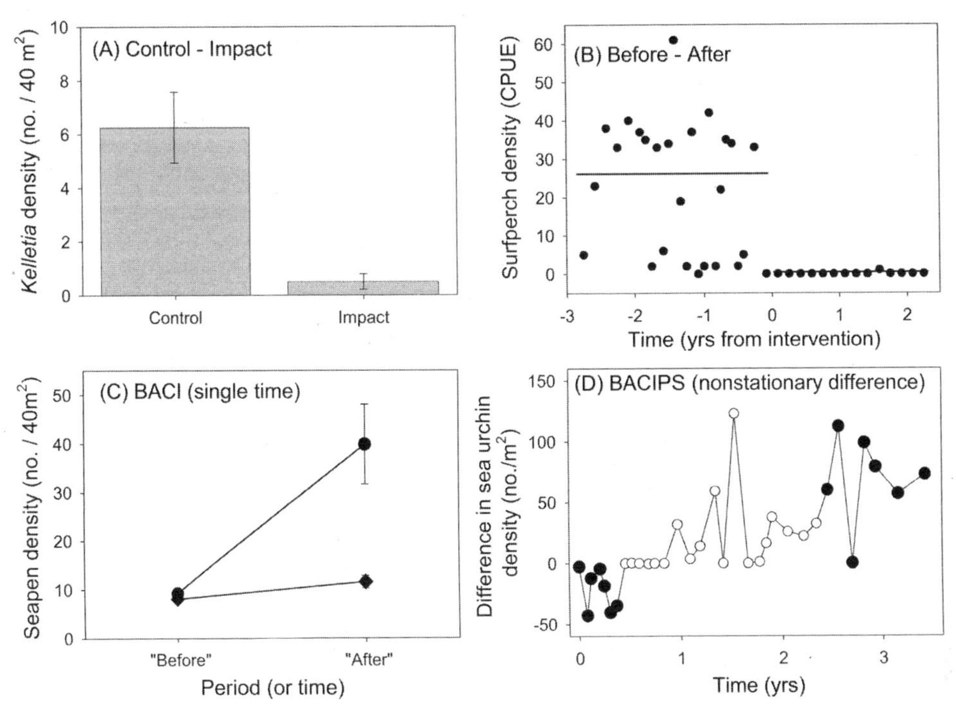

FIGURE 13.1 Empirical examples of assessment designs in which erroneous inferences would be drawn due to confounding of natural variability with effects of an intervention. (A) A Control-Impact design investigating effects of oil and gas production on a benthic mollusc (*Kelletia kelletii*) (see Osenberg et al. 1992, 1994; Osenberg and Schmitt 1996). The data were taken in a Before period and therefore represent preexisting spatial variation in density and not an effect of the oil production activity. (B) A Before-After design investigating effects of the cooling tower effluent of a nuclear power plant on the abundance (catch per unit effort = CPUE) of pink surfperch, *Zalembius rosaceus* (see Murdoch et al. 1989). Time = 0 indicates the date on which power was first generated following expansion of the power plant. However, these data came from a Control site and indicate natural temporal variability, not effects of the power plant. (C) A BACI design (without a time series) studying effects of oil production on the density of seapens (*Acanthoptilum* sp.). Production did not begin when expected, so this relative change in the Control and Impact sites represents a natural space-by-time interaction and not an effect of oil production. (D) A BACIPS study showing a time series of differences in sea urchin (*Lytechinus anamesis*) between an Impact and Control site, illustrating the possible confounding of an effect with long-term natural changes in density (e.g., if the two time periods indicated by filled circles happened to define the Before and After periods). The data come from a Before period and indicate a long-term trend in the differences independent of the intervention.

rerouting of traffic following the opening of the 405 freeway (these two were considered together due to their temporal confluence); and (2) redesign of the engines of new cars. The first intervention was predicted to produce a step-change reduction in ozone, and the second was expected to gradually reduce ozone as new cars replaced older versions. Box and Tiao framed a stochastic model of the interventions, defined an analytic approach based on that model, ran diagnostics to determine model inadequacies, and, barring the latter, derived inferences about the response of ozone to the interventions. They concluded that both interventions had demonstrable effects (Figure 13.2).

Box and Tiao's success was, in part, due to (1) the long and dense time series (monthly averages of ozone from an 18-year period); (2) the well-behaved temporal dynamics of ozone; and (3) the simple expectations about plausible effects of the interventions on ozone. These advantages are unlikely to exist for most ecological studies (perhaps with the possible exception of some epidemiological studies: e.g., Earn et al. 2000). Figures 13.1b and 13.1d offer examples of ecologically "long" time series (five years) that were too short to capture relevant background temporal dynamics. We return to this issue in the next section.

Before-After–Control-Impact (BACI) Designs

BACI designs attempt to deal with both spatial and temporal variation by sampling at one or more Control site(s) and the Impact site both Before and After the intervention. A variety of permutations on the basic theme have been proposed.

BACI (Single Time)

Green (1979) proposed a BACI design in which a Control and Impact site were sampled once Before and once After an intervention. A site-by-time interaction indicates an effect of

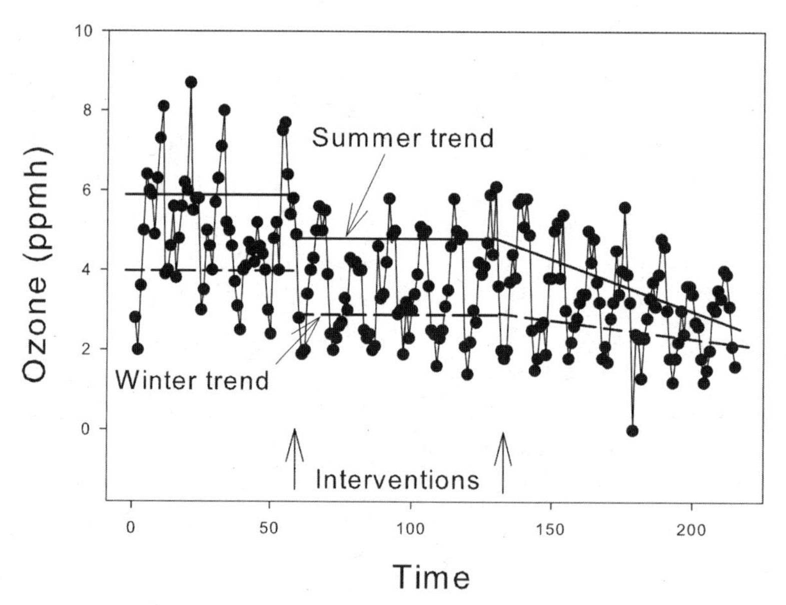

Figure 13.2 Summary of the results of Box and Tiao's (1975) Before-After study of ozone in the Los Angeles basin. Points give monthly ozone concentrations. Arrows indicate the timing of the two interventions: one hypothesized to result in a step change and one hypothesized to result in a gradual reduction in ozone. The solid line gives the estimated trend for summer conditions and the dashed line gives the estimated trend for winter conditions (other seasonal trends are excluded for clarity).

the intervention. However, no two sites show the same temporal dynamics. Thus, we expect site-by-time interactions when two sites are sampled intensively on two different dates (Figure 13.1c). This BACI design therefore confounds effects of the intervention with other factors that cause site-by-time interactions.

BACI-Paired Series (BACIPS)

In the basic BACIPS design, a Control (or set of Controls) and an Impact site are sampled simultaneously several times Before and After the perturbation (Stewart-Oaten et al. 1986). The parameter of interest is the *difference* in a chosen variable (e.g., density of a target species) between the Control and Impact sites estimated on each sampling date. Each difference from the Before period provides an estimate of the spatial variation between the two sites and thus is an estimate of the expected difference that should exist in the After period in the absence of an effect of the intervention. The difference between the average Before and After differences provides an estimate of the magnitude of the effect of the intervention. The simplest design assumes that there is no serial correlation (or temporal trend) in the differences between the Control and Impact sites (serial correlation will result if the sampling within a site is done at too short an interval). If there is serial correlation in the differences, then an autoregressive approach can be used to account for the correlation structure (see Stewart-Oaten et al. 1992; Stewart-Oaten and Bence 2001).

If sampling is done too close together for too short a time period, the serial correlation structure cannot be detected and may be confounded with the "effect" (Figure 13.1d); instead of indicating a true effect of the intervention, the change in the difference from Before to After may be the result of oversampling during a single, short-lived, local perturbation in each period, or sampling over a time interval in which the true difference was changing naturally and gradually through time. Indeed, the Before period is critical for developing diagnostic tests of the patterns of covariation between the Control and Impact sites (but see Murtaugh 2002, 2003). Especially important are the pattern of serial correlation and the additivity of site and time effects (Stewart-Oaten et al. 1986, 1992; Bence 1995; Stewart-Oaten 1996b; Bence et al. 1996). We return to this below.

Predictive BACIPS

The BACIPS design uses the Control site to predict the Impact's state (Bence et al. 1996): the Impact site's state in the After period (and assuming no effect of the intervention) can be predicted as the sum of the Control's state in the After period plus the mean difference between the Control and Impact site estimated during the Before period. Bence et al. (1996) have advocated a more flexible approach in which the relationship between the Control and Impact values is compared Before and After the intervention (Figure 13.3). This approach is intuitively appealing and has the advantage of allowing the effect size to vary (e.g., with the overall environmental conditions, as indexed by the Control value), but it has the disadvantage that the independent variable (the Control value) is measured with error and thus violates a standard assumption of Model I regression models. This problem has not been clearly resolved in the predictive-BACIPS approach.

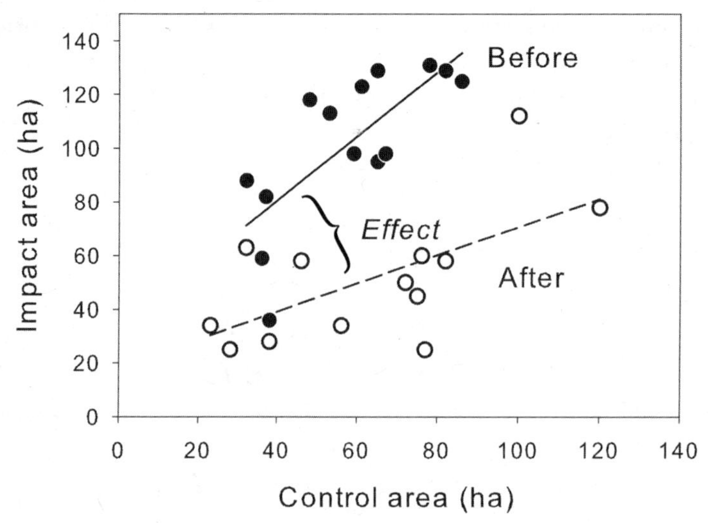

FIGURE 13.3 Illustration of the predictive-BACIPS design using the Bence et al. (1996) study of the effect of a nuclear power plant on the areal extent of kelp (*Macrocystis pyrifera*) in southern California. The difference between the relationships between the Impact and Control site from Before to After gives an estimate of the effect of the intervention (operation of the power plant). In this case, the effect ranges from a reduction in kelp cover of ~40–80 ha, with the largest effects expected when conditions are good (i.e., when there is more kelp at the Control site).

BEYOND BACI

Underwood (1991, 1992, 1994) promoted a different elaboration of BACI that uses an "asymmetrical design" in which there are multiple Control sites. The data are not paired in time (i.e., the samples at the Controls and Impact sites do not share a common time effect) and thus the differencing approach of BACIPS is not relevant. Stewart-Oaten and Bence (2001) have critiqued this approach in depth, so we concentrate on the BACIPS designs.

Why Have a Control? What Makes a Good One?

Recall that Box and Tiao (1975) successfully used a BA design to examine effects of interventions on ozone in downtown Los Angeles. Yet we dismissed BA designs above as confounding effects of the intervention with other sources of temporal variation. However, Box and Tiao had a very long-time series of data, from which they were able to construct (and evaluate) plausible models of ozone dynamics with and without the interventions. In essence, their model of ozone dynamics from the Before period could be extrapolated to the After period and contrasted with the observed behavior to infer effects of the intervention. In ecological assessments we usually lack long-time series and well-defined temporal dynamics. Thus, a predictive ecological model from the Before period is not likely to provide an accurate null expectation for the After Period (Figure 13.1b). This is where the Control site helps.

Imagine that the variable of interest at the Impact site varies considerably (and possibly erratically) through time. Developing a predictive model of these dynamics may be very difficult. However, if another site (the Control) exhibits similar temporal dynamics in the absence of the intervention, then the Control site can be used to develop a more accurate model of the Impact's dynamics. Indeed, this is the key feature of a good Control site: it is not necessarily a site that is most like the Impact site, but rather it is one that changes through time in a way comparable to the Impact site in the absence of an intervention (Figure 13.4) (Magnuson et al. 1990; Osenberg et al. 1994). If the Control and Impact sites track one another through time (show high coherence), then there will be low variation in the differences through time, and BACIPS and predictive-BACIPS will have high power and will give rise to more accurate estimates of the effect sizes (Figure 13.4).

To illustrate this more specifically, let the parameter of interest be the difference (hereafter referred to as "delta," Δ, or D for its estimate) in density or other suitable variable between the Control and Impact sites as estimated on each sampling date (e.g., $D_{P,i} = N_{I,P,i} - N_{C,P,i}$), where $N_{I,P,i}$ and $N_{C,P,i}$ are sampled densities (often log-transformed) at the Control and Impact sites on the ith date of Period P (i.e., Before or After). Each difference Before provides an estimate of the spatial variation between the two sites (Δ_B), which is the expected difference that should exist in the After period in the absence of an effect of the intervention. The difference between the average Before and After differences ($\bar{D}_B - \bar{D}_A$) provides an estimate of the effect of the intervention. Confidence in this estimate is determined by the variance in differences pooled across periods (s^2), as well as the number of sampling dates (i.e., replicates) in each of the Before and After periods (n_B, n_A). In the absence of serial correlation in the time series of differences (see also Stewart-Oaten and Bence 2001):

$$\text{Effect Size: } E = \bar{D}_B - \bar{D}_A = \frac{\Sigma D_{B,i}}{n_B} - \frac{\Sigma D_{A,i}}{n_A} \quad (1)$$

$$\text{Variance: } s^2 = \frac{s_B^2}{n_B} + \frac{s_A^2}{n_A} \quad (2)$$

$$95\% \text{ Confidence Interval: } E \pm s \cdot t_{n_B+n_A-2,0.025} \quad (3)$$

where for period P,

$$s_P^2 = [\Sigma (D_{P,i} - \bar{D}_P)^2]/n_P - 1 \quad (4)$$

In a standard null-hypothesis testing context, low variability (s^2, Equation 2 or 4) will lead to a more powerful test of the intervention effect and more accurate estimates of the effect (i.e., smaller confidence limits, Equation 3): see Osenberg et al. (1994). By taking differences between the Control and Impact sites (E, Equation 1), BACIPS removes the effects of background sources of variation that are common to both sites (e.g., responses to climatic events). By emphasizing differences and using a time-series approach, the BACIPS design accounts for some sources of spatial and temporal variation ignored in the BA and CI and BACI designs (Stewart-Oaten et al. 1986; Stewart-Oaten 1996a, 1996b).

Notice that the two main sources of variation in a BACIPS design are quite different from those used in other designs. The estimate of the effect (E, Equation 1) is derived from the Period-by-Location term (in standard ANOVA terms), which indicates how much the response variable at the Impact site (*relative* to the Control site) changed from the Before to After periods (i.e., $\bar{D}_B - \bar{D}_A$). The error component (Equation 2 or 4) measures how much the difference between the response variable at the Control and Impact sites varies in the

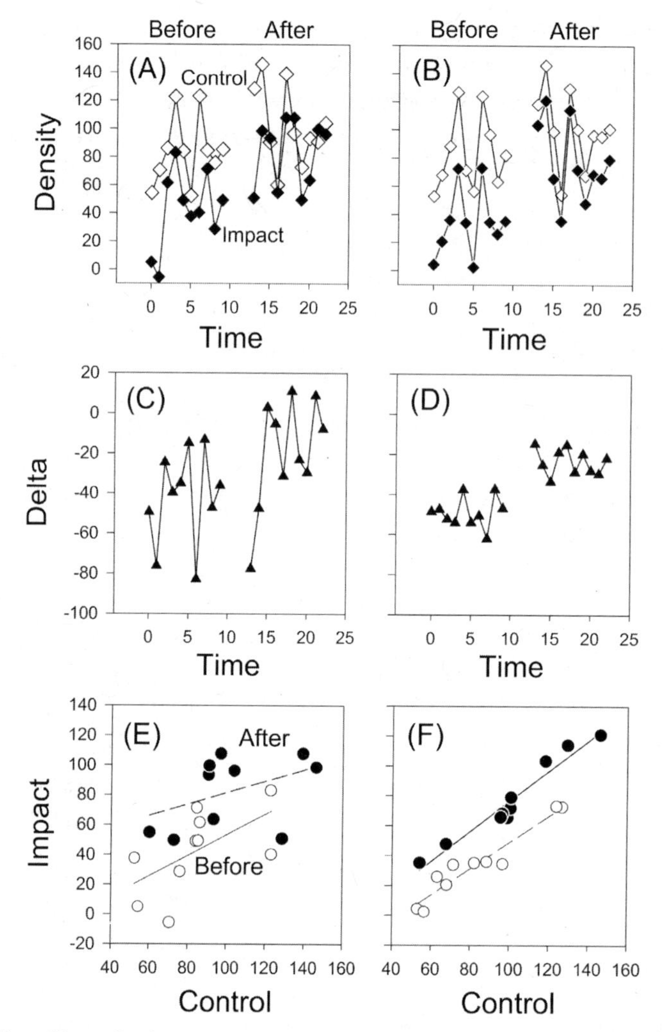

FIGURE 13.4 The effect of coherence between the Control and Impact site on the ability of the BACIPS and predictive-BACIPS designs to detect effects of an intervention. Coherence is the degree of strength of the correlation between the Control and Impact sites through time in the absence of a change in the status of the intervention (Magnuson et al. 1990; Osenberg et al. 1994). The panels on the left (A, C, and E) are for a system with relatively low coherence, whereas the panels on the right (B, D, and F) apply to a system with relatively high coherence. The data were simulated by constructing a time series from a random distribution and imposing a temporal trend in densities at both sites with a sine function. The variance in densities is the same under low and high coherence; the effect size is also identical (25). The only difference is the correlation between the two sites (high: $r = 0.99$; low: $r = 0.63$). The influence on inferences from BACIPS analyses is potentially dramatic. Effect sizes were estimated to be 25.6±6.6 (95% CI) for high coherence versus 21.9±24.9 for low coherence. Notice that under low coherence, the CI was very wide and included positive effects as well as deleterious effects; a *t*-test failed to reject the null hypothesis of no effect ($t_{18} = 1.91$, $P = 0.07$). For predictive-BACIPS, estimated effects and uncertainty were similarly affected (note difference in elevation and scatter in panels E and F, which give separate regression lines for the Before and After periods).

absence of a change in the intervention (i.e., the interaction between site and time *within* a period). Other designs use error terms based on the within-site sampling variation (Control-Impact and BACI designs) or temporal variation (Before-After design). Of course, inferences about cause and effect can be increased with ancillary studies of the mechanisms that might elicit change at the sites (e.g., Stewart-Oaten et al. 1992; Schroeter et al. 1993).

Case Studies: Marine Restoration Using Reserves

Marine reserves, or marine protected areas (MPAs), have been touted as a powerful tool to restore degraded marine systems, improve fisheries management, and conserve biodiversity. By limiting human activities, MPAs are thought to produce long-lasting increases in the density, size, diversity, and productivity of marine organisms within reserve boundaries due to decreased mortality and habitat destruction, as well as indirect ecosystem effects (e.g., Halpern 2003). Importantly, the effects of MPAs are hypothesized to extend beyond the boundaries of the MPA by "spillover": that is, via the density-dependent migration of juveniles or adults from inside to outside the MPA, or via increased production of planktonic larvae (spawned within the MPA), which are then exported outside of the MPA (e.g., Sanchez Lizaso et al. 2000). Thus, we expect both local and regional effects of MPAs. Indeed, it is the regional effect that is often used to motivate the designation of MPAs to the fishing community: reserves must enhance fisheries enough to compensate for the loss of fishing habitat (Palumbi 2000). Similar regional effects are expected in other conservation contexts, for example, by protecting the wintering grounds of a migratory bird or butterfly, effects should also arise in the breeding grounds.

Given the potential importance of MPAs as a restoration tool, many studies have examined effects of marine reserves on fishes and invertebrates and a recent meta-analysis by Halpern (2003) summarized those effects. We evaluated the designs of studies reviewed by Halpern and added additional studies by searching Web of Science for papers with the key words "marine protected area" or "marine reserve." We maintained the criteria for inclusion used by Halpern (2003): (1) data had to allow an inference about effects of the MPA; (2) measured variables had to include ecological responses (e.g., density or biomass); and (3) MPAs had to be "no-take" reserves. In total, we found 118 studies of MPA effects. Each study was conducted under various constraints (both political and scientific) and therefore the studies used different designs (e.g., CI versus BA versus BACI) and examined different scales of effects (i.e., local versus regional).

The majority of studies (70%) used a Control-Impact design to study local effects (Table 13.1). Fewer than 8% of the studies explored regional effects. No studies used a full BACIPS design with time series in both the Before and After periods (although some studies had time series in the After period and a single sample date in the Before period). Thus, not a single study used the most powerful assessment design (BACIPS) to study the regional effects that are of most interest to managers and often promoted by the scientific community.

Below, we look at several different approaches that have been taken, and highlight their limitations based on our previous generic discussions of assessment designs. We do this to emphasize the differences among the various study designs and their ability to look at appropriate scales of effects, and to inform future restoration studies, especially of marine reserves.

TABLE 13.1

Designs and scales of effects examined in studies of marine protected areas. Studies were obtained from Halpern's (2003) review and supplemented with further searches of the literature.

	Scale of study	
Design	Local	Regional
Control-Impact	82	3
Before-After	17	5
BACI	10	1
BACIPS	0	0

Control-Impact Studies

Because most of the studies that Halpern (2003) tabulated used a CI design to evaluate local effects (Table 13.1), we discuss Halpern's results in that context. Halpern achieved replication by combining the results from many unreplicated studies. Indeed, he observed strikingly consistent responses across the studies: for example, densities in reserves were 91% (95% CI: ~ 35–147%) greater than outside the reserves. He concluded that this consistent pattern was the result of a beneficial effect of MPAs on the densities of marine organisms. Increases also were observed for species richness (23%), organismal size (31%) and biomass (192%). Is there a reasonable alternative explanation to the appealing interpretation that the designation of MPAs has these beneficial effects?

In any single CI design the MPA effect is confounded with other factors whose effects vary spatially. Thus, we would expect the MPA to sometimes be placed in a "better" site and other times that the Control would go in the "better" site. On average, however, there should be no difference between the MPA and Control in the absence of an effect (assuming the MPA was assigned at random). Thus, the meta-analysis, which achieved replication by looking across studies, is comparable to a large-scale experiment (with MPA systems representing blocks, but lacking replication within blocks).

Of course, MPAs and Controls are not usually assigned randomly. Instead, MPAs are typically established following a laborious site selection process. Controls are rarely if ever discussed in the process; indeed, planning for a scientific assessment is rare. This is why CI designs are so common—the assessments are done after the fact, and the Control sites are often chosen by the investigator in a post hoc attempt to find sites that are otherwise "identical" to the MPA. Of course this is impossible. In most cases, MPAs (like most restoration sites) are put in specific sites—for example, the best remaining shallow coral reef habitat.

Thus, an alternative explanation for Halpern's result is that it reflects differences between the MPA and Control site that existed *prior* to the establishment of the MPA. Indeed, other meta-analyses indicate that the size of the reserve effect does not increase with time since the establishment of the MPA (Cote et al. 2001; Halpern and Warner 2002), suggesting a large role of initial conditions (but see Halpern and Warner 2002 for an alternative explanation). The problem, of course, is that the data cannot distinguish between the two alternatives. Hence, we are left either "believing" that MPAs are good and are in no better position than we were before the study was conducted or being skeptical and arguing that we need better data.

To further complicate inferences derived from such approaches, note that in the presence of regional effects, CI designs will underestimate true local effects because the Impact (MPA) site response will cause a concordant response at the Control site (i.e., they are not independent). Our hope is that by understanding the limits of even the best studies, such as Halpern's, we can ultimately obtain more defensible and less ambiguous interpretations. This requires Before data using designs conducted at appropriate scales.

Before-After Studies

Given the problems with site selection and possible non-independence between Control and Impact sites, why not simply avoid the use of Control sites all together and attempt to emulate the success of Box and Tiao (1975)? To explore this approach, we have extracted data from the studies of Russ and Alcala (1996, 2003) in the central Philippines. Although Russ and Alcala had Control sites, many of their inferences were based on patterns of change at two sites on Sumilon Island where fishing was "turned on and off" through time. As with most ecological studies, the data set is relatively sparse. We used these data (Equation 5a–b) to fit a model of fish dynamics that allowed us to estimate the effect of fishing:

$$N(t+1) = N(t) + (r + \varepsilon_r(t)) - (a + fF(t))N(t) \tag{5a}$$

$$N_{obs}(t) = N(t) + \varepsilon_{obs}(t) \tag{5b}$$

where $N(t)$ was the sampled density in year t; r was the average recruitment of new settlers into the local population and $\varepsilon_r(t)$ represents independent, normally distributed error with mean 0 and standard deviation σ_r; a is the background (nonfishing induced) mortality; f is the effect of fishing when it was allowed; $F(t)$ is the fraction of the year during which fishing was allowed (between 0 and 1); and $\varepsilon_{obs}(t)$ represents independent, normally distributed observation error with mean 0 and standard deviation σ_{obs}. We specified $N(0)$, the starting density of the population, as a parameter. When $\sigma_r = 0$, the other parameters can be estimated by simple least-squares fitting of the estimated population densities over time to the observed population densities, with σ_{obs} estimated from the residual sum of squares. To fit the model with process error ($\sigma_r > 0$), we ran many (up to 50,000) realizations of the population dynamics for a given set of parameters and used these realizations to compute the theoretical mean vector $\mathbf{m}$ of the observations as well as the variance-covariance matrix $\mathbf{V}$ among the observations. We then calculated the log-likelihood of the observed data given a multivariate normal distribution with mean $\mathbf{m}$ and variance-covariance matrix $\mathbf{V}$, and used a nonlinear fitting routine to maximize the log-likelihood. In practice, since estimates of standard errors were available for individual measurements, we determined σ_{obs} from the estimated sample standard error for a given census rather than trying to estimate this parameter from data. We used published data from 1983–2000 for the Sumilon Nonreserve (SNR) and from 1983–1994 for the Sumilon Reserve (SR). Limited fishing was permitted from 1995–2000 at SR, so we excluded this period of partial protection. Over these time periods SNR was opened for fishing except for 1987–1992 and therefore had a pattern of open-closed-open. SR was opened for fishing during two, approximately two-year, periods, and therefore had a closed-open-closed-open pattern of exploitation. These repeated "on-off" patterns potentially provide greater ability to detect interventions than the more standard single switch in Box and Tiao's study. We used estimates of f (the fishing effect) to infer effects of the MPA on fish dynamics.

When we included process error, we obtained estimates of $f = -0.15 \pm 0.20$ yr^{-1} (95% CI) for SNR, which overlapped zero and failed to distinguish between beneficial and deleterious effects, and $f = 0.60 \pm 0.25$ yr^{-1} for SR, which provided good evidence for a demonstrable effect of the MPA (i.e., an increase of ~60% per year in the growth of the fish population released from fishing). Indeed, the confidence intervals of the fishing effect at the two reserves do not overlap, suggesting heterogeneity in the efficacy of the reserves. However, estimates of σ_r were large (e.g., 3.4 for SNR), suggesting a major role of environmental variability due to recruitment, r. Without process error, it was difficult to reconcile the data from SNR with a biologically plausible model, due in part to the large fluctuations in density that occurred when the site was continually fished (Figure 13.5). In contrast, the fit of the SR data was quite good, even in the absence of process error (Figure 13.5).

Our approach assumes that all fluctuations in the growth rate r are independent (and that there is no variation in the effect of fishing, f) and thus ignores serial correlation in the process error (in Box and Tiao's terms, we are fitting the autoregressive part of the model and ignoring the moving-average terms). Accounting for the serial correlation should be done but would only make our estimates even more uncertain. Despite estimating a significant effect of fishing for one of the sites, we are dangerously short on data. We are trying to fit a model

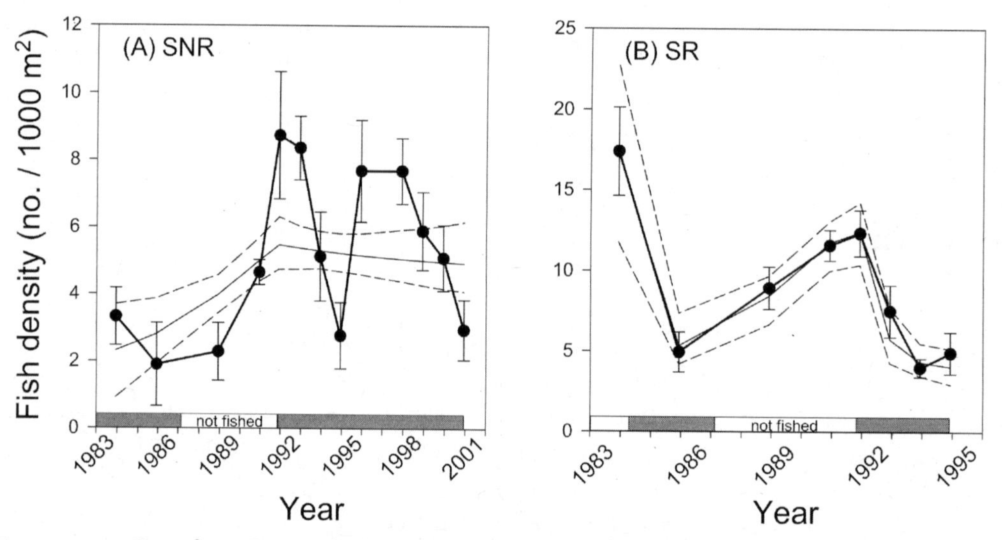

FIGURE 13.5 Data from Russ and Alcala's (2003) study of the response of large predatory fishes to the implementation of marine reserves in the Philippines. SNR and SR are two sites in which fishing was allowed or prohibited at different times between 1983 and 2000. Data points with error bars (±SE) give the observed fish densities. Solid lines without points give the predicted dynamics based on the mean of 1,000 simulations using parameter values drawn from the sampling distribution of the parameters (estimated from the curvature of the likelihood surface at the MLE): see equation 5a–b. The dashed lines bound 95% of all simulations. The simulation did not include process error (i.e., we set $\varepsilon_r = 0$) and thus the confidence bands reflect uncertainty in the parameter estimates and not temporal variation in the recruitment parameter.

with five parameters (two of them variances, which are notoriously hard to estimate) to 8 (or 13) data points in a time-series, which are not even independent of one another (and hence represent less than three, or eight, degrees of freedom). Indeed, most ecological data will not be sufficient in these regards. Furthermore, it will be difficult to develop detailed diagnostic checks and to compare alternate model formulations (e.g., functional forms, as well as error structure and serial correlation).

Unfortunately, Russ and Alcala's study is one of the best available with a fairly extensive time series by ecological standards. It helped that we were able to use a semimechanistic model, based on at least a caricature of a population growth model, that we had relatively detailed data on the interventions (=fishing intensity), and that the intervention fluctuated more than once (providing a stronger signal to pull out from the noise). Despite doing better than we initially expected (being able to pull out a signal at all), the estimates of fishing effects were uncertain. Can we do better?

Before-After–Control-Impact Studies and Spatial Scale

Although there are not any well-designed BACIPS regional studies, there are several that have elements of a BACIPS design. Here we discuss an important study by Roberts et al. (2001) on the St. Lucia reserve network in the Caribbean. Roberts et al. focused on regional effects on fisheries. Although they did not present a formal model, they did present data and results from a statistical analysis relevant to assessment designs. They used data collected once prior to the establishment of the MPA and four times during the subsequent five years. Data on fish biomass were taken inside and outside the reserve (Figure 13.6). They analyzed the data using a statistical model that included time, location (MPA versus outside of MPA),

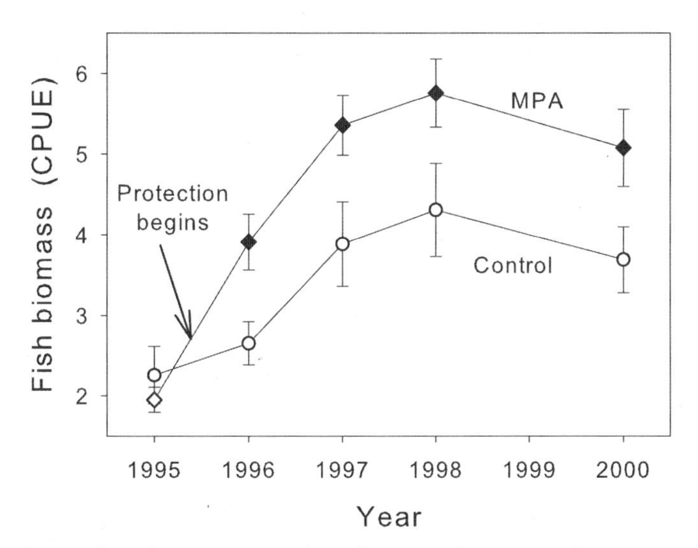

FIGURE 13.6 Relative abundance (CPUE, based on visual counts) of commercially important fishes from the Roberts et al. (2001) study of the MPA network on St. Lucia in the Caribbean. One survey date was available prior to, and four after, establishment of the MPA. Samples were taken inside (MPA) and outside (Control) the protected areas.

and a time-by-location interaction. The interaction was nonsignificant, suggesting little evidence for a differential change with time at the two sites. Taken alone, these data would argue against a local enhancement, but Roberts et al. were interested primarily in the regional effects, so their working hypothesis was that densities both inside and outside the reserve would increase, which they observed.

However, the fish responses observed by Roberts et al. apparently occurred within a year of establishment of the MPA network (Figure 13.6). This rapid response may be plausible for local effects where migration of fishes into MPAs may exaggerate local effects (e.g., note the relatively quick response suggested in Figure 13.5). It seems implausible, however, that *regional* effects would be manifest in a year's time, because they arise primarily through enhanced larval production from increased *adult* stocks or density-dependent movement, probably of older life stages (Russ and Alcala 2003; Russ et al. 2004).

An alternative explanation for this result, as was noted by the authors, is that there was another factor that caused the regional increase in fish stocks. Because fish stocks fluctuate for many reasons, and each Control site was within ~1 km of the nearest reserve, a common response of the Control and Impact sites to another factor is plausible.

The two competing hypotheses (regional effects of the MPA versus other factors) cannot be distinguished with the available data. Interviews suggested that fishers thought the MPA had worked; however, the fishers might have reasonably, but perhaps falsely, inferred an effect based on the general increase in fish biomass (no matter the cause). The authors observed in a footnote that they had no evidence for similar increases in fish abundance on nearby islands, although quantitative data were not collected. These interviews illustrate a more suitable approach: what is needed is an *appropriate* Control instead of nearby sites that are expected to be influenced by the MPA (see below and Box 13.1).

Box 13.1

Case Study: The Application of BACIPS to Lagoonal Fisheries

Local and Regional BACIPS: BACIPS designs can be used to assess both local (Figure A) and regional (Figure B) effects of marine reserves (MPA) on lagoonal fisheries. Local and regional designs are distinguished by the location of the Control site(s). The regional assess-

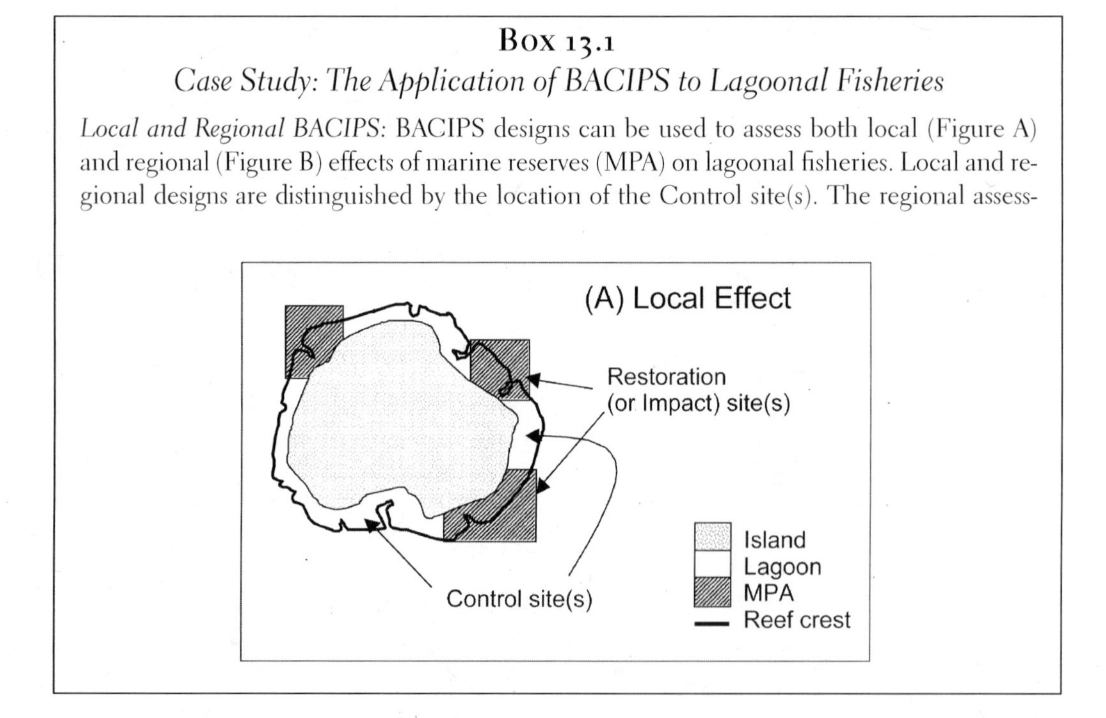

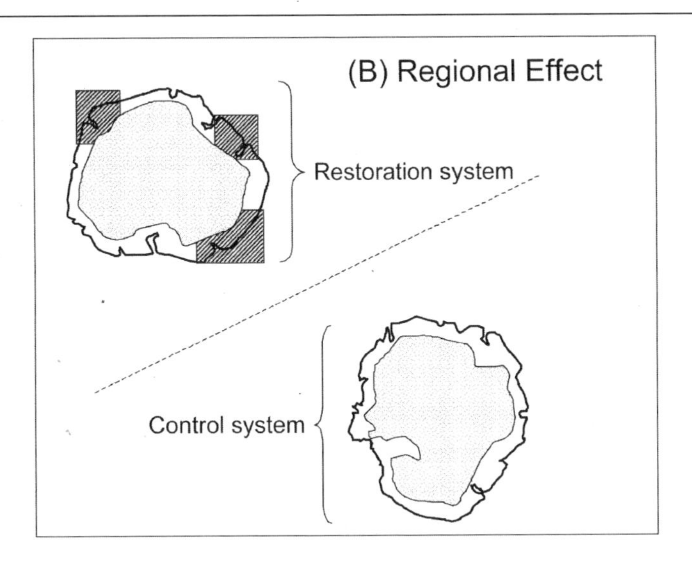

ment design (B) has never been used in any system but is feasible for lagoonal fisheries because local retention of larvae is thought to be high (Planes et al. 1993, 1998; Bernardi et al. 2001) and fishing effort localized. Thus, two adjacent islands may have fairly independent lagoonal fisheries yet may be close enough to one another to be affected similarly by oceanographic and weather conditions. One island could serve as the Impact (receiving an MPA network within its lagoons) and one island could serve as a Control (lacking MPAs). Monitoring of population densities, size-structure, and fisheries yields in the two islands, Before and After implementation of the MPA, would provide a test of regional effects of MPAs on fisheries. The expectation is that the fishing yields (and stocks) outside of the MPA, but on the island with an MPA network, would increase relative to the Control island, despite removal of habitat from the fishers.

Similar opportunities likely exist in other restoration projects that may have regional-scale effects, such as prairie restorations designed to rescue other nearby habitats; habitat protection programs for migrating birds or butterflies that affect dynamics on other continents; or fire management regimes that promote local diversity and thus enhance the regional pool and therefore the richness of sites outside of the managed areas.

Coordinated Assessments: X-BACIPS. Consider ten islands in the Indo-Pacific, with five receiving MPA networks and five others remaining as Controls (as in Figure B). If MPAs were assigned at random and each site sampled after MPA enforcement, the resulting data could be analyzed using a standard experimental approach. Although this design could discern average effects, it could not estimate effects of MPAs on any particular island. Instead, if each island pair was sampled Before and After MPA enforcement, then inferences could be made about individual islands (as in assessment designs) and about the population of MPAs as a whole (as with a standard experimental design). Mean effects and variances could be estimated, for example, using mixed-model meta-analysis (Gurevitch and Hedges 1999; Osenberg et al. 1999) or maximum likelihood. Due to the combination of experimental and assessment approaches, we term this design "X-BACIPS."

Regional Assessments: BACIPS with an Appropriate Control

Studies of MPAs highlight the need to better match the spatial scale of interest to the assessment design. To study regional effects of MPAs (or any other restoration effort), we require not a comparison of inside versus outside the MPA, but instead comparison of a region *with* an MPA network with a region *lacking* an MPA (Box 13.1; see Russ 2002 for an alternate design). At least two major scientific problems are likely to arise in implementing such a study: (1) the spatial scale of movement of organisms can be large, suggesting that spillover effects will be fairly dispersed in time and will be hard to detect; and (2) large-scale movement will also require that the Control site be located a sufficient distance from the region with the MPA, thus reducing coherence and the power of the resulting analysis. Fortunately, there are some systems in which such a design is feasible (Box 13.1), and it is imperative that we take advantage of these opportunities.

Summary of Lessons Learned from MPAs

Roberts et al. (2001) and Halpern (2003) are among the best of all available studies of MPA effects. They were constrained by the available data and absence of key design features (such as random assignment, suitable Controls, and Before data). However, if these studies of restoration effects lead, at best, to equivocal results, then it is clear that additional studies (with poorer designs) will lead to even greater equivocation. Thus, we need a better approach (not just more of the same). This is not unique to restoration of marine systems but constrains the assessment of most large-scale restoration projects.

Future Directions

Future assessments will be enhanced through the use of better designs, such as BACIPS, and improved statistical tools. Although statistical tools are important, we believe design issues and the increased use of better designs are even more critical. We conclude with a brief discussion of one analytic tool (Bayesian approaches combined with meta-analysis) that we believe is relevant and provide a final discussion about the application of BACIPS.

Bayesian analyses are increasingly common in the ecological literature, although many ecologists avoid them because of concerns about the subjectivity of the prior distribution of effect sizes. As better assessment studies accumulate (with unbiased estimates of effect sizes and variances), mixed model meta-analyses can be used to quantify the distributions of effect sizes (Gurevitch and Hedges 1999; Osenberg et al. 1999). This distribution, useful in its own right, also can be used to define the prior distribution in a later Bayesian analysis of a new restoration project. Of course, each restoration setting is unique, and the number of potential BACIPS studies will likely be small for many types of restoration activities. Thus, the prior distribution itself will be estimated poorly and should probably carry relatively little weight (obviating one advantage of the Bayesian method).

Crome et al (1996) took a different Bayesian approach and used interviews with different parties involved in forestry practices to assess how much their initial opinions (the priors) would change (as reflected in the posteriors) by a scientific study based on a BACIPS design. This was an innovative way to incorporate an assessment into a public policy arena, attempting to gauge the interaction between public opinion and scientific information. If opposing sides of an environmental issue are sufficiently intransigent, then even a well-designed scientific study may do little to bring the groups to consensus. Of course, we already may be in such a position today, because past studies used poor designs and led to debate among scientists. In such instances, the public has had little choice but to ignore the science and base their opinions on other matters, like economics.

In far too many cases (as illustrated for marine reserves) assessments of restoration projects are post hoc and lack Before data and, therefore, are open to alternate interpretations. As a result, these scientific studies do little to inform the science or public policy. Moving beyond BA, CI, and BACI-single-sample designs and toward greater reliance on BACIPS designs will not be trivial. Successful application of BACIPS requires planning. For politically charged projects (like MPAs), the science often takes a back seat to social considerations. Sites may be relocated several times during the planning phases. This may prevent the collection of Before data from appropriate sites. However, in many cases, candidate sites are known. Sampling can be done at several sites during the planning phase. One of these will likely become the restoration site (e.g., MPA); the others could be used as Control(s). Some sites may not track the restoration site well and can be dropped later from the study (for discussion of some of these issues, see Stewart-Oaten 1996b). Of course, conducting such a "risky" study requires foresight on the part of funding agencies. This is sometimes possible (Piltz 1996 and Ambrose et al. 1996 give two nice examples). However, if the scientific community does not aspire to conduct BACIPS studies, then regulatory and funding agencies will never support them. By recognizing the limitations of existing studies, we hope to facilitate the execution of better-designed and more informative studies that will lead to the development of more effective, large-scale, restoration activities.

Summary

Rigorous statistical evaluation and sound inference of restoration efforts is difficult to achieve. As a result, quantitative assessments are often missing, incomplete, or misinterpreted. Appropriate analyses must be applied within the broader context of the study design and the limitations of these designs evaluated within the context of the restoration goals (including spatial scale). We presented the central statistical concepts relevant to restoration evaluation and contrasted the strengths and weaknesses of possible approaches, including the Control-Impact, Before-After, and Before-After–Control-Impact designs. We advocate the use of Before-After–Control-Impact Paired Series design because it can be applied to spatially unreplicated interventions, is site-specific, does not require random assignment of sites, and yields defensible estimates of the effect of the restoration activity (rather than *P*-values from null-hypothesis tests). We illustrated advantages and limitations of different approaches through a discussion of studies of marine protected areas and closed by proposing future directions, including the use of more appropriate designs that will address current shortcomings and enhance the practice of restoration ecology.

Acknowledgments

We thank Ray Hilborn, René Galzin, and Caroline Vieux for helpful discussion; Tom Adam and Jackie Wilson for help compiling the MPA literature; Margaret Palmer, Don Falk, and

Joy Zedler for useful comments; and the Minerals Management Service, Florida and National Sea Grant programs, and the NSF (OCE-0242312) for support that contributed to the development of these ideas.

Literature Cited

Allison, G. W., J. Lubchenco, and M. H. Carr. 1998. Marine reserves are necessary but not sufficient for marine conservation. *Ecological Applications* 8:S79–S92.

Ambrose, R. F., R. J. Schmitt, and C. W. Osenberg. 1996. Predicted and observed environmental impacts: Can we foretell ecological change? In *Detecting ecological impacts: Concepts and applications in coastal habitats*, ed. R. J. Schmitt and C. W. Osenberg, 343–367. San Diego: Academic Press.

Anderson, D. R., K. P. Burnham, and W. L. Thompson. 2000. Null hypothesis testing: Problems, prevalence, and an alternative. *Journal of Wildlife Management* 64:912–923.

Bence, J. R. 1995. Analysis of short time series: Correcting for autocorrelation. *Ecology* 76:628–639.

Bence, J. R., A. Stewart-Oaten, and S. C. Schroeter. 1996. Estimating the size of an effect from a before-after–control-impact paired series design: The predictive approach applied to a power plant study. In *Detecting ecological impacts: Concepts and applications in coastal habitats*, ed. R. J. Schmitt and C. W. Osenberg, 133–149. San Diego: Academic Press.

Bernardi, G., S. J. Holbrook, and R. J. Schmitt. 2001. Gene flow at three spatial scales in a coral reef fish, the three spot dascyllus, *Dascyllus trimaculatus*. *Marine Biology* 138:457–465.

Box, G. E. P., and G. C. Tiao. 1975. Intervention analysis with applications to economic and environmental problems. *Journal of the American Statistical Association* 70:70–79.

Cote, I.M., I. Mosqueira, and J. D. Reynolds. 2001. Effects of marine reserve characteristics on the protection of fish populations: A meta-analysis. *Journal of Fish Biology* 59 (suppl. A): 178–189.

Crome, F. H. J., M. R. Thomas, and L. A. Moore. 1996. A novel Bayesian approach to assessing impacts of rain forest logging. *Ecological Applications* 6:1104–1123.

Earn, D. J. D., P. Rohani, B. M. Bolker, and B. T. Grenfell. 2000. A simple model for complex dynamical transitions in epidemics. *Science* 287:667–670.

Englund G., and S. D. Cooper. 2003. Scale effects and extrapolation in ecological experiments. *Advances in Ecological Research* 33:161–213.

Green, R. H. 1979. *Sampling design and statistical methods for environmental biologists*. New York: John Wiley & Sons.

Gurevitch, J., and L. V. Hedges. 1999. Statistical issues in ecological meta-analyses. *Ecology* 80:1142–1149.

Halpern, B. S. 2003. The impact of marine reserves: Do reserves work and does reserve size matter? *Ecological Applications* 13:S117–S137.

Halpern, B. S., and R. R. Warner. 2002. Marine reserves have rapid and lasting effects. *Ecology Letters* 5:361–366.

Johnson, D. M. 1999. The insignificance of statistical significance testing. *Journal of Wildlife Management* 63:763–772.

Lubchenco, J., S. R. Palumbi, S. D. Gaines, and S. Andelman. 2003. Plugging a hole in the ocean: The emerging science of marine reserves. *Ecological Applications* 13:S3–S7.

Magnuson, J. J., B. J. Benson, and T. K. Kratz. 1990. Temporal coherence in the limnology of a suite of lakes in Wisconsin, U.S.A. *Freshwater Biology* 23:145–159.

Mapstone, B. D. 1995. Scalable decision rules for environmental impact studies—Effect size, type-I, and type-II errors. *Ecological Applications* 5:401–410.

Melbourne, B. A., and P. Chesson. 2005. Scaling up population dynamics: Integrating theory and data. *Oecologia* 145:179–187.

Michener, W. K. 1997. Quantitatively evaluating restoration experiments: Research design, statistical analysis and data management considerations. *Restoration Ecology* 5:324–337.

Murdoch, W. W., B. Mechalas, and R. C. Fay. 1989. *Final report of the Marine Review Committee to the California Coastal Commission on the effects of the San Onofre Nuclear Generating Station on the marine environment*. San Francisco: California Coastal Commission.

Murtaugh, P. A. 2002. On rejection rates of paired intervention analysis. *Ecology* 83:1752–1761.

Murtaugh, P. A. 2003. On rejection rates of paired intervention analysis: Reply. *Ecology* 84:2799–2802.

Norse, E. A., C. B. Grimes, S. V. Ralston, R. Hilborn, J. C. Castilla, S. R. Palumbi, D. Fraser, and P. Kareiva. 2003. Marine reserves: The best options for our oceans (Forum). *Frontiers in Ecology and the Environment* 1:495–502.

Osenberg, C. W., O. Sarnelle, S. D. Cooper, and R. D. Holt. 1999. Resolving ecological questions through meta-analysis: Goals, metrics and models. *Ecology* 80:1105–1117.

Osenberg, C. W., and R. J. Schmitt. 1996. Detecting ecological impacts caused by human activities. In *Detecting ecological impacts: Concepts and applications in coastal habitats*, ed. R. J. Schmitt and C. W. Osenberg, 3–16. San Diego: Academic Press.

Osenberg, C. W., R. J. Schmitt, S. J. Holbrook, K. E. Abu-Saba, and A. R. Flegal. 1994. Detection of environmental impacts: Natural variability, effect size, and power analysis. *Ecological Applications* 4:16–30.

Osenberg, C. W., R. J. Schmitt, S. J. Holbrook, and D. Canestro. 1992. Spatial scale of ecological effects associated with an open coast discharge of produced water. In *Produced water: Technological/environmental issues and solutions*, ed. J. P. Ray and F. R. Englehardt, 387–402. New York: Plenum Publishing.

Osenberg, C. W., C. M. St. Mary, R. J. Schmitt, S. J. Holbrook, P. Chesson, B. Byrne. 2002. Rethinking ecological inference: Density dependence in reef fishes. *Ecology Letters* 5:715–721.

Palumbi, S. R. 2000. The ecology of marine protected areas. In *Marine ecology: The new synthesis*, ed. M. D. Bertness, 509–530. Sunderland, MA: Sinauer Associates.

Piltz, F. M. 1996. Organization constraints on environmental impact assessment research. In *Detecting ecological impacts: Concepts and applications in coastal habitats*, ed. R. J. Schmitt and C. W. Osenberg, 335–345. San Diego: Academic Press.

Planes, S. 1993. Genetic differentiation in relation to restricted larval dispersal of the convict surgeonfish *Acanthurus triostegus* in French-Polynesia. *Marine Ecology Progress Series* 98 (3): 237–246.

Planes, S., P. Romans, and R. Lecomte-Finiger. 1998. Genetic evidence of closed life-cycles for some coral reef fishes within Taiaro Lagoon (Tuamotu Archipelago, French Polynesia). *Coral Reefs* 17:9–14.

Roberts, C. M., J. A. Bohnsack, F. Gell, J. P. Hawkins, and R. Goodridge. 2001. Effects of marine reserves on adjacent fisheries. *Science* 294:1920–1923.

Russ, G. R. 2002. Yet another review of marine reserves as reef fishery management tools. In *Coral reef fishes: Dynamics and diversity in a complex ecosystem*, ed. P. F. Sale, 421–443. San Diego: Academic Press.

Russ, G. R., and A. C. Alcala. 1996. Marine reserves: Rates and patterns of recovery and decline of large predatory fish. *Ecological Applications* 6:947–961.

Russ, G. R., and A. C. Alcala. 2003. Marine reserves: Rates and patterns of recovery and decline of predator fish, 1983–2000. *Ecological Applications* 13:1553–1565.

Russ, G. R., A. C. Alcala, A. P. Maypa, H. P. Calumpong, and A. T. White. 2004. Marine reserve benefits local fisheries. *Ecological Applications* 14:597–606.

Sanchez Lizaso, J. L., R. Goni, O. Renones, J. A. Garcia Charton, R. Galzin, J. T. Bayle, P. Sanchez Jerez, A. Perez Ruzafa, and A. A. Ramos. 2000. Density dependence in marine protected populations: A review. *Environmental Conservation* 27:144–158.

Scheiner, S. M., and J. Gurevitch, editors. 2001. *Design and analysis of ecological experiments*. New York: Oxford University Press.

Schindler D. W. 1998. Replication versus realism: The need for ecosystem-scale experiments. *Ecosystems* 1:323–334.

Schmitt, R. J., and C. W. Osenberg, editors and contributing authors. 1996. *Detecting ecological impacts: Concepts and applications in coastal habitats*. San Diego: Academic Press.

Schmitz, O. 2005. Scaling from plot experiments to landscapes: Studying grasshoppers to inform forest ecosystem management. *Oecologia* 145:225–234.

Schreuder, H. T., R. Ernst, and H. Ramirez-Maldonado. 2004. *Statistical techniques for sampling and monitoring natural resources*. General Technical Report RMRS-GTR-126. Rocky Mountain Research Station, Fort Collins: USDA, Forest Service. 111 pp. http://www.treesearch.fs.fed.us/.

Schroeter, S. C., J. D. Dixon, J. Kastendiek, R. O. Smith, and J. R. Bence. 1993. Effects of the cooling system for a coastal power plant on kelp forest invertebrates. *Ecological Applications* 3:331–350.

Stewart-Oaten, A. 1996a. Goals in environmental monitoring. In *Detecting ecological impacts: Concepts and applications in coastal habitats*, ed. R. J. Schmitt and C. W. Osenberg, 17–27. San Diego: Academic Press.

Stewart-Oaten, A. 1996b. Problems in the analysis of environmental monitoring data. In *Detecting ecological impacts: Concepts and applications in coastal habitats*, ed. R. J. Schmitt and C. W. Osenberg, 109–131. San Diego: Academic Press.

Stewart-Oaten, A., and J. R. Bence. 2001. Temporal and spatial variation in environmental impact assessment. *Ecological Monographs* 71:305–339.

Stewart-Oaten, A., J. R. Bence, and C. W. Osenberg. 1992. Detecting effects of unreplicated perturbations: No simple solution. *Ecology* 73:1396–1404.

Stewart-Oaten, A., W. W. Murdoch, and K. R. Parker. 1986. Environmental impact assessment: "Pseudoreplication" in time? *Ecology* 67:929–940.

Underwood, A. J. 1991. Beyond BACI: Experimental designs for detecting human environmental impacts on temporal variations in natural populations. *Australian Journal of Marine and Freshwater Research* 42:569–587.

Underwood, A. J. 1992. Beyond BACI: The detection of environmental impacts on populations in the real, but variable, world. *Journal of Experimental Marine Biology and Ecology* 161:145–178.

Underwood, A. J. 1994. On beyond BACI: Sampling designs that might reliably detect environmental disturbances. *Ecological Applications* 4:3–15.

Underwood, A. J. 1997. *Experiments in ecology: Their logical design and interpretation using analysis of variance*. New York: Cambridge University Press.

Yoccoz, N. G. 1991. Use, overuse and misuse of significance tests in evolutionary biology and ecology. *Bulletin of the Ecological Society of America* 72:106–111.

Chapter 14

Ecological Restoration from a Macroscopic Perspective

Brian A. Maurer

Restoration implies returning an ecological system to a configuration that approximates its state prior to its alteration by human activities (SER 2002). Determining what this state is can be difficult for several reasons. First, ecological systems are constantly in flux, with changes occurring on a variety of spatial and temporal scales. These changes make it difficult to delineate natural boundaries that existed prior to human impact. Furthermore, ecological dynamics and change occur at different rates for different processes. For example, some population processes might operate on very different temporal and spatial scales than biogeochemical processes. Second, the original state that an ecological system occupied previous to human incursions might have existed at a much larger spatial extent then is available to the restored system. This means that the biotic and abiotic context within which the restored system must operate may be very different than the original context. Third, ecological systems may have many alternative steady states at which they can operate, some of which may not be accessible to the restored system due to historical contingencies or changed boundary conditions. Finally, for many of the reasons just listed, the original components that made up the restored ecological system may be missing. The resulting configuration of the "restored" system may be quite different from its original state in terms of ecological functioning and biodiversity. The restored system is also likely to be simpler both ecologically and genetically for reasons that will become clearer below.

Many of these concerns can best be understood by taking a macroscopic perspective regarding how the ecological systems function. An emerging paradigm, often referred to as "macroecology," is an attempt to deal explicitly with the structure and functions of ecosystems on geographic spatial scales (Brown 1995; Gaston and Blackburn 2000). By examining the properties of large collections of species on continents, macroecology provides the empirical basis for developing new insights into the structure of biological diversity. Because of its focus on processes operating on large spatial scales, macroecology explicitly assumes that the spatial and temporal scales of ecological systems extend far beyond political, geographical, and functional boundaries within which these systems are often managed. In this chapter, I'll examine the implications of this paradigm for the practical aspects of ecological restoration. Macroecology can provide a guide for understanding the context within which ecological restoration efforts must operate and the limitations that are imposed by restricting the spatial extent of restoration efforts. One of the major questions currently confronting macroecology is the extent to which large-scale patterns are determined by simple stochastic processes that assume all species are ecologically equivalent (Hubbell 2001; Maurer and McGill 2004). This question strikes at the heart of restoration ecology, and accumulated results from studies of successful restoration efforts may shed light on how equivalent species really are.

Macroecology began as an attempt to explain patterns in geographical distribution, abundance, and body size among many species inhabiting continents (Brown and Maurer 1987, 1989). It was quickly realized that the explanations for such patterns required an expansion of perspective from the focus on local ecological processes to larger, global-scale processes (Ricklefs 1987; Brown 1995; Brown 1999; Maurer 1999; Gaston and Blackburn 2000; Price 2003; Ricklefs 2004). These insights were reinforced by advances in a number of other fields, including studies of species diversity (Rosenzweig 1995), community ecology (Ricklefs and Schluter 1993), biogeochemistry (Schlesinger 1997), global ecology (Rambler et al. 1989; Peters and Lovejoy 1992; Kareiva et al. 1993; Southwick 1996), and biogeography (Brown and Lomolino 1998; Lomolino 2000b, 2000c; Hubbell 2001). Essentially, ecologists found that it was necessary to expand the temporal and spatial scales at which they viewed ecological systems in order to understand what processes were important in determining patterns in distribution, abundance, ecosystems processes, and species diversity.

This focus on the need for a global perspective on ecological systems has been mirrored in conservation efforts that focus on continental- and global-scale biological diversity (Scott et al. 1993; Heywood et al. 1995; Soulé et al. 1999). Such assessments and protocols call for an expanded agenda for the preservation of ecological systems. Within such a context, restoration of individual ecological systems is clearly an integral part of global conservation. But, in focusing on the local scale, it is possible that the large-scale context, constraints, and process requirements that are needed to truly restore a local system may be neglected. Hence, it is important to examine the nature of large-scale effects on local-scale ecological structures.

Given these concerns with the widespread nature of ecosystem changes, it is important to examine the concept of ecological restoration from a large-scale perspective. In this chapter, I address three issues from such a perspective. First, I will examine the effects of system size and ecological processes. As the area occupied by an ecosystem becomes smaller and more isolated from similar ecosystems, a number of changes occur in the inherent dynamics of the system. I will review what some of these changes are thought to be and how they might influence restoration goals. This discussion leads into a consideration of the role of external processes in determining local ecosystem structure and function. If an ecosystem becomes sufficiently isolated from other ecosystems, the exchanges of materials, organisms, and energy that normally would occur are disrupted, and these disruptions should have important consequences for the way the ecosystem behaves. Finally, I will consider the question of the scale at which long-term progress in the restoration of ecological systems might be expected.

How System Size Affects Ecological Processes

Biological systems of all kinds vary functionally with size (Brown and West 2000; Brown et al. 2002). It is fairly straightforward to understand the consequences of size variation when examining the properties of individual organisms (Peters 1983; Calder 1984; Schmidt-Nielsen

1984). As organisms grow larger, physiological processes often change in a nonlinear fashion as a consequence of the fractal nature of organismal structure (West et al. 1997, 1999a). These changes in physiological processes with size have a number of profound implications for the structure of ecological systems (Enquist et al. 1998, 1999a; Enquist et al. 1999b; West et al. 1999b; Gillooly et al. 2001). Although much is known about organismal scaling, much less is known about the way that population, community, and ecosystem processes change with ecosystem size. The seminal contribution to our understanding of such processes was the publication of the equilibrium theory of island biogeography (MacArthur and Wilson 1963, 1967). After considering this theory and its extensions, I will expand the discussion to consider a variety of other effects that ecosystem size has on ecological processes.

Ecosystem Size and Species Diversity

Since the dawn of modern ecology in the past century, ecologists have sought to understand how the size of an ecosystem determines the number of species inhabiting it (Rosenzweig 1995). The pioneering work of MacArthur and Wilson (1963, 1967) explained the number of species in an ecosystem of a given size as a dynamic balance between immigration of new species into the system and local extinction of species already residing in the system. Although capable of predicting some aspects of species-area relationships (SAR), the MacArthur-Wilson model seems to be too simplistic to explain adequately how species richness varies with ecosystem size (Brown 1986; Lomolino 1994, 2000a, 2000b, 2000c; Lomolino et al. 1995). Clearly, the mechanisms underlying SARs on continents and island archipelagos must include both dispersal and population viability, as MacArthur and Wilson (1967) envisioned. But the processes that generate the population rates that determine viability and dispersal depend on a large number of complexly interacting mechanisms. These mechanisms include, among other things, the ecological attributes of individual organisms, the abundance and variety of resources, the spatial patterns of those resources, and the spatial context in which the ecosystem exists.

Taking a large-scale perspective on the processes that generate SARs provides a different perspective that in some ways overcomes the problems with more mechanism-based explanations. The basic idea is that SARs are generated as a consequence of the overlapping distributions of species in geographic space (Maurer 1999; McGill and Collins 2003). To explain why each species has a characteristic geographic distribution is difficult if one focuses on the particular mechanisms limiting each species. However, if the patterns in demography of species across their ranges are examined, the myriad causes underlying SARs can be condensed into simpler models describing population mechanisms responsible for SARs (Maurer 1999; Hubbell 2001; Maurer and Taper 2002; McGill and Collins 2003). Consider the following simple model for distributions of species in space (Maurer 1999; McGill and Collins 2003). Suppose species are distributed across space in a unimodal manner (Figure 14.1). Each species has a different-sized geographic range, some with larger ranges, others with smaller ranges. At any given point in space, this results in a skewed distribution of abundance. The SAR resulting from this pattern is similar to empirical patterns seen in many collections of species (Figure 14.2). This approach can be expanded to examine SARs at different geographic scales, leading to the prediction that SAR exponents will vary with geographical scale (Rosenzweig 1995).

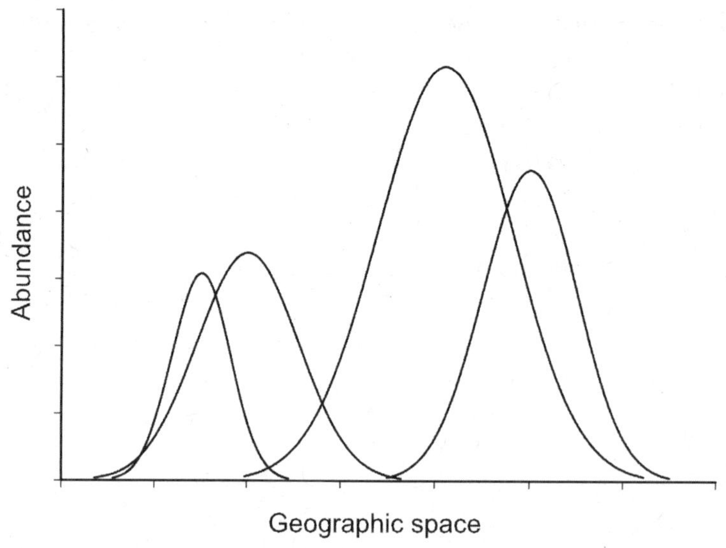

FIGURE 14.1 Schematic representation of the distribution of species in geographic space hypothesized to be responsible for species-area relationships.

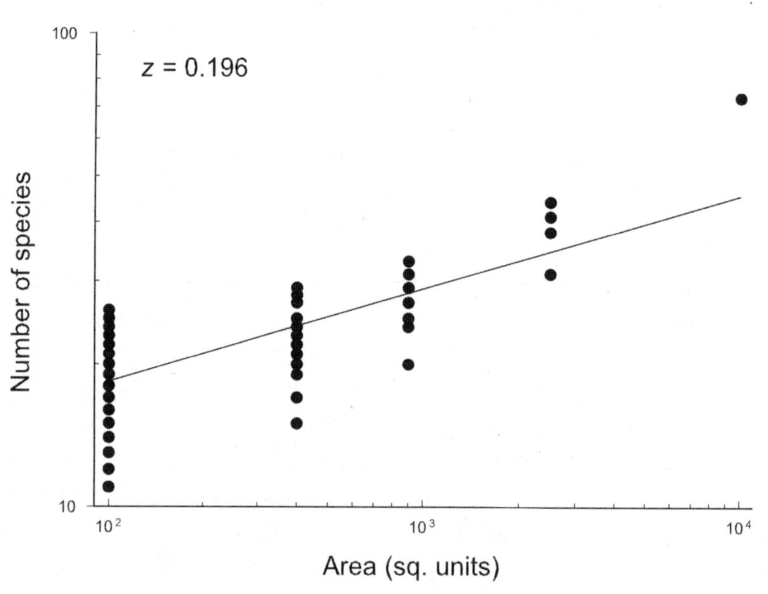

FIGURE 14.2 Species-area relationship derived from a simulation of a two-dimensional version of the model represented in Figure 14.1.

Of particular relevance to ecological restoration is the observation that the smallest islands often depart from the SAR for an archipelago (MacArthur and Wilson 1967; Brown and Lomolino 1998). Many such islands are too small to maintain viable populations of any species and therefore must be solely maintained by immigration alone. This observation has important implications for ecological restoration projects. Small areas that are being restored may require close proximity to a source of colonization, if the goal is to maintain the species diversity of a larger ecosystem. Many local populations persist as parts of larger metapopulations (Hanski 1998a, 1998b, 1999). If the size of the area to be restored is too small, or the area is too isolated from colonization sources, then the likelihood of maintaining the original species richness and diversity of the restored ecosystem may be low (Maschinski, this volume). These consequences regarding species diversity have been known and debated for many years (Brown and Lomolino 1998; Whittaker 1998). What is less clear is whether there are additional properties of ecosystems that are affected by ecosystem size.

Ecosystem Size and Ecosystem Function

If ecosystem size has important consequences for species diversity, then it follows that the ecosystem processes in which those species participate might be altered as well by ecosystem size. This raises the question, however, of how species diversity impacts ecosystem function. The relationship between species richness and ecosystem function has been a controversial topic that has not yet given rise to many clear-cut principles. On the one hand, since species are each unique, one might expect that the ecosystem function that the species served would be unique. If this were true, then loss of one species could not be offset in any way by other species. On the other hand, species may fill redundant functions in an ecosystem, as suggested by recent theories of ecosystem dynamics (Bell 2000, 2001; Hubbell 2001; Hubbell and Lake 2003; Volkov et al. 2003). Experiments designed to distinguish between these two alternatives have been equivocal (Huston 1997, 1999; Kinzig et al. 2001; Loreau et al. 2001; Naeem, this volume). Moreover, the assumption of functional equivalence of species in ecosystems, as required by neutral theories, remains problematic (Chave 2004).

One of the most ambitious attempts to examine ecosystem structure and function is the Biosphere 2 experiment (Marino and Odum 1999). The idea behind the experiment was to enclose eight humans, four small ecosystems, and an intensive agricultural plot in an airtight compartment for an extended period of time. The facility covered 1.27 ha and had roughly 180,000 m^3 of atmosphere. Each ecosystem was inoculated with a specified number of species of plants and animals and then left to "self-organize" over time. The major result of this experiment was a significant deviation of gas concentrations in the biosphere's atmosphere from its original state. Carbon dioxide concentrations increased to high levels, while oxygen levels dropped. Eventually oxygen concentrations in the atmosphere were increased by injection of pure oxygen from outside. The basic lesson from this part of the experiment is that there was not enough atmosphere in Biosphere 2 to buffer the system against relatively small perturbations (Maurer 1999). The missing oxygen turned out to have been absorbed by concrete structures within the biosphere (Dempster 1999). The sheer size of the Earth's atmosphere ensures that such oxygen sinks have relatively minor effects on the amount of oxygen available for ecosystem respiration. But at the scale of Biosphere 2, the reaction of oxygen with concrete became a major limitation on the oxygen cycle of the ecosystem, so much so that it was not self-sustaining.

A second important result from the Biosphere 2 experiment was that species richness in most of the ecosystems was reduced (although ecosystems were "overstocked" to allow "self-organization"). For example, in the tropical rainforest biome of Biosphere 2, species richness of plants declined 39‰ from over 282 species to 172 species after two years (Leigh et al. 1999). Similar reductions occurred in other biomes in the structure. Although apparently there was reproduction of some plant species within the tropical rain forest biome, it is not clear that sustainable populations would grow there over long periods of time. Intensive human intervention was also required to maintain forest structure and function.

External Transport Processes and Ecosystem Function

The results of the Biosphere 2 experiment suggest that much of what happens in a local ecosystem is caused by fluxes and flows across its boundaries. When those boundaries are closed, as occurred in Biosphere 2, the resulting system operates in a very different manner than a similar-sized ecosystem that is open to flows from the rest of the Earth's biosphere. It is clear that the scientists and engineers involved in Biosphere 2 never anticipated all that happened in the ecosystem, despite considerable expertise available during the construction and operation of the facility. An important consideration for ecological restoration is the manner in which the ecosystem being restored is connected with other ecosystems.

Population biologists have developed extensive theoretical ideas regarding the way in which flows (migration) into and out of populations influence the behavior of population systems (both genetically and demographically). The most intriguing set of theories are those that deal with metapopulations (Hanski 1998b). The way a metapopulation works underscores the importance of external transport processes in maintaining a viable ecological system. A metapopulation is an aggregate of local populations connected together by immigration. Each local population in the metapopulation has a short lifespan; hence, all populations will go extinct in a relatively short period of time. The metapopulation can persist only if there is enough exchange of individuals among local populations to offset extinctions with establishment of new local populations. That is, all local populations can experience negative growth rates, yet the metapopulation is maintained indefinitely by transport of colonists among local populations. Within the context of ecosystem restoration, this means that the species diversity of a restored ecosystem may depend heavily on the amount of immigration the ecosystem receives from beyond its borders (Maschinski, this volume).

The principle that ecosystem functions depend on external transport processes generalizes to other types of ecosystem attributes, such as material cycles and energy flows. The carbon and oxygen cycles of Biosphere 2 dramatically illustrate this. The only way the oxygen cycle of Bioshpere 2 could operate within limits able to support the ecosystem was by external transport of oxygen from outside. This principle can be understood by considering an analogy from economics (Maurer 1999). In most markets, a corporation (conglomeration of local businesses) can almost always outcompete a small business that manufactures and sells the same product for at least two reasons. First, the corporation need only derive a small profit from each local business, because the accumulated profit across many local businesses makes up for the small margin of profit expected from each business. In addition, many costs of conducting business do not scale linearly with corporation size (e.g., raw materials can be bought in bulk at reduced prices by a corporation). Second, if the local economy experiences a downturn, a corporation can sustain a longer period of financial losses from a local business than a comparable small business. Essentially, corporate-wide profits can underwrite local losses, while a small business has no such resources. If the Earth's biosphere is analogous to a large corporation and Biosphere 2 to a small business, then it should be clear that for any local ecosystem, having access to the biogeochemical products of the Earth's biosphere more than makes up for local deficits in ecosystem function over the short term.

Temporal and Spatial Scales Needed for Long-Term Restoration

A comprehensive discussion of how big and connected restored ecosystems need to be to preserve target species and ecosystem functions is beyond the scope of this chapter. For the present purpose, we can say confidently that any restoration project will need to take into account practical issues regarding both size and connectedness of the ecosystem being restored. Since ecosystem functions often require input from processes not physically contained within the boundary of the ecosystem, a fundamental principle of ecosystem restoration should be to ensure that the restored ecosystem resides within a spatial context that is conducive to providing adequate flows of individual organisms, energy, and materials to maintain ecosystem function over a specified period of time.

Consider the following example of the importance of spatial context in determining the course a restored ecosystem might take. I obtained data on land cover for the state of Michigan (USA) from land-cover maps derived for the early 1800s (Albert 1995) and contemporary satellite imagery (Donovan et al. 2004). The land-cover classifications used in each of these data maps were crosswalked at a very general level of classification (Figure 14.3) in order to determine large-scale changes in land cover over the past two hundred years in Michigan. Interestingly, the relative amounts of deciduous and coniferous forest cover for the state has changed little in the past two centuries. The most dramatic changes have been the net loss of

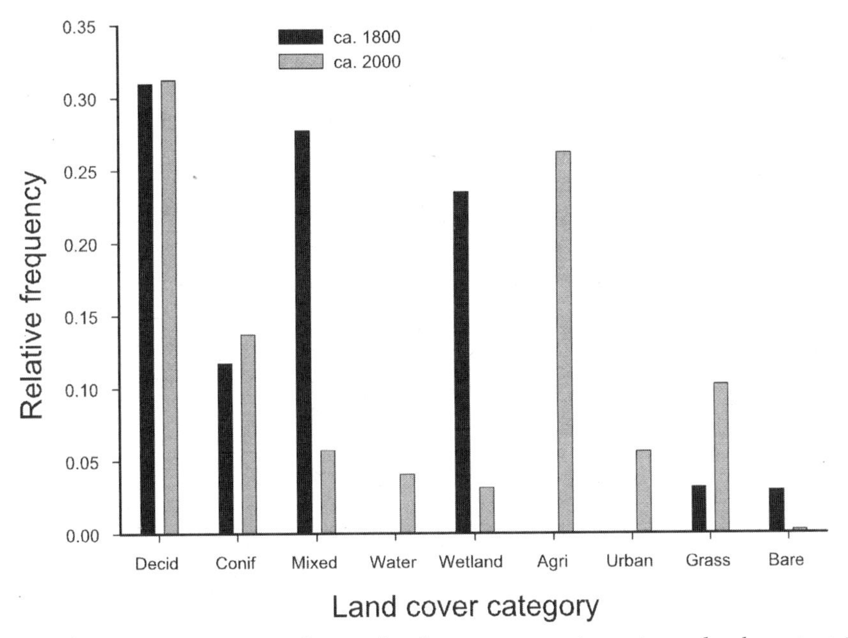

FIGURE 14.3 Relative frequencies of major land-cover categories estimated to be extant in 1800 and in 2000. Notice the major decline in the relative frequencies of wetlands and mixed forests.

mixed forests and wetlands, with a corresponding increase in agricultural, urban, and grassland cover types.

Now, consider the problem of wetland restoration in such an altered landscape. Nearly 25‰ of Michigan was covered with wetlands 200 years ago. That figure has shrunk to less than 5‰ today. A local wetland ecosystem in Michigan 200 years ago would have had a high likelihood of being connected or relatively close to another wetland in that landscape (Figure 14.4). The same ecosystem in a contemporary landscape would exist in a very different ecological context. The average size of a wetland ecosystem in modern Michigan landscapes looks to be much smaller than the average 200 years ago. Furthermore, the ecosystem is likely to be more isolated and surrounded by land uses not conducive to wetland ecosystem functions. For these reasons, a restored wetland ecosystem in a contemporary Michigan landscape will probably function very differently than it would have in the same landscape 200 years ago. If the goal of the restoration is to reestablish the original ecosystem function and species composition, it is likely that intensive management will be needed to replace the external transport processes that operated in the ecosystem 200 years ago.

Restoration Ecology and Neutral Macroecology

In attempting to understand the underlying causes for patterns such as SARs, the neutral theory of biodiversity (Hubbell 2001) posits a specific population mechanism responsible for these patterns. As discussed above, this mechanism is based on the assumption of functional equivalence among species. From the perspective of restoration ecology, this means that a complete, functional ecosystem can be constituted from an arbitrary set of species from the pool of species that could occupy a given site. Since one species is substitutable for another, the relative abundances and identities of species in a community should have little impact on the final structure and function of the ecosystem.

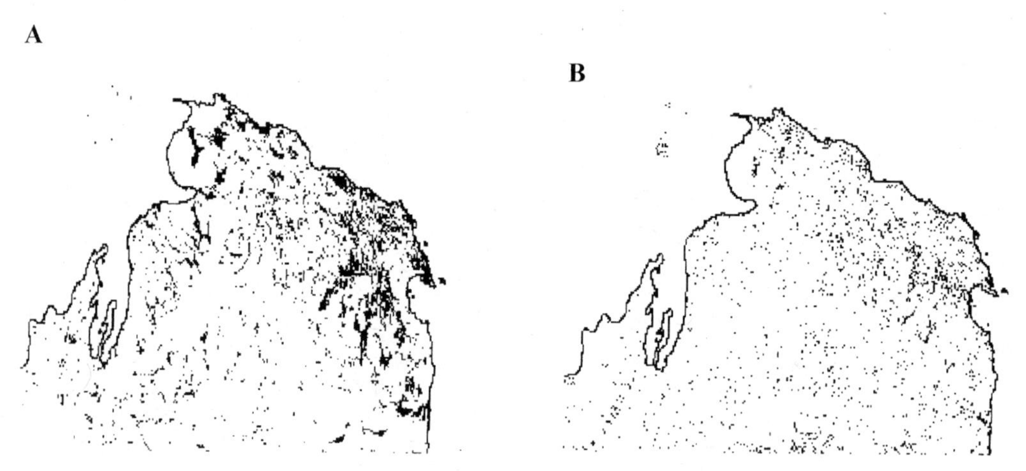

FIGURE 14.4 Estimated spatial distributions of wetlands in the northern lower peninsula of Michigan in 1800 (A) and 2000 (B). Note that the same major wetland complexes can be seen at both times. The sizes of these complexes were much larger and more clustered in 1800 than they are in 2000.

To what degree is this assumption met in nature? The initial response to this question by many ecologists would be that differences among species are ecologically important. But demonstrating that differences among species have a cumulative impact on the structure and function of ecosystems has not been straightforward (Kinzig et al. 2001; Loreau et al. 2001). Successful ecological restoration, however, often may depend on bringing together the right combination of species to generate and maintain a functional ecosystem. In this respect, it may be possible to view ecological restoration projects as experiments that can provide tests for the assumption of functional equivalence among species. If this assumption is true, then there may be a relatively large number of species combinations that might produce a persistent, functional ecosystem. If the assumption is false, then there may be only a few appropriate combinations of species that will produce an ecosystem that can persist and function appropriately over time. There are a number of ways that these hypotheses might be tested. A survey of restoration activities that were evaluated based on whether or not they led to an appropriately functioning ecosystem might provide a test of functional equivalence, if it was found that species composition was an important factor in determining the success of the restoration attempt. Better yet, adaptive management of restored ecosystems might provide opportunities to design experiments to test the degree to which ecosystem integrity depends on species composition.

Because of the importance of understanding how species composition of an ecosystem affects its structure and function to ecological restoration, it is imperative that restoration projects be carefully monitored after they are completed. Such monitoring will serve the dual purpose of establishing criteria to judge the degree to which restoration objectives are met and providing data to test models of community assembly that make assumptions about the functional equivalence of species. In this way, ecological restoration can become not only a practical field that deals with the how of restoring ecosystems but it can also provide a fertile field to test scientific theories that provide answers about why and how ecosystems must be reconstructed in order to maintain their integrity in space and time.

Summary

The large-scale perspective I have discussed in this chapter suggests that ecosystem restoration cannot be successfully carried out without careful consideration of the spatial context within which the ecosystem will exist. An ecosystem is defined not only by the species composition, edaphic conditions, and interaction networks that exist within its boundaries but also by the fluxes and flows across its boundaries that connect its internal processes with larger systems. A restoration project that fails to consider these flows may be unable to meet some restoration objectives, because the restored system may be governed by flows across its borders from landscapes not conducive to its functional integrity. One important way to maintain the integrity of a restored ecosystem may thus be to replace external transport processes with intensive management activities that serve the same function. Data on the relationship between species composition and the success of restoration activities can be used to test the important assumption of functional equivalence of species that underlies neutral models of macroecological patterns. In particular, carefully planned restoration projects can be used as experiments to test this important assumption for different kinds of ecosystems.

Literature Cited

Albert, D. A. 1995. Regional landscapes of Michigan, Minnesota, and Winsconsin. Lansing, MI: Michigan Natural Features Inventory, Report Number 1995-01.

Bell, G. 2000. The distribution of abundance in neutral communities. *American Naturalist* 155:606–617.

Bell, G. 2001. Neutral macroecology. *Science* 293:2413–2418.

Brown, J. H. 1986. Two decades of interaction between the MacArthur-Wilson model and the complexities of mammalian distributions. *Biological Journal of the Linnean Society* 28:231–251.

Brown, J. H. 1995. *Macroecology*. Chicago: University of Chicago Press.

Brown, J. H. 1999. Macroecology: Progress and prospect. *Oikos* 87:3–14.

Brown, J. H., V. K. Gupta, B. L. Li, B. T. Milne, C. Restrepo, and G. B. West. 2002. The fractal nature of nature: Power laws, ecological complexity and biodiversity. *Philosophical Transactions of the Royal Society of London, Series B* 357:619–626.

Brown, J. H., and M. V. Lomolino. 1998. *Biogeography*. Sunderland, MA: Sinauer Associates.

Brown, J. H., and B. A. Maurer. 1987. Evolution of species assemblages: Effects of energetic constraints and species dynamics on the diversification of the North American avifauna. *American Naturalist* 130: 1–17.

Brown, J. H., and B. A. Maurer. 1989. Macroecology: The division of food and space among species on continents. *Science* 243:1145–1150.

Brown, J. H., and G. B. West. 2000. *Scaling in biology*. New York: Oxford University Press.

Calder, W. A. 1984. *Size, function, and life history*. Cambridge: Harvard University Press.

Chave, J. 2004. Neutral theory and community ecology. *Ecology Letters* 7:241–253.

Dempster, W. F. 1999. Biosphere 2 engineering design. *Ecological Engineering* 13:31–42.

Donovan, M. L., G. M. Nesslage, J. J. Skillen, and B. A. Maurer. 2004. The Michigan Gap Analysis Project final report. Lansing, MI: Wildlife Division, Michigan Department of Natural Resources.

Enquist, B. J., J. H. Brown, and G. B. West. 1998. Allometric scaling of plant energetics and population density. *Nature* 395:163–165.

Enquist, B. J., J. H. Brown, and G. B. West. 1999a. Plant energetics and population density. *Nature* 398:573.

Enquist, B. J., G. B. West, E. L. Charnov, and J. H. Brown. 1999b. Allometric scaling of production and life-history variation in vascular plants. *Nature* 401:907–911.

Gaston, K. J., and T. M. Blackburn. 2000. *Pattern and process in macroecology*. Malden, MA: Blackwell Science.

Gillooly, J. F., J. H. Brown, G. B. West, V. M. Savage, and E. L. Charnov. 2001. Effects of size and temperature on metabolic rate. *Science* 293:2248–2251.

Hanski, I. 1998a. Connecting the parameters of local extinction and metapopulation dynamics. *Oikos* 83:390–396.

Hanski, I. 1998b. Metapopulation dynamics. *Nature* 396:41–49.

Hanski, I. 1999. Habitat connectivity, habitat continuity, and meta-populations in dynamic landscapes. *Oikos* 87:209–219.

Heywood, V. H., R. T. Watson, and United Nations Environment Programme. 1995. *Global biodiversity assessment*. New York: Cambridge University Press.

Hubbell, S. P. 2001. *The unified neutral theory of biodiversity and biogeography: Monographs in population biology*. Volume 32. Princeton: Princeton University Press.

Hubbell, S. P., and J. K. Lake. 2003. The neutral theory of biodiversity and biogeography, and beyond. In *Macroecology: Concepts and consequences*, ed. T. M. Blackburn and K. J. Gaston, 45–63. Oxford, UK: Blackwell Science.

Huston, M. A. 1997. Hidden treatments in ecological experiments: Re-evaluating the ecosystem function of biodiversity. *Oecologia* 110:449–460.

Huston, M. A. 1999. Microcosm experiments have limited relevance for community and ecosystem ecology: Synthesis of comments. *Ecology* 80:1088–1089.

Kareiva, P. M., J. G. Kingsolver, and R. B. Huey. 1993. Biotic interactions and global change. Sunderland, MA: Sinauer Associates.

Kinzig, A. P., S. W. Pacala, and D. Tilman. 2001. The functional consequences of biodiversity: Empirical progress and theoretical extensions. *Monographs in population biology*. Volume 33. Princeton: Princeton University Press.

Leigh, L. S., T. Burgess, B. D. V. Marino, and Y. D. Wei. 1999. Tropical rainforest biome of Biosphere 2: Structure, composition and results of the first two years of operation. *Ecological Engineering* 13:65–93.

Lomolino, M. V. 1994. Species richness of mammals inhabiting nearshore archipelagoes—Area, isolation, and immigration filters. *Journal of Mammalogy* 75:39–49.

Lomolino, M. V. 2000a. A call for a new paradigm of island biogeography. *Global Ecology and Biogeography* 9:1–6.

Lomolino, M. V. 2000b. Ecology's most general, yet protean pattern: The species-area relationship. *Journal of Biogeography* 27:17–26.

Lomolino, M. V. 2000c. A species-based theory of insular zoogeography. *Global Ecology and Biogeography* 9:39–58.

Lomolino, M. V., J. H. Brown, and R. Davis. 1995. Analyzing insular distribution patterns: Statistical approaches and biological inferences. *Annales Zoologici Fennici* 32:435–437.

Loreau, M., S. Naeem, P. Inchausti, J. Bengtsson, J. P. Grime, A. Hector, D. U. Hooper, M. A. Houston, D. Raffaelli, B. Schmid, D. Tilman, and D. A. Wardle. 2001. Ecology—Biodiversity and ecosystem functioning: Current knowledge and future challenges. *Science* 294:804–808.

MacArthur, R. H., and E. O. Wilson. 1963. An equilibrium theory of insular zoogeography. *Evolution* 17:373–387.

MacArthur, R. H., and E. O. Wilson. 1967. *The theory of island biogeography* Volume 1. Princeton: Princeton University Press.

Marino, B. D. V., and H. T. Odum. 1999. *Biosphere 2: Research past and present*. Amsterdam: Elsevier Science B.V.

Maurer, B. A. 1999. *Untangling ecological complexity: The macroscopic perspective*. Chicago: University of Chicago Press.

Maurer, B. A., and B. J. McGill. 2005. Neutral and non-neutral macroecology. *Basic and Applied Ecology* 5:413–422.

Maurer, B. A., and M. L. Taper. 2002. Connecting geographical distributions with population processes. *Ecology Letters* 5:223–231.

McGill, B., and C. Collins. 2003. A unified theory for macroecology based on spatial patterns of abundance. *Evolutionary Ecology Research* 5:469–492.

Peters, R. H. 1983. *The ecological implications of body size: Cambridge studies in ecology*. Cambridge, UK: Press Syndicate of the University of Cambridge.

Peters, R. L., and T. E. Lovejoy. 1992. *Global warming and biological diversity*. New Haven: Yale University Press.

Price, P. W. 2003. *Macroevolutionary theory on macroecological patterns*. Cambridge, UK: Cambridge University Press.

Rambler, M. B., L. Margulis, and R. Fester. 1989. *Global ecology: Towards a science of the biosphere*. Boston: Academic Press.

Ricklefs, R. E. 1987. Community diversity: Relative roles of local and regional processes. *Science* 235:167–171.

Ricklefs, R. E. 2004. A comprehensive framework for global patterns in biodiversity. *Ecology Letters* 7:1–15.

Ricklefs, R. E., and D. Schluter. 1993. *Species diversity in ecological communities: Historical and geographical perspectives*. Chicago: University of Chicago Press.

Rosenzweig, M. L. 1995. *Species diversity in space and time*. Cambridge, UK: Cambridge University Press.

Schlesinger, W. H. 1997. *Biogeochemistry: An analysis of global change*. San Diego: Academic Press.

Schmidt-Nielsen, K. 1984. *Scaling, why is animal size so important?* Cambridge, UK: Cambridge University Press.

Scott, J. M., F. Davis, B. Csuti, R. Noss, B. Butterfield, C. Groves, H. Anderson, S. Caico, F. Derchia, T. C. Edwards, J. Ulliman, and R. G. Wright. 1993. Gap analysis—A geographic approach to protection of biological diversity. *Wildlife Monographs* 121:1–41.

Society for Ecological Restoration International (SERI). 2002. *The SERI primer on ecological restoration*. www.ser.org/.

Soulé, M. E., J. Terborgh, and Wildlands Project. 1999. *Continental conservation: Scientific foundations of regional reserve networks*. Washington, DC: Island Press.

Southwick, C. H. 1996. *Global ecology in human perspective*. New York: Oxford University Press.

Volkov, I., J. R. Banavar, S. P. Hubbell, and A. Maritan. 2003. Neutral theory and relative species abundance in ecology. *Nature* 424:1035–1037.

West, G. B., J. H. Brown, and B. J. Enquist. 1997. A general model for the origin of allometric scaling laws in biology. *Science* 276:122–126.

West, G. B., J. H. Brown, and B. J. Enquist. 1999a. The fourth dimension of life: Fractal geometry and allometric scaling of organisms. *Science* 284:1677–1679.

West, G. B., J. H. Brown, and B. J. Enquist. 1999b. A general model for the structure and allometry of plant vascular systems. *Nature* 400:664–667.

Whittaker, R. J. 1998. *Island biogeography: Ecology, evolution, and conservation*. Oxford, UK: Oxford University Press.

Chapter 15

Climate Change and Paleoecology: New Contexts for Restoration Ecology

CONSTANCE I. MILLAR AND LINDA B. BRUBAKER

In this chapter, we explore linkages between two fields that have been little acquainted yet have much to say to one another: restoration ecology and climatology. The limited discourse between these fields is surprising. In the last two decades there have been significant theoretical breakthroughs and a proliferation of research on historical climate and climate-related sciences that have led to an overhaul of our understanding of Earth's climate system (Smith and Uppenbrink 2001). These new insights are relevant to restoration and ecology—so much so that fuller understanding could trigger rethinking of fundamental principles.

Climate Variability as an Ecosystem Architect—In Perspective

Conceptual views of the natural world influence tactical approaches to conservation, restoration, and resource management. The phrase *climate change* usually connotes global warming, greenhouse gas impacts, novel anthropogenic threats, and international politics. There is, however, a larger context that we must begin to understand and assimilate into restoration ecology theory—that is, the role of the natural climate system as a pervasive force of ecological change.

Advances in environmental sciences during the mid-to-late twentieth century on ecological succession, disturbance, and spatial and temporal variability motivated a shift from viewing nature as static and typological to dynamic and process driven. In turn, restoration ecology and practice matured from emphasis on museum-like nature preservation to maintaining variability and natural function (Jordan et al. 1990). As a result, prescribed fires and managed floods, for instance, became important restoration tools, and recovery of ecosystem function, composition, and structure was added to restoration goals.

Important as these changes have been, static concepts still constrain our understanding of natural dynamism and limit our conservation successes. The recent advances in climate-system sciences characterize recurrent climate change as a central physical force on Earth and significant agent of physical, ecological, and even cultural change at micro- to macroscales. From this perspective, climate is a macrodisturbance element, or the background stage of change on which evolutionary and successional dynamics play out. Such dynamism has only begun to be incorporated into evolutionary and ecological theory, and remains largely untranslated into conservation and restoration ecology. As a result, resource analyses and prescriptions, such as evaluation and diagnoses of ecological change, determination of baselines and evaluation of change in monitoring, and development of targets for restoration, may have limited applicability.

In this chapter, we bring forward new ideas in paleoclimatology and paleoecology that are relevant to restoration ecology. In so doing, we hope to foster discussion about fundamental goals and purposes in restoration that result from dialog between the fields. Our examples draw from western North American plant communities, but generalities from these extend to other areas and are supported by theoretical treatments (e.g., Jackson and Overpeck 2000).

Earth's Climate System: A Paleoclimatology Primer

Changes in weather are familiar features of Earth's surface, readily recognizable as diurnal variations, seasonal cycles, and annual differences that irregularly include extremes of drought, wet, heat, and cold. All forms of life are influenced by this variability in how and where they live, and mitigate adverse weather effects through conditioned responses and evolved adaptations. Cycles of climate change occur also over periods of decades to millennia, although these fluctuations have been little known and poorly understood. Until recently our knowledge of past climates came mostly from interpreting their indirect effects on the Earth's surface—for example, glacial moraines as evidence of past ice ages, coastal terraces as clues to former sea levels. Collectively these led to early interpretations of the Pleistocene (0.01–2.5 million years ago) as a long, cold interval—the "Great Ice Age" of Agassiz (1840). By the late nineteenth century, evidence for multiple glaciations accumulated and led to widespread description of four major glacial periods in the Pleistocene bracketed by brief warm intervals. The ice ages were regarded as ending about 10,000 years ago with the arrival of novel warmth of our present epoch, called the Holocene, or Recent, to signify its difference from the Pleistocene. Because the climate of the Holocene was interpreted as distinct from the past, Pleistocene climate processes were viewed as having little relevance to the present.

In the past two decades, new tools with high precision and resolution, new theory reliant on high-speed computing capacity, and a critical mass of empirical research have revolutionized understanding of Quaternary (the last 2.5 million years) climate. Quaternary climates are now understood as being far more variable and complex than previously imagined (Bradley 1999; Cronin 1999; Ruddiman 2001). The most widely applied and useful proxies first derived from long ice cores retrieved in polar ice caps (Cuffey et al. 1995). Gases and atmospheric particles trapped in ice faithfully record atmospheric conditions at the time of deposition. Due to annual layering and the ability to date layers accurately, analysis of thin sections at regular intervals yields high-resolution historic climate information for continuous time series. Cores drilled to the bottom of continental ice sheets (e.g., Greenland) have yielded highly resolved information on more than 40 climate variables that extend over 200,000 years (Lorius et al. 1990). The most important are isotopes of oxygen. Ratios of heavy to normal oxygen isotopes ($\delta^{18}O$) quantify the relative amount of oxygen stored in land ice relative to seawater, and provide robust indicators of surface air temperature at the time the isotopes were trapped in the ice. Analysis of these and other climate-related isotopes are now routinely extracted from other situations where undisturbed deposition occurs, such as lake beds, coral reefs, and sea floors sediments. Other climatologically important indicators retrievable from ice and sediment cores include greenhouse gases (CO_2, CH_4) and atmospheric aerosols that indicate dust and volcanic ash. Studies of varying time depth around the world, from a few decades to over 60 million years ago (Zachos et al. 2001), have led to detailed global and regional reconstructions of historic climate, which cumulatively provide new insight on the causal nature of climate variability.

Multimillennial Climate Cycles: Glacial/Interglacial or Orbital Cycles

Taken together, these long, highly resolved records collectively document the repeating, cyclic nature of climate over the past 2.5 million years (Figure 15.1) (Wright 1989; Raymo and Ruddiman 1992). Unlike earlier assumptions of persistent Pleistocene ice, oxygen-isotope records show a repeating pattern of over 40 glacial/interglacial cycles. A startling insight revealed by the oxygen-isotope records is the overall similarity of the Holocene to previous interglacials in length, trends, and relative temperatures; our Recent is not wholly novel after all. From the many oxygen-isotope curves now available around the world, it has become clear that these major warm-cold oscillations of glacial/interglacial phases were expressed globally and more-or-less synchronously. Global temperature differences between glacial and interglacial periods averaged 10°–20°C (Petit et al. 1997). Compare this to 0.7°C, the twentieth-century increase (IPCC 2001).

The oxygen-isotope curves further reveal a repeating structure of climate variability within glacial and interglacial phases (Lorius et al. 1990). Extensive cold glacial periods (*stades*) of the past were interrupted by warm phases (*interstades*) of about one-third the warmth of interglacials, and they terminated abruptly into interglacials. At a coarse scale, interglacials, including the Holocene, began abruptly, peaked in temperature in early to middle cycle, and terminated in a series of steps, each with abrupt transitions into cold stades of the subsequent glacial period. The cumulative effect is a sawtooth pattern typical of Quaternary climate records around the world (Figure 15.1).

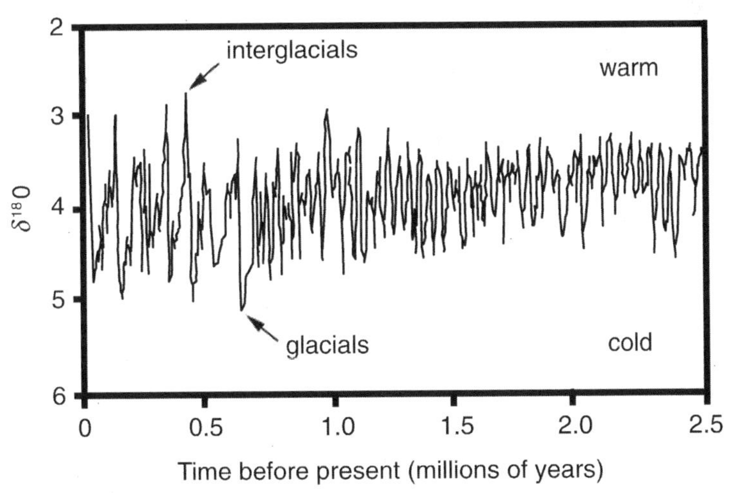

FIGURE 15.1 Primary temperature fluctuations between glacial and interglacial periods of the past 2.5 million years derived from oxygen-isotope analysis of ice cores from the Greenland ice sheet. High values of $\delta^{18}O$ indicate cold temperatures (glacial periods), and low values indicate warm temperatures (interglacial periods). Our current interglacial period (Holocene) is at the far left, from 0 to 10,000 years ago. From Wright 1989.

A mechanistic cause for climatic oscillations was proposed by Serbian mathematician Milatun Milankovitch (1941) long before detailed paleoclimate variability had been documented. Milankovitch integrated knowledge about Earth's orbit around the sun into a unified theory of climate oscillations. This has been revised subsequently into a modern orbital theory that is widely accepted as the pacemaker for the ice ages (Imbrie et al. 1992, 1993). Three major cycles of orbital variability recur over time (Figure 15.2) (Hays et al. 1976): (1) change in the shape of Earth's orbit around the sun from elliptical to circular (100,000 years);

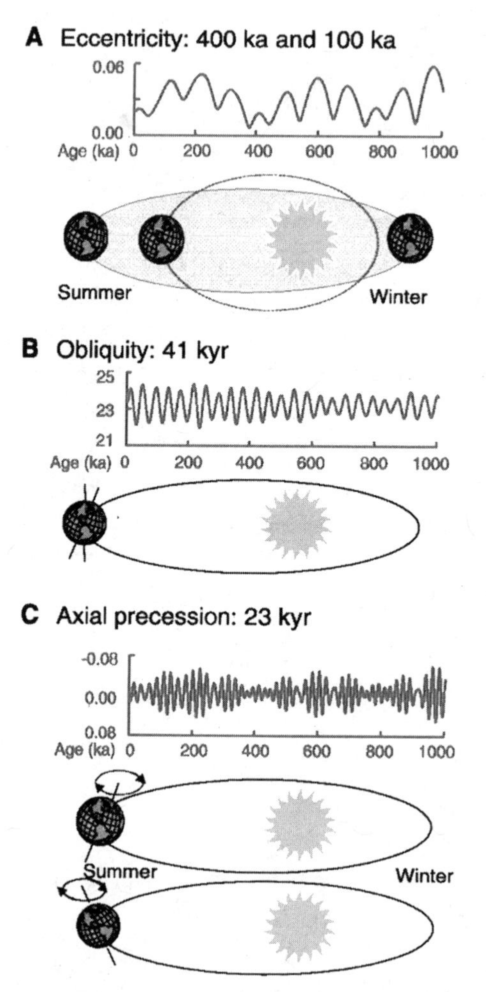

FIGURE 15.2 Primary orbital cycles of the Earth, the fundamental mechanism for oscillating climates of the past 2.5 million years. Temperatures on Earth vary depending on how much heat from the sun (solar insolation) reaches Earth's surface. This in turn varies, depending on the exact position of Earth within each of three orbital cycles. Mathematical integration of the three curves produces a graph of temperature over time that closely matches temperature reconstructions from $\delta^{18}O$ (e.g., Figure 15.1). (A) Eccentricity cycle, or changes in shape of the Earth's orbit from elliptical to circular (100,000-year cycle). (B) Obliquity cycle, or change in tilt of the Earth on its axis (41,000-year cycle). (C) Axial precession cycle, or change in time of year of perihelion (when Earth is closest to the sun; 23,000-year cycle). From Zachos et al. 2001.

(2) change in the angle of Earth's tilt on its axis (41,000 years); and (3) change in time of year when the Earth is closest to the sun (23,000 years). The amount of heat from the sun reaching the Earth (solar insolation) at any point in time varies with the Earth's position in each cycle. Integrating the three cycles mathematically results in a curve over time of predicted temperature on Earth that corresponds to the observed oxygen-isotope curves (e.g., Figure 15.1).

Century- to Millennial-Scale Climate Cycles

Analyses of oxygen-isotope variation at finer temporal resolution further reveal century to millennial length oscillations nested within orbitally driven climate cycles. These were known first from a few well-studied climate events, such as the interval known as the Younger Dryas (Kennett 1990), a 1,000-year return to ice age conditions that interrupted warming at the end of the last glacial period (11,500 years–12,500 years ago); Heinrich events (Heinrich 1988), a series of short (100 years–1,000 years), extremely cold intervals within the last glacial period; and Dansgaard/Oeschger interstadials (Dansgaard et al. 1993), brief, abrupt, warm intervals during the last glacial period. These climate events are increasingly understood as part of a pervasive oscillation pattern, called "Bond cycles," documented for at least the last 130,000 years (Bond et al. 1997). Bond cycles average 1,300 years–1,500 years, meaning that for each warm or cold phase (each ca. 700 years), the warmest and coldest half-phases last 300 years–400 years (Figure 15.3). Climate intervals during the Holocene that exemplify Bond cycles include the Little Ice Age (LIA), a minor ice advance and global cold period from A.D. 1450 to 1920 (Grove 1988; Overpeck et al. 1997); the Medieval Climate Anomaly, a warm, dry interval with regional variability from A.D. 900 to 1350 (Hughes and Diaz 1994; Stine 1994; Esper et al. 2002); and the 8,200-year cold event (Alley et al. 1997).

Painstaking analysis at high resolution of several well-known Bond intervals has documented that oscillations often begin and end extremely abruptly. Annual analysis, for example, of 150 years centered on the major collapse of ice at the end of the Younger Dryas cold

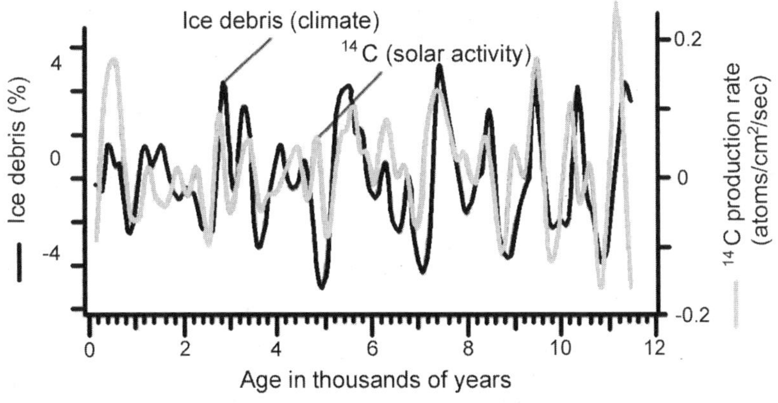

FIGURE 15.3 Century-millennial scale oscillations, or Bond cycles, have been pervasive at least through the Holocene and last major glacial age. Individual events have long been recognized, such as the Little Ice Age (A.D. 1450–1920) and the Younger Dryas (11,500–12,500 years ago), but these are only recently understood as part of a systematic cycle. Correlations with solar activity suggest a sun-driven mechanism for this climate pattern. From Bond et al. 1997, 2001.

event revealed that 15°C warming occurred in two 10-year periods (ca. 7°–8°C each) separated by a 20-year plateau of no detectable change (White et al. 2001).

Of particular interest is the warming of the twentieth century. During the preceding four-century-long Little Ice Age, temperatures in western North America were on average 1°C colder than present; glaciers in many western North American mountain ranges were at their greatest extent since the end of the Pleistocene, over 10,000 years ago (Clark and Gillespie 1997). Warming since the late 1800s has been ca. 0.7°C globally with much of the increase occurring due to increases in minimum temperature (IPCC 2001). Increases in the early part of the century are now widely accepted as natural climate forcing, whereas continued warming since mid-twentieth century can be explained only by recent anthropogenic-induced greenhouse gases (IPCC 2001).

The natural mechanisms driving climate oscillations at the century-millennial scale are a topic of great current interest. The relationship of extremely cold intervals within glacial periods to sudden surges of polar ice into high-latitude oceans, and resulting abrupt changes in global ocean salinity, first led climatologists to believe these intervals were driven by ice and ocean-circulation dynamics (Broecker et al. 1990; Clark et al. 2001). Recently, however, millennial cycles in the sun's intensity (mediated by sun spots and other changes on the sun's surface) have been shown to match the timing of the Bond cycles over the last 130,000 years with high precision (Figure 15.3) (Bond et al. 2001). This has led climatologists to speculate that a trigger for century-millennial climate changes comes from outside the Earth—that is, changes in the sun—and that resulting changes in ocean circulation subsequently regulate and abruptly communicate solar signals worldwide.

Interannual- to Decadal-Scale Climate Change

In recent years, climatologists have defined high-frequency climate cycles operating on scales from a few years to several decades. The best known of these is the El Niño pattern, called the El Niño/Southern Oscillation (ENSO) for its interhemispheric expression and ocean-based cause (Diaz and Markgraf 2000). Every several years, hemispheric trade winds that typically blow warm tropical ocean water westward across the Pacific Ocean stall, and warm water, instead, accumulates in the eastern Pacific Ocean. This leads to the presence of unusual water temperatures offshore from North and South America. Each year there is some degree of El Niño or its opposite effect, La Niña. Extreme events cycle on a 2-year to 8-year basis (Figure 15.4). El Niño events bring different conditions to different parts of the world. For instance, they portend unusually warm and wet falls and winters in central and southern California, and unusually cold and dry weather in the Pacific Northwest. The reverse occurs during La Niña events.

Climate oscillations on multidecadal (20-year to 60-year) periodicities have also been described recently. Like ENSO, these act regionally but have effects on distant locations. The Pacific Decadal Oscillation (PDO) is a recently characterized multidecadal cycle affecting western North America. It appears to be regulated by decadal changes in ocean circulation patterns in the high-latitude Pacific Ocean (as opposed to ENSO's tropical locus) and yields climate effects and regional patterns similar to extended ENSO effects (Mantua et al. 1997; Zhang et al. 1997). Warm (or positive) phases are extensive (10-year to 25-year) periods of El Niño-like conditions that alternate with cool (or negative) phases of La Niña-like conditions.

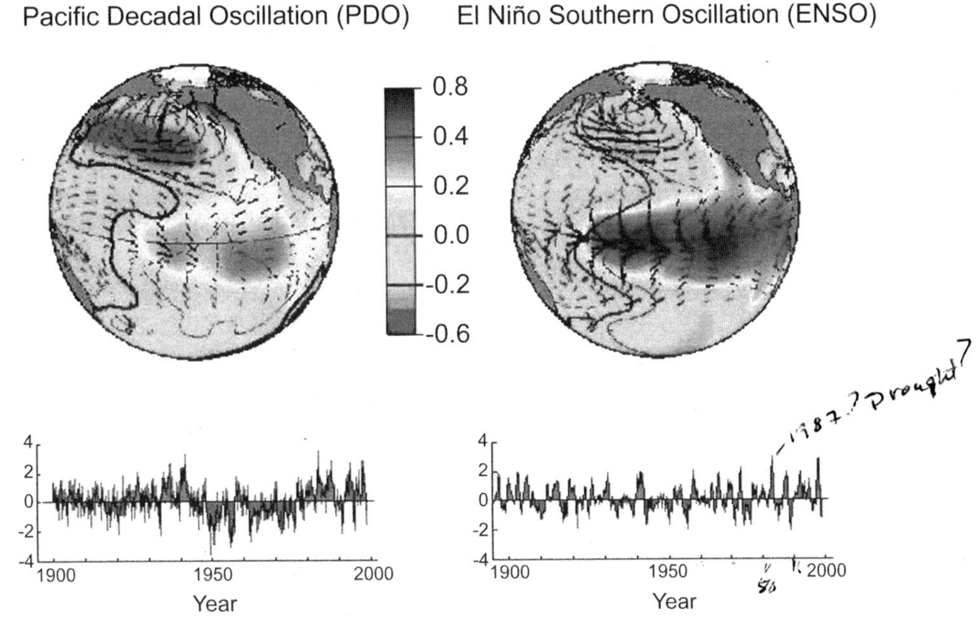

FIGURE 15.4 The El Niño/Southern Oscillation (right) and Pacific Decadal Oscillation (left) are internally regulated, ocean-atmospheric patterns that affect climate on interannual (ENSO) and multidecadal (PDO) scales. Positive ENSO (El Niño) and PDO periods bring systematically warm, wet conditions to certain parts of the world, while negative ENSO (La Niña) and PDO bring cool, dry conditions. PDO and ENSO interact such that during times of positive PDO, El Niño signals are enhanced, and the reverse is true during negative PDO decades. PDO regime shifts, such as in 1976, can be abrupt, with dramatic physical and ecological effects. Reproduced with permission from the University of Washington's Joint Institute for the Atmosphere and Oceans. Figure created by Dr. Stephen Hare, International Pacific Halibut Commission.

Other ocean-mediated multidecadal patterns affect other parts of the world, such as the North Atlantic Oscillation, and the Arctic Oscillation (Cronin 1999).

Climate Change as an Ecosystem Architect

Abundant evidence worldwide indicates that life on Earth has responded to climate change at each of these scales. Changes in biota over time can be measured in many ways, such as from sediment cores taken from wet areas including meadows, bogs, lakes, and ocean bottoms. In dry environments, packrat middens preserve macrofossils, while in temperate forests tree-ring records archive annual tree growth. In ocean environments, annual coral layers record ecosystem responses.

At *multimillennial* scales, paleoecological records collectively document that, at any one place, compositions of flora changed significantly in correspondence with major climate phases, often showing complete species turnover and recurring patterns of similar groups of species or species with similar adaptations alternating between glacial and interglacial periods. In relatively flat terrain, such as in the northeastern United States, eastern Canada, parts of Scandinavia, and northern Asia, species shifted latitudinally north and south many hundreds of kilometers, as modeled, for example, for spruce (*Picea*) in eastern North America (Figure 15.5) (Jackson et al. 1987). In mountainous regions, by contrast, species responded primarily by elevational shifts, as indicated by conifers of the Great Basin and southwestern desert region, which shifted as much as 1,500 m (Figure 15.6) (Thompson 1988, 1990; Grayson 1993). In regions where habitats were highly patchy, with steep and discontinuous gradients, species responded primarily by fluctuations in population size and minor geographic shifts in location, as exemplified by oaks in California (Adam 1988; Heusser 1995). Areas occupied by continental ice caps were often revegetated via rapid colonizations from refugia (Brubaker and McLachlan 1996).

Significant and rapid response of vegetation to *century*-scale climate change is also well documented. Before temperature proxies such as oxygen isotopes provided independent measures of historic climate, millennial-scale abrupt climate events were inferred from changes in flora and fauna. For instance, the Younger Dryas cold interval was known from changes in abundance of the arctic tundra plant *Dryas octopetala* (Jensen 1935). This species dominated paleofloras of western Europe during the coldest ice ages and was being replaced by warm temperate vegetation as climate warmed at the end of the last major ice age. An abrupt, short-lived reversal to full-glacial abundances of *D. octopetala* became known as the Younger Dryas (in contrast to an earlier interval known as the Older Dryas), now recognized as a phase of the Bond cycles.

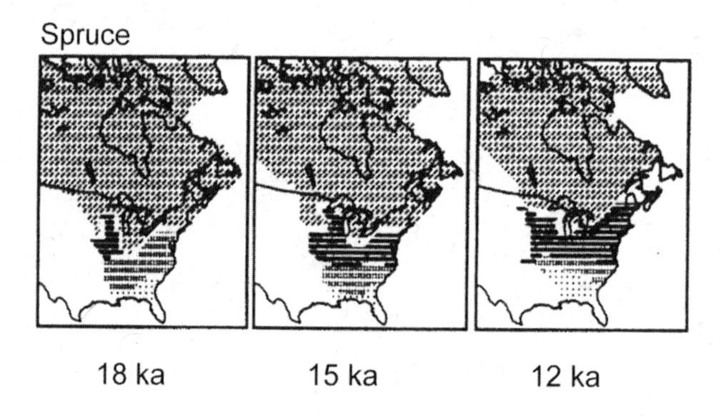

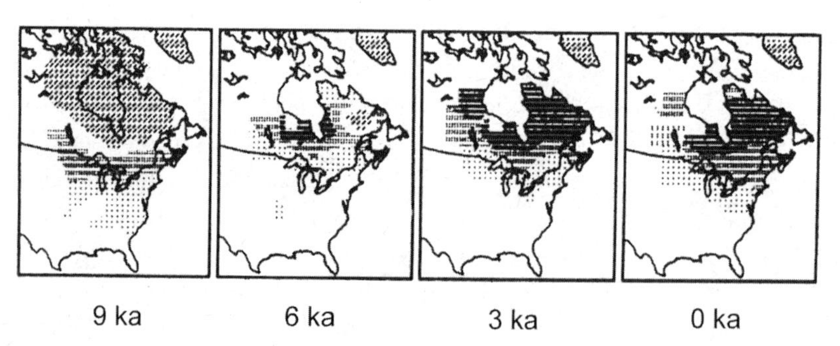

FIGURE 15.5 Shift in ranges of spruce (*Picea*) forests in eastern North America as they track changing temperatures from the Last Glacial Maximum to present. Reconstructed from pollen abundances in lake sediments for intervals of 3,000 years. Dots indicate range of spruce; other stippling represents glacial and periglacial environments. From Jackson et al. 1987.

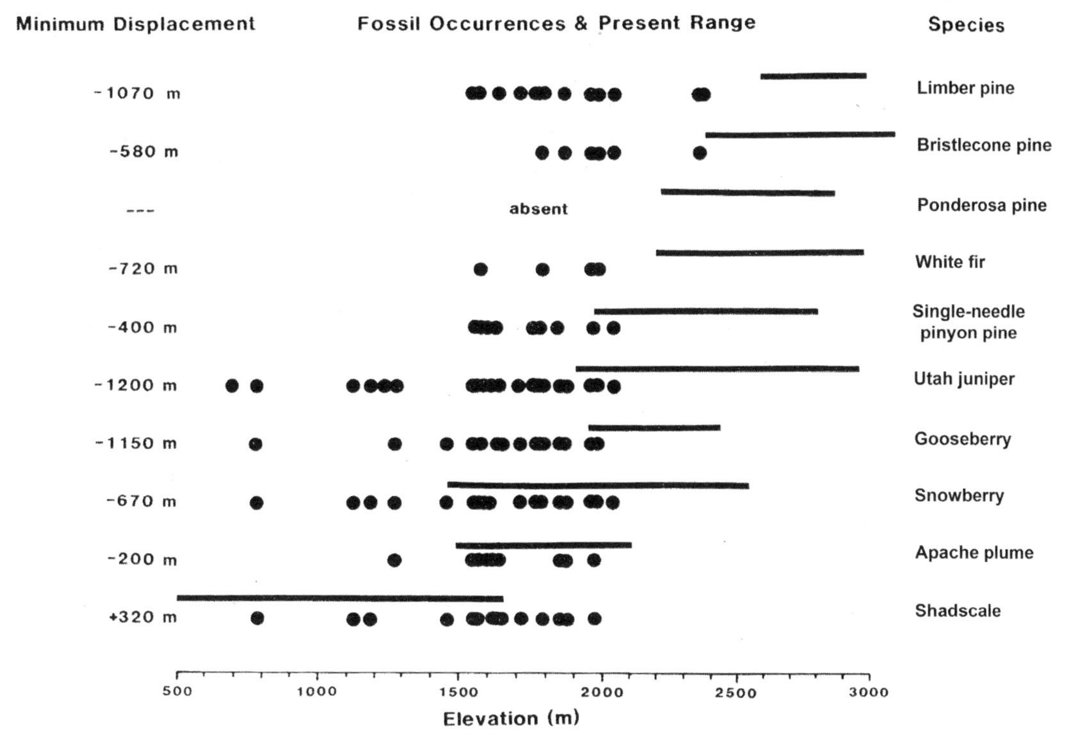

FIGURE 15.6 Glacial/interglacial shifts in elevation for plant species of the Sheep Range, southern Nevada, showing current (interglacial, solid line) and past (glacial, pre-11,000 years ago, dots) elevation limits, and individualistic responses of species. From Thompson 1990.

Many examples now show fluctuating changes of vegetation corresponding to Bond cycles. An illustrative example is the abrupt change in pine versus oak vegetation in southern Florida that corresponds to Heinrich events (Figure 15.7) (Grimm et al. 1993). Another example from the California region comes from the work of Heusser (2000), who demonstrated that abrupt changes in the dominance of oak versus juniper corresponded to rapid climate oscillations of the last 160,000 years. In the Great Basin of North America, major changes in population size and extent of pinyon pine (*P. monophylla*), and changes in floristic diversity, correspond to Bond-scale, century-long cycling (Tausch et al. 2004). Whereas recurring patterns emerge at coarse scales, species responses are individualistic, lags are common, and non-analog patterns frequent, so that population increases or decreases may not appear to be "in synch" with climate change, especially when climate changes are extreme and abrupt (Jackson and Overpeck 2000).

Vegetation responds also to *interannual* and *decadal* variability. At ENSO scale, changes occur primarily in plant productivity and abundance within populations. The oscillations contribute to regional fire regimes, where fuel loads build during wet years and burn during dry years. These lead to mesoscale vegetation changes as ENSO itself cycles, and thus fire regimes change over time (Swetnam and Betancourt 1998; Kitzberger et al. 2001). Decadal climate and vegetation oscillations have been well documented in secondary growth of trees, such as recurring droughts over the past 400 years that led to reduced ring-widths in ponderosa pine in New Mexico (Figure 15.8) (Grissino-Mayer 1996), and the recurring pattern

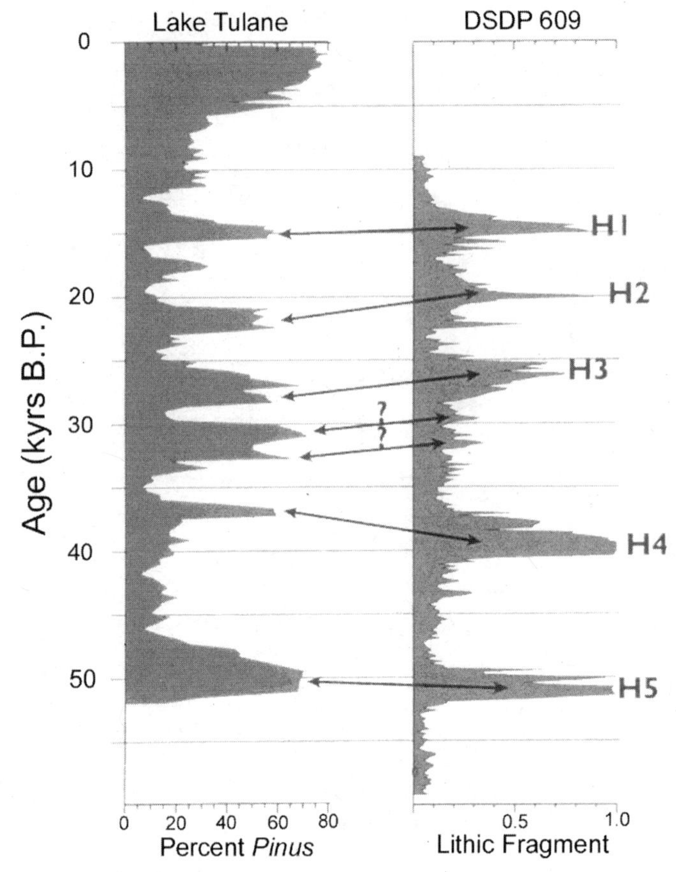

FIGURE 15.7 Correspondence in abundance of pine from Lake Tulane, Florida (indicated by pollen %, left panel) with millennial scale cold, or Heinrich, events of the last glacial period (indicated by % lithics, or ice-rafted rock debris, right panel). Data from Grimm et al. 1993; figure first produced by NOAA National Geophysical Data Center's Paleoclimatology Program (T. G. Andres, J. T. Andrews, and L. M. Lixey).

of ring-widths in big-cone Douglas-fir (*Pseudotsuga macrocarpa;* Biondi et al. 2001), mountain hemlock (*Tsuga mertensiana;* Peterson and Peterson 2001) and subalpine fir (*Abies lasiocarpa;* Peterson et al. 2002) that correlate with PDO for up to 400 years in the past. Vegetation type conversions from meadow to forest, changes in species growth rates and crown morphology, and changes in forest density were associated with PDO cycles in conifer forests of the Sierra Nevada, California (Millar et al. 2004).

Summary of Climate Change and Vegetation Response

This brief overview yields several conclusions: First, climate has *oscillated* between warm and cold, wet and dry regimes over the last 2.5 million years rather than being dominantly directional or stochastic. In broad terms, our present warm period (Holocene) is similar to interglacials of the past, and the last glacial period had many antecedents before it. Second, climate has oscillated simultaneously at *multiple* and *nested temporal scales*, including interan-

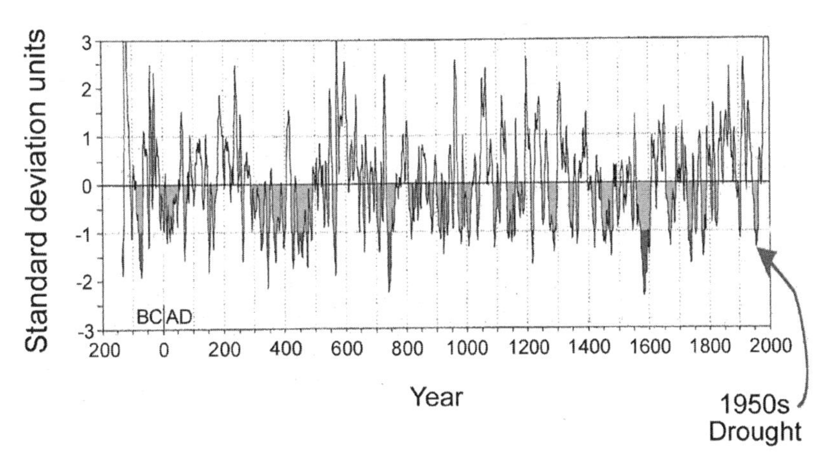

FIGURE 15.8 Decadal variability in precipitation for the past 2,200 years as indicated in ponderosa pine/Douglas-fir (*Pinus ponderosa/Pseudotsuga menziesii*) tree-ring reconstruction of annual rainfall from western New Mexico. The drought of the 1950s is shown, as well as droughts of equal and greater magnitude. Data from Grissino-Mayer 1996; graphic modified and first produced by Connie Woodhouse, NOAA National Geophysical Data Center Paleoclimatology Program.

nual, decadal, century, millennial, and multimillennial; mechanisms and the nature of expression differ depending on the scale, although the effects interact. Third, transitions between major and minor climate phases often occurred *abruptly* (a few years to decades), and were accompanied by significant changes in climate (e.g., 3°–15°C). Finally, *vegetation responded to climate change* at each scale. Vegetation responses to annual/decadal variability were mostly in productivity, abundance, and local shifts in community composition, whereas responses at century/millennial scales involved major and often recurring colonization and extirpation (migration and range shift) events. Repetitive climate changes at each scale thus exert significant recurring evolutionary and ecological force on vegetation. Modern species have been exposed to fluctuating climates and rapid transitions for at least two million years, and they have likely been exposed to similar phases that punctuated the past 20 million years (Zachos et al. 2001).

A key characteristic of Quaternary paleoecology is that species respond individually to particular climate cues with unique rates and sensitivities to individual climate variables. Individual species follow their own ecological trajectories as climates cycle, leading to changes in community compositions that themselves form, dissolve, and may reform over time. Often non-analog communities form, that is, assemblages not observed in modern vegetation. From this perspective, plant communities exist as transient assemblages of species; species move individually through time and space following favorable climates and environments. The apparent recoalescence of vegetation assemblages results from recurrence of similar climate conditions, although lags and differences in individual species responses, as well as stochastic events, give variation to the exact structure of plant communities at any one time (Davis et al. 1986; Webb 1986).

Implications for Restoration Ecology

Compelling evidence of climate variation across all timescales has implications for the theoretical bases and practice of restoration ecology. In particular, our awareness of the dominant effect that climate has in driving ecological change, and of the dynamic nature of climate and vegetation change, prompts us to evaluate assumptions about ecological sustainability, native species range, and restoration targets.

Concepts of Sustainability

Ecological sustainability is a dominant paradigm in restoration ecology. Variously defined, sustainability and related concepts of ecological heath and integrity imply species, community, and ecosystem endurance and persistence over time, and they are often used as implicit or explicit restoration goals (e.g., Jordon et al. 1990; Lele and Norgaard 1996). In practice, sustainability has been difficult to describe or to recognize. Sustainability is generally accepted to pertain when natural species diversity is maintained, species are abundantly distributed throughout their recent historic native range, community associations are maintained, natural processes occur at reference intervals and conditions, and human disturbance is minimized (Lackey 1995; Hunter 1996).

The complex and recurring cycles of ecological change in response to climate cycling challenge the conceptual bases of this interpretation of ecological sustainability. Species ranges have and will—even in the absence of human influence—shift naturally and individualistically over small to large distances as species follow, and attempt to equilibrate with, changes in climate. In the course of adjustment, plant demography, dominance, and abundance levels change, as do vegetation associates and wildlife habitat relations. A major conclusion from the paleorecord is that, at scales from years to millennia, ecological conditions are not in equilibrium, do not remain stable, nor are they sustained, but, by contrast, are in ongoing flux (Jackson and Overpeck 2000). The flux is not random, chaotic, or unlimited, but it is driven by regional climate and modified stochastically by local conditions.

It is important to note that the timescales under discussion are short relative to the lifespans of most extant plant species. Notwithstanding recently speciated taxa, many native North American plant species originated during the Tertiary or earlier, commonly 20 to 40 million years ago. Thus, many extant species have been subjected to the demands of shifting climates, in both large tempo and small, throughout their histories. This implies that adaptation to abrupt climate changes has had many opportunities to evolve. Resilience and sustainability, at least in terms of species persistence, appear to have been met through the capacity of plants to track favorable environments as they shift over time and through adjustment in range distribution, habitat, associates, and population characteristics.

Paleorecords in areas where abundant information exists can be used as a test of what has been sustained naturally over time. When Quaternary vegetation records from the Sierra Nevada were assessed in this way, Millar and Woolfenden (1999a) found that only a few conditions often associated with ecological sustainability concepts pertained. These included (1) relative stability of the Sierra Nevada ecoregion, that is, persistence of a distinct ecoregion over time; and (2) persistence of overall species diversity at the scale of the entire Sierra Nevada ecoregion, with only one species, a spruce (*Picea* spp.), disappearing from the region about 500,000 years ago.

Beyond these two features, however, other conditions associated with ecological sustainability did not occur. At subregional scales within the Sierra Nevada, species diversity changed at timescales of centuries to millennia. Similarly, individual species ranges and population abundances shifted, often drastically. An example is giant sequoia (*Sequoiadendron giganteum*). Currently limited to small and disjunct groves between 1,500 and 2,100 m in the southwestern Sierra Nevada, giant sequoia's range over the past 10,000 to 26,000 years included the eastern Sierra Nevada (Mono Lake, Davis 1999a), and locations in the western Sierra Nevada that are both well above (2,863 m, Power 1998) and below (1,000 m in current chaparral shrubland, Cole 1983; and 54 m at Tulare Lake in the California Central Valley, Davis 1999b) its current range. Giant sequoia did not appear in its current range until 4,500 years ago and did not reach modern abundance there until about 2,000 ago, that is, the age of the oldest living individuals (Figure 15.9) (Anderson and Smith 1994).

Several other conditions often considered elements of ecological sustainability did not pertain in the California paleorecord. Movement of individual species meant that vegetation assemblages changed over time and/or shifted locations as individual species followed climate gradients (Woolfenden 1996). Vegetation communities appeared sometimes to shift locations, when individual species tracked climate coincidentally, and in other cases, changed composition and dominance relations. Non-analog communities occurred transiently, such

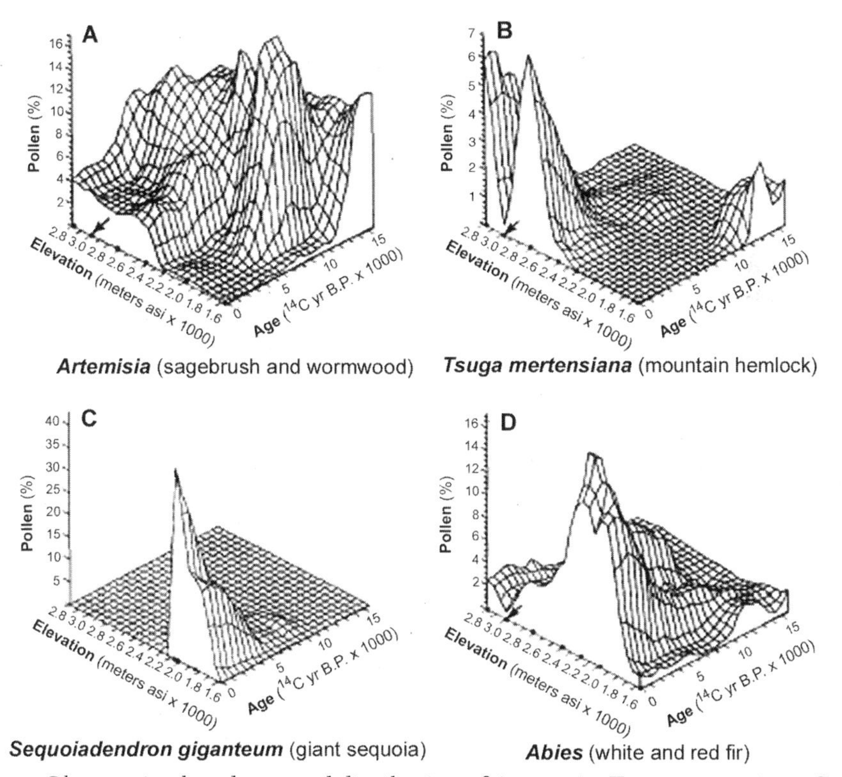

FIGURE 15.9 Changes in abundance and distribution of *Artemesia*, *Tsuga mertensiana*, *Sequoiadendron giganteum*, and *Abies* over the last 15,000 years as summed from pollen analyses in western Sierra Nevada meadows. *Sequoiadendron* pollen reached its present abundance and native range at Giant Forest only in the last 2,000–3,000 years. From Anderson and Smith 1994.

as the co-occurrence 20,000 to 30,000 years ago in the southern Sierra Nevada of yucca (*Yucca brevifolia*) and Utah juniper (*Juniperus osteosperma*) with an understory of *Artemesia tridentata, Purshia tridentata,* and *Atriplex concertifolia* (Koehler and Anderson 1995).

Historic fire regimes reconstructed from charcoal analyses in paleorecords also changed over time at multiple scales. Over the last 10,000 years, for instance, fire in mid-elevations of the western Sierra Nevada was a minor ecosystem architect. Beginning about 4,000 years ago, charcoal records indicate increased local fires and effect on regional vegetation (Anderson and Smith 1994, 1997). At scales of decades to centuries, Swetnam (1993) showed that fire regimes in giant sequoia forests shifted from frequent, light, and localized fires to infrequent, intense, and widespread fires in the last 1,000 years following climate changes.

These and similar records challenge interpretations of ecological sustainability that have emphasized persistence of species and stability of communities within current ranges. As widely used, such concepts of sustainability do not adequately accommodate natural dynamics and promote misinterpretations about behavior of natural systems.

Population Size, Population Abundance, and Native Species Range

Declines (or increases) in population size and abundance—observed through monitoring or other measures—and reductions (or increases) in overall range often are assumed to be anthropogenic, whereas these may be instead natural species' responses to climate change. Two examples in California illustrate adaptation at millennial to decadal scales. Species of oak (*Quercus*) and juniper (*Juniperus*) expand and contract in complementary fashion: oak population abundances and total range distribution expanded repeatedly during warm climates and, as often, contracted during cool climates, while the opposite occurred for juniper species (Figure 15.10) (Adam and West 1983; Heusser 1995). Although oaks in general are widespread and common in California now, during repeated long glacial periods, they were rare in the region. Although these changes are most obvious between glacial and interglacial times, significant changes in abundance tracked climate at scales as short as a decade (Heusser and Sirocko 1997).

Coast redwood (*Sequoia sempervirens*) is another example. Currently rare, it has fluctuated in population extent and abundance following both long (millennial) and short (century-decadal) cold/warm cycles. Redwood was even more sparsely distributed than at present during climate periods when coastal fog did not develop and temperatures were hotter or cooler relative to present (Heusser 1998; Poore et al. 2000). Redwood expanded during mild, equable parts of interglacials when ocean temperature and circulation influenced development of coastal fog.

This perspective of Quaternary vegetation dynamics compels us to evaluate causes for changes in population size, abundance, and native range more carefully. Rather than interpreting changes as resulting from undesired anthropogenic threat, we might investigate instead whether these are natural species' adaptations. For instance, *Juniperus* expanding in Great Basin rangelands has been considered an exotic invasive, and measures have been taken to remove plants. These changes appear, rather, to be adaptive responses to climate change (Nowak et al.1994). Other things being equal, an ecologically informed conservation action would be to encourage, not thwart, juniper expansion.

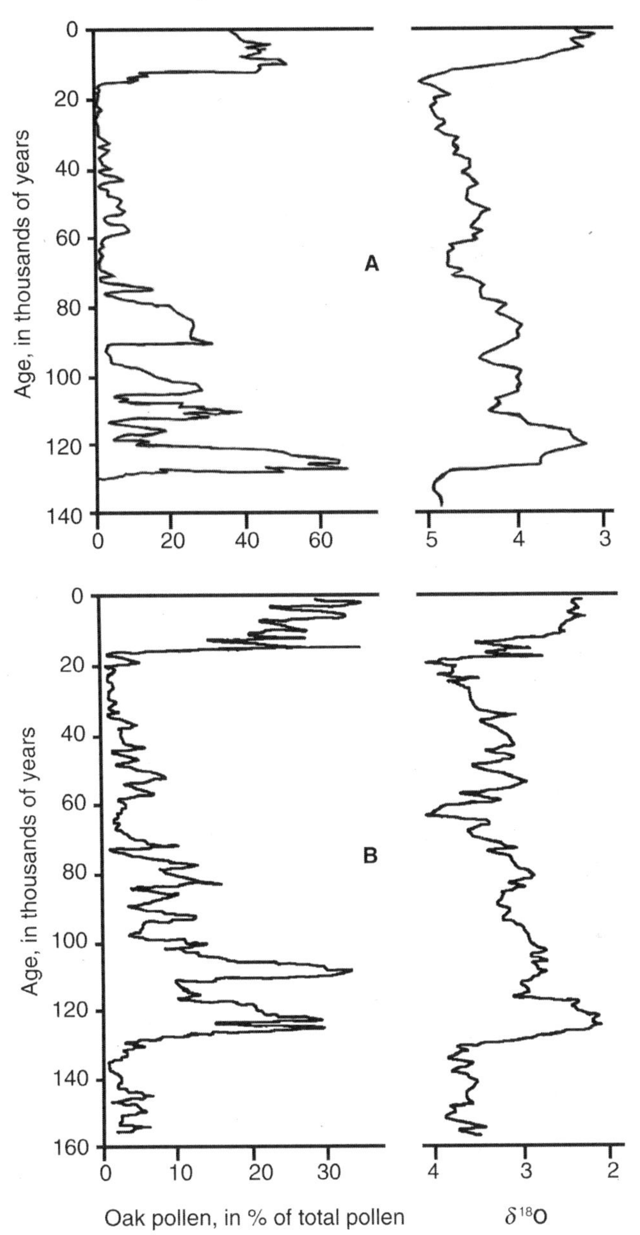

FIGURE 15.10 Correspondence of California oak (*Quercus*) abundance (pollen ‰) and temperature fluctuations recorded as variability in oxygen-isotope ratios indicating fluctuations between high and low abundance. Similar and synchronous patterns throughout the California oak ranges document species fluctuations between rare and widespread condition. (A) 140,000-year record from Clear Lake, Lake County, CA, in the north-central Coast Ranges (from Adam 1988). (B) 160,000-year record from Santa Barbara Basin, Santa Barbara County, CA (ODP 893A; from Heusser 1995).

Although changes in population size and distribution may be natural responses to climate change, causes are often difficult to discern in practice. Lags in adjustment and other disequilibria between population distributions and climate mean that population increases or decreases may not be synchronous with climate change, especially during periods when rapid climate changes occur over short periods, making difficult the search for mechanistic causes (Jackson and Overpeck 2000). Because individual plants, unlike animals, cannot "pick up and move" (intragenerational), they migrate and shift in distribution by dying in some areas while expanding in others (intergenerational). These may be messy on the landscape—with patchiness and irregularity characteristic—making the effects difficult to evaluate while they're happening. Causes may be attributed readily to other proximal factors, such as to insects and pathogens, or anthropogenic effects, such as fire suppression, even where climate is the ultimate underlying factor.

A challenging question for restoration ecology becomes, "what is the native range of a species?" To define the range of a species is the basis for monitoring its condition, understanding favorable habitat and ecological interactions, diagnosing threats and risks, determining restoration targets, and indicting species as "exotic" (Jackson 1997). Viewed against historic changes in distribution and natural flux, the native range of a species must be considered a transient and dynamic process itself, readily capable of moving in space as climate shifts over the landscape. Recognizing that non-equilibrium conditions exist and vegetation lags occur means that, like Lewis Carroll's Red Queen, vegetation chases a target (climate) that is itself changing. Population abundances and species' distribution ranges may be relatively stable whenever climate is in a more stable phase and/or if the environment of a species offers considerable local heterogeneity (Thompson 1988; Jackson and Overpeck 2000; Williams et al. 2001). In these cases, shifts in climate may be tracked with relatively minor overall geographic changes. By contrast, in regions that are inelastic to change, for instance, landscapes with little topographic diversity, even small shifts in climate may bring large changes in population condition. Given that the climate has been undergoing rapid changes with high variability during the twentieth and twenty-first centuries, we would expect population demographics and species ranges to also be highly unstable.

Genetics and Restoration in a Climate-Change Context

During the last 15 years, increasing attention has been given to the importance of genetics in restoration ecology (Millar and Libby 1989; Falk and Holsinger 1991; Lesica and Allendorf 1999; Guerrant et al. 2004; Rice and Emory 2004). Theory and guidelines have been developed regarding provenance (genetic origin); allelic and genotypic diversity; effective population size, gene flow, and genetic drift relevant to material for restoration; and long-term maintenance of reintroduction populations (Falk et al. 1996). These have been based on population-genetic assumptions that local populations are best adapted, that genetic contamination and inbreeding or drift decrease fitness and are to be avoided, and that safeguarding diverse local gene pools over time provides the best option for adaptedness in reintroduced populations.

New understanding of climate variability and its influence on plant population dynamics and biogeography raises questions about assumptions concerning local adaptation and genic diversity and suggests a need to rethink the role of genetics in restoration (Westfall and Millar 2004). Genetic theory on selection, drift, and genic diversity rests heavily on equilibrium conditions in regard to population size, gene flow, and drift. If, by contrast, rates of fragmentation, coalescence, population growth and mortality, and selection coefficients are changing within or between generations, then local populations will almost always remain in disequilibrium (Pease et al. 1989; Bürger and Gimelfarb 2002; García-Ramos and Rodríguez 2002). This leads to the likelihood that local populations are not necessarily best adapted, as has been found in common-garden experiments (Matyas 1996; Rehfeldt et al. 1999; Rehfeldt et al. 2001). Further, lags and persistent disequilibria could accumulate over time, creating a new kind of genetic load. Whenever climates are relatively stable, progress toward equilibrium may occur. In most species, especially perennial species, and during times of high climate variability, however, this is unlikely to be the case. Such theoretic possibilities suggest that restoration guidelines regarding appropriate germplasm collection zones, requirements for genic diversity in restoration, and germplasm transfer rules be reevaluated.

Reference Conditions and Restoration Targets

"Predisturbance" or "pre-Euro-American impact" conditions are used routinely as reference models and descriptions of desired targets for ecological restoration. This assumes, however, that climate hasn't changed between the historic target time and the present, and that human influence hasn't confounded historic conditions. These assumptions are tenuous, and the likelihood of their validity decreases with time between the historic target and present.

In western North America the disturbance period is regularly assumed to start at European/Asian contact with native peoples and their landscape, about 1840–1860, and the prior centuries are used as predisturbance reference conditions. As that period coincides with the coldest part of the Little Ice Age, however, it makes a poor model for twenty-first century restoration. Even in eastern North America, where European contact with the landscape was several centuries earlier, the dominant climate was Little Ice Age, with ecological conditions very different from present. Although "premodern contact" times differ around the world, the point remains: because of climate change, historic conditions are likely to be very different from present and are poor models for restoration.

The use of historic "predisturbance" landscapes is made further tenuous by confounding of human influence on environmental conditions. A bold new hypothesis offers an extreme example. Compiling several lines of evidence, Ruddiman (2003) suggests that humans have significantly altered world climates for over 8,000 years as a result of the spread of agriculture, thus radically changing the physical and ecological trajectory of Earth systems for millennia. Ruddiman's model postulates that cultural practices associated with agriculture triggered increases in wetlands and clearing and burning of forests, which in turn elevated carbon dioxide and methane far above natural levels. The cumulative effect is that global temperature is 1.5°C above what it would be without the anthropogenic input, and that Holocene climates are more stable than they would have been (W. Ruddiman, pers. comm.). Such long-term confounding of human with nonhuman influences challenges use of historic conditions as models for "pristine" or natural conditions in restoration.

Restoration or Realignment?

The discussions above prompt reevaluation of restoration assumptions and goals. If sustainability is to remain a guiding concept in restoration ecology, its interpretation would better focus on sustaining future options for flexibility and adaptation to changing conditions, rather than attempting to recreate stable conditions that resist change. In practice, rather than emphasizing historic ranges, or predisturbance species assemblages, compositions, structures, and landscape patterns, sustainability might instead embrace landscape macrodynamics that have characterized populations and species over long timeframes. These include, for instance, the ability to shift locations significantly, fragment into refugia, expand or contract in range, coalesce with formerly disjunct populations, alter dominance relations, foster non-equilibrium genetic diversities, and accommodate population extirpations and colonizations—all in response to changing regional and global conditions.

Biotas increasingly respond to more than change in climate at these scales. Although a fundamental goal of ecological restoration may be to remove direct manipulative effects of humans, in many cases this is impossible. Air pollution, alterations in landscapes surrounding the restoration site, presence of exotic invasives from distant continents, and changes in disturbance regimes all are imposed on restoration populations with little hope of mitigation. Sustainability in this context implies encouraging successful adaptation to conditions that cannot be turned back. Restoration ecology would better minimize the focus on restoring predisturbance historic structures and functions, and, instead promote efforts that foster natural macrodynamics as processes of adapting to inevitable change.

This does not imply, however, that "anything goes" in restoration. Adaptation is not chaotic, although stochastic processes play important roles. Populations and species respond and adapt to external forces of climate, invasives, and disturbance regimes with definable relationships and patterns; these can be better defined by restoration science, and mimicked in restoration practice. Rather than *restoring* past conditions, the challenge may be *realigning* systems to present and anticipated future conditions in such a way that they can respond adaptively to ongoing change (Millar and Woolfenden 1999b).

Realignment will require an understanding of relevant prehistories as well as changing influences on population dynamics over time in the restoration region. Modeling (quantitatively or qualitatively) these conditions as a trajectory forward into the present and future, including known or anticipated changes in climate as well as other environmental changes, allows target conditions for a realigned population to be developed. An example comes from the Mono Lake, California, ecosystem. A former pluvial lake at the western edge of the Great Basin, Mono Lake receives its water from four Sierra Nevada streams. It has been documented that surface elevation has naturally fluctuated with climate for the last 3,700 years (Stine 1990). In 1941, when the natural elevation of Mono Lake was 1,956 m, the city of Los Angeles began diverting water from Mono's tributaries for municipal use. This caused Mono Basin rivers to dry, aquatic communities and riparian forests to disappear, the lake level to drop, salinity to increase; diversion also triggered significant declines in floral and faunal populations. At the low point, Mono Lake stood 14 m below the 1941 elevation (Stine 1990, 1991). Rather than adopting the predisturbance lake elevation of 1,956 m as a restoration goal, scientists sought to determine a level based on current and anticipated future climate and water conditions. Using historic relationships of surface elevation, snowpack, stream flow, and climate, a water balance model was developed that allowed estimation of the current elevation level, if diversions didn't exist, incorporating antecedent climates (Vorster 1985). Then, estimates of future climate and water trends, extremes, and fluctuations were used to estimate input needed to keep the lake at or above a level considered adequate to sustain aquatic and riparian biota. The resulting lake level modeled was lower than the 1941 level, but this can be understood given that the lake was in rebound at the time from an extremely wet period in the early twentieth century. The court set the level at 1,948 m, incorporating the scientific approach in its decision. A lawsuit involving restoration advocates and the City of Los Angeles resulted in a court decision that incorporated scientific realignment, and the level for Mono Lake was set at 1,948.

Another example of realignment comes from a conservation assessment of Monterey pine (*Pinus radiata*), currently a rare species with three small California coastal populations and two Mexican island populations. Each population is suffering significant declines from human threats and conservation plans have been designed to restore these populations. All plans focus on improving conditions of the extant populations. Another approach to restoration derives from examining Quaternary dynamics of the species (Millar 1999). Analysis of Monterey pine paleorecords and paleoclimatologic data suggests that the species has a repeating metapopulation behavior that responds sensitively to fluctuations in climate. Under favorable climates, Monterey pine responds via colonization of many small, disjunct populations extensively along the California and Mexican coast, while during unfavorable climates, the species contracts to small networks of few populations. This process appears to have been repeated many times in Monterey pine's history.

Monterey pine occurred in the past in coastal northern California locations, as far as 600 km from the closest current native population. In this region, Monterey pine has been planted for landscaping, where it has naturalized widely, eventually spreading into parks and nature reserves. In these locations, the species is considered an unwanted exotic and is aggressively removed as part of restoration projects.

Based on analysis of Monterey pine's paleoecology, a realignment strategy was proposed as a supplemental restoration approach for the species (Millar 1998). The core of the idea is that Monterey pine would be encouraged to persist in certain areas on the north coast rather than being removed as an exotic pest. These locations are defined as areas where Monterey pine has naturalized, overlaps its historic range under similar climates at present, and includes floristic associates found in Monterey pine fossil assemblages. Such realignment locations are considered "neonative" sites for Monterey pine.

Opportunities for Climatological Research from Ecological Restoration

Whereas we have been describing implications from climate sciences to ecological restoration, there are opportunities for the reciprocal, that is, situations where ecological restoration research could advance climate science, as the following sections suggest.

Paleoclimate and Modern Climate Monitoring

Restoration projects that incorporate concepts of climate variability and realignment will, by design, include assessments of prehistory and paleoecological relationships of the restoration ecosystems. When this is done for the benefit of designing a restoration project, it will also contribute to paleoclimatology and paleoecology broadly. Restoration sites are often in locations that would not be selected for paleoclimatological research, and thus their addition to databases can be valuable. A similar benefit derives from archeology, where inferences about cultural sites and ancient human behavior depend on understanding the paleoenvironmental context of settlement sites. The need for this has motivated many excellent new paleoecological and paleoclimatic analyses that would not have been undertaken otherwise. The cumulative effect of such contributions is to saturate regions with local information, which in turn provides new understanding of well-resolved spatial relationships in paleoclimate and paleoecology.

Similarly, assessment of existing ecological conditions for a restoration project may provide information about modern climate relationships. In restoration ecology, causes for impacts are sought. Because many aspects of species biology are sensitive indicators of climate change, such analysis may bring to light important climate effects that would otherwise be unmonitored. For instance, investigation of conifer invasion into mountain meadows revealed that the primary correlations with invasions were not human disturbance (grazing or fire suppression) as had been suspected, but instead, multidecadal climate patterns related to the Pacific Decadal Oscillation (PDO) (Millar et al. 2004). Prior to this assessment, it had not been known whether PDO was expressed in that region or the nature of its effect. Similarly, restoration concerns throughout southwestern United States, where massive forest diebacks are occurring, provide an opportunity to understand spatial patterning of climate. Forest mortality is a sensitive indicator of persistent drought, and so it can be used to delineate resolved maps of the affected climate landscape. Local areas within a general drought zone where mortality is low may indicate microspatial patterns, such as anomalies in storm tracks that wouldn't be recognized otherwise. In this way, restoration projects can contribute to understanding the spatial and temporal influences of current climate and ongoing changes in climate.

Biotic Feedback to the Climate System

Global circulation models are the workhorses of modern climate change analysis. As models become more sophisticated, they are able to accommodate more information and thus become better estimators of future change. Model improvement focuses on greater spatial resolution of climate, incorporation of background climate variability (e.g., ENSO, PDO), and role of biota as feedback to climate. The latter is poorly understood and little integrated into climate change models. Biotic feedback occurs when changes in climate induce changes in biota, which in turn trigger further changes in climate. Examples include climate-mediated changes in vegetation life form (e.g., tree to shrub to grass), or changes in fire regimes or wetland extant, which initiate changes in albedo and carbon storage and eventually feedback to further climate changes. Ruddiman's (2003) Anthropocene is a millennial-scale hypothesis of biotic feedback to climate. Recent experiments in Rocky Mountain flora demonstrate carbon sequestration and feedbacks in meadow ecosystems (Saleska et al. 2002).

Restoration, when conducted with strategic experimental design, provides opportunities to investigate changes in variables related to climate feedback. For instance, introduction of restoration materials presents a chance to compare, at one location, the introduced system with the original, control condition. Monitoring changes in climate-responsive variables, such as carbon storage and albedo for specific environments, ecosystems, species, and climate zones, would contribute valuable input into the nature of biotic feedback and enable calculations of the significance of cumulative effects.

"Space-for-Time" Studies Versus Experimental Manipulation

A challenge in environmental sciences is how to study long-term processes in short, available (grant-determined) timescales. An option is to substitute space for time in research design. Experimental manipulation studies are also surrogates for time. Because studies about climate change implicitly invoke time-series analysis, these two approaches are often used—either one, or the other, or interchangeably, with the assumption that they yield similar results.

What is rarely studied is a comparison of methods and their value or appropriateness for different ecological or physical processes. Long-term studies by both space-for-time (gradient-analysis studies) versus experimental manipulation of Rocky Mountain plant communities to climate change document that, depending on trait, the two approaches can yield similar or different results. For some traits (e.g., flowering phenology) the two experimental approaches were comparable, while for others (e.g., carbon storage) they were discordant (Shaw and Harte 2001).

Because experimental manipulation is expensive, it is less often chosen in ecological or climatological research. Ecological restoration projects, by contrast, are almost always manipulative by design and provide an opportunity to do active experimentation. In such cases, opportunities exist to provide valuable information to climate and climate-related sciences about anticipated future responses to climate change under contrasting situations of experimentation, gradient analysis (e.g., contrasting elevations approximates space-for-time studying climate change), and *in situ* controls.

Global Warming and Restoration Ecology

The specter of global warming has raised much concern in conservation communities. As we now understand, this is not something coming in the future, but something we already are experiencing. At one extreme, the "Anthropocene" era of human-induced climate began 8,000 years ago with the spread of agriculture and its cumulative biotic feedback effects. Nested within this background, warming observed in the last 120 years is partly rebound in the Bond cycling events, superimposed on the longer periods of internal and orbital cycling to which Earth is inextricably bound, and partly induced by modern human effects. Abrupt climate change and vegetation response have been common in Earth's history. On the one hand, this is comforting in that most species, whose roots extend into the Tertiary, must be at least somewhat adapted to the rates of change occurring now. Certain responses, such as massive landscape mortality events, range expansions, minor and major population extirpations, shifts in native ranges, or changes in community composition, may be expressions of landscape-scale resilience and realignment to changing external forces. Accommodating these factors—if we choose to accept them— will require rethinking our concepts about what and where native habitat is, what are "healthy" population sizes, what are causes of changes in population size, and when is change acceptable and appropriate. Society may choose not to accept such consequences and manage instead for other desired conditions. In such cases we will benefit by knowing that our management and conservation efforts may run counter to natural process, and thus restoration efforts may require continuing manipulative input to maintain the desired conditions. The lessons implied from paleoclimatology and paleoecology suggest that making friends with physical and ecological change is an important prerequisite to effective stewardship. Incorporating these ideas into new restoration ecology science and practice will require considerable thought, discussion, experimentation, and research in coming years.

Summary

New information from climate sciences and paleoecology increasingly challenges our ability to grasp dynamic nature. Key concepts for restoration include that natural (without human influence) climate *oscillates* regularly, at *multiple and nested temporal scales*, including interannual, decadal, century, millennial, and multimillennial. In addition, *transitions between climate phases often occur abruptly*, and *vegetation responds to climate change*. Repetitive climate changes at each scale exert significant recurring evolutionary and ecological force on vegetation, and species have evolved mechanisms to adapt despite ongoing environmental change. These include changes in population size, abundance, and productivity, population migration, colonization, and extirpation. Plant communities exist as transient assemblages as species move individually through time and space following favorable climates and environments.

Such conclusions suggest a rethinking of concepts of sustainability and restoration targets. Rather than restoring historic, "pre-human-disturbance" conditions, we may better help species persist into the future by realigning populations with current and future anticipated conditions, and providing options to cope with uncertain futures with certain high variability. The capacity for populations to grow, decline, migrate, colonize, even extirpate, has determined species survival under past conditions of rapid change. Many situations thwart this capacity at present, including fragmentation, urbanization and development, static land-use policies (including conservation measures such as reserves, easements, etc.), and even rigid conservation philosophies that hold species hostage to specific locations and conditions. Understanding that species have coped with change in the past suggests that restoration sciences have opportunities to assist species cope with the dynamics of the current world.

Literature Cited

Adam, D. P. 1988. Palynology of two upper Quaternary cores from Clear Lake, Lake County, California. *USGS Professional Paper* 1363:1–86, plates and figures.

Adam, D. P., and G. J. West. 1983. Temperature and precipitation estimates through the last glacial cycle from Clear Lake, CA. Pollen data. *Science* 219:168–170.

Aggasiz, L. 1840. *Etudes sur les Glaciers*. Neuchatel, Switzerland.

Alley, R. B., P. A. Mayewski, T. Sowers, M. Stuiver, K. C. Taylor, and P. U. Clark. 1997. Holocene climatic instability: A prominent, widespread event 8,200 years ago. *Geology* 25 (6): 483–486.

Anderson, R. S., and S. J. Smith. 1994. Paleoclimatic interpretations of meadow sediment and pollen stratigraphies from California. *Geology* 22:723–726.

Anderson, R. S., and S. J. Smith. 1997. The sedimentary record of fire in montane meadows, Sierra Nevada, California, USA: A preliminary assessment. In *Sediment records of biomass burning and global change*, ed. J. S. Clark, H. Cachier, J. G. Goldammer, and B. Stocks, 313–327. Berlin: Springer-Verlag.

Biondi, R., A. Gershunov, and D. R. Cayan. 2001. North Pacific decadal climate variability since 1661. *Journal of Climate* 14:5–10.

Bond, G., B. Kromer, J. Beer, R. Muscheler, M. Evans, W. Showers, S. Hoffmann, R. Lotti-Bond, I. Hajdas, G. Bonani. 2001. Persistent solar influence on North Atlantic climate during the Holocene. *Science* 294:2130–2136.

Bond, G., W. Showers, M. Cheseby, R. Lotti, P. Almasi, P. deMenocal, P. Priore, H. Cullen, I. Hajdas, and G. Bonani. 1997. A pervasive millennial-scale cycle in North Atlantic Holocene and glacial climates. *Science* 278:1257–1266.

Bradley, R. S. 1999. *Paleoclimatology. Reconstructing climates of the Quaternary*. 2nd Edition. San Diego, CA: Academic Press.

Broecker, W. S., G. Bond, M. Klas, G. Bonani, and W. Wolfli. 1990. A salt oscillator in the glacial Atlantic? I. The concept. *Paleoceanography* 4:469–477.

Brubaker, L. B., and J. S. McLachlan. 1996. Landscape diversity and vegetation response to long-term climate change in the eastern Olympic Peninsula, Pacific Northwest, USA. In *Global change and terrestrial ecosystems*, ed. B. Walker and W. Steffen, 184–203. London: Cambridge University Press.

Bürger, R., and A. Gimelfarb. 2002. Fluctuating environments and the role of mutation in maintaining quantitative genetic variation. *Genetical Research* 80:31–46.

Clark, D. H., and A. R. Gillespie. 1997. Timing and significance of late-glacial and Holocene cirque glaciation in the Sierra Nevada, California. *Quaternary Research* 19:117–129.

Clark, P. U., S. J. Marshall, G. K. Clarke, S. Hostetler, J. M. Licciardi, and J. T. Teller. 2001. Freshwater forcing of abrupt climate change during the last glaciation. *Science* 293:283–287.

Cole, K. 1983. Late Pleistocene vegetation of Kings Canyon, Sierra Nevada, California. *Quaternary Research* 19:117–129.

Cronin, T. M. 1999. *Principles of paleoclimatology*. New York: Columbia University Press.

Cuffey, K. M., G. D. Clow, R. B. Alley, M. Stuvier, E. D. Waddington, and R. W. Saltus. 1995. Large Arctic temperature change at the Wisconsin-Holocene glacial transition. *Science* 270:455–458.

Dansgaard, W., S. J. Johnsen, H. B. Clausen, D. Dahl-Jensen, N. S. Gundestrup, C. U. Hammer, C. S. Hvidberg, J. P. Steffensen, A. E. Sveinbjornsdottir, J. Jouzel, and G. Bond. 1993. Evidence for general instability of climate from a 250-kyr ice-core record. *Nature* 364:218–220.

Davis, O. K. 1999a. Pollen analysis of a late-glacial and Holocene sediment core from Mono Lake, Mono County, California. *Quaternary Research* 52:243–249.

Davis, O. K. 1999b. Pollen analysis of Tulare Lake, California: Great Basin-like vegetation in central California during the full-glacial and early Holocene. *Review of Palaeobotany and Palynology*. 107:249–257.

Davis, M. B., K. D. Woods, S. L. Webb, and R. P. Futyma. 1986. Dispersal versus climate: Expansion of *Fagus* and *Tsuga* into the Upper Great Lakes region. *Vegetatio* 67:93–103.

Diaz, H. F., and V. Markgraf, editors. 2000. *El Niño and the Southern Oscillation: Multiscale variability, global, and regional impacts*. Cambridge, UK: Cambridge University Press.

Esper, J., E. R. Cook, and F. H. Schweingruber. 2002. Low-frequency signals in long tree-ring chronologies for reconstructing past temperature variability. *Science* 295:2250–2253.

Falk, D. A., and K. E. Holsinger. 1991. *Genetics and conservation of rare plants*. New York, NY: Oxford University Press.

Falk, D. A., C. I. Millar, and M. Olwell. 1996. Guidelines for developing a rare plant reintroduction plan. Part 5. In *Restoring diversity: Strategies for reintroduction of endangered plants*, ed. D. A. Falk, C. I. Millar, and M. Olwell, 453–490. Washington, DC: Island Press.

García-Ramos, G., and D. Rodríguez. 2002. Evolutionary speed of species invasions. *Evolution* 56:661–668.

Grayson, D. K. 1993. *The desert's past. A natural prehistory of the Great Basin*. Washington, DC: Smithsonian Institution Press.

Grimm, E. C., G. L. Jacobson, W. A. Watts, B. C. Hansen, and K. A. Maasch. 1993. A 50,000-year record of climate oscillations from Florida and its temporal correlation with Heinrich events. *Science* 261:198–200.

Grissino-Mayer, H. D. 1996. A 2,129-year reconstruction of precipitation for northwestern New Mexico, USA. In *Tree rings, environment, and humanity*, ed. J. S. Dean, D. M. Meko, and T. W. Swetnam, 191–204. Tucson: Radiocarbon.

Grove, J. M. 1988. *The little ice age*. London: Methuen Publishing.

Guerrant Jr., E. O., K. Havens, et al., editors. 2004. *Ex situ plant conservation: Supporting species survival in the wild*. Washington, DC: Island Press.

Hays, J. D., J. Imbrie, and N. J. Shackleton. 1976. Variations in the Earth's orbit: Pacemaker of the ice ages. *Science* 194:1121–1132.

Heinrich, H. 1988. Origin and consequence of cyclic ice rafting in the northeast Atlantic Ocean during the past 130,000 years. *Quaternary Research* 29:142–152.

Heusser, L. 1995. Pollen stratigraphy and paleoecologic interpretation of the 160 k.y. record from Santa Barbara Basin, Hole 893A. In *Proceedings of the ocean drilling program, scientific results*, ed. J. P. Kennet, J. G. Baldauf, and M. Lyle, 265–277. Volume 146, Part 2. Ocean Drilling Program. College Station, TX.

Heusser, L. 1998. Direct correlation of millennial-scale changes in western North American vegetation and climate with changes in the California current system over the past ~60 kyr. *Paleogeograph, Paleoclimatology, and Palaeoecology* 13:252–262.

Heusser, L. E. 2000. Rapid oscillations in western North America vegetation and climate during oxygen isotope stage 5 inferred from pollen data from Santa Barbara Basin (Hole 893A). *Palaeogeography, Palaeoclimatology, and Palaeoecology* 161:407–421.

Heusser, L. E., and F. Sirocko. 1997. Millennial pulsing of environmental change in southern California from the past 24 k.y.: A record of Indo-Pacific ENSO events? *Geology* 25:243–246.

Hughes, M. K., and H. F. Diaz. 1994. Was there a "Medieval Warm Period" and, if so, where and when? *Climate Change* 26:109–142.

Hunter, M. L. 1996. *Fundamentals of conservation biology*. Cambridge, MA: Blackwell Science.

Imbrie, J., A. Berger, and E. Boyle. 1993. On the structure and origin of major glaciation cycles. 2. The 100,000 year cycle. *Paleoceanography* 8:699–735.

Imbrie, J., E. Boyle, S. C. Clemens, A. Duffy, W. R. Howard, G. Kukla, J. Kutzbach, D. G. Martinson, A. McIntyre, A. C. Mix, B. Molfino, J. J. Morley, L. C. Peterson, N. G. Pisias, W. L. Prell, M. E. Raymo, N. J. Shackleton, and J. R. Toggweiler. 1992. On the structure and origin of major glaciation cycles. 1. Linear responses to Milankovitch forcing. *Paleoceanography* 7:701–738.

International Panel on Climate Change (IPCC). 2001. *Climate change 2001. Third assessment report of the intergovernmental panel on climate change*. Three reports and overview for policy makers. Cambridge, UK: Cambridge University Press.

Jackson, G., T. Webb III, E. C. Grimm, W. F. Ruddiman, H. E. Wright Jr., editors. 1987. North America and adjacent oceans during the last deglaciation. *Geological Society of America* 3:277–288.

Jackson, S. T. 1997. Documenting natural and human-caused plant invasions using paleoecological methods. In *Assessment and Management of Plant Invasions*, ed. J. O. Luken and J. W. Thieret, 37–55. New York, NY: Springer-Verlag.

Jackson, S. T., and J. T. Overpeck. 2000. Responses of plant populations and communities to environmental changes of the late Quaternary. *Paleobiology* 25:194–220.

Jensen, K. 1935. Archaeological dating in the history of North Jutland's vegetation. *Acta Archaeologica* 5:185–214.

Jordan, W. R., M. E. Gilpin, and J. D. Aber, editors. 1990. *Restoration ecology: A synthetic approach to ecological research*. Cambridge, UK: Cambridge University Press.

Kennett, J. P. 1990. The Younger Dryas cooling event: An introduction. *Paleoceanography* 5:891–895.

Kitzberger, T., T. W. Swetnam, and T. T. Veblen. 2001. Inter-hemispheric synchrony of forest fires and the El Niño-Southern Oscillation. *Global Ecology and Biogeography* 10:315–326.

Koehler, P. A., and R. S. Anderson. 1995. Full-glacial shoreline vegetation during the maximum high stand at Owens Lake, California. *Great Basin Naturalist* 54 (2): 142–149.

Lackey, R. T. 1995. Seven pillars of ecosystem management. *Landscape and Urban Planning* 40 (1): 21–30.

Lele, S., and R. B. Norgaard. 1996. Sustainability and the scientist's burden. *Conservation Biology* 10:354–365.

Lesica, P., and F. W. Allendorf. 1999. Ecological genetics and the restoration of plant communities: Mix or match? *Restoration Ecology* 7 (1): 42–50.

Lorius, C., J. Jouzel, D. Raynaud, J. Hansen, and H. LeTreut. 1990. The ice-core record: Climate sensitivity and future greenhouse warming. *Nature* 347:139–145.

Mantua, N. J., S. R. Hare, Y. Zhang, J. M. Wallace, and R. C. Francis. 1997. A Pacific interdecadal climate oscillation with impacts on salmon production. *Bulletin of the American Meteorological Society* 78:1069–1079.

Matyas, C. 1996. Climatic adaptation of trees—Rediscovering provenance tests. *Euphytica* 92:45–54.

Milankovitch, M. 1941. Canon of insolation and the ice-age problem. *Royal Serbian Academy, Special Publication Number 132*. Translated from the German by the Israel Program for Scientific Translations, Jerusalem, 969.

Millar, C. I. 1998. Reconsidering the conservation of Monterey pine. *Fremontia* 26 (3): 12–16.

Millar, C. I. 1999. Evolution and biogeography of *Pinus radiata*, with a proposed revision of its Quaternary history. *New Zealand Journal of Forestry Science* 29 (3): 335–365.

Millar, C. I., and W. J. Libby. 1989. Disneyland or native ecosystem: The genetic purity question. *Restoration Management* 7 (1): 18–23.

Millar, C. I., and W. B. Woolfenden. 1999a. Sierra Nevada forests: Where did they come from? Where are they going? What does it mean? In *Natural resource management: Perceptions and realities*, ed. R. E. McCabe and S. E. Loos, 206–236. Transactions of the 64th North American Wildlife and Natural Resource Conference. Washington, DC: Wildlife Management Institute.

Millar, C. I., and W. B. Woolfenden. 1999b. The role of climate change in interpreting historical variability. *Ecological Applications* 9 (4): 1207–1216.

Millar, C. I., R. D. Westfall, D. L. Delany, J. C. King, and L. J. Graumlich. 2004. Response of subalpine conifers in the Sierra Nevada, California, USA, to twentieth-century warming and decadal climate variability. *Arctic, Antarctic, and Alpine Research* 36 (2): 181–200.

Nowak, C. L., R. S. Nowak, R. J. Tausch, and P. E. Wigand. 1994. Tree and shrub dynamics in northwestern Great Basin woodland and shrub steppe during the Late-Pleistocene and Holocene. *American Journal of Botany* 81:265–277.

Overpeck, J., K. Hughen, D. Hardy, R. Bradley, R. Case, M. Douglas, B. Finney, K. Gajewski, G. Jacoby, A. Jennings, S. Lamoureux, A. Lasca, G. MacDonald, J. Moore, M. Retelle, S. Smith, A. Wolfe, and G. Zielinski. 1997. Arctic environmental change of the last four centuries. *Science* 278:1251–1256.

Pease, C. M., R. Lande, and J. J. Bull. 1989. A model of population growth, dispersal and evolution in a changing environment. *Ecology* 70:1657–1664.

Peterson, D. W., and D. L. Peterson. 2001. Mountain hemlock growth responds to climatic variability at annual and decadal timescales. *Ecology* 82 (12): 3330–3345.

Peterson, D. W., D. L. Peterson, and G. J. Ettl. 2002. Growth responses of subalpine fir to climatic variability in the Pacific Northwest. *Canadian Journal of Forest Resources* 32:1503–1517.

Petit, J. R., I. Basile, A. Leruyuet, D. Raynaud, C. Lorius, J. Jouzel, M. Stievenard, V. Y. Lipenkov, N. I. Barkov, B. B. Kudryashov, M. Davis, E. Saltzman, and V. Kotlyakov. 1997. Four climate cycles in Vostok ice core. *Nature* 387:56–58.

Poore, R. Z., H. J. Dowsett, J. A. Barron, L. Heusser, A. C. Ravelo, and A. Mix. 2000. Multiproxy record of the last interglacial (MIS 5e) of central and northern California USA, from Ocean Drilling Program sites 1018 and 1020. *USGS Professional Paper* 1632:1–19.

Power, M. J. 1998. Paleoclimatic interpretations of an alpine lake in southcentral Sierra Nevada. MS Thesis, Northern Arizona University. Flagstaff.

Raymo, M. E., and W. F. Ruddiman. 1992. Tectonic forcing of late Cenozoic climate. *Nature* 359:117–122.

Rehfeldt, G. E., W. R. Wykoff, and C. C. Ying. 2001. Physiologic plasticity, evolution, and impacts of a changing climate on *Pinus contorta*. *Climatic Change* 50:355–376.

Rehfeldt, G. E., C. C. Ying, D. L. Spittlehouse, and D. A. Hamilton. 1999. Genetic responses to climate in *Pinus contorta:* Niche breadth, climate change, and reforestation. *Ecological Monographs* 69:375–407.

Rice, K. J., and N. C. Emory. 2003. Managing microevolution: Restoration in the face of global change. *Frontiers in Ecology* 1 (9): 469–478.

Ruddiman, W. F. 2001. *Earth's climate: Past and future*. New York: W.H. Freeman & Co.

Ruddiman, W. F. 2003. The Anthropocene greenhouse era began thousands of years ago. *Climatic Change* 61:261–293.

Saleska, S., M. Shaw, M. Fischer, J. Dunne, M. Shaw, M. Holman, C. Still, and J. Harte. 2002. Carbon-cycle feedbacks to climate change in montane meadows: Results from a warming experiment and a natural climate gradient. *Global Biogeochemical Cycles* 16 (4): 1055.

Shaw, M., and J. Harte. 2001. Control of litter decomposition under simulated climate change in a subalpine meadow: The role of plant species and microclimate. *Ecological Applications* 11 (4): 1206–1223.

Smith, J., and J. Uppenbrink. 2001. Earth's variable climate past: Special issue on paleoclimatology. *Science* 292:657–693.

Stine, S. 1990. Late Holocene fluctuations of Mono Lake, eastern California. *Palaeogeogaphy, Palaeoclimatology, Palaeoecology* 78:333–382.

Stine, S. 1991. Geomorphic, geographic, and hydrographic basis for resolving the Mono Lake controversy. *Environmental Geology and Water Sciences* 17:67–83.

Stine, S. 1994. Extreme and persistent drought in California and Patagonia during Medieval time. *Nature* 369:546–549.

Swetnam, T. 1993. Fire history and climate change in giant sequoia groves. *Science* 262:885–889.

Swetnam, T. W., and J. L. Betancourt. 1998. Mesoscale disturbance and ecological response to decadal climatic variability in the American Southwest. *Journal of Climate* 11:3128–3147.

Tausch, R, S. Mensing, and C. Nowak. 2004. Climate change and associated vegetation dynamics during the Holocene—The paleoecological record. In *Great Basin riparian ecosystems: Ecology, management and restoration*, ed. J. C. Chambers and J. R. Miller, 24–48. Washington, DC: Island Press.

Thompson, R. S. 1988. Western North America. In *Vegetation History*, ed. B. Huntley and T. Webb III, 415–459. Dordrecht, The Netherlands: Kluwer Academic Press.

Thompson, R. S. 1990. Late Quaternary vegetation and climate in the Great Basin. In *Packrat middens. The last 40,000 years of biotic change*, ed. J. L. Betancourt, T. vanDevender, and P. S. Martin, 201–239. Tucson: University of Arizona Press.

Vorster, P. 1985. A water balance forecast model for Mono Lake, California. *Earth Resources Monograph* 10. San Francisco: USDA, Forest Service, PSW Region.

Webb, T. 1986. Is vegetation in equilibrium with climate? How to interpret Late-Quaternary pollen data. *Vegetatio* 67:75–91.

Westfall, R. D., and C. I. Millar. 2004. Genetic consequences of forest population dynamics influenced by historic climatic variability in the western USA. *Forest Ecology and Management* 197:159–170.

White, J. W., T. Popp, S. J. Johnsen, V. Masson, and J. Jouzel. 2001. Clocking the speed of climate change: The end of the Younger Dryas as recorded by four Greenland ice cores. Abstract. Fall Meeting of the American Geophysical Union, 10–12 December. San Francisco. Pg. F22.

Williams, J. W., B. N. Shuman, and T. Webb III. 2001. Dissimilarity analyses of late-Quaternary vegetation and climate in eastern North America. *Ecology* 82:3346–3362.

Woolfenden, W. B. 1996. Quaternary vegetation history. *Sierra Nevada ecosystem project: Final report to Congress.* Volume II. Assessments and scientific basis for management options. Wildland Resources Center. Report No. 37, pp 47–70.

Wright, H. E. 1989. The Quaternary. In *The Geology of North America*, ed. A. W. Bally and A. R. Palmer, 513–536. Boulder, CO: Geologic Society of America.

Zachos, J., M. Pagaini, L. Sloan, E. Thomas, and K. Billups. 2001. Trends, rhythms, and aberrations in global climate, 65 Ma to present. *Science* 292:686–693.

Zhang, Y., J. M. Wallace, and D. S. Battisti. 1997. ENSO-like interdecadal variability: 1900–1993. *Journal of Climate* 10:1004–1020.

Chapter 16

Integrating Restoration Ecology and Ecological Theory: A Synthesis

DONALD A. FALK, MARGARET A. PALMER, AND JOY B. ZEDLER

Restoration ecology would be easier in a world of linear, deterministic, ordered, predictable change tending toward stable equilibria. In such a world, many restoration projects would require only that the restorationist give a degraded or damaged ecosystem an initial push, and then stand back and watch the system heal itself.

But this is not the world that most ecologists believe we inhabit (Botkin 1990; Wu and Loucks 1995). Contemporary ecology describes a world characterized largely by nonlinear, stochastic, imperfectly predictable processes where historical contingencies, spatial context, and initial conditions are strong determinants of change following perturbation, and in which equilibria, if they exist at all, are likely to be unstable (Maurer, Menninger and Palmer, Suding and Gross, this volume). Contemporary ecology sees constant interactions between intrinsic or endogenous dynamics (for example, population cycles) and a nonstationary physical environment with multiple frequencies and amplitudes of change. What we now understand about climate variability suggests that the physical environment is nowhere near as stable—even on "ecological" time scales—as was once supposed (Cayan et al. 1998; McCabe et al. 2004; Millar and Brubaker, this volume). Indeed, ecological and evolutionary adaptation to spatial and temporal variability is a powerful new line of ecological inquiry (Chesson 2000; Clauss and Venable 2000; Reed et al. 2003).

These emerging views of how the world works pose a fundamental challenge for restoration ecology (Pickett and Parker 1994; Hobbs and Norton 1996; Anand and Desrosiers 2004): Given that ecosystems are in a constant state of dynamic flux, what state should be restored?

The contributors to this volume offer some novel and important answers, if only as working hypotheses. On the whole they emphasize *ecological processes* that underlie the visible composition and structure of ecological communities. Although "saving the parts" (Leopold 1953) is often used as shorthand for restoration, restoration ecology shows that *how* the pieces are assembled, and how they work together, are at least as critical (Naeem, this volume).

Retaining all the individual components (species) of communities and ecosystems remains important, however. Restoration is becoming more attuned to underappreciated keystone functional groups, such as soil microflora and microfauna, cryptobiotic crusts, and dispersal agents. Uncommon and rare species may also play unknown ecological roles at small spatial scales. Nonetheless, there is a world of difference between having all the parts of an automobile laid out neatly on the garage floor and an assembled machine that can take you down the highway. Restoration requires having all the right pieces, even if the real interest is how they will function once reassembled.

Perhaps the most important lesson from these fifteen chapters is the reciprocal, mutually beneficial relationship between ecological theory *sensu latu* and restoration ecology (Hobbs 1998; Palmer et al., this volume). We see many compelling reasons for closer connections between restoration ecology and ecological theory, two of which emerge as central themes in this book.

Ecological Theory Can Help Inform and Improve the Science and Practice of Restoration

The idea that ecological theory can be of significant value to restoration science and practice runs through every chapter in this book. The science of restoration was motivated initially by practical applications rather than by theoretical inquiry (Jordan et al. 1987). Increasingly, however, restoration ecology is defining itself as a scientific discipline in the sense that it strives not only to observe, but to explain (Palmer et al. 1997; Ginzburg and Jensen 2004). This is reflected in the growth of journals such as *Restoration Ecology*, as well as academic and research programs in restoration ecology around the world (www.seri.org).

In principle, there are important differences between restoration science and restoration practice. Science is a means of inquiry, which progresses by asking questions, collecting data, and forming interpretations that help us to understand the world around us. The aim of research is ultimately understanding and the ability not only to quantify but, more importantly, to offer coherent explanations for how the world works (Weiner 1995). In science, to learn is to succeed.

Restoration practice typically begins with a different goal, which is to accomplish specific objectives. Clients might want to reestablish a species in a particular place; reduce rates of soil erosion; bring the pH of a lake within its natural range; reestablish a natural disturbance regime, such as fire; eliminate an aggressive invading species; or create vegetation structure that will provide nesting habitat for a species of interest.

In reality, the line between restoration science and practice is often fuzzy, and both can advance simultaneously if each capitalizes upon the other. Even when a restoration project has a limited objective, the practitioner usually tries out a few alternative treatments to evaluate "what works." We assert throughout this book that even applied restoration practice offers many opportunities for learning and testing of scientific ideas. For example, Callaway et al. (2003) accomplished restoration of the species-rich canopy in a degraded salt marsh plain while simultaneously testing predictions of biodiversity-ecosystem function theory.

Of course, not all important insights begin with a theoretical question; ecologists sometimes begin by being good natural historians, observing and processing what they see in a synthetic, holistic mode of thinking. In complex systems, the best questions—and the most challenging problems—may not be amenable to a simple reductionist paradigm (Pickett et al. 1994). Depending on one's training and research focus, testable hypotheses can emerge from good natural history at least as often as the reverse (Weiner 1995).

Restoration ecology can also benefit from closer integration with ecological theory in the area of research design and statistical analysis (Michener 1997). Restoration experiments are often constrained by practical considerations that limit replication, the use of balanced factorial designs, and the range of experimental conditions, especially at large spatial scales. New research designs and statistical methods can help restorationists deal with these contingencies and, in so doing, help solidify restoration ecology as an empirical science (Osenberg et al., this volume). Likewise, mathematical and simulation models are becoming more widely recognized in restoration ecology as valuable tools for anticipating and, in many cases, simulating the responses of complex systems to a variety of perturbations (Anand and Desrosiers 2004; Urban, this volume). Broader application of ecological modeling could help restoration ecology grow beyond trial-and-error experimentation.

Restoration Ecology Can Help Test Basic Elements of Ecological Theory

While we contend that restoration will benefit from closer integration with ecological theory, a parallel tenet of this book is that restoration ecology has a great deal of reciprocal value to offer (Jordan et al. 1987; Hobbs 1998). The contributing authors of this book highlight many interesting opportunities for restoration ecology to contribute to the development of ecological theory. It is hardly an exaggeration to suggest that restoration ecology offers some of the most promising prospects for advancements in our understanding of how ecosystems work.

We find examples of such potential at all levels of biological hierarchy. The simple act of augmenting or reintroducing a population of a single species provides opportunities for controlled, empirical tests of concepts in population and ecological genetics, such as founder events, effective population size, inbreeding and outbreeding depression, metapopulation genetics, and temporal changes in gene frequencies (Falk et al., this volume). At the population level, restoration ecology offers the opportunity to test predictions about dispersal and establishment limitation, demographic variability, intra- and interspecific competition, and the contribution of metapopulation dynamics to persistence and resilience in changing environments (Maschinski, this volume).

Restorationists have already learned a great deal about the influence of spatial variability of resources, such as water and limiting nutrients, and how fine-scale heterogeneity influences species interactions and community structure (Larkin et al., this volume). Similarly, it is the large extent of manipulation needed to restore land (and water) that allows community and ecosystem ecologists to test ideas at the large scale. Restoration of whole communities gives ecologists unparalleled opportunities for detailed and controlled experimentation with higher-order processes, such as community assembly, food-web organization, diversity-stability relationships, and successional pathways, under controlled, repeatable circumstances (Menninger and Palmer, Van der Zanden et al., Naeem, Suding and Gross, this volume).

Disturbed or altered communities and ecosystems, including those that have been invaded by exotic species, are a central domain of ecological restoration (D'Antonio and Chambers, this volume). Restoration ecology overlaps substantially with disturbance ecology and invasive species control efforts, partly because species invasions are often a critical factor triggering the call for restoration. Degraded and restored settings offer a chance to examine the properties of invasive species, invaded communities, and the effects of removal at large scales under controlled conditions.

By its very nature, restoration exposes species to novel environmental conditions. In the short term, controlled *in situ* experimentation in a restoration context can reveal the ecophysiological responses of organisms to stress, and phenotypic tolerance of extreme conditions (Ehleringer and Sandquist, this volume). In the longer term, restoration creates empirical tests of the ability of species to adapt to novel evolutionary environments (Stockwell et al., this volume). The evolutionary response to changing climate, biogeochemical cycles, and landscape configuration may be the most pervasive outcome, not only of our globally altered environments, but also of our efforts to restore them.

Good restoration practice and science both require continual observation and data collection. To realize their full scientific potential, restoration projects need to acquire adequate baseline (pretreatment) data, establish treatments as replicated experiments, and monitor outcomes systematically (Zedler and Callaway 2003; Zedler 2005). Unfortunately, this is still not practiced consistently; for example, Bernhardt and colleagues (2005) found that only 10% of more than 37,000 river restoration projects in the United States had documentation and monitoring protocols in place. Although some responses to restoration actions are visible immediately after treatments, others may take years to unfold. If we do not monitor consistently to decadal scales, we run the risk of basing adaptive management decisions only on the short-term component of ecological response. We would then miss important slow changes in species composition, competitive and coexistence interactions, soil properties, hydrologic regimes, and community structure (e.g., Friederici 2003; Temperton et al. 2004; Packard and Mutel 2005). If we want to learn how best to restore the dynamics of ecological systems, even in an applied context, we need to follow the outcomes of representative projects over decades, with preference given to well-documented, replicated experiments (Larkin et al., this volume).

Ecology *sensu latu* embodies a wide domain of subjects and subdisciplines and, in this first attempt at integration, we have not covered them all. Belowground ecology, species interactions, social organization, quantitative spatial ecology, ecosystem ecology, biosphere-atmosphere couplings, and ecological time-series analysis are among the areas within ecology that merit further exploration from a restoration perspective. Ecology's allied peer disciplines—such as soil science, hydrology, geomorphology, and biogeochemistry—are equally deserving of a careful treatment of links to restoration theory and practice. We find ample room for fuller exploration of the potential to join restoration ecology to all of these fields.

In the meantime, we hope this book will lead more restoration ecologists to look to ecological theory for a unifying framework for their work, and more ecologists to look to restoration as an opportunity to test their most basic ideas.

Literature Cited

Anand, M., and R. E. Desrosiers. 2004. Quantification of restoration success using complex systems concepts and models. *Restoration Ecology* 12 (1): 117–123.

Bernhardt, E. S., M. A. Palmer, J. D. Allan, G. Alexander, K. Barnas, S. Brooks, J. Carr, S. Clayton, C. Dahm, J. Follstad-Shah, D. Galat, S. Gloss, P. Goodwin, D. Hart, B. Hassett, R. Jenkinson, S. Katz, G. M. Kondolf, P. S. Lake, R. Lave, J. L. Meyer, T. K. O'Donnell, L. Pagano, B. Powell, and E. Sudduth. 2005. Synthesizing U.S. river restoration efforts. *Science* 308:636–637.

Botkin, D. B. 1990. *Discordant harmonies: A new ecology for the twenty-first century*. New York: Oxford University Press.

Callaway, J. C., G. Sullivan, and J. B. Zedler. 2003. Species-rich plantings increase biomass and nitrogen accumulation in a wetland restoration experiment. *Ecological Applications* 13:1626–1639.

Cayan, D. R., M. D. Dettinger, H. F. Diaz, and N. Graham. 1998. Decadal variability of precipitation over western North America. *Journal of Climate* 11:3148–3166.

Chesson, P. 2000. Mechanisms of maintenance of species diversity. *Annual Review of Ecology and Systematics* 31:343–366.

Clauss, M. J., and D. L. Venable. 2000. Seed germination in desert annuals: An empirical test of adaptive bet hedging. *American Naturalist* 155 (2): 168–186.

Friederici, P., editor. 2003. Ecological restoration of southwestern Ponderosa pine forests. *Science and practice of ecological restoration*. Washington, DC: Island Press.

Ginzburg, L. R., and C. X. J. Jensen. 2004. Rules of thumb for judging ecological theories. *Trends in Ecology & Evolution* 19 (3): 121–126.

Hobbs, R. J. 1998. Managing ecological systems and processes. In *Ecological scale: Theory and applications*, ed. D. L. Peterson and V. T. Parker, 459–484. New York: Columbia University Press.

Hobbs, R. J., and D. A. Norton. 1996. Towards a conceptual framework for restoration ecology. *Restoration Ecology* 4 (2): 93–110.

Jordan, W. R. I., M. E. Gilpin, and J. D. Aber, editors. 1987. *Restoration ecology: A synthetic approach to ecological research*. Cambridge, UK: Cambridge University Press.

Leopold, A. 1953. *The Round River*. New York: Oxford University Press.

McCabe, G. J., M. A. Palecki, and J. L. Betancourt. 2004. Pacific and Atlantic ocean influences on multidecadal drought frequency in the United States. *Proceedings of the National Academy of Sciences, USA* 101 (12): 4136–4141.

Michener, W. K. 1997. Quantitatively evaluating restoration experiments: Research design, statistical analysis, and data management considerations. *Restoration Ecology* 5 (4): 324–337.

Packard, S., and C. F. Mutel. 2005. *Tallgrass restoration handbook for prairies, savannas, and woodlands*. Washington, DC: Island Press.

Palmer, M. A., R. F. Ambrose, and N. L. Poff. 1997. Ecological theory and community ecology. *Restoration Ecology* 5 (4): 291–300.

Pickett, S. T. A., J. Kolasa, and C. D. Jones. 1994. *Ecological understanding: The nature of theory and the theory of nature*. San Diego: Academic Press.

Pickett, S. T. A., and V. T. Parker. 1994. Avoiding the old pitfalls: Opportunities in a new discipline. *Restoration Ecology* 2 (2): 75–79.

Reed, D. H., E. H. Lowe, D. A. Brisco, and R. Frankham. (2003). Fitness and adaptation in a novel environment: Effect of inbreeding, prior environment, and lineage. *Evolution* 57:1822–1828.

Temperton, V. M., R. J. Hobbs, T. Nuttle, and S. Halle, editors. 2004. Assembly rules and restoration ecology. In *Science and practice of ecological restoration*. Washington, DC: Island Press.

Weiner, J. 1995. On the practice of ecology. *Journal of Ecology* 83:153–158.

Wu, J., and O. L. Loucks. 1995. From balance of nature to hierarchical patch dynamics: A paradigm shift in ecology. *Quarterly Review of Biology* 70 (4): 439–466.

Zedler, J. B. 2005. Restoring wetland plant diversity: A comparison of existing and adaptive approaches. *Wetlands Ecology & Management* 13 (1): 5–14.

Zedler, J. B., and J. C. Callaway. 2003. Adaptive restoration: A strategic approach for integrating research into restoration projects. In *Managing for healthy ecosystems*, ed. D. J. Rapport, W. L. Lasley, D. E. Rolston et al., 164–174. Boca Raton: Lewis Publishers.

EDITORS

Don Falk is an associate professor in the School of Natural Resources and the Laboratory of Tree-Ring Research at the University of Arizona. His research focuses on fire history, fire ecology, dendroecology, and restoration ecology, including multiscale studies of fire as an ecological and physical phenomenon. He has received research support and fellowships from the National Science Foundation and the USDA Forest Service and has been honored for his work by the Pinchot Institute, the Australian Fulbright Foundation, the McGinnies Scholarship in Arid Lands Studies, and the International Association of Landscape Ecology. In 2003 he received the Edward S. Deevey Award from the Ecological Society of America for outstanding graduate work in paleoecology; he is also a fellow of the American Association for the Advancement of Science (AAAS) and a member of the Arizona Forest Health Advisory Council. Dr. Falk was cofounder and executive director of the Center for Plant Conservation, and served subsequently as the first executive director of the Society for Ecological Restoration (now SER International). His publications include two books on conservation genetics and restoration of endangered species. Falk currently serves as associate editor for the Island Press–SER series, *Science and Practice of Restoration Ecology*.

Margaret A. Palmer is a professor at the University of Maryland, College Park, and director of the Chesapeake Biological Laboratory of the University of Maryland Center for Environmental Sciences. Her research has focused on coastal and freshwater ecosystems with a current emphasis on restoration of biodiversity and ecosystem processes in urbanizing watersheds. She is an international leader in riverine science and restoration ecology, with more than 100 publications. She recently led efforts to establish the National River Restoration Science Synthesis, the first comprehensive national database on river and stream restoration. Her research has been supported by the National Science Foundation, the U.S. Environmental Protection Agency, the C. S. Mott Foundation, CALFED, and the David and Lucile Packard Foundation. Dr. Palmer serves on numerous scientific advisory boards, including American Rivers; the Center for Watershed Protection; the Grand Canyon Research and Monitoring Center; and the National Science Foundation's National NEON Design Consortium and National Center for Earth-surface Dynamics. Her awards include AAAS Fellow, Aldo Leopold Leadership Fellow, Lilly Fellow, and the University of Maryland Distinguished Scholar-Teacher Award.

Joy B. Zedler is the Aldo Leopold Professor in Restoration Ecology and professor of botany at the University of Wisconsin–Madison. She directs research at the UW–Madison Arboretum and trains graduate students in restoration ecology and adaptive restoration involving Wisconsin herbaceous wetlands and southern California tidal marshes. Dr. Zedler has served as a board member for The Nature Conservancy (both nationally and for the Wisconsin chapter) and Environmental Defense. Her publications concern adaptive restoration, degradation of wetlands by invasive plants, indicators of wetland integrity, relationships between diversity and ecosystem function, shortcomings of mitigation policy, and methods of improving the effectiveness of ecological restoration. Her professional awards include a Career Achievement Award from her alma mater, Augustana College, in 2004; the William Niering Outstanding Educator Award, from the Estuarine Research Federation; the First Award for Achievements in Regional Wetland Conservation from the Southern California Wetlands Recovery Project, in 2000; the National Wetlands Award for Science Research, in 1997; and the Society for Ecological Restoration's Theodore Sperry Award, in 1995. She served on three panels for the National Research Council concerning wetland policy and compensatory mitigation, including chairing the panel that wrote *Compensating for Wetland Losses Under the Clean Water Act.*

CONTRIBUTORS

Benjamin M. Bolker is an associate professor in the Department of Zoology at the University of Florida. Dr. Bolker is a theoretical ecologist with expertise in the dynamics of spatially structured systems. He has contributed to our understanding of a wide range of problems, including measles outbreaks, forest dynamics, reef fish recruitment and management, and sea turtle conservation.

Linda B. Brubaker is professor of forest ecology in the College of Forest Resources, University of Washington, with primary emphasis in dendrochronology and palynology. Her research interests focus on the responses of vegetation to climate change and disturbance at different temporal and spatial scales in the Pacific Northwest, Alaska, and the Siberian Far Northeast. She has served on numerous advisory committees for arctic research.

Jeanne C. Chambers is a research ecologist with the USDA, Forest Service, Rocky Mountain Research Station, and an adjunct professor at the University of Nevada, Reno. Her research focuses on the restoration of disturbed or degraded ecosystems in the western United States. She has served as the team leader of the Great Basin Ecosystem Management Project for Restoring and Maintaining Sustainable Watersheds and Riparian Ecosystems since 1993.

Carla M. D'Antonio is the Schuyler Chair of Environmental Studies and professor of ecology in the Department of Ecology, Evolution, and Marine Biology at the University of California, Santa Barbara. Her research has focused on terrestrial shrubland and grassland ecosystems and processes affecting the success and impacts of invading species in these habitats. After serving as a faculty member at UC Berkeley for 12 years, she was the lead scientist with the USDA's Exotic and Invasive Weeds Research Unit in Reno, Nevada, for three years.

James R. Ehleringer is a Distinguished Professor of biology at the University of Utah and serves as director of the Stable Isotope Ratio Facility for Environmental Research (SIRFER). His research focuses on understanding terrestrial ecosystem processes through stable isotope analyses. Dr. Ehleringer's foci include gas exchange and biosphere-atmosphere interactions; water relations in arid land, riparian, and urban ecosystems; and stable isotopes applied to homeland security issues. He serves as editor-in-chief for *Oecologia* and chairs the Biosphere-Atmosphere Stable Isotope Network (BASIN).

Claudio Gratton is an assistant professor of entomology at the University of Wisconsin–Madison. His interests are in understanding the role of top-down and bottom-up forces and landscape interactions in terrestrial arthropod food webs in salt marshes and, more recently, in agroecosystems.

Katherine L. Gross is a professor in the Department of Plant Biology and acting director of the W. K. Kellogg Biological Station, Michigan State University. Her research interests are in plant community ecology, particularly the causes and consequences of variation in species diversity across community types. Much of her work has focused on successional processes in midwestern old-fields, but she is now also working in row-crop agricultural systems, and native and restorable grasslands.

Andrew P. Hendry is assistant professor in the Redpath Museum and Department of Biology at McGill University. He studies the evolution of biological diversity, particularly in relation to natural selection, gene flow, and speciation. His empirical systems include Pacific salmon, Trinidadian guppies, threespine stickleback, and Darwin's finches.

Richard J. Hobbs is professor of environmental science at Murdoch University in Western Australia. His research interests span basic ecology, restoration ecology, landscape ecology, and conservation biology, and he is particularly interested in the management and repair of degraded and altered ecosystems. Dr. Hobbs has worked mostly in the fragmented agricultural landscapes of Western Australia, but he also has ongoing, long-term research interests in the dynamics of serpentine grasslands in northern California. He is currently editor-in-chief of the journal *Restoration Ecology*.

Michael T. Kinnison is assistant professor of biological sciences at the University of Maine. He specializes in the contemporary evolution of fish populations in the wild, including local adaptation of introduced species and the interaction of natural selection with other evolutionary factors, such as gene flow and population decline. Dr. Kinnison serves as an advisor on conservation genetics and related issues in the restoration of endangered salmon populations.

Eric E. Knapp is a research ecologist with the USDA, Forest Service, Pacific Southwest Research Station. While a postdoctoral researcher at the University of California, Davis, he studied the population biology of native grass species used in restoration. His current research focuses on the restoration of fire to forested ecosystems.

Daniel Larkin is a doctoral student in the Department of Botany, University of Wisconsin–Madison. His research on restoration of southern California salt marshes focuses on the effects of topographic heterogeneity on trophic dynamics and ecosystem development.

Joyce Maschinski is the conservation ecologist leading the South Florida conservation program at Fairchild Tropical Botanic Garden. Her current research interests center on rare plant populations, especially understanding factors that limit their reproduction, growth, and expansion; the impact of human activities; and the potential management solutions for their conservation. Dr. Maschinski has also worked as a conservation officer at the Center for Plant Conservation for more than 15 years.

Brian A. Maurer is associate professor in the Department of Fisheries and Wildlife, Michigan State University. His research focuses on understanding the structure and dynamics of species assemblages across geographic space. Recently, he collaborated with the Michigan Department of Natural Resources (MDNR) on wildlife habitat modeling for the Michigan Gap Analysis project, including investigation of the accuracy of models at different spatial scales, and on developing refinements to be used in the ecosystem MDNR management-decision support system. Research in his lab focuses on spatial dynamics of ecological communities at different spatial and temporal scales.

Holly L. Menninger is a doctoral candidate in the Behavior, Ecology, Evolution, and Systematics program at the University of Maryland. She studies terrestrial-aquatic linkages in streams adjacent to human-altered landscapes.

Constance I. Millar is a senior research scientist with the Sierra Nevada Research Center, Pacific Southwest Research Station, USDA, Forest Service. Her research has involved forest-population, evolutionary, and conservation genetics. Currently her team focuses on climate change and the effects of historic climate variability on high-elevation forests of the Sierra Nevada and the Great Basin. She was coleader of the Sierra Nevada Ecosystem Project and currently cochairs the Consortium for Integrated Climate Research in Western Mountains.

Arlee M. Montalvo is a plant restoration ecologist at the Riverside-Corona Resource Conservation District and associate of the Department of Botany and Plant Sciences, University of California, Riverside. Her research has focused on plant population biology and restoration genetics, including experimental studies on the effect of translocation and hybridization on population fitness. She has written extensively on the importance of considering genetic information in making informed choices of plant materials for restoration and is working to establish standards for the collection and use of native plants for restoring biodiversity.

Shahid Naeem is professor of ecology at Columbia University, New York City. His theoretical and empirical research focuses on the ecosystem consequences of biodiversity loss. His studies include experimental manipulation of plant, animal, and microbial diversity in laboratory, greenhouse, and field studies. Much of his field work is conducted in the old-fields and grasslands of Cedar Creek, Minnesota; Black Rock Forest in New York; and the grasslands of Inner Mongolia in China. He is director of the Research Coordinating Network BioMERGE, which coordinates data-driven studies of the consequences of biodiversity loss in rainforest, marine, and grassland ecosystems.

Julian D. Olden is a Nature Conservancy David H. Smith Postdoctoral Fellow at the Center for Limnology, University of Wisconsin–Madison. His research interests include using life-history theory to predict fish invasions and extinctions, exploring the mechanisms driving the biological and functional homogenization of fauna and floral communities, and studying the ecological effects of river regulation in aquatic systems.

Craig W. Osenberg is a professor in the Department of Zoology at the University of Florida. He works on the dynamics of stage-structured populations (especially fishes) and has contributed to the development of assessment designs and the application of meta-analysis. His interests in restoration and conservation include studies of gas and oil production, coastal springs restoration, artificial reefs, and marine reserves.

Christopher M. Richards is a population geneticist at the USDA's central gene bank in Ft. Collins, Colorado. Richards received his Ph.D. in botany and genetics at Duke University, working with Janis Antonovics on metapopulation genetics and gene flow dynamics, which he continued to study at Vanderbilt University with David McCauley. His current research centers on the theoretical and empirical aspects of *ex situ* conservation, both as a source of novel variation for plant breeding and for applied restoration projects.

Colette M. St. Mary is an associate professor in the Department of Zoology at the University of Florida. Her work in marine management and conservation combine her interests in behavioral ecology, reproductive biology, and mathematical modeling. In addition, she has contributed to marine conservation via her studies of habitat restoration and marine reserves.

Darren R. Sandquist is associate professor of biological science at California State University, Fullerton, where he is a member of the biodiversity, ecology, and conservation concentration. His research focuses on water-use ecophysiology in tropical and arid land plants, their evolution and their response to changing climate and biotic interactions.

Jeffrey S. Shima is a senior lecturer in the School of Biological Sciences at Victoria, University of Wellington New Zealand. Dr. Shima is a population ecologist specializing in marine systems and the effects of species interactions. His work on spatial and temporal variation in density-dependence has influenced our views of marine systems. He serves as director of the Victoria University Marine Lab.

Craig A. Stockwell is associate professor in the Department of Biological Sciences at North Dakota State University. He works in the area of evolutionary conservation biology. His primary research interests consider the evolutionary implications of active management of rare fish species. Dr. Stockwell currently serves as scientific advisor to the White Sands Pupfish Conservation Team.

Katharine N. Suding is assistant professor in the Department of Ecology and Evolutionary Biology at University of California, Irvine. Her research bridges plant community and ecosystem ecology, particularly in grassland and tundra systems. Her interests include plant-soil feedbacks, functional traits, and species invasions.

Dean L. Urban is associate professor and director of the Landscape Ecology Laboratory in the Nicholas School of the Environment and Earth Sciences at Duke University, where he teaches landscape ecology, spatial analysis, and multivariate methods for ecological assessment. His research focuses on extrapolating ecological understanding from the small scale of field studies to the regional scales of management and policy.

M. Jake Vander Zanden is assistant professor at the Center for Limnology and the Department of Zoology at the University of Wisconsin–Madison. He was a Nature Conservancy David H. Smith Postdoctoral Fellow at the University of California, Davis, from 1999 to 2001. Dr. Vander Zanden works on aquatic ecosystems, with research interests that include invasive species, food webs, stable isotopes, benthic ecology, and salmonid conservation and restoration.

Gabrielle Vivian-Smith is a senior scientist (weed ecology) with the Queensland Government and is a member of the Cooperative Research Centre for Australian Weed Management. She has 12 years of experience in ecological restoration. Her current research focuses on the ecology of invasive plants in southeast Queensland, a region where invasive species present a major restoration barrier. She has also participated in restoration research in New York, New Jersey, California, Costa Rica, and Mexico.

Jada-Simone S. White is a graduate student in the Department of Zoology at the University of Florida. She works on the dynamics of open populations and is interested in applying her interests in marine ecology to marine management, including the implementation of marine reserves.

INDEX